# Welfare of production animals: assessment and management of risks

ECVPH

efsa

European Food Safety Authority

# Food safety assurance and veterinary public health

## – volume 5 –

## *Welfare of production animals: assessment and management of risks*

edited by:
Frans J.M. Smulders
Bo Algers

Wageningen Academic
Publishers

**ISBN 978-90-8686-122-4**



**First published, 2009**

# Preface

The International Committee of the World Organisation for Animal Health (OIE) has recognised that animal welfare is a complex, multifaceted public policy issue which includes scientific, ethical, economic and political dimensions. Determinants of animal welfare may influence the occurrence of animal diseases and have a positive or negative effect on food safety. For example, stress may increase susceptibility to infections, which in turn may lead to illness in animals or increased shedding of pathogens and contaminants into the food chain. On the other hand, diseases often decrease the welfare of animals. Whereas outdoor farming systems may improve animal welfare, they may also increase the exposure of animals to certain zoonotic hazards. These interlinking aspects between animal welfare and animal health highlight how important animal welfare is within the area of veterinary public health.

Both the European College of Veterinary Public Health (ECVPH) and the European Food Safety Authority (EFSA) provide unique European platforms to collect and distribute the skills and knowledge on animal welfare. EFSA provides independent scientific advice to the European Commission, European Parliament and Member States on food and feed safety, animal health and welfare, plant protection products, plant health and nutrition. Although the main focus of ECVPH has been in the area of animal diseases and food safety, animal welfare represents an important element within the college, too.

ECVPH aims at the improvement and promotion of: (a) the quality of animal health care and welfare by making available specialized knowledge and skills in the subspecialties of population medicine and food science to the benefit of animals; (b) the quality of veterinary practice through contacts of general practitioners with registered specialists; (c) the structure of animal health care through enhancing the application of formal risk assessment procedures, quantitative problem analysis methods, systems of monitoring and surveillance at population level, food safety and process quality management systems; (d) the structure of population medicine, risk management and risk communication by improving the knowledge and perception of veterinarians, livestock owners, food processing industries and the general public; (e) consumer protection with regard to prevention and control of foodborne hazards and to food hygiene procedures; (f) the further development of veterinary public health and its subspecialties, population medicine and food science; and, (g) the integrated, multidisciplinary approach towards analysis, control and prevention of hazards to human and animal health.

EFSA is on the forefront for developing risk assessment methodologies, including also methods for the assessment of animal welfare. EFSA's activities in this area are carried out by the Panel on Animal Health and Welfare (AHAW). Its scientific opinions focus on helping risk managers to identify methods to reduce unnecessary pain, distress and suffering of animals and to increase welfare where possible. EFSA is not mandated to give advice on ethical or cultural issues related to animal welfare. Since the initiation of EFSA in 2002, the AHAW Panel has adopted several Scientific Opinions on animal welfare, dealing – amongst others – with laboratory animals, stunning and killing methods, animal transport, welfare of calves, welfare risks of the import of captive birds as well as welfare of pigs, fish and dairy cows. In addition, EFSA's Scientific Committee has, in 2009, adopted an opinion on experimental animals examining the possibility of replacement, reduction and refinement of animal testing in the food and feed sector.

Given the close link between the welfare conditions of farmed animals on the one hand, and disease occurrence and food safety on the other, animal welfare questions have also been looked at by other Scientific Panels of EFSA. For instance, in relation to pigs the Panel on Biological Hazards (BIOHAZ) has delivered opinions on issues such as the food safety aspects of pig housing and husbandry systems. The Panel on Additives and Product Substances Used in Animal Feed (FEEDAP) issues scientific opinions on feed additives used in animal nutrition, including a number of substances for intended use in pig feed.

International organisations have issued recommendations and guidelines concerning animal welfare, such as the World Organisation for Animal Health (OIE) and the Council of Europe. The EU is signatory to the European Convention for the Protection of Animals kept for Farming Purposes, which has been adopted by the Council of Europe. As the International reference organisation for animal health, OIE started in 2002 its activities in animal welfare, i.e. elaborating recommendations and guidelines covering animal welfare practices, reaffirming that animal health is a key component of animal welfare. A permanent Animal Welfare Working Group was formed in May 2002 and the first recommendations of the Working Group were adopted one year later. The OIE Guiding Principles on animal welfare were included in the OIE Terrestrial Animal Health Code ('Terrestrial Code').

The EU has one of the world's highest standards of animal welfare. Harmonised EU rules are in place, covering a range of animal species and welfare-affecting issues. Council Directive 98/58/EC lays down the minimum standards for the protection of all farmed animals, while other EU legislation sets welfare standards for farmed animals during transport and at the time of stunning and slaughter. Specific Directives cover the protection of individual animal categories such as calves, pigs and laying hens. In addition to farmed animals, animals used in laboratory tests and wild animals kept in zoos are also protected by harmonised EU standards. The overall framework for EU action on animal welfare is set out in rolling action plans, currently the Community Action Plan on the Protection and Welfare of Animals 2006-2010. On the 19$^{th}$ of September 2007, the European Commission (EC) adopted a Communication setting out the EU's Animal Health Strategy (AHS) for 2007-2013.

Although the importance of animal welfare is recognised both at the international and EU level, there is still a lack of harmonization of indicators for animal welfare. EFSA, in line with mandates received from the EC, will start to assess the indicators for the welfare of food-producing animals with a view to developing a harmonised European monitoring system assessing the quality of farming systems and their impact on the diseases and welfare of animals according to the outcome-based approach described in the Community Action Plan on the Protection and Welfare of Animals 2006-2010. Such indicators could be integrated into future legislative proposals and should provide the basis for the development of an EU label for animal welfare proposed by the EC.

Both the European and international organisations responsible for ensuring animal welfare need scientific advice. In this context it is important to appreciate the complexity of animal welfare, which is impacted on by a range of factors including, for instance, housing and bedding, space and crowding, transport conditions, stunning, slaughter and killing methods, castration of males and tail docking. This complex nature of animal welfare challenges our competence and knowledge in many ways. Whilst animal welfare science is developing fast, many basic questions are still waiting for their answers. The application of the risk analysis framework, particularly the development of methodologies for the assessment of risk for animal welfare, has started, but their implementation is still in its infancy as compared with the longer tradition within the food safety and disease prevention area. Moreover, in addition to its complex nature, the fact that many animal welfare problems are introduced by human beings, make it even more difficult to separate purely scientific facts from the perceptions and ethical issues of individuals. Finally, the risks for animal welfare considered acceptable by society are continually re-appreciated, as an inevitable result of which decisions taken a long time ago may not necessarily still be considered ethically valid in modern day society.

Animal welfare should be taken into account not only by welfare scientists, but should represent a natural part of the daily activities of other professionals – e.g. those dealing with food-borne hazards and animal diseases – as decisions taken in these areas may indeed have an impact on animal welfare. Also, one must realize that animal welfare is an important issue when evaluating the appropriateness of methods for transport and slaughtering of production animals, or when the acceptability of animal testing related to safety evaluations and diagnostics has to be considered. Research in animal welfare has greatly advanced our ability to understand the behaviour of animals and which factors can cause unnecessary pain and suffering. This knowledge needs to be taken into account in many practical situations, e.g. in planning new stables for animals, building new slaughterhouses, defining transport conditions or for purposes of handling diseases outbreaks. Consequently, a basic understanding of welfare concepts needs to be acquired by farmers as well as individuals in charge of transport, slaughter and killing of animals.

A sustainable high level of welfare for food producing animals cannot be achieved without costs. In addition, other legitimate factors need to be taken into account when making decisions. Since daily decisions on handling animals are to be based both on a proper understanding of animal behaviour and on an awareness of the practical constraints, it does not suffice if only those active in Academia or governmental bodies understand the essential aspects of animal welfare. In general, when one sets out to make sustainable political decisions on risks, it is crucial that the definition of acceptable levels of risks is established through an intensive dialogue with stakeholders, so as to ensure that such definitions are supported by the society in which we live. This is particularly challenging for an area such as animal welfare. There seems to be no single commonly acceptable level of risks within a single country, let alone within the entirety of the European Union when it comes to taking a stance on animal welfare issues. Consequently, equally challenging will be the area of risk communication. In this context major questions include (1) how to measure animal welfare objectively? (2) what level of risks are considered acceptable by society? And (3) which specific risk management options should be contemplated, so as to avoid unnecessary pain and suffering? It is not uncommon that the various stakeholders' standpoints range from supporting a total ban of the use of animals for human interest to allowing animal production on the current commercial level. Moreover, as long as many primarily associate animal welfare issues with illicit activities or even criminal cases of animal abuse, it is extremely complicated to mobilise the various stakeholders to join a fair and scientifically based dialogue with the ultimate aim to safeguard or even improve the welfare status of animals.

I hope this book finds its way to the hands of many professionals. Not only should the contributions included in this Volume serve to enhance discussions among welfare scientists, they would seem to represent equally relevant information for individuals active in other areas of veterinary public health, i.e. where legitimate considerations of another than animal welfare nature play an important role. Ultimately, ensuring animal welfare is a job of humans, and, unless the different actors and factions in the animal production field keep abreast of the latest information and thus stay well-informed about essentials, animal welfare issues will not be adequately addressed.

**Riitta Maijala, DVM, PhD, Dipl. ECVPH**

Director of Risk Assessment at EFSA
Senior Vice-President of ECVPH

# Contents

## Part 2 – The assessment of animal welfare risks

## Part 4 – Reconciling animal welfare and food safety

# Introduction

# *Hominum animaliumque saluti*, or 'giving welfare proper consideration in animal production'

**Frans J.M. Smulders[1] and Bo Algers[2]**
[1]Department of Production Animal Medicine and Veterinary Public Health, University of Veterinary Medicine, Veterinärplatz 1, 1210 Vienna, Austria; frans.smulders@vu-wien.ac.at
[2]Department of Animal Environment and Health, Swedish University of Agricultural Sciences, P.O. Box 234, SE-532 23 Skara, Sweden

## 1. Introduction

In Volume 14 of his *magnum opus* '*Noctes Atticae*' the 2nd century Roman author Aulus Gellius fables an event, which has since been an inspiration to philosophers, playwrights and the veterinary profession alike. Androclus, a runaway slave hiding in what turns out to be a lion's den, removes a giant spine from the animal's paw and after both are captured and subsequently confronted with each other in the arena, the engagement of the two – rather than turning into a mortal battle – results in the lion being extremely excited to see his benefactor again. Androclus is pardoned by the Emperor, even is given the lion as a reward, whereupon both wander through the streets of Rome being showered with flowers and gifts, and Roman citizens exclaiming: '*Hic est leo hospes hominis, hic est homo medicus leonis*' (this is the lion which is a man's friend, this is the man who is the lion's doctor). The fable illustrates in a poetical way that compassion and proper care for animals is not only a decent thing to do, but can also be entirely rewarding in an existential and material sense.

One would have hoped that such a realisation might have carried over to the following centuries. However, the use of animals in agricultural production, the appreciation of animals' innate physiological functions and behaviour and, in consequence, of their needs has over the centuries not always been met with appropriate actions from animal keepers (or human society in general for that matter). In fact, and sadly so, man's attitude towards animals has rather been tainted by a largely anthropocentric stance towards animals and how they should be treated.

In Greek (particularly Platonian and Aristotelian-), as well as in Judean-Christian and Islamic philosophy the position of animals has been frequently debated. In Arabic sources, notably in one of the 52 philosophical epistles of the Ichwan as-safa (Brethren of purity; issued in Baghdad in the late 10th century) '*The case of the animals versus man before the king of the Jinn*' is described, with the king's final ruling being that man is far superior to animals because human beings are God's chosen creatures and consequently that animals are to be led by man and are subject to man's precepts and prohibitions (see Goodman, 1978). Man's superiority has been equally stressed in Christian tradition. For instance, justified by biblical texts (notably Exodus 21: 28-32), from the middle ages up to the late 17th century, animals were even convicted to cruel deaths in court cases for having behaved in no other way than one would expect them to (Berriat-Saint-Prix, 1829). It was not until the end of the 18th century that the realisation started to sink in that, provided animal welfare is given proper consideration, human beings will fare equally well.

## 2. Animal welfare – a growing concern

Arguably, the tradition of human concern about animal welfare really began with animal rights advocate Jeremy Bentham's famous query from 1789: *'The question is not, can they reason? Nor, can they talk? But, can they suffer?'* (Bentham, 1789). More recently, the debate was rekindled by Ruth Harrison's book *'Animal machines'* in which she describes production conditions that sent shock waves through the public (Harrison, 1964). Nowadays the European Union on a regular basis investigates peoples attitudes to animal welfare through the Eurobarometer (EC, 2009). It shows that, although the concern for animal welfare varies among countries, if people have visited farms their awareness and concern for animal welfare increases.

The growing concern of the public in the western industrialised world has in recent years even led to the foundation of a number of political parties, e.g. the 'Human Environment Animal Welfare Party' (Germany, 1993), the 'Party against Bull Fighting, Cruelty and Mistreatment of Animals' (Spain, foundation date unknown), the 'Animals Count' party (UK, 2002), the 'Animal Alliance Environment Voters Party' (Canada, 2005), and most notably the 'Party for the Animals' (the Netherlands, 2002). The latter party was the first to actually gain a parliamentary status in the 2006 elections and they currently (2009) hold two (out of a total of 150) seats in the Dutch House of Representatives, their self-proclaimed major (some argue only) goal being to represent a 'testimonial' party fighting for animal rights and welfare, rather than seeking political power (Wikipedia, 2009).

## 3. Animal welfare science

What then is animal welfare? The OIE definition of animal welfare is: '... *how an animal is coping with the conditions in which it lives. An animal is in a good state of health if (as indicated by scientific evidence) it is healthy, comfortable, well nourished, safe, able to express innate behaviour, and if it is not suffering from unpleasant states such as pain, fear and distress. Good animal welfare requires disease prevention and veterinary treatment, appropriate shelter, management, nutrition, humane handling, and humane slaughter/killing. Animal welfare refers to the state of the animal; the treatment the animal receives is covered by other terms, such as animal care, animal husbandry, and humane treatment*' (OIE, 2009). This means that the environment does not stress or overtax any demands on the animal, so that it adapts easily and its welfare is good. Or if the environment is stressful, that the animal is given ample opportunity to protect itself, so that for example free-range animals are able to seek protection from inclement weather. If, on the other hand, the animal exhibits frustration, anxiety, pain, injury, disease or stress, its welfare is impaired. However, this is not the only definition of animal welfare used, and also some aspects of animal welfare, when it regards our judgement of how to prioritize the various components of welfare, refer to underlying values that may guide decisions on priorities. Therefore, animal welfare science can be divided into the science about various elements of welfare together with the methodologies to measure and validate such measures on one hand and on the other the science about which weight we put on various aspects of welfare and why.

## 4. The risk of poor animal welfare

According to the 1997 Amsterdam Treaty, the EU's member states are supposed to pay full regard to the welfare requirements of animals. In various surveys, the citizens of the EU have called for better animal protection standards to improve food safety and quality. In their Whitepaper on Food Safety (EC, 2000), the European commission explains the relationships between food safety and animal protection through the entire food processing chain.

The purpose of the new European Food Safety Authority, EFSA, is to serve as an independent body and carry out risk assessments for food safety, animal health and animal welfare. A distinction is now made between EFSA's scientifically based risk assessment for animal welfare and the Commission's risk management, which takes political aspects and legislation into consideration.

Scientific methods exist for risk assessments of food safety and assessing the risk of spreading contagious diseases between countries and regions. But standardised methods for scientifically assessing the risk of impaired animal welfare are lacking. It is explained in the reports published by EFSA that animal welfare includes how the animal feels, but also how well the animal copes with its environment and the conditions under which it lives. Methods are therefore under development for risk assessments of animal welfare, which is opening up a new research front.

## 5. The management of animal welfare risks

In January 2006 the EU adopted a new action plan for animal welfare. The goals include upgrading existing minimum standards for animal protection and welfare, giving a high priority to promoting policy-orientated future research on animal protection and welfare, and introducing standardized animal welfare indicators for scientific assessment of animal welfare in various production systems, livestock herds and slaughterhouses. Rules are lacking for dairy cows and handling of animals in connection with slaughter. Fish also await rules governing animal welfare, such as how densely fish may be kept in the tanks and pens of fish farms, as well as requirements on slaughter methods. More recent research can provide guidance on how fish are affected by stress.

Internationally, a consensus is being sought regarding what consideration should be given to animal protection and welfare in global trade. Another goal is to increase awareness and promote better understanding of the conditions of animal production among the public and animal keepers. At the same time, producers have to pay more attention to prevailing views on what an animal can reasonably be expected to endure.

The handling of animals in connection with transport and killing are urgent questions for research, but also for application of the knowledge that exists today.

## 6. Veterinarians and animal welfare

The mission statement of the World Veterinary Association (WVA), which represents the veterinary profession, veterinary medicine and science in all their aspects on a global international level, states *expressis verbis* as one of its major missions to: '… promote animal and human health and well-being through sustainable and humane use and management of animals' (WVA, 2009). Similarly the European College of Veterinary Public Health (ECVPH) lists as a main aim to improve and promote '... the quality of animal health care and welfare by making available specialised knowledge and skills in the subspecialties of population medicine and food science to the benefit of the animals' (ECVPH, 2009). In essence, both statements reflect what the Latin motto of this contribution [*hominum animaliumque saluti* (towards welfare of man and animal alike)] signifies for the veterinary profession. Such an 'extended' Hippocratic oath represents a promise to appropriately treat animals both through curative and preventive medicine, and expresses the veterinarian's creed to ensure public health.

Whereas companion animal clinicians are inherently confronted with animal owners, who generally have a more emotional bond with their pets, production animal practitioners face animal owners, who – for existential reasons – have to consider profitability issues as well as consequences for the safety of consumers ingesting foods of animal origin. It would appear that particularly the latter category of veterinarians have to weigh the occasionally conflicting interests of humans and animals and have to do so with proper consideration of animal welfare issues. Finally, veterinary professionals in a control function ('official veterinarians') are held (and supported) by European and national legislation to enforce that such is achieved in production animal practice.

## 7. The aim of this book

Today no one questions the possibility of assessing animal welfare by scientific methodology. Much has lately been published on this topic. This book aims at bringing the interested reader up to date not only on the science of animal welfare, but also on how this science may be used to help risk managers to prioritize their resources when taking actions to improve welfare for animals that we use for food production.

## Acknowledgements

We take this opportunity to express our gratitude to the authorship of this Volume. Being able to rely on the expertise of the various contributors has allowed to adequately sketch what achievements have recently been made in animal welfare science, and all of them should be acknowledged for delivering high quality scripts and for strictly complying with the Instructions for Authors.

A special word of thanks is due to the Directorate of the European Food Safety Agency (EFSA) at Parma, Italy, for their permission to utilize sections of relevant EFSA documents previously issued as EFSA Scientific Opinions and for their willingness to join the European College of

Veterinary Public Health (ECVPH) in issuing this book. In addition, we are particularly grateful to the Staff members of EFSA associated with its Scientific Panels on Animal Health and Welfare (AHAW) and Biohazards (BIOHAZ) for their assistance in our endeavours.

Last but not least, we wish to acknowledge the efforts of our assistant-editor Alexandra Bauer BSc, DVM, and thank her for her exemplary support in ensuring that formatting issues were properly addressed and for keeping us sharp throughout this most rewarding exercise.

## References

Bentham, J., 1789. Introduction to the principles of morals and legislation. Imprint 1996, Clarendon Press, Oxford, UK.

Berriat-Saint-Prix, J., 1829. Rapport et recherches sur les procès et jugements relatifs aux animaux. In: Mémoires de la Société Royale des Antiquaires de France, T VIII, Publ. Selligue, Paris, France, pp. 403-450.

European College of Veterinary Public Health (ECVPH), 2009. Available at: www.ecvph.org/index.aspx. Accessed 4 June 2009.

European Commission (EC), 2000. Whitepaper on Food Safety, COM (99) 719 Final, 12th January 2000, 52 pp. Available at: http://ec.europa.eu/dgs/health_consumer/library/pub/pub06_en.pdf

European Commission (EC), 2009. Attitudes of consumers towards the welfare of farmed animals. Available at: http://ec.europa.eu/food/animalwelfare/eurobarometer25_pdf_en.pdf. Accessed 10 June 2009.

Goodman, L.E., 1978. The case of the animals versus man before the king of the Jinn. An English translation of a major portion of the second section of the Rasa'il with extensive introduction and notes. Twayne Publishers, Boston, MA, USA.

Harrison, R., 1964. Animal machines: the new factory farming industry. Stuart, London, UK, 186 pp.

Office International des Épizooties (OIE), 2009. Terrestrial animal health code, chapter 7.1, article 7.1.1. Available at: http://www.oie.int/eng/normes/mcode/en_chapitre_1.7.1htm. Accessed 10 June 2009.

Wikipedia, 2009. Party for the Animals. Available at: http://en.wikipedia.org/wiki/Party_for_the_Animals. Accessed 4 June 2009.

World Veterinary Association (WVA), 2009, Vision/mission statement. www.worldvet.org/manuals/01missio.stm.pdf. Accessed 4 June 2009.

# Part 1 – Animal welfare – theoretical basis

# The concept of animal welfare: a philosopher's view

**Lennart Nordenfelt**
Department of Medical and Health Sciences, Linköping University, 58183 Linköping, Sweden; lennart.nordenfelt@liu.se

## Summary

In this chapter the major interpretations of the expression 'animal welfare' that have been proposed in the contemporary animal science literature are scrutinised, with a particular focus on the following conceptions: (1) freedom to perform natural behaviour, (2) health and natural function, (3) coping with the environment, and (4) positive subjective experiences. It is argued that these conceptions, although basically different, are interrelated both from a logical and empirical point of view. Particularly striking in this context are some causal relationships, such as an animal's natural behaviour, health and coping which all partly determine its sustainable positive experiences. It is discussed why 'positive subjective experiences' are crucial to animal welfare. The two main arguments for this include: (1) the existence of subjective experiences often being cited as supporting evidence that other, associated, factors are components of welfare, and (2) the main ethical motive for enhancing animal welfare, and indeed preventing cruelty to animals, being that welfare relates to animal feelings. It is acknowledged that there is a special reason for giving 'freedom to perform natural behaviour' a separate place in the set of proposed animal welfare definitions since this conception refers to the dignity of an animal. Thus, one may say that – assuming the same level of positive subjective experience of animals A and B, whilst A has a greater freedom of behaviour than B – animal A has a higher degree of welfare than does animal B. It is argued that animal welfare is an evaluative concept, albeit that it does not necessarily include ethical evaluation, i.e. it should be possible to address welfare without having to adhere to a particular moral attitude. Finally, this author claims that once a level of welfare has been established (by evaluation), a number of purely scientific procedures, for instance measurement and causal analysis, can be performed.

**Keywords**: natural behaviour, health, coping, subjective experiences, evaluation

## 1. Introduction

An extensive, largely ethically motivated, discussion on the nature of animal welfare in animal- and veterinary science and practice has followed the lively debate in the 1960's about the treatment of animals both in the food industry and in medical research. The starting point for the discussion was the publication of Ruth Harrison's *Animal machines* (1964), where she particularly criticised modern farming methods.

As a result, a great number of definitions of animal welfare have appeared in the scientific literature. Some of these are quite distinct and would, if applied, indeed have significant consequences for the treatment of animals in general and for the organisation of research in particular. Therefore, it is more than timely to thoroughly analyse the various proposals and consider how reasonable they are as theoretical platforms in animal science and veterinary practice.

On the other hand, even though some of the proposals are quite different, they still have crucial connections to each other, both logically and empirically. Whilst some elements of the proposals are identical, the factors included in the various definitions may stand in causal relationship to each other. For instance, health, coping ability (an integral part of good health) and freedom of behaviour are typical causes of positive subjective experiences. Hence, the conceptual area of animal welfare is complex and difficult to disentangle, not in the least because various theorists use their terms with diverging connotations.

It is crucial to distinguish between at least the following questions:

- What is the definition of the concept of animal welfare?
- What are the criteria and indicators of animal welfare (biological, psychological and ethological)?
- What are the methods for determining animal welfare?
- What are the causes and conditions of animal welfare?

It would, in the framework of this contribution, be impossible to answer all these questions systematically. However, the distinctions indicated by the above questions will appear throughout this chapter.

In the following section some of the most influential definitions of animal welfare that have been presented over the last 30 years are introduced. Subsequently, their conceptual and empirical relations are briefly considered and arguments in favour of using one particular definition are provided. Finally, an attempt is made to determine whether or not the concept of welfare is evaluative.

## 2. Some contemporary definitions of welfare

### 2.1 Organic definitions of welfare

Some biologically oriented theorists claim that welfare must be defined in evolutionary terms (Barnard and Hurst, 1996). According to this idea welfare should be interpreted in terms of what natural selection has designed an organism to do and how circumstances impinge on its functional design. In order to assess a state of welfare one should try to understand the organism's naturally selected performance patterns.

The organism's priority, then, is to maximise reproductive success through efficient self-expenditure; the priority is not necessarily the preservation of the organism itself. The welfare of animal A is identical with the maximisation of A's reproductive contribution. In many instances the behaviour and development leading to such success are equivalent to self-preservation and growth. In these cases welfare in this sense might coincide with other ideas of welfare. However, since natural selection may have designed an organism for suffering and self-expenditure in certain contexts, suffering and self-expenditure may in these contexts constitute a high level of welfare. Thus, it is by no means self-evident that the (evolutionary) welfare of animals should consist of self-preservation or positive feelings in those animals where such can occur. In the extreme case 'the best thing' for A to do in order to maximise its reproductive contribution is to die. Its dying process could thus, paradoxically, constitute its welfare. An individual's health and subjective well-being can clearly be jeopardised in its struggle for maximum reproduction.

A different but still evolutionarily based theory of welfare focuses on the notion of *coping* (Broom, 1988 and later). Other authors use the term 'adaptation' to signify a similar concept (Fraser, 2008). The idea is that the welfare of an individual is its state as regards its attempts to *cope* with its environment. The mental and bodily stability that is the successful result of coping defines the animal's biological fitness. Fitness in its turn is characterised and measured in terms of survival, growth and reproduction (Broom, 1988). Since survival and growth are central elements in this conception, self-expenditure could never count as welfare even if it were to increase reproduction. (NB, observe that Broom's conception of welfare also comprises further elements. His full conception is therefore of a conglomerate kind; see below, section 2.6.)

Some theorists with a biological inclination underline how similar the notions of *animal health* and *animal welfare* are. Hughes and Curtis (1997) express this as follows: '*The relationship between health and welfare is self-evident. Some people regard them as almost synonymous*'. It is even more common to say that welfare, if not the same as health, at least includes health (Broom, 1996).

## 2.2 A theory of welfare in terms of feelings

A common and highly influential conception of animal welfare focuses on the animal's feelings or, in general, its subjective experiences.

Animal welfare involves the subjective feelings of animals. The growing concern for animals in laboratories, farms and zoos is not just about their physical health, important though that is. Nor is it just to ensure that animals function properly, like well-maintained machines, desirable though that may be (Dawkins, 1990).

Dawkins (1980, 1990) emphasises that almost all arguments about the treatment of animals focus on the experience of pain or suffering in the most general sense (including fear, depression and boredom). There is also a strong relation between this idea of welfare and the animal rights' movement, the basic common concept being that animals are morally the equals of humans and would consequently have rights comparable to those of humans. This idea is based on the belief that humans and animals have an equal capacity to suffer and that both have other similar qualities which influence subjective experience (De Grazia, 2002).

Dawkins's approach is shared by Ian Duncan, who notes that we speak of welfare only in the case of organisms that are capable of experiencing subjective feelings, or in his words that '*neither health nor lack of stress nor fitness is necessary and/or sufficient to conclude that an animal has good welfare. Welfare is dependent on what animals feel'* (Duncan, 1993). [Further well-known proponents of a feelings-based theory are Simonsen (1996), Singer (1990), McMillan (2000, 2005)]

## 2.3 The idea of welfare as fulfilment of preferences and wants

A popular notion for the characterisation of both the quality of human life and animal welfare is the idea of fulfilment of preferences. It is central in many accounts of human happiness (McGill, 1967; Tatarkiewicz, 1976; Veenhoven, 1984; Nordenfelt, 2000) and has been employed in the discussion of animal welfare by for instance Dawkins (1990), Sandøe (1996) and Haynes (2008). Two different ways in which the fulfilment of preferences (or the satisfaction of wants or desires) can play a role need to be considered:
1. Welfare is *assessed* via the person's or animal's choices in certain situations. One could say that the person or animal through these choices determines the conditions for its high degree of welfare. Here preference functions as an indicator of welfare.
2. Welfare is *defined* in terms of fulfilment of preferences. To say that A has a high degree of welfare means that most of A's preferences have been fulfilled.

In animal science it is a popular idea – most clearly set out and practised by Dawkins – that one can determine the level of animal welfare by studying various situations where the animals make choices. Here preference is used as a method for investigating, not as an element in a definition of, welfare. Several sophisticated methods are now available to determine the strength of preferences of animals. Dawkins suggests that in order to determine which preferences are most important to animals (in the sense that not fulfilling them will create the highest degree of suffering) one could essentially follow the model which economists use when they rank preferences. Other researchers have, for instance, measured the strength of the preference of pigs for earth floors by forcing them to make a large number of responses in order to be allowed access to such a floor.

However, preference satisfaction is not a completely reliable method if one is interested in the animal's subjective well-being in the long run. When, for example, an animal shows preference for a certain food or level of eating, it may be mistaken in its choice and its welfare may become poor. One can make an analogy with humans: *'Anorexic girls have a very strong preference not to eat, to the detriment of their welfare'* (Broom, 1991). Discovering the preferences of animals is here a *method* by which we can determine the welfare of animals. Neither Dawkins nor Broom explicitly define welfare in terms of preferences.

Let us now turn to *definitions* of human and animal welfare in terms of fulfilment of preferences or satisfaction of wants and desires. In the philosophy of human welfare (happiness) it is customary to offer such a definition as follows: P is completely happy if, and only if, P wants at t that x1... xn shall be the case at t; x1... xn constitutes the totality of P's wants at t; and P finds that x1... xn is the case. Thus P has the best welfare if, and only if, P observes that all his or her present wants are being satisfied (Nordenfelt, 2000, 2006). A similar definition applied both to the human and to the animal case has been suggested by Sandøe (1996) and particularly to the animal case by Haynes (2008).

Although preference refers to a subjective state, a definition in terms of preferences is not equivalent to the feeling-state definition mentioned above. Welfare, in the preference sense, is not identical with pleasure or any other collection of positive sensations. Instead, welfare is defined as a relation between the subject's preferences and the subject's perception of the present state of affairs. However, the fulfilment of preferences normally *results in* a feeling of satisfaction. For the sake of simplicity this position is considered in the following as being similar to the feelings-based definition presented in section 2.2.

### 2.4 Animal welfare as the satisfaction of needs: a spurious definition

Many animal scientists talk about welfare in terms of the satisfaction of the animal's *needs*. This is a proposal that requires much further specification to become useful. 'Need' is basically a relational term, more precisely a four-place predicate. This is most easily seen when considering locutions where 'need' is a verb: Person A needs y in order to attain the goal G. Example: John needs a hammer in order to repair his house. Essentially this phrase means that John's using a hammer is a necessary condition for his repairing his house. In such locutions we can detect the most general and clearest sense of 'need'. 'Need' here simply stands for any necessary condition for the attaining of a goal. The term 'goal' is so far not specified; it can be a goal set by a person or an animal but it can also be a goal in an abstract sense. This analysis shows that 'need' presupposes that there is a goal; otherwise, the need analysis is nonsensical.

For further illustration of this reasoning one should consider one significant theory of human needs (Maslow, 1968):

a. There is a limited set of basic human needs universal to all human beings. Maslow suggests the following six goals to which these needs are related: survival, safety, belongingness, love, esteem and self-actualisation.
b. The basic needs are ordered in a hierarchy where the physiological needs are the most, and the need for self-actualisation the least basic. This hierarchy entails that a more basic need must be fulfilled before a less basic need can be recognised by the agent in question.

For comparison, an example from the use of the term 'needs' in animal science is illustrative. Dawkins (1983) distinguishes between two kinds of needs: ultimate and proximate needs. Animals need food and water because without them they would die. Dawkins calls this kind of need 'ultimate needs': i.e. death and reproductive failure result if they are not met. Hens in battery cages on the other hand do not die when they are unable to dust-bathe. Still, it is reasonable to say that a hen has a need to do so. If the hen does not get her dust-bath she will experience considerable suffering. Thus the hen has a proximate need of dust baths.

In this context it should be considered what goal is served by the various needs. Dawkins refers to death and suffering if the needs are not met. Thus the goal of need must, self-evidently, involve life itself, but also well-being in some feeling sense. The need analysis therefore primarily refers to the goals of life and well-being. To say that welfare is the fulfilment of needs is to say that welfare is tantamount to attaining the goals of the needs, which are, as we have seen, normally described in terms of survival, positive well-being or avoidance of suffering. Hence the characterisation of welfare in terms of needs can be analysed in terms of other welfare definitions.

## 2.5 Theories of welfare in terms of natural behaviour

A different idea for characterising welfare is to focus on the behaviour of the animal. An influential proposal is to say that an animal has welfare if it can perform its natural behaviour.

The idea behind this (Rollin, 1992; Alröe *et al.*, 2001) is that every animal has – through evolution – received an innate genetically coded nature. This nature has sometimes been called the animal's *telos,* i.e. the goal towards which the animal's life is directed (Rollin, 1992, and later). According to this nature the animal behaves in certain ways, unless it is prevented to do so. It seeks food in its peculiar ways, it provides for its shelter and breeding in certain ways, etc. The natural behaviour presupposes a 'natural' environment. In order to behave in a natural way the animal must have a background in terms of an environment that has not been changed in artificial ways.

The philosophy of natural behaviour, then, is quite closely tied to the idea of a genetically coded nature. According to this idea, if human beings interfere with this behaviour there is a great risk that they interfere with the welfare of the animals in a negative way.

The idea of natural behaviour does not, however, presuppose determinism. Animals can normally freely choose between various alternatives. Birds of the same species typically choose different geographical sites for their nesting. They can have variations in their food-intake. Some chaffinches eat corn but some eat insects, and the same individual can change its menu over time. And almost all individuals constantly introduce variations into their lives. They vary their movements and choose to see new locations. Like humans, many animals have a lot of real freedom. There is not much that is predetermined in how they live their lives in detail. Given these facts, what is natural is the freedom of the animals, or rather the potentiality for freedom, in the realisation of the coded goals.

The interpretation of welfare in terms of freedom to perform natural behaviour is sometimes coupled with and supported by the observation that such behaviour normally contributes to the happiness of the animal (Rollin, 2005). It may, however, also stand as an independent idea (see under section 5 of this contribution).

## 2.6 On conglomerate definitions of animal welfare

Several authors are not content with any single one of the mentioned definitions of welfare. Instead, they propose complex, so-called *conglomerate* definitions (Nordenfelt, 2006), according to which welfare is the sum of states along a number of dimensions, for instance coping function, feelings and natural behaviour. In order to have a high degree of welfare one must, according to these authors, score high on all dimensions. The theories of Fraser *et al.* (1997), Broom (1998 and later) and Alrøe *et al.* (2001) provide conglomerate concepts of welfare. Fraser combines health (or natural function) with natural behaviour and positive subjective experiences. Broom combines coping with health and positive feelings. Alrøe *et al.* claim that a systemic view, which is central to the idea of organic farming, calls for a more complex notion of welfare than most theorists propose. They claim that one cannot simply consider fulfilment of the animal's basic needs, without also considering such factors as the richness of the animal's life and its relation to the whole eco-system. [For a further analysis and critical evaluation of complex definitions of welfare, see Nordenfelt (2006) and Lerner (2008)].

## 3. The major definitions of animal welfare under scrutiny

In spite of the apparent variety of proposals for the characterisation of animal welfare there seems to be a reasonably strong consensus that some ideas are more prominent than others, not in the least as foci from an ethical point of view. The following attempt to compare and evaluate these various ideas is partly inspired by Fraser *et al.* (1997) and Fraser (2008).

Among animal scientists and practising veterinary surgeons there is a large degree of consensus to the effect that a high level of welfare implies (1) freedom to perform natural behaviour (2) a high level of health and biological functioning (3) success in coping with (or adapting to) the environment, and (4) freedom from suffering and pain as well as having positive experiences. [NB it should be noted that Fraser *et al.* concentrate on (1), (2) and (4).]

The first animal welfare component to be considered is the idea of *freedom to perform natural behaviour*. This conception is for the most part well in harmony with feelings- and functioning-based interpretations. An animal can be prevented from building a nest on the spot it has chosen. Animals can be caught and taken to zoos and hens and turkeys can be kept in small cages awaiting slaughter, measures that normally cause suffering and are often detrimental to health. However, sometimes freedom to perform natural behaviour is a more autonomous conception. Concern with regard to a bird's opportunity to fly is mainly motivated by the view that animals should be allowed to live according to their natural adaptations. But such concern is not so obviously covered by an emphasis on health and subjective feelings.

Living a natural life is no guarantee that all ethical concerns will be satisfied. Even though an animal is kept in a natural environment, it can still suffer and become ill. Wild animals are continually exposed to hazards such as disease and predation and indeed, as a result, captive animals may often be healthier and in this sense better off than are wild ones. It must be kept in mind that it is not only humans that can interfere with the freedom of particular animals. There is interference within the animal world between the animals themselves, even within the same species. Notably, the predator continuously interferes with the predated animals. Such interference is continuous also in the sense that the predated animals must always be aware of dangers and behave cautiously even when the predator happens to be absent. Moreover, natural forces, such as earthquakes and sandstorms, can interfere with the intentions and behaviours of animals. In summary: in a 'natural' environment all this happens and this is also associated with suffering and a great lack of freedom.

Thus the wild and natural life is not always a happy life. Some animals, that are lucky and have few natural enemies, will probably live a reasonable life. Others clearly live a horrid and in many cases extremely short life. Nature has no mercy and the notion of help in nature is almost non-existent. Help, on the other hand, is a notion in the human world that can become relevant to such animals as they enter the circle of humans: livestock, zoo-animals and pets are fed, taken care of and frequently provided with medical care by their human owners.

This consideration inspires a preliminary assessment of the idea of natural behaviour as a candidate for a definition of welfare. It is questionable whether natural behaviour as such (i.e. behaviour not interfered with by humans) can *define* welfare. If this were the case, then behaviour aborted by an earthquake or by a leopard attack would constitute welfare. This is not what is intended. The root of the philosophy of natural behaviour normally is that an animal flourishes and is happy when it has the freedom to fulfil its potential. In this interpretation natural behaviour is a cause of happiness. But then the definition of welfare is not the natural behaviour as such, but is rather based on the notion of freedom to fulfil one's preferences and thereby become happy.

A second welfare concept is based on *diseases and other disturbances of normal biological functioning*. These are often the concern of veterinary surgeons. Thus some theorists would characterise welfare in terms of health or normal functioning (Hughes and Curtis, 1997). Undoubtedly, health is crucial for animal welfare whichever ultimate definition one chooses. However, is it questionable if it is reasonable to *identify* welfare with health, as this would suggest that a healthy animal cannot suffer a low degree of welfare. In considering humans one normally makes a clear distinction between the two concepts: a person with perfect health[1] but who has lost his entire fortune in the global recession presumably has a low degree of welfare.

It is quite clear that (ill) health does not always reflect an animal's subjective experiences. There are salient cases of suffering in animals where there are no apparent changes in physiological functioning. An example of such a case is where an animal has a strong inherited motivation for certain behaviour, such as the sucking of a calf in order to drink. So, whereas a calf with access to a bucket of water can satisfy its thirst, it will nevertheless have an innate desire to suck that is not fulfilled and will thus suffer, albeit such is not associated with any obvious malady or physiological abnormality. Conversely, the functioning of an animal can be impaired because of the presence of pathogens, even though the associated patho-physiological changes are not necessarily sensed by the animal until a very late stage of the disease. In short, there is no simple one-to-one correspondence between health or normal functioning and the subjective experience of animals.

A third welfare concept, *coping* or *adapting* is central in welfare discussions particularly in Europe. Coping denotes the ongoing activity of an animal to remain in a stable position with regard to its health (as well as its growth and survival). Consequently, it relates to (but is not identical with) both health and natural behaviour.

---

[1] Note that a conventional concept of health is assumed here, i.e. the absence of diseases. In my own analysis of health (Nordenfelt, 2000) the following holistic conception of health is proposed: A is completely healthy if, and only if, A is able to realise all his or her vital goals in standard circumstances.

For the most part, coping is also connected with positive feelings: an animal that copes is normally a happy animal. However, this is not universally true. For instance, a horse that is basically strong and fit but is faced with a harsh climatic environment must work hard to find enough food for its subsistence and consequently the horse will often be hungry and will suffer from the cold. Yet, *ex hypothesi*, the horse copes well, it struggles successfully to find food and in spite of the cold climate it does not contract a disease. However, the conclusion (suggested by the coping conception of welfare) that the horse has a high degree of welfare is hardly tenable.

This example illustrates that suffering is sometimes a necessary part of successful coping, as it may alert the animal to the fact that something has to be done to retain stability, however painful the struggle. [For the distinction between coping as a process-concept and coping as a result-concept, see Nordenfelt (2006).] This is the reason why a conglomerate notion of welfare that includes both coping and positive feelings has an inherent tension.

Finally, most authors addressing animal welfare maintain that *positive feelings* play a part in, or indeed represent the essence of welfare. As mentioned above, much of the ethical concern for animals is motivated by the supposition that animals, in particular those in captivity, can and do suffer significantly. The whole idea of cruelty to animals is hard to understand without the presupposition that most animals (i.e. those possessing some neurological complexity) have subjective feelings. Moreover, many of the arguments proposed in favour of the other conceptions are related to an understanding of welfare in terms of positive feelings: freedom of behaviour is crucial because it brings happiness or other positive feelings; health and coping normally result in positive experiences, etc.

The problem with considering feelings as being relevant for animal welfare has basically been the fact that this concept has for a long time not been seen as scientifically viable. The scientific school of *positivism* has had a considerable impact in this respect. In positivists' terms only such sentences as refer to observable events are scientifically meaningful. Sentences referring to unobservable events, such as mental states or values, are deemed meaningless and should be expelled from science. The impact of positivism remained quite powerful in animal science until a few decades ago. Many animal scientists have claimed (and still claim) that ethically important questions about subjective experience cannot normally be answered empirically. One might, for example, contemplate whether or not keeping a bird in a cage reduces its welfare by depriving it of the pleasure of flying. But this question cannot be answered by empirical science. In the positivists' view there is no acceptable method to quantify the pleasure or pain experienced by the animal.

Although, currently, this has become an unusual position, some theorists still argue that there is no such thing as animal mentation. Their main arguments for this stance include the following. Animal minds cannot be observed. At best, one must rely on indirect indications of an animal's mental state, for instance by studying their behaviour or physiological functions which might be similar to the reactions of humans experiencing pain or pleasure. Yet, these observations cannot prove the existence of an animal's subjective experiences.

However, one must confront the fact that one cannot prove anything in empirical science. 'Proof' is a concept that belongs to mathematics and logic. This truth has been particularly emphasised in modern philosophy of science (Popper, 1959; Kuhn, 1962). In empirical science one makes observations based on hypotheses, which are generally inspired by theoretical (unobservable) presuppositions. Among plausible hypotheses are the ones that concern animal mentation. Most human and animal behaviour is explained in terms of intentions, motives or – in general – feelings. Such hypotheses have been frequently tested and 'corroborated'. There are a large number of both pharmacological and neurological data, usually involving comparisons with humans, which speak in favour of animal mentation (Rowan, 1988; Panksepp, 1998; Uvnäs-Moberg, 1998).

It is crucial to add that we do not directly observe the subjective experiences of our fellow human beings either. The existence of such experiences in other human beings is also a hypothesis that one needs to continuously corroborate. However, obviously, nobody is questioning that all humans have a subjective mental life.

Are feelings or the subjective experiences the *chief candidate*, then, for characterising animal welfare? Some authors indeed argue they are and we have already seen several arguments for this thesis. Others authors still maintain a sceptical attitude, as discussed in the following section.

## 4. Are subjective feelings the central component in animal welfare?

### 4.1 The relation between health and subjective feelings

As we have already seen, there is no one-to-one correspondence between ill health and negative subjective experiences. Given the supposition that ill health in an animal must entail a low level of welfare, we should by consequence have to conclude that subjective experiences are not all that there is to welfare.

There are *two reasonable replies* to this statement. One is to deny that reduced health necessarily causes reduced welfare. This is a strong argument in the discussion of human welfare. People with a low level of health can have a high level of welfare (or quality of life), and *vice versa*. Also, a person with a chronic disease can live quite well and in general enjoy life.

The other argument is that ill health primes the animal for a *disposition* for negative experiences. Even though, admittedly, health is not always accompanied by negative subjective experiences (notably in the initial stage of a disease), ill health or disease will nearly always – directly or indirectly (e.g. through disability) – result in such a disposition. Only rarely a disease (e.g. some trivial skin disorder) has no bearing at all on the animal's welfare, but in such cases one would probably not even consider including such a disease in our conception of what is a reduced level of welfare.

By consequence, none of the above considerations represent arguments against putting the subjective experiences at the centre of animal welfare.

### 4.2 The relation between natural behaviour and positive experiences

We have already seen that there is not a one-to-one correspondence between natural behaviour and positive experiences. Natural life can involve a lot of painful struggle. However, if we restrict ourselves to interpreting natural life as freedom to cope and to fulfil one's preferences we certainly come closer to a one-to-one correspondence, albeit still not completely. One might argue that an animal can have a high degree of welfare without necessarily having full freedom to fulfil its preferences, as is the case in the situation of many pets and livestock, where – although freedom is restricted – animals are kept under very good conditions and their basic needs are cared for. In this case humans compensate for the animal's lack of freedom by seeing to its needs and thereby maintaining its positive experiences, i.e. its welfare.

### 4.3 Summary with regard to the definition of welfare

The aforementioned considerations essentially suggest that (particularly sustainably having) positive subjective experiences represent a very strong candidate for constituting the kernel of animal welfare. This is also illustrated by the fact that arguments in favour of the competing alternatives are often made with reference to animal's feelings. The value of natural behaviour is highly dependent on its relation to the long-term subjective comfort of the animal in question. The value of health, natural function and coping is highly dependent on its relation to the long-term subjective well-being of the animal. In conclusion, the realisation that there is a causal relation between, on the one hand, natural behaviour, health and coping and, on the other hand, positive subjective experiences, has in the scientific discussion been considered crucial for selecting the former factors as essential in the definition of animal welfare.

There is, however, no symmetry here. The value of positive feelings is rarely dependent on their relation to natural living, coping or health. For instance, whereas positive feelings induced by drugs and alcohol are not valued greatly since they are dependent on unnatural or even detrimental physiological function, such positive feelings are not enduring. Long-term positive feelings seem to retain their high value irrespective of natural living, health or coping.

It is important to once more stress that the existence of negative experiences is the chief ethical motivational factor for paying attention to animal welfare. For many it is only the animal's suffering that ultimately is of ethical relevance (McMillan, 2005; Rollin, 2005).

## 5. A word of caution

Notwithstanding this preliminary conclusion in favour of subjective experiences, there may be a feeling among many practitioners and theoreticians, that animal welfare is not properly addressed by merely following the 'subjective experience' approach. For instance, one is inclined to concur with the statement that a horse, kept in a large outdoor area in which it can stroll around and forage in the company of other horses, has a higher degree of welfare than another horse kept in a stable and being restricted in its movements most of the time, even if both have the same degree of pleasant experiences. This is an idea that has an Aristotelian origin. Aristotle thought that welfare for human beings entails virtuous *activity* and not just good *feelings*. It belongs, one could say, to the *dignity* of both human beings and animals that they should be able to behave and act in accordance with their innate or learned preferences. This natural life of unimpeded activity is in some respect better than the restricted life even if they were to involve the same amount of pleasure. This is a powerful argument.

In addition to this philosophical argument a more pragmatic consideration needs to be mentioned. Even if we were to accept a 'subjective experience' conception of animal welfare as ultimately the most reasonable option (i.e. for animals possessing some neurological complexity), we might for practical reasons wish to choose a much broader approach in our research. Animal welfare ethics/science must in practice deal with other things than the subjective experiences of animals. The reason for this is both methodological and empirical: there is not yet any accepted method to quantify and measure the pleasures and sufferings of animals. Indeed, a lot is known about the causal relationship between on the one hand coping, natural behaviour and health, and on the other sustainable positive feelings. Our knowledge, however, is incomplete and we cannot measure these relations. Therefore, much of the concrete welfare work must be directed towards such observable phenomena, along different dimensions, as we know they are related to sustainable subjective feelings. (For similar conclusions, see Fraser, 2008)

However, a multiple approach, be it based on a conglomerate concept of welfare or just on pragmatic methodological considerations, is not uncomplicated. Questions on the balance between the dimensions have arisen, for instance whether a high score on one dimension is in some way more crucial than a high score on another one. The conceptual analysis *per se* does not provide the answer. The assessors must judge whether one of the four dimensions is more important than the other. This means that when a conglomerate concept of welfare is put to use a number of difficult choices have to be made to ensure that this approach is operational.

## 6. Is welfare an evaluative concept?

Let us now turn to a philosophical issue which has plagued the contemporary animal welfare discussion, i.e. the question whether or not animal welfare is a straight-forward concept as is the concept of a stone or an atom, which scientists can define objectively without having to rely on evaluative terms. Or, alternatively, is animal welfare a concept like excellence or virtue which are inherently evaluative, and what theoretical consequences does an answer to these questions have?

Proponents of biologically oriented theories, including Barnard and Hurst (1996) and Broom, claim that the concept of welfare can be given a purely scientific characterisation without any moral or otherwise evaluative component, in other words: the process of assessing welfare is quite different from making a moral evaluation. Other theorists (in particular the supporters of the 'feeling' concept) deny that such assessment is possible without a moral evaluation. This controversy has a counterpart in the philosophy of human health and welfare, and the arguments in favour or against this notion should be considered in an attempt to take a stand on the issue.

Tannenbaum (1991) is one of the theorists who argue that '*the pure science model fundamentally misconstrues the nature of animal welfare*'. Tannenbaum states that it is impossible to study animal welfare scientifically without considering, or taking any positions with respect to, ethical issues. Tannenbaum also analyses the following position: Scientists might agree that the decision of an acceptable level of welfare is an ethical decision. However, they may still insist that once this decision is made one can carry out the study without reference to values. Tannenbaum objects also to this position. He notes that goodness is the most central evaluative concept and claims that welfare necessarily refers to what is good for somebody.

In order to settle this issue one has to distinguish between an ethical decision and a value judgement in general. The philosophy of values covers a much wider ground than ethics and includes e.g. aesthetic, religious and intellectual evaluations that have little or nothing to do with ethics. To say that something is beautiful, sublime or well-organised represents an evaluation, but not one in an ethical sense. Moreover, there are certain evaluations that are highly relevant to ethics without *ipso facto* being ethical. Central among such evaluations are those related to welfare. One can say that a certain state of affairs has a high welfare value without making an ethical judgment. The ethical judgment comes in when one decides that it is morally good or right to aim at this state of affairs.

Tannenbaum's position can hence be qualified by saying that a welfare evaluation can be made without taking any *ethical* decision. However, the pure scientists may still contend that determining that state X is a state of welfare is a purely scientific matter, not involving any evaluations, ethical or otherwise.

This author's viewpoint is that primary and secondary assessments of welfare need to be clearly distinguished. A primary assessment must involve an evaluation. A secondary assessment, although relying on a primary assessment, does in itself not express an evaluation. For instance, when a theorist has decided that coping to maintain fitness (in terms of survival, growth and reproduction) is *what should be meant* by animal welfare, this decision (or this choice of definition) involves an evaluation. It proposes that coping to maintain fitness is a *good* state to be in for an animal: it is the faring well of this animal. With this statement of Tannenbaum this author concurs completely: the scientist who supports this definition for his or her further scientific purposes has made a primary assessment and is committed to an evaluation.

However, once this primary assessment has been performed there are many things to follow of a non-evaluative, i.e. 'purely scientific', nature. One can, for instance, ask many causal questions concerning what environments, or what physiological states, contribute to welfare. In a secondary way the scientist can thereby assess that such contributing states belong to the states of welfare. However, in making such an assessment the scientist need not be committed to the primary evaluation. He or she need not even agree with this evaluation.

It should, however, be noted that it is difficult to avoid evaluations even on the secondary level. This relates to the fact that it is often difficult to precisely answer the causal (scientific) questions, and in such a case the less cautious scientist may unjustifiably assume that a state X is conducive to welfare or a hazard to welfare. The following is an example. A certain condition has long been included in a conventional list of diseases. Although it is often assumed that *all* diseases compromise welfare [for instance the ability to cope (see above)], the mere fact that the condition is listed under the heading 'disease' does not justify qualifying the condition as a hazard to welfare. To make such an assessment is to evaluate and not to make a scientific judgment.

It can thus be argued that 'welfare' is an evaluative term. In the lexical meaning of the word, for an animal to 'fare well' is to live a good life. When some theorists should deny that this is so, it is questionable why they choose the term 'welfare' in the first place and not 'coping for survival or reproduction' or any of the other notions that appear in the *analysans* of their investigations. To designate coping for survival as welfare is to say – unescapably so – that coping for survival is something *good*, something that we should appreciate. By consequence, one has then admitted that using the term 'welfare' commits one to an evaluation, although not necessarily a moral one. [Consider for further valuable distinctions concerning this matter, Sandøe and Simonsen (1992) and Fraser (1995)].

## 7. Conclusions

### 7.1 What has been achieved?

After having scrutinised some of the major interpretations of the expression 'animal welfare' it can be concluded that the various, essentially different, conceptions of animal welfare (freedom to perform natural behaviour – health and natural function – coping with the environment – positive subjective experiences) are to a significant degree both logically and empirically interrelated. This author has singled out 'positive subjective experiences' as the most central conception of welfare, firstly because these are often highlighted in arguments on which are essential components of welfare (e.g. natural behaviour for being conducive to an animal's happiness), and secondly because animal feelings represent the main ethical motive for enhancing animal welfare, and indeed preventing cruelty to animals.

Although animal welfare is an evaluative concept, one can discuss welfare without necessarily taking an ethical stance. Once a level of welfare has been established (by evaluation), then a number of purely scientific procedures, regarding measurement and causal analysis, can be performed.

### 7.2 What has been neglected?

Although a distinction between definitions, criteria for their application, methods for assessment, and causes and conditions of animal welfare should be made – as indeed has been made by some prominent authors (e.g. Broom and Kirkden, 2003; Fraser, 2008) – such has repeatedly been neglected in much of the literature on the subject. Consequently, it is not uncommon that an apparent conflict of opinion stems from the mere fact that – e.g. in characterising welfare – one author refers to observable criteria of welfare, whilst the other refers to underlying concepts. It is crucial, however, to acknowledge the differences between these different approaches to allow making progress in the area of animal welfare research.

### 7.3 What needs to be done?

An extended further analysis is necessary to assure that the afore-mentioned distinctions are identified and applied systematically, e.g. by establishing a map of relations between all the concepts surrounding animal welfare as discussed in this chapter. Such an analysis should also consider alternative mappings, i.e. alternative definitions of the relevant concepts. Lerner (2008) has initiated a systematic study of this kind. He proposes the construction of alternative clusters of concepts (in his case the concepts of health, well-being and welfare). He envisages that there is more than one reasonable way of reconstructing the relations between these concepts where the following preconditions are still fulfilled:
1. The definitions of the concepts are consistent with each other.
2. The definitions of the concepts are closely related to the use in ordinary language and the language in the science of animal health and welfare.
3. The definitions of the concepts are useful for specified purposes within the science of animal health and welfare.

By using these criteria Lerner has constructed different *clusters* of the triad of the concepts health, well-being and welfare. All of these proposed clusters are both logically possible and reasonable. In future research such an approach should be pursued further, and a larger number of relevant concepts, e.g. some of the ones discussed in this chapter (notably coping, natural behaviour and subjective experiences) should be included.

## References

Alrøe, H.F., Vaarst, M. and Kristensen, E.S., 2001. Does organic farming face distinctive livestock welfare issues? A conceptual analysis. Journal of Agricultural and Environmental Ethics 14, 275-299.

Barnard, C.J. and Hurst, J.L., 1996. Welfare by design: the natural selection of welfare criteria. Animal Welfare 5, 405-433.

Broom, D.M., 1988. The scientific assessment of animal welfare. Applied Animal Behaviour Science 20, 5-19.

Broom, D.M., 1991. Animal welfare: concepts and measurement. Journal of Animal Science 69, 4167-4175.

Broom, D.M., 1996. Animal welfare defined in terms of attempts to cope with the environment. Acta Agriculturae Scandinavica. Section A. Animal Science Supplementum 27, 22-28.

Broom, D.M., 1998. Welfare, stress and the evolution of feelings. Advances in the Study of Behavior 27, 371-403.

Broom, D.M. and Kirkden, R.D., 2003. Welfare, stress, behaviour and pathophysiology. In: Dunlop, R.H. (Ed.). Veterinary pathophysiology. Iowa State University Press, Ames, IA, USA.

Dawkins, M.S., 1980. Animal suffering: the science of animal welfare. Chapman and Hall, London, UK, 149 p.

Dawkins, M.S., 1983. Battery hens name their price: consumer demand theory and the measurement of ethological 'needs'. Animal Behaviour 31, 1195-1205.

Dawkins, M.S., 1990. From an animal's point of view: motivation, fitness and animal welfare. Behavioral and Brain Sciences 13, 1-9.

De Grazia, D., 2002. Animal rights: a very short introduction. Oxford University Press, Oxford, UK, 131 p.
Duncan, I.J.H., 1993. Welfare is to do with what animals feel. Journal of Agricultural and Environmental Ethics 6, Suppl. 2, 8-14.
Fraser, D., 1995. Science, values and animal welfare: exploring the 'inextricable connection'. Animal Welfare 4, 103-117.
Fraser, D., 2008. Understanding animal welfare: the science in its cultural context. Wiley-Blackwell, Oxford, UK, 324 p.
Fraser, D., Weary, D.M., Pajor, E.A. and Milligan, B.N., 1997. A scientific conception animal welfare that reflects ethical concerns. Animal Welfare 6, 187-205.
Harrison, R., 1964. Animal machines: the new factory farming industry. Stuart, London, UK, 186 p.
Haynes, R.P., 2008. Animal welfare: competing conceptions and their ethical implications. Springer, Dordrecht, the Netherlands, 160 p.
Hughes, B.O. and Curtis, P.E., 1997. Health and disease. In: Appleby, M.C. and Hughes, B.O. (Eds.). Animal welfare. CABI Publishers, Wallingford, UK, pp. 109-125.
Kuhn, T.S., 1962. The structure of scientific revolutions. University of Chicago Press, Chicago, IL, USA, 172 p.
Lerner, H., 2008. The concepts of health, well-being and welfare as applied to animals: A philosophical analysis of the concepts with regard to the differences between animals, Linköping Studies in Arts and Sciences No. 438, Linköping University, Linköping, Sweden, 212 p.
Maslow, A., 1968. Toward a psychology of being. Van Nostrand Co., New York, NY, USA, 240 p.
McGill, V.J., 1967. The idea of happiness. Fredrik A. Jaeger Publishers, New York, NY, USA, 360 p.
McMillan, F.D., 2000. Quality of life in animals. Journal of the American Veterinary Association 216, 1904-1910.
McMillan, F.D., 2005. The concept of quality of life in animals. In: McMillan, F.D. (Ed.). Mental health and wellbeing in animals. Blackwell Publishing, Oxford, UK, pp. 183-200.
Nordenfelt, L., 2000. Action, ability and health. Kluwer Academic Publishers, Dordrecht, the Netherlands, 173 p.
Nordenfelt, L., 2006. Animal and human health and welfare: a comparative philosophical analysis. CABI Publishers, Wallingford, UK, 190 p.
Panksepp, J., 1998. Affective neuroscience: the foundations of human and animal emotions. Oxford University Press, Oxford, UK, 466 p.
Popper, K.R., 1959. The logic of scientific discovery. Hutchinson, London, UK, 480 p.
Rollin, B.E., 1992. Animal rights and human morality. Prometheus Books, Buffalo, NY, USA, 248 p.
Rollin, B.E., 2005. Animal happiness: a philosophical view. In: McMillan, F.D. (Ed.). Mental health and wellbeing in animals. Blackwell Publishing, Oxford, UK, pp. 235-241.
Rowan, A.N., 1988. Animal anxiety and animal suffering. Applied Animal Behaviour Science 20, 135-142.
Sandöe, P., 1996. Animal and human welfare – are they the same kind of thing? Acta Agriculturae Scandinavica, Section A, Animal science, Suppl. 27, 11-15.
Sandøe, P. and Simonsen, H.B., 1992. Assessing animal welfare: where does science end and philosophy begin? Animal Welfare 1, 257-267.
Simonsen, H.B., 1996. Assessment of animal welfare by a holistic approach: behaviour, health and measured opinion. Acta Agriculturae Scandinavica. Section A, Animal Science, Suppl. 27, 91-96.
Singer, P., 1990. The significance of animal suffering. Behavioral and Brain Sciences 13, 9-12.
Tannenbaum, J., 1991. Ethics and animal welfare: the inextricable connection. Journal of the American Veterinary Medical Association 198, 1360-1376.
Tatarkiewicz, W., 1976. Analysis of happiness. Martinus Nijhoff, The Hague, the Netherlands, 356 p.
Uvnäs-Moberg, K., 1998. Oxytocin may mediate the benefits of positive social interaction and emotions. Psychoneuroendochrinology 23, 819-835.
Veenhoven, R., 1984. Conditions of happiness. D. Reidel Publishing Company, Dordrecht, the Netherlands, 461 p.

# Animal welfare in science and society

**Ron S. Broglio**
School of Liberal Arts, Georgia Gwinnett College, 1000 University Center Lane, Lawrenceville GA 30043, USA and Visiting Scholar, Department of English Language and Literature, University of Waterloo, Waterloo, Ontario N2L 3G1, Canada; rbroglio@charter.net

## Summary

Science serves as but one player within a larger public sphere that has competing players exerting influence on the discourse regarding animal welfare. The multiple voices show the interest in animal welfare yet unresolved nature of the topic. In part debates played out in the public sphere are indicative of social values but are also a result of competing social discourses including interests of: food industry, agricultural industry, government, environmental science, animal advocacy groups, 'foodies' concerned citizens all alongside animal welfare science. Fundamental in this conversation are (a) competing images of food producing animals and (b) production of food which remains opaque to the public eye. Further complicating the image of livestock animals and food is the general public's relationship to non-food domesticated animals, particularly pets. Conflicting images of domesticated animals complicates animal welfare discourse. Additionally, the audience in this dialogue is fragmented and conflicted in itself as the public takes on roles as consumer, citizen, and moral agent. Within each segmentation of the public's role there are degrees of activity, interest, or investment. Public discourse on animal welfare fits within an analytic and legal-political model of language and logic. Alternative language for thinking animal welfare and vulnerability could provide a larger ethical ground for thinking about the similarities and differences of animal from humans. Finally, the expert information and assessment by scientists are much sought after by a range of social players. The details of scientific findings can help shape the public discourse on animal welfare.

**Keywords**: advocacy, public sphere, social discourse, media, animal welfare

## 1. Introduction

There is an ongoing social concern as to who speaks for animals and their welfare. Scientists have developed a range of welfare policies for food producing animals (Fraser, 2008). However, scientists are not the only voice which must contend to be heard in the public sphere and in the spheres of policy. This essay will examine the players involved within the social dialogue of animal welfare along with their competing voices: food industry, agriculture industry, government, sciences, animal advocacy groups, 'foodies', and concerned citizens. The variety of forces does not present a single view of animal welfare but rather gives conflicting information and rhetorical presentation of welfare. The result of such conflicts is a dissonance for the general public regarding the welfare of food producing animals. An added problem for public perception of these animals is that welfare must contend with the social image of food which itself has a variety of players staking claims in the public sphere.

There is an increasing need for mediators and translators who can present the current social discourse in a reliable and unbiased manner.

## 2. Competing voices for animal welfare

Contact with animals other than domestic pets has decreased rapidly in the 20$^{th}$ century and now into the 21$^{st}$ century. This is particularly true in the developed world where, for example, 41% of Americans worked on farms in 1900 and less than 2% by 2000 (Bittman, 2009). As a result of less direct engagement with livestock, the general population relies upon mediated images of animals and, by extension, mediated images for animal welfare.

Science is one of several prominent forces competing for a voice in how society views food producing animals. Each group has its own goals and standards which results in competing paradigms for animal welfare. Currently under construction is an 'ethics in the making' rather than an already formed social ethics toward livestock animals (Latour, 1988). Animal behaviourists and psychologists have developed a picture of animal psychology and physiology sufficient for a basic understanding of when animals are in pain and suffering. Animal welfare scientists have translated some of these findings into practical ethical prescriptions for handling animals. Certainly there are controversies within the fields of animal behaviour and animal ethics but for the broader picture of how these fields connect to society, it can be said that science presents a sufficient picture of animals to provide a foundation for an animal welfare. Nevertheless, such an understanding of animals and implications for their handling have not solidified into a clear social view nor demand of how livestock should be treated.

### 2.1 Participants in the conversation

In addition to animal welfare scientists, active agents in creating a social language for animal welfare include the following:

- Food industry. The goal for industry is to maximize profits while minimizing expenses. In processing animals for food, the factory model predominates. Under such a model, the standard for evaluating production is maximized order and efficiency (Dreyfus, 2006). In factory farming, animals are a material to be fashioned in accord with machine efficiency. Animal welfare is secondary to the final product as defined by public policy parameters (Broom, 2000). In this food industry model animal welfare matters as a corporate public relations concern rather than as a primary concern within production.
- Agriculture industry. Agriculture is often closely associated with the food industry as part of agrobusiness and as such their model of animals and their end goal of order and efficiency are the same. Certainly agriculture has found some models of small scale production which separate it from the food industry; however, the predominant methods accord with the goals of the food industry.
- Government. Because governments set policies and regulations, they are fundamental agents in determining the limits and permissions of other players within the discourse of animal welfare. Recently government policies have come under scrutiny as they relate directly and indirectly to animal welfare. As an example of recent direct policy changes

in the United States, the state of California's Proposition 2 'Standards for confining farm animals' changes the space requirements for farm animals. As an example of government policy indirectly effecting animals, Pollan and others have criticized US subsidies for maize and soy which, because of their low costs, become a primary source of animal feed. Such feed then effects care and welfare of livestock (Pollan, 2007). Bittman points out that the 'US Food Pyramid' as a ratio of foods for daily consumption influences public perception of meats, dairy, and eggs (Bittman, 2009).
- Environmental science. Recent concerns over climate change as well as water quality have caused environmental scientists to voice concern over the rising population of livestock and their negative effects on the environmental (Steinfeld *et al.*, 2006). Prominent concerns include: the increased demand for meat among developing countries as they see increased prosperity, effects of extensive grazing as forests are converted into pastures and the effects of intensive animal feeding (in confined animal feed operations, CAFOs).
- Animal advocacy groups. Such groups are generally concerned with the health of animals in agrobusiness. Beyond general health, advocacy groups fracture around dietary proscriptions. For example, PETA promotes vegan and vegetarian diets while the Humane Society is less proscriptive by focusing their concern on the welfare of animals raised for food but without advocating particularly dietary regimes (Humane Society of the United States, 2006). A small separate but related set of animal advocates promote heritage breeds for the sake of genetic diversity, historical novelty and taste differences.
- 'Foodies'. Recently chefs, gourmands, and food writers have had a large impact on the popular and social image of food producing animals. Noted among these are Mark Bittman, Michael Pollan, Jamie Oliver, and Hugh Fearnley-Whittingstall. Because these figures have ready access to media, they are able to provide readily accessible and easily understood images of animal life and welfare. Recent trends in food writing and cooking shows give visibility to the methods of production from farm to table (see section 3.1). A prominent part of such methods is the rearing of animals.
- Concerned citizens. Where people have a choice in food consumption, they are able to voicing their preferences by their purchases. Carlo Petrini, a 'slow food' advocate, calls consumers 'co-producers' since their choices help shape the directions of animal husbandry (Petrini, 2006).

### 2.2 Public perceptions and overlapping sciences

Rarely do scientific studies reach the public unfiltered. Of concern is how an information saturated public can comprehend and process study results and their implications. The global demand for meat is rising. Since 1980 there has been a quadrupling of pig and chicken production and a doubling of cattle, sheep, and goat production (Bittman, 2009). In the next 40 years, meat consumption globally is targeted to double (Bittman, 2009). Livestock production is the fastest growing sector of agriculture. The food industry has continued a factory farming model to meet growing demand. The marketing end of the food industry continues to promote inexpensive meat, particularly in the developed world where it is in large supply. This signal from food and agriculture does not correspond to information from environmental and animal welfare sciences.

Environmental sciences are concerned about the release of methane from livestock. For example, methane emission from cattle effects climate change at a rate equal to the $CO_2$ emissions from cars (Steinfeld *et al.*, 2006). Additionally, sewage runoff from CAFOs has a negative effect on the surrounding rivers and streams (Environmental Protection Agency, 2000). Extensive grazing replaces forests with pastures, as for example, in the Amazon. Intensive rearing in CAFOs creates detrimental pools of waste that have a greater adverse impact on the environment than waste left in fields (Steinfeld *et al.*, 2006). CAFOs produce a number of welfare concerns as well since animal environment and behaviour is modified to meet conditions infelicitous to the proclivity of animals.

Given the food and agricultural industry message and the message from environmental and welfare sciences, the public image of food producing animals is fractured and incongruous. The result is a public which holds at the same time competing images of animals and competing values regarding their welfare (Pollan, 2007). Some popular writers have tried to reconcile the demand for meat with interest in the environment such as Bittman's *Food matters* which is subtitles 'A guide to conscious eating'. Yet, in such instances, the solution is eating less meat and using extensive grazing. Eating less is a message contrary to the message of the food industry and agribusiness. Extensive grazing, as noted above, is not without problems depending on what types of landscapes the pasture replaces and the effects of livestock on water supplies and water quality. So, again, the public is left with competing values and interests.

## 2.3 Static and dynamic images of the animal

With fewer people labouring on farms, particularly in industrialized countries, the public image of food producing animals is incongruous with the actual state of affairs on farms. For example, since 2002 the California Milk Advisory Board has disseminated the slogan 'Good cheese comes from happy cows. Happy cows come from California. Real California cheese'. The accompanying ad campaign shows cattle relaxing in verdant pastures. Bulls lasciviously eye cows and comment on their beauty. One wonders why bulls would have a prominent role in a dairy farm and ads for cheese. In this and many other instances, the food industry relies upon an antiquated and bucolic image of farming particularly one without intrusive technologies which would call attention to contemporary methods of food production.

### *2.3.1 Static and dynamic images defined*

The livestock industry has developed a number of technological interventions in the lives of animals from birth to death, or from artificial insemination to stun guns and irradiated meats. From even a cursory look at the livestock industry and scientific advancement of the industry, it is evident that the contemporary state of rearing livestock is different from the image commonly held by the public. At issue here is the difference between a static image of animals and a dynamic image of animals.

In a static image, animals have inherent characteristics which do not change over time. A cow is a cow, for example. In a dynamic image, the animal's traits change over time according to variables at least in part controlled by breeders and handlers. So, for example, because of an increased number of animals in CAFOs, it is possible to breed animals which more adequately adapt to 'hot rations' and other environmental conditions encountered by most livestock today (Pollan, 2007). Because of an increased demand and higher price for Angus cattle, it is possible to increase the number of cattle with Angus markings (51% black hide) through increased breeding of Angus and cross-breeds which include sufficient Angus characteristics. So, for agribusiness and for science, the image of animals is dynamic.

### *2.3.2 Interior and exterior of animals as related to dynamic image*

Animal breeding provides a look at the dynamic image of animals and the influence of that view on actual individual animals as well as breeds. Traditionally livestock breeders read the outward appearance of prized animals, and the only glance at the animal's interior happened at slaughter when the inside of the animal becomes a visible outer surface of meat. In the case of prized breed cattle, the animal itself would not be slaughtered; instead, its traits are discerned outwardly and by the meat produced by its progeny. However, today with functional genomics the outer surface of cattle hair and skin are transformed into genetic information which yields a look 'inside' the beast. Given this look inside, computer programs which run genomic inputs now allow cattlemen to match theirs cow and bulls to achieve 'targets' or ideal offspring.

The challenge for animal husbandry both in the past and today is to see beyond the surface. If one could see inside the animal and into the structure of its flesh and intramuscular fat or 'marbling', then it would be easier to make decisions on which animals to mate in order to achieve desired goals. Current ultrasound technology is helping breeders to see the insides of breeding cattle. Yet, to be even more ambitious, if one could see into the future, if one could not just imagine but also produce images of projected progeny from a particular dam and sire pairing, then animal husbandry would have taken a revolutionary turn. It is just such a turn that is underway today as technology has created new means of breeding through the combination of artificial insemination and genetic selection. Genetic selection remains in its infancy compared to many other breeding technologies but promises the most intensive predictive power and chance both to look below the surface of the animal and to project and alter future progeny. Animals are breed according to traits desired by the market and by the method of rearing them.

### *2.3.3 Interior and exterior of animals as related to a static image*

Genetics provide new means of targeting particular characteristics of livestock and making demands upon the animal to 'beef up' these traits. While ranchers work to know and modify the inside or interior of the animal, what remains under question (particularly for issues of welfare) is the animal's 'interiority'. Do animals have a 'world view', a *gestalt*, and how might genetic modification through breeding effect this view from within? (Uexküll, 1957) To a great degree, this is an unanswerable question (Nagel, 1974).

Yet, the very futility of asking what it is like to experience the world as various livestock do is the usefulness of this question since it requires one to create a new means of evaluating animals. This question allows for a particular interiority of the animal and precisely to the degree that humans can not access it, such interiority remains just that, interior and not an externalized and accessible commodity (see section 5). Essentially, such an interiority evades representation and the accompanying technological refashioning of the interior through breeding. Whether static or not, the concept of an interiority of the animal – the cow as cow, for example – has an effect on the limits of what could be ethically done in breeding.

The Committee on Managing Global Genetic Resources has reported their concern over the reduction of genetic difference within a species through modern breeding practices; however, this has done nothing to affect the economy of industrial breeding. Instead, it merely reserves a place for genetic difference in databanks and cryogenic sperm banks. Concerns for any 'interiority' of the animal are also noticeably absent from the US National Research Council report on the health issues related to biotechnological farm and research animals (Davis, 2002). What is lacking here and elsewhere in scientific discourse is a conversation about breed and species 'mutation' which takes place through [. Lassen *et al.* (2006) states the case quite plainly:

'Often they [concerns regarding genetic manipulation of animals] are expressed in terms of boarders and limits. They indicate that to some people there are as it were freestanding limits to the degree of control that humans can ethically exercise over other living beings. Concepts such as integrity and dignity are often evoked to describe these limits. There is a tendency in the [scientific] academic literature to write off such limits as irrational and/or religious, but lack of conceptual clarity should not lead to dismiss them too quickly.'

Unfortunately the authors do not address these concerns but rather state them as a problem of public opinion, to be corrected or negotiated by scientists. At stake, however are larger questions outside the boarders of typical scientific discourse, questions which address the role of technology and anthropological questions about the role of human behaviour in shaping domesticated animals. Knowing the animal's insides is not knowing its interiority or 'what it is like' to be livestock. Genetics has extended and is poised to supersede phenotypical methods of breeding and shaping animals inside and out, in what appears to be a totalizing knowledge; this data that proposes 'insider knowledge' refuses to engage with the static image of animals and with the question of interiority, the possibility that there is an animality of the animal outside the horizons of human knowledge.

When a dynamic image of the animal and static image of the animal are not reconciled, or to the degree that interiority is not a functional concept, the public remains confused about the image of livestock. Such an image is necessary in order to advance a conversation about animal welfare. Again, the question is raised as to who speaks for these animals. Furthermore, what part of these animals – exteriors, interiors, interiorities – are part of the conversation?

## 2.4 Role of the social citizen

Competition in speaking on behalf of food producing animals is matched by the segmented role of the public. At various times, individuals are called upon to play the role of consumers, citizens of a particular country or region, moral agents, connoisseur of taste, and biological beings. Each role pitted against the other creates competing values. These roles may have competing demands on the individual which is further complicated by the number of participants speaking on behalf of animals and their welfare (section 2.1). So, for example, a consumer may want to pay as little as possible for meat or dairy but products from humanely raised animals may cost more. An individual may want to support local agriculture only to find that prices for local animal products are higher than those of large scale national or international brands. In another example, the culturally developed taste for a particular sort of animal product – for example 'marbled' or fatty meat – may run at odds with biological health. The welfare of food producing animals occupies only a segment of the individual's biological, culinary, and lifestyle choices as he assumes the role of consumer, citizen of a particular country or region, moral agent, connoisseur of taste, and biological being.

#### *2.4.1 Consumer as co-producer*

Carlo Petrini's concept of consumer as co-producer is an attempt to make individuals aware of these conflicts (Petrini, 2006). Petrini makes a rhetorical shift in defining the consumer. The consumer as co-producer means that purchases are an investment in or funding for particular farming and agricultural practices. The consumer becomes co-responsible with the producer in animal welfare since the consumer as co-producer is funding particular methods of production. Petrini assumes here that the consumer or co-producer has sufficient knowledge upon which to make judgments about what would constitute humane treatment of animals and that consumers would in fact care to pursue this concern.

Behind Petrini's concept is a particular Enlightenment understanding of social democracy: that objects and situations can be clearly conceived, that citizens act from reason, and that the citizen has sufficient knowledge for judgment. Petrini assumes that one can gain sufficient knowledge about animal production and welfare at the point of purchase or through some casual inquiry. He also assumes that through *sensus communus* or common sense – a sensibility held in common by all in the civic society – one can make proper choices as to what constitutes animal welfare. So, while for ethicists and for veterinary scientists, animal welfare is a particular disciplinary study, for Petrini such study is only useful as it informs the public of its practical choices.

#### *2.4.2 Varied investment of consumers*

Petrini as a founder of the slow food movement has as his model a citizenship that is invested in knowing food production and capable of obtaining such knowledge. While this is a particular construction of the citizen, at best it represents a small segment of the population. More practically, social roles are variously segmented as: consumer, citizen, moral agent. Within each segmentation there are degrees of activity, interest, or investment from highly involved, to somewhat, casually involved, or not at all interested. So, for example, a highly involved consumer but with low interest in ethical debates regarding food production would buy according to cost or taste. An active citizen may buy according to community or locale but not be as concerned for price, taste, or methods of production. In short, animal welfare becomes one of the competing values for understanding food choices.

## 3. The social image of food producing animals contends with images of food

One of the fundamental issues of food production is transparency. How much information should the consumer be able to obtain at the point of sales and how much should be readily accessible for inquiring consumers? Technological scaling of production has increased the distance between consumer and methods of production. Such distance is both literal and metaphorical. That is, fewer people participate in or see food production. 'Distance' means consumers are removed not only physically but conceptually from understanding and inquiry into food production.

For developed countries, the rise of food sciences and engineering for mechanical production of food has helped produce more food and at lower consumer costs than any moment in recent history. Two prominent characteristics of this model are assembly production and food systems:

- The assembly model of agriculture remains prominent. Fundamental to this mode of production is thinking of biological systems within the framework of mechanical or assembly line systems with limited and containable variables and inputs (Shukin, 2009).
- Food systems also known as processed foods are more profitable for agribusiness than less processed or raw foods (Pollan, 2006). A food system breaks down raw components of food and reconfigures them in processed form as ready to eat food items. The 'chicken finger' is an example of a food system. Such processed chicken parts are more profitable when configured as a food item than if sold as raw products.

The lack of transparency within current methods of production benefits business through commodity fetish (Pollan, 2006; Shukin, 2009). The food item as commodity can be valued above its cost of production because the methods of production are not sufficiently understood by the consumer. The 'fetish' quality of the food item is its value above and beyond its contents and production values.

What allows the obfuscation of food production is the black boxing of methods of production. A black box, as defined by Latour, is a complex set of actions simplified by bracketing out the details and only representing the input and output actions (Latour, 1988). By black boxing the production of food derived from livestock, ethical questions regarding production methods are inumbrated. In place of detailed production history is the 'fetish' or image value of the animal or animal product (see section 3.3 below).

### 3.1 Recovery of the methods of production and production history

Over the past years there has been an increased interest in the methods of food production. Such interest does not seem fully organized as a coordinated and sustained social movement and may only be a passing fad for the media and the public. Nevertheless, it is worth noting that commensurate with the distancing between consumers and food production is an interest in how food gets produced. The result is an *unscientific but popular* set of investigations into production and animal welfare advocacy. Exemplary of such interest are the following:

- Cooking shows. British celebrity chef Hugh Fearnley-Whittingstall was featured as part of BBC 4's cooking show *Big Food Fight*. Fearnley-Whittingstall's contribution consisted of three episodes, showing consecutively intensive chicken farming, commercial free range, and a local community farm. The methods of production in each farm were discussed in some detail. Following the show, Fearnley-Whittingstall began a campaign called 'Chicken Out!' to advocate alternatives to intensively raised chickens (Compassion in World Farming, 2008). The campaign focused on consumers as co-producers and asked consumers to buy only free range chickens (see section 2.4.1). Fearnley-Whittingstall used the guidelines established by Great Britain's Farm Animal Welfare Council to determine welfare in methods of production.
- Talk shows. Oprah Winfrey has used her talk show along with its considerable audience as a base for taking up issues of animal welfare. In April of 1996, she addressed the issue of mad cow disease (BSE) and rendered animal parts in animal feed. More recently, in October of 2008, she aired a full episode on 'How We Treat The Food We Eat'. The episode focused on housing and care of food producing animals just prior to the California Proposition 2 vote on animal confinement in November of 2008 (Ling, 2008). As a matter of media dissonance, in April 2009 Oprah offered discount coupons for Kentucky Fried Chicken *via* her website and so provided positive media attention for the very companies that the 'How We Treat The Food We Eat' argues against. Such contrasting animal welfare values within the same program are indicative of the conflicting sentiments held simultaneously by the public regarding food producing animals (see section 2.4).
- Journalism. An increasing number of journalists are writing on issues that connect public policy, human health, and animal welfare. Most notable and exemplary among these is Michael Pollan in *Omnivore's dilemma* and in *Defense of food*. The gourmet and food writer Mark Bittman has followed with *Food matters*. These authors are concerned with meeting animal needs or proclivities alongside market demands for animal products.

These examples, which are indicative of the social discourse on animal welfare, fail to employ the work of veterinary scientists and animal welfare science. It is worth noting the disconnect between science and media outlets by which most citizens and consumers get information. Perhaps the one 'media star' among animal scientists has been Temple Grandin. It may be that her personal story of autism coupled with her relationship to animals has allowed a particular notoriety not accessible by other scientists working with livestock. Grandin's media success is an exception that proves the rule that overall scientists have not had direct impact on the public through media outlets.

### 3.2 Niche food markets which incorporate part of the methods of production

Increasingly niche food markets have developed to take advantage of public interest in the methods food production. Perhaps best known in this category is milk from cows not given growth hormones (such as rBGH). Despite the US Federal Drug Administration approval of the hormone, there has been a steady and increasing demand for organic milk. Scientific study and government approval are trumped by consumer anxiety regarding the purity of milk (Corey, 2009).

There are a number of fast food establishments which have marketed and profited from particular traits of livestock. For example, Chipotle sells meat that is 'naturally raised' without growth hormones or antibiotics. From its growth in 2001 until 2006 McDonalds held a majority stake in Chipotle. Outside of welfare issues but included in methods of production are restaurants which sell Angus beef. The successful marketing campaign regarding the flavour of Angus beef has led to increased profit for this group of cattlemen. It is commonly held that the National Cattleman's Beef Association wants to do for beef what Perdue did for the chicken, i.e. create consistent and marketable name brands.

As stated previously, the public is given a variety of messages with conflicting values and interests. For example, Angus cattle may provide flavourful meat, but so do other breeds and the specialty breed label does not address animal welfare. Chipotle's niche food market for 'humane' meat supplemented McDonald's larger market. While beyond the scope of this chapter, it is worth considering how animal welfare is used as a marketing tool.

## 4. The family pet and the family food

The pet population has been on the rise in developed nations. 58.3% of all American households have at least one pet. There are now 81.7 million pet cats and 72.1 million dogs in the United States (AVMA, 2007). Following affluence in the human society, there has been a growing market for pet care. Veterinary medicine for pets has become more affordable and more far reaching it its ability to sustain the lives of pets (Kay, 2008). *Speaking for spot: be the advocate your dog needs to live a happy, healthy, longer life* by Nancy Kay gives advice and examples of the complex issues involved in how to be an advocate for the welfare of one's pet. There is a rise in animal products and pet specialty stores. A number of television shows develop and reinforce human bonds with pets.

It is unclear how the affection and care for pets has affected the welfare for food producing animals. It is occasionally noted with some irony that the family dog or cat has a privileged position in the home while intelligent and socially adept animals such as pigs have a place at the family table as food. Haraway has adopted a language for discussing the co-evolution of humans with working animals but focuses primarily on 'companion species' such as the dog. For Haraway the harmony achieved with some species such as dogs is at the cost of other species such as livestock. For her, there is a violence and competition among species that may include animal welfare but does not preclude animal death. Consequently, for Haraway, there are degrees of welfare and advocacy that are coordinated with an animal's role within our social system. The public sphere has competing values or different values for different domesticated animals (Haraway, 2007).

## 5. Continental philosophy, animal rights, and welfare advocacy

In general, when ethical arguments are invoked in media or personal debates regarding livestock animal welfare, the parties involved explicitly or implicitly work from an analytic philosophy framework. This is in part because Peter Singer's *Animal liberation* from 1975 and Tom Regan's *The case for animal rights* from 1985 are well known and have had time to circulate. More particularly, the social and legal framework is sympathetic to analytic arguments in a rights perspective. The pragmatics of rights fits with liberal jurisprudence as both are inheritors of similar philosophical, historical and ideological modes of thinking (Wolfe, 2009). Singer and Regan seek positive characteristics and capabilities of animals and from these abilities argue for their rights.

Singer wishes to take into account the interests of each being. Equal consideration for animals should be based upon consideration of their interests. Each being is treated differently according to its interests and proclivities. To deny animals their interests is to make an artificial divide between humans and other species (Singer, 2009). Regan argues from animal autonomy defined as preferences and ability to act upon these preferences. Consideration of autonomy includes interests, benefits, and harms according to physical and psychological constraints and capacities of various animals (Regan, 2004).

Recent work in strands of continental philosophy shifts the grounds for ethics away from rights arguments. While admittedly rights arguments are most productive for social-political action, for Cary Wolfe and Cora Diamond, they fall short of a more radical ethics of care (Cavell *et al.*, 2008). Following Niklas Luhmann, Wolfe argues that every mode of identification holds a particular blind spot to difference. Otherwise said, every mode of seeing the world entails a particular blindness. Where analytic and rights philosophy argues according to a difference within sameness (different needs of various animals and humans but sameness in having needs and preferences), Wolfe argues for an ethics of difference (Wolfe, 2009). An ethics of difference argument makes the following claims:

- Any social discourse will be blind to a discourse that is not similar to itself. We can only think and understand based on similarities.
- Laws established through a particular social discourse will become a self-instantiating tautology since of necessity the social and legal system cannot think from outside of its limits and fundamental laws which found it. It is under this rubric that analytic philosophy and rights arguments make their claims of continuity between humans and nonhuman animals.
- Analytic philosophy and rights arguments assume that people make choices based on consciousness, reason, and transparency. Moral decisions include conceptual clarity, information, and rationality (Regan, 2004). This comes close to an ethics of calculability. That is, one could calculate and rationally dictate moral decisions. Continental philosophers discussing animal rights are generally skeptical in this regard. According to continental philosophy (broadly conceived), moral and legal discourse is situated within a particular social framework that already has cultural standards and power relations established. Such standards and relations inform the decision making process and so tip the scales toward particular social prejudices.
- Wolfe and Diamond turn from a language of rights and capabilities back to Jeremy Bentham's argument of suffering: animals should be given rights because like humans they can suffer. For these authors, suffering indicates a vulnerability which humans are aware of and can share with other animals. As defined by these authors, vulnerability is not an argument of capabilities and powers but rather the very lack of such power. Animals and humans share the negative capacity of not having power, i.e. being vulnerable.

Wolfe and Diamond do not arrive at a particular set of welfare practices nor legal proscriptions but admittedly this is not their aim. Instead, they are interested in the limits of ethics within the social discourse. It is possible that the multiple discourses discussed previously in this chapter provide disjunctive narratives and ethical claims that allow for competing ways of understanding the lives and roles of food producing animals within a particular culture. In the competing languages, the insights and blind spots of any particular approach is confronted by other approaches. The result is not necessarily a clarity of moral concepts.

Wolfe and Diamond's argument opens the possibility of thinking of animal lives in-themselves rather than animals as they relate to and are understood by humans. That is, to recall terminology used earlier, Wolfe and Diamond consider animal interiority. By considering the possible blindness of human judgment as relates to animals and by considering vulnerability as a category of concern, these authors have created a mode of thinking animal welfare that, perhaps radically, is outside how humans interact with and use livestock animals. Singer and Regan are concerned with the interests and abilities of animals in order to make decisions regarding proper treatment and handling of animals. Wolfe and Diamond could be seen as admitting there are some ways of animal being (what Singer might call 'interests') to which humans remain blind. Such was of being would constitute animal interiority. These authors would prefer to maintain a space for considering that which humans cannot know or perceive regarding the animal.

This may seem like a radical or idealist concept but it may have practical application when considering future models for livestock breeding and overall valuation of animals as 'fit' for the parameters of current large scale management operations. In other words, the question becomes: are there genetic, physical, and psychological factors that are not items of concern for breeders or handlers but over years of breeding could change the animal in-itself (interiority) in ways that remain unknown until many generations later (if at all)? Such speculative thinking may be outside the normal range of practical scientific inquiry but the conversation between ethics and science at this level constitutes an ethical concern that is wide and far reaching for animals and the environments they inhabit. By way of examples, one may think of models for environmental effects of importing Norwegian Red Cattle into Iceland and displacing the more 'native' Icelandic cattle population. Or, one can consider how diet and breeding of cattle might affect the prevalence and virility of *E. coli* 0157:H7 and so have effects on human health. The range of possible effects of such blind spots is, of course, by definition not wholly conceivable.

## 6. Conclusions

### 6.1 What has been achieved?

Particularly in developed countries animal welfare has become an item of media attention and public discussion. The multiple discourses show the interest yet unresolved nature of the topic. In part debates played out in the public sphere are indicative of social values but are also a result of competing social discourses including interests of: food industry, agricultural industry, government, environmental science, animal advocacy groups, 'foodies', concerned citizens all alongside animal welfare science. Among each group there is a growing awareness that animal welfare is part of the larger social discourse. In some cases, animal welfare science plays a role in shaping the opinion of these various groups and interests.

## 6.2 What has been neglected?

For most citizens, [ is a legitimate and respected authority which is often but not always trusted. Recent food born illnesses have weakened social trust in science and agricultural sciences, not always for rational reasons. Furthermore, most citizens are not aware of the heterogeneous discourse in science regarding animal welfare. Competing social discourses, such as the food industry pitted against animal advocacy groups, may selectively choose scientific statements according to their own agendas. The result for the general public is a confused message as to scientific standards and directions for animal welfare.

While scientists talk to each other and to the industries which they study, there has been little visibility of scientists in the public discourses on animal welfare. From the public perspective, many players weigh in to the debate but notably absent is the voice of scientists. This may be in part because scientists speak through various social players – such as the agricultural and food industries. Outside the conversation between science and industry, these social players fashion their own message regarding animal welfare. Speculatively, for science to speak against industry would have a negative effect on animal welfare as it would sever the dialogue between scientific findings and industrial standards. In general, animal welfare science seems to have found its most impact in speaking to social players rather than to the public in general.

## 6.3 What needs to be done?

Animal welfare scientists would do well to consider themselves as social players in the conversations regarding animal welfare. What remains absent and needed is a non-political, non-profit and unbiased space for translating animal welfare science to the public. As a model for such a space, one could consider the work of climate scientists at Climate Central. Because of conflicting interests in the number of players in the public sphere regarding climate change, the general public has a mixed and sometimes confused understanding of the state of affairs. Climate Central is designed as a one stop, timely, relevant, and informative internet site from which the public and media can gather reliable material regarding climate change. The center is non-profit and non-partisan. It employs a regular staff and has a board of respected scientists in the fields of climate change.

The history of Climate Central serves as a model for constructing an engaged space for animal welfare science. Climate Central emerged from several high level scientific conferences both large and small. Funds were gathered from a non-profit interested in this issue. The group maintains high visibility both within the scientific and social arenas.

Regardless of whatever space scientists make for themselves within the public sphere, it remains important to understand that voice and visibility of animal welfare sciences is not in conflict to scientific objectivity. Rather, science is always already a player within the public sphere. The expert information and assessment by scientists are much sought after by a range of social players. The details of scientific findings can help shape the public discourse on animal welfare. Such information allows the public to think through the particularities of ethics and welfare.

## References


American Veterinary Medical Association, 2007. U.S. Pet ownership and demographics sourcebook 2007, 159 pp.

Bittman, M., 2009. Food matters: a guide to conscious eating. Simon and Shuster, New York, NY, USA, pp. 325.

Broom, D.M., 2000. Does present legislation help animal welfare? Sustainable animal production: workshops, discussion, online resources. Available at: http://www.agriculture.de /acms1/conf6/ws5alegisl.htm. Accessed 25 April 2009.

Cavell, S., Diamond, C., McDowell, J., Hacking, I. and Wolfe, C., 2008. Philosophy and animal life. Columbia University Press, New York, NY, USA, 172 pp.

Corey, B., 2009. Bovine growth hormone: harmless for humans. FDA Consumer, April 2009. Available at: http://www.fda.gov/bbs/topics/consumer/con00068.html. Accessed on 3 May 2009.

Compassion in World Farming, 2008. Chicken Out! Available at: http://www.chickenout.tv/. Accessed 10 May 2009.

Davis, D., 2002. National Research Council, and Committee on Defining Science-based Concerns Associated with Products of Animal Biotechnology. Animal Biotechnology: Science-Based Concerns. National Academies Press, Washington, DC, USA, 201 pp.

Dreyfus, H., 2006. Heidegger on the connection between nihilism, art, technology and politics. In: Guignon, C.B. (Ed.). The Cambridge companion to Heidegger. Cambridge University Press, Cambridge, pp. 289-316.

Environmental Protection Agency, 2000. Proposed regulations to address water pollution from concentrated animal feeding operations. Available at: http://www.p2pays.org/ref/01/00830.pdf. Accessed on 1 March 2009.

Fraser, D., 2008. Understanding animal welfare: the science in its cultural context. Wiley-Blackwell, Oxford, UK, 324 pp.

Haraway, D., 2007. When species meet. Minnesota University Press, Minneapolis, MN, USA, 360 pp.

Humane Society of the United States, 2006. An HSUS report: the welfare of animals in the meat, egg, and dairy industries. Available at: http://www.hsus.org/farm/resources/research /welfare/welfare_overview.html. Accessed 3 March 2009.

Kay, N., 2008. Speaking for Spot: be the advocate your dog needs to live a happy, healthy, longer life. Trafalgar Square Books, London, UK, 384 pp.

Latour, B., 1988. Science in action: how to follow scientists and engineers through society. Harvard University Press, Cambridge, MA, USA, 288 pp.

Lassen, J., Gjerris, M. and Sandøe, P., 2006. After Dolly: ethical limits to the use of biotechnology. Theriogenology 65, 992-1004.

Ling, L., 2008. How we treat the food we eat. Available at: http://www.oprah.com/dated/oprahshow/oprahshow_20081008_animals. Accessed 1 May 2009.

Nagel, T., 1974. What is it like to be a bat? Philosophical Review 83, 435-450.

Petrini, C., 2006. Slow food revolution: a new culture for eating and living. Rizzoli, New York, NY, USA, 312 pp.

Pollan, M., 2007. The omnivore's dilemma: a natural history of four meals. Penguin, New York, NY, USA, 464 pp.

Regan, T., 2004. The case for animal rights. University of California Press, Los Angeles, CA, USA, 425 pp.

Steinfeld, H., Gerber, P., Wassenaar, T., Castel, V., Rosales, M. and C. De Haan, 2006. Livestock's long shadow: environmental issues and options. Food and Agriculture Organization, Rome, Italy. Available at: ftp://ftp.fao.org/docrep/fao/010/A0701E/A0701E00.pdf. Accessed 3 March 2009.

Shukin, N., 2009. Animal capital: rendering life in biopolitical times. Minnesota University Press, Minneapolis, MN. USA, 306 pp.
Singer, P., 2009. Animal liberation. Harper, New York, NY, USA, 311 pp.
Uexküll J., 1957. A stroll through the world of animals and men. In: Schiller, C.H. (Ed.). Instinctive behavior: the development of a modern concept. International Universities Press, New York, NY, USA, pp. 5-80.
Wolfe, C., 2009. On a certain blindness in human beings. In: Cavalieri, P. (Ed.). The death of the animal. Columbia University Press, New York, NY, USA, pp. 123-134.

# Animal welfare components


## Xavier Manteca[1], Antonio Velarde[2] and Bryan Jones[3]

[1]School of Veterinary Science, Universitat Autònoma de Barcelona, Spain; xavier.manteca@uab.cat
[2]Animal Welfare, Institute for Food and Agricultural Research and Technology (IRTA), Spain
[3]Animal Behaviour & Welfare Consultant, Edinburgh, Scotland, United Kingdom



## Summary

Animal welfare can be defined in a number of different ways, but there is a growing consensus that whatever the definition, it has to include three elements: the emotional state of the animal, its biological functioning and its ability to show normal patterns of behaviour. These three elements are by no means contradictory; in fact they are closely interrelated. The Five Freedoms developed by the Farm Animal Welfare Council of the UK provided an elegant framework for doing this. These freedoms, which represent ideal states rather than actual standards for animal welfare, include: freedom from hunger and thirst, freedom from discomfort, freedom from pain, injury and disease, freedom to express normal behaviour, and freedom from fear and distress. More recently, the WelfareQuality® project built on and extended the Five Freedoms to four principles (good housing, good feeding, good health and appropriate behaviour). Each of these four principles comprises several criteria, with an overall total of 12 criteria. These 12 animal welfare criteria are discussed in this chapter as they provide a very useful framework for understanding the components of animal welfare. The next step is to establish reliable and practical ways of measuring the many components of welfare.


**Keywords**: animal welfare, assessment, Five Freedoms, WelfareQuality®

## 1. Introduction

Animal welfare can be defined in a number of different ways, but there is a growing consensus that whatever the definition, it has to include three elements: the emotional state of the animal, its biological functioning and its ability to show normal patterns of behaviour (Duncan and Fraser, 1997; Mendl, 2001). Indeed, it is now widely accepted that an animal's welfare embraces its physical and mental state and that good animal welfare implies both fitness and a sense of well-being.

Since the subjective feelings of the animal are an essential part of its welfare (Dawkins, 1990) a logical argument is that welfare will be reduced by negative subjective states such as pain and chronic fear, for example, and that it will be improved by positive states such as comfort and play.

Measures based on biological functioning and on the animals' ability to cope with the environment also provide relevant information on welfare. It was proposed that both a possible failure to cope and the fact that the animal has difficulty in coping indicate poor welfare (Broom, 1986, 2001). Researchers following this approach have been greatly influenced by the concept of stress (Duncan and Fraser, 1997). In 1929 Cannon described stress as the sympatho-adrenomedullary (SA) system's attempt to regulate homeostasis when threatened by a variety of aversive stimuli. Later on, Selye (1936) conducted some of his classic studies on the response of the hypothalamic-pituitary-adrenal (HPA) axis to noxious stimuli; Selye suggested that the organism reacted in a nonspecific manner to a wide variety of aversive stimuli, and this stress reaction was termed 'General Adaptation Syndrome'. More recently, Mason (1971) suggested that the psychological component of the aversive stimuli is the main determinant of the stress response. For example, animals that could control and/or predict the occurrence of an electric shock showed less pronounced stress responses than counterparts with no control or warning signals (Weiss, 1970, 1971). Most researchers agree that the animal's appraisal of the situation is a major determinant of the stress response (Terlouw *et al.*, 1997).

Farm animals kept in semi-natural environments have been show to express most of the behavioural patterns of their wild ancestors (Jensen, 1986). There is evidence that the performance of some of these behaviours is important for the welfare of the animals because the achievement of their functional consequences, in the absence of the behaviours themselves, is not sufficient to reduce their motivation (Petherick and Rushen, 1997).

The three elements mentioned above are by no means contradictory; in fact they are closely interrelated. For example, when animals are prevented from performing a particular behaviour pattern a stress response may follow (Mason *et al.*, 2001). Also, negative emotional states often result from the animal's inability to show appropriate behavioural responses and thereby failing to cope with the situation (Broom, 2001).

Before we can fully understand and improve animal welfare we need to firmly establish its influential components as well as the major welfare problems faced by each species and the 'conditions of good welfare' that they should be provided with. The Five Freedoms developed by the Farm Animal Welfare Council of the UK provided an elegant framework for doing this. These freedoms, which represent ideal states rather than actual standards for animal welfare, include: freedom from hunger and thirst, freedom from discomfort, freedom from pain, injury and disease, freedom to express normal behaviour, and freedom from fear and distress. The next step is to establish reliable and practical ways of measuring the many components of welfare. As one of the first steps in its efforts to develop standardised assessment systems the WelfareQuality® project built on and extended the Five Freedoms to the four principles and 12 criteria described immediately below.

## 2. The WelfareQuality® project's principles and criteria of good welfare

WelfareQuality® (2004) is an integrated research project funded by the European Commission in the sixth framework programme, priority 5: food quality and safety. Forty-four institutes and universities (representing thirteen European countries and four Latin American countries) with expertise in animal welfare participate in this project which runs from May 2004 to

December 2009. One of its main objectives is to develop an integrated, standardised welfare assessment system for cattle, pigs and poultry from farm to slaughter (www.welfarequality.net). The awareness that welfare is multidimensional and that its overall assessment requires a multicriteria evaluation resulted in a decision to base the WelfareQuality® assessment system on four principles of animal welfare: good housing, good feeding, good health and appropriate behaviour. Each of these four principles comprises several criteria, with an overall total of 12 criteria (Botreau *et al.*, 2007; Table 1). These 12 animal welfare criteria will be discussed in the following sections as they provide a very useful framework for understanding the components of animal welfare.

## 2.1 Good feeding

Good feeding includes two elements or criteria: absence of prolonged hunger and absence of prolonged thirst. Hunger may result from malnutrition, undernutrition or both. Malnutrition occurs when nutritients are not balanced, whereas undernutrition reflects insufficient supply. There are several reasons why prolonged hunger results in poor welfare. First, both malnutrition and undernutrition cause stress and, if sufficiently prolonged or severe, this can lead to debilitation, loss of body condition, immunosuppression and disease. Consequently, prolonged hunger results in inadequate biological functioning and it is likely to be an unpleasant emotional state (Webster, 1995; Kyriazakis and Savory, 1997).

Prolonged hunger also plays a key role in the development of stereotypies in farm animals. Stereotypies are considered an indicator of poor welfare because they tend to develop in apparently aversive and stressful situations. In general, farming systems associated with a high prevalence of stereotypies are consistently ranked as 'welfare unfriendly' when other indicators are used (Lawrence *et al.*, 1993).

*Table 1. Principles and criteria of animal welfare as developed in the WelfareQuality® project (Botreau* et al.*, 2007).*

| Principles | Criteria |
|---|---|
| Good feeding | Absence of prolonged hunger |
| | Absence of prolonged thirst |
| Good housing | Comfort around resting |
| | Thermal comfort |
| | Ease of movement |
| Good health | Absence of injuries |
| | Absence of disease |
| | Absence of pain induced by management procedures |
| Appropriate behaviour | Expression of social behaviours |
| | Expression of other behaviours |
| | Good human-animal relationship |
| | Positive emotional state |

Malnutrition may be deliberately caused in certain cases, for example, when veal calves are fed a diet deficient in iron to produce 'pale meat' (Van Putten, 1982). More often however, it may result from a mismatch between an individual animal's nutritional requirements (which are a consequence of its sex, age, stage of growth or reproduction, and previous nutritional history), and the common farming practice of providing a single diet designed to satisfy the needs of the 'average' animal (Manteca *et al.*, 2008)

Undernutrition may be a consequence of neglect, poor husbandry and/or circumstance. In extensive conditions, grazing ruminants may suffer undernutrition when forage conditions are very poor. Competition with conspecifics may also lead to undernutrition when access to food is limited due, for example, to insufficient feeding space (see below).

Undernutrition is intentionally imposed in some production systems, usually to prevent reproductive and health problems and/or to reduce food costs. Firstly for example, broiler breeders are offered 25 to 50% of what they would eat if fed *ad libitum*. This is probably the highest level of quantitative food restriction imposed on any farm animal. Secondly, pregnant sows are also usually food restricted and they experience prolonged hunger as a result (Lawrence *et al.*, 1993). Thirdly, forced moulting in laying hens is often induced by withholding food temporarily, sometimes for several days. Feeding recommences when the birds have lost up to 30% of their body weight (Brake *et al.*, 1982).

Hunger may also be a problem during transport and lairage. For example, pigs are fasted to reduce gut content during the pre-slaughter period and to prevent the release and spread of bacterial contamination in the faeces during transport and lairage as well as through the spillage of gut contents during carcass evisceration (Faucitano and Schaefer, 2008). Fasting before slaughter, within reasonable limits, is beneficial for the welfare of pigs as it prevents vomiting and hyperthermia. However, a prolonged fasting period causes hunger, aggressiveness (Warriss *et al.*, 1994), weakness, lethargy and sensitivity to cold (Gregory, 1998). In pigs, fat mobilization, as the main source of energy, starts after about 16 h of starvation.

In high-producing animals, particularly dairy cows and laying hens, food intake may not be sufficient to compensate for the strong production demands. This may lead to a severe loss of body condition. Starvation is also a hazard for newborn animals, particularly piglets, and is a main cause of neonatal mortality (Mellor and Stafford, 2004). Food intake can also be insufficient when animals are exposed to stressful conditions; this is often the case with weaning pigs.

Thirst is the sensation that accompanies dehydration. Prolonged thirst causes stress and, if long-lasting or severe, leads to debilitation, loss of body condition and disease. For example, low water intake in pregnant sows may cause urinary infections. Thirst also reduces food intake which, in turn, may lead to the welfare problems associated with prolonged hunger (see above). Prolonged thirst can occur when animals are given water of poor quality or when drinking facilities are insufficient or inadequate, mainly due to neglect or poor husbandry. The latter condition is exacerbated when there is competition with conspecifics. Water availability may also be inadequate in extensive conditions.

Dehydration might appear in animals that are transported over long distances during dry hot weather and when airflow through the moving truck is high. At lairage, the ability to cope with dehydration varies between species and age (Gregory, 1998). Suckling animals are particularly susceptible to dehydration because they have not learnt how to drink from a trough and thereby fail to drink the water provided at the abattoir. Poultry and rabbits are transported in containers and supplying water under these conditions is difficult.

## 2.2 Good housing

Good housing includes three elements: comfort around resting, thermal comfort and ease of movement. Lack of comfort around resting is likely to reduce resting time. This can lead to at least two major welfare problems. First, the risk of lameness increases if animals receive inadequate rest and this is particularly important in dairy cattle. Second, animals are often strongly motivated to rest and preventing them from doing so may cause them distress.

When housing is inadequate or inappropriate, the animals may have to use abnormal sequences of movements to lie down and get up, thereby increasing the risk of injury. Insufficient resting space may also elicit increased competition and aggression. Lying behaviour is an important element of behavioural thermoregulation. For example, in pigs the adoption of sternal recumbency and huddling mean that it is too cold, i.e. pigs attempt to reduce heat loss (Velarde, 2008). Pigs adopt a lateral recumbency position when the effective temperature is high and may experience heat stress if prevented from doing so due to lack of space (Petherick, 1983).

Lack of comfort around resting may be a consequence of an excessive stocking density or inadequate housing facilities, particularly inadequate flooring or an inappropriate number or design of cubicles on dairy farms. For instance, pigs prefer to separate their dunging area from their lying area. If the stocking density is too high, some pigs may have to lie in the dunging area, among faeces and urine. To satisfy its need for comfort around resting, each animal should have enough space to stand up, lie down and turn around. Minimal lying area requirements can be calculated based on body weight.

Movement patterns and durations of lying down and standing up might also affect welfare. Altered behaviour patterns can be seen in negative conditions e.g. when space is restricted in sow crates or when slatted pens have hard floors. For example, standing up with forelegs first and lying down with the hind legs first reveals in cattle an adaptation to hard ground to avoid pain in carpal joints coming into contact with the floor. The average duration of lying down and of standing up is much longer for sows in crates than for group housed sows (20.4 vs. 9.3 sec; Marchant and Broom, 1996), and is prolonged on hard slatted floors. A longer duration of lying down could be interpreted as an attempt to avoid pain and injuries (Hörning, 2007).

Farm animals are homoeothermic, which means they are able to maintain a relatively constant deep body temperature that differs from the environmental temperature within certain limits. A relatively constant deep body temperature means that heat production and heat loss are equal. Lower environmental temperature leads to higher heat losses, which have to be compensated by a higher heat production. Thermal comfort and the relationship between animals and their thermal environment are explained using the concept of thermoneutral zone. This is defined as the range of ambient temperatures that provides a sensation of comfort and minimises stress.

Temperatures which are too low or too high cause cold and heat stress respectively. The temperatures that define the thermoneutral zone depend on the species and may also vary among breeds of the same species. Even animals of the same breed may respond differently to ambient temperatures if they have been raised in different environments. Furthermore, the level of production and the amount and type of food given to the animals can all influence their response to the thermal environment. The effects of the thermal environment are not solely dependent on air temperature but on 'effective temperature', which is the end-result of the interaction between air temperature, relative humidity, ventilation and flooring. Solar radiation is also important. Temperatures which are too low or too high cause stress, which can lead to disease and even death if it is severe or prolonged. Heat stress also increases the amount of water required and can therefore increase the risk of prolonged thirst if water supply is limited (NRC, 1981).

Cold stress is a particular hazard for newborn animals and, together with starvation, plays an important role in neonatal mortality, particularly in piglets (Mellor and Stafford, 2004). Pigs and poultry have great difficulty in losing heat and may therefore suffer heat stress at ambient temperatures close to the upper limit of their thermoneutral zone and at high humidity. Heat stress may result from poor ventilation, inadequate housing and an overly high stocking density. Heat stress is a very common and important welfare problem for dairy cows kept in warm countries (West, 2003). Under extensive conditions, particularly in the tropics, non-adapted, exotic breeds of animals may suffer an increased risk of heat stress.

Ease of movement, i.e. the ability of animals to turn round, groom, get up, lie down and stretch their legs or wings has long been considered a basic requisite for good welfare (Brambell Committee, 1965). These movements are part of the behavioural repertoire of all species, and animals are highly motivated to perform them. They are also important to maintain the adequate functioning of the body. Difficulty of movement may be caused by a lack of space in the home environment, for example in hens kept in battery cages or sows housed in farrowing crates. Too high a stocking density may also prevent animals from moving normally, as is often evident in broiler chickens towards the end of their lives. Inadequate design of housing facilities may prevent animals from lying down and getting up normally. The presence of dominant individuals, particularly when stocking density is high or housing facilities are inadequate, may further curtail the movement of subordinate animals.

### 2.3 Good health

Good health is an important component of animal welfare and it can be defined as the absence of injuries, disease and pain. These negative states can have many causes, including certain management procedures. Injuries can cause acute and/or chronic pain. Pain is defined as an aversive emotional experience and is therefore a welfare problem (see section 1 of this chapter).

The legs and the feet are the parts of the body that are most frequently injured in farm animals. These injuries interfere with normal behaviour and locomotion, and may have a debilitating effect by preventing the animal from feeding normally. Mouth lesions may also hamper feeding. Wounds can become infected and, under some circumstances, may lead to systemic disease. Infectious, systemic diseases secondary to injuries as well as the debilitating effect of some injuries may result in the animals being culled.

Pigs show a high prevalence of neonatal mortality. Data from the UK indicates that 11.85% of all live-born pigs die within the 72 h post-parturition period (Meat and Livestock Commission, 2006). Besides constituting an important economic problem, piglet mortality is also becoming an increasingly significant welfare concern. Neonatal mortality in pigs is a complex multi-factorial problem, which involves elements related to piglet health status and behaviour, the behaviour of the sow and the characteristics of the physical environment (Baxter *et al.*, 2008). The most common event preceding live-born death is crushing or overlying by the sow. Thus, the traditional approach to preventing neonatal mortality in pigs has been the use of a farrowing crate that reduces the likelihood of crushing by restricting the movement of the sow. Various studies have confirmed the efficacy of such sow restraint systems in reducing crushing (Edwards and Fraser, 1997), but confinement at the time round parturition seems to be very stressful for the sow. Levels of distress may be particularly high for sows that are accustomed to being loose-housed during gestation, a practice that will be mandatory for all EU farms in 2013. In addition, the use of farrowing crates might compromise some aspects of maternal behaviour that could promote offspring survival (Baxter *et al.*, 2008). In any case, public concerns regarding animal welfare may limit the use of restraint systems in the future, thereby highlighting the need to develop alternative welfare-friendly strategies for preventing crushing (Edwards, 2002).

Injuries may be caused by abuse or rough handling, the latter being more common when animals are loaded and unloaded during transport. Injuries can result from accidents, such as when animals become entangled in wire or run into a fence or other obstacle. Hens can also be crushed if several try to re-enter the poultry house through the pop holes at the same time (Jones, 1997). Such accidents are often seen if animals are frightened and become panicked. Poor flooring and inadequate design or maintenance of housing facilities (e.g. slippery floors, sharp protrusions) may also cause injuries. Fighting with other animals can also cause injury; this is more common when animals are mixed with unacquainted individuals (particularly in pigs and to some degree cattle) and when animals have to compete for access to feed, water or resting space.

Tail-biting and feather pecking and cannibalism are important causes of injuries in pigs and laying hens respectively. According to the most widely accepted hypothesis, tail biting is a form of redirected behaviour derived from the thwarting of normal exploratory, feeding, social and sexual motivations. Tail-biting is a welfare problem because of the pain and suffering experienced by the bitten animal (not only due to the biting but also to secondary infections), the stress caused to the group (restlessness), and the likely frustration of the biting animal. As is true for other behaviour problems in intensive pig production, tail biting is a multi-factorial problem involving both internal and environmental risk factors; these include genetic background, sex, age, health status, diet, feeding management and different characteristics of the pen (SchrØder-Petersen and Simonsen, 2001; Moinard *et al.*, 2003; Schrøder-Petersen *et al.*, 2003, 2004).

Feather pecking (FP) can be defined as pecking at or pulling out and eating feathers of another hen (Bilcík and Keeling, 2000). The term cannibalism refers to the behaviour of pecking and pulling at the skin and the underlying tissue of another hen (Keeling, 1994). There are several forms of cannibalism in laying hens, including vent pecking, toe pecking and the pecking of defeathered body regions. Although the latter can develop as a consequence of the injuries caused by feather pecking, vent and toe pecking are independent from feather pecking. Both feather pecking and cannibalism are major welfare and economic problems in laying hens and turkeys (Jones *et al.*, 2004). FP has also been observed in broiler chickens reared to 12 weeks or more before slaughter. It imposes an economic burden because denuded birds lose heat faster and have to eat more to maintain body temperature. In terms of reduced welfare, pecking at another hen is painful and stressful for the victim (Gentle and Hunter, 1990) and it can lead to cannibalism and the painful death of target birds (McAdie and Keeling, 2000). FP can also be particularly problematic in alternative systems, such as percheries and free range, because it is more difficult to control when birds are kept in large flocks

Broken bones are relatively common in laying hens and, according to some studies, may occur in up to 12% to 25% of hens kept in free-range units or battery cages, respectively. Injuries caused by crash landings are also common in hens housed in percheries and aviaries. Laying hens develop osteoporosis because of their high turnover of calcium in producing eggshells and their lack of movement; this makes their bones more likely to break, especially when they are removed from their cages. In broiler chickens, hock burn, breast burn and breast blisters may occur when animals are kept on wet or inadequate litter.

Absence of disease is a basic requisite for good welfare. Diseases can cause pain and may interfere with normal behaviour. Chronic diseases often have a debilitating effect on the animal and may lead to it being culled. Control of infectious diseases, (e.g. BSE, foot and mouth, avian influenza), can cause major welfare problems when large numbers of animals have to be killed to avoid transmission of the disease.

Transport poses major challenges to the animals, as they have to cope with a variety of physical, psychological, social and climatic stressors over a relatively short period of time. Poor conditions during transport may cause injury, thirst, debilitation or even death, particularly in pigs and poultry. These problems can be exacerbated during longer journeys.

The gastrointestinal tract of most pigs can act as a reservoir for enterohaemorrhagic *Escherichia coli*, *Campylobacter* and *Salmonella* species. These enterobacteria usually become established and multiply within the caecum, and long feed withdrawal times during transport lead to a more favourable environment for their growth (Martin-Pelaez *et al.*, 2008). Furthermore, stress can result in increased emptying of the caeca into the colon and faster evacuation of the digest. Therefore, long fasting periods and stress increase *Salmonella* shedding in infected animals and even unleash excretion in silent carriers. Animals exposed to *Salmonella* can start to shed it in the faeces within 2 h.

Some of the diseases that are more relevant from an animal welfare standpoint are called 'multifactorial diseases', meaning that they are caused by the interplay of several factors. Lameness in dairy cows and broilers are good examples of multifactorial diseases. Lameness resulting from leg disorders is commonly regarded as one of the main welfare problems in broiler chickens. The aetiology of leg disorders in broilers includes factors such as genetic background, growth rate, feed conversion efficiency, body conformation, exercise, nutrition and stocking density. These categories are not mutually exclusive as one factor may affect another (Bradshaw *et al.*, 2002). As many as 90% of birds in some flocks show at least some degree of lameness by slaughter age (Kestin *et al.*, 1992), and some studies report up to 30% of birds moderately to severely lame (Sanotra *et al.*, 2001). However, prevalence of lameness in broilers varies to a large extent between farms. In a large scale study Dawkins *et al.* (2004) reported a mean percentage of severely lame birds of 9%, with a range of 0 to 20. As intensive broiler chicken production now exceeds $2 \times 10^{10}$ birds worldwide (Dawkins *et al.*, 2004), lameness in broilers is likely to be one of the most widespread farm animal welfare problems in modern agriculture. Leg problems have serious welfare consequences as lame birds may suffer pain (Pickup *et al.*, 1997) and their behaviour is significantly altered (Weeks *et al.*, 2000). Lameness has can also result in negative economic effects, as lame birds may lose weight and are more likely to be downgraded at slaughter (Kestin *et al.*, 1999).

Lameness in dairy cattle is also an important and mounting problem in the modern dairy industry. Actual levels of clinical lameness show a very large variability across farms, for example, they range between 5 and 70 cases/100 cows per year in the United Kingdom (Green *et al.*, 2003). Lameness compromises the animals' welfare by causing long-term pain and impairing their normal behaviour. In addition, it can result in significant economic losses. For example, Green *et al.* (2003) found that in clinically lame cows, milk yield was reduced up to 4 months before a case was diagnosed and treated and for the 5 months after treatment; (the total mean estimated reduction in milk yield per 305 days lactation was about 360 kg). Diseases of the claw (hoof) account for about 90% of all lameness incidents (Weaver, 2000). As in broilers, lameness in dairy cattle is a multifactorial condition resulting from an interaction between several factors (Clarkson *et al.*, 1993).

Poor housing and husbandry may in some cases predispose the animals to infection, even when the disease is caused by a microorganism. The background genome can also be influential, for instance some diseases are more common in animals that have been selected for improved production. Examples include lameness in broiler chickens (see above) and metabolic diseases in dairy cows which may result from selection for very high milk production. In pigs, transport related mortality varies depending on the animals' genetic background, the transport conditions and ambient temperature. Mortality during transport in broilers is also very much dependent on ambient temperature, transport conditions and journey duration.

Several procedures that are routinely carried out in farm animals can cause pain. These include tail docking and castration in pigs and less frequently in cattle; beak trimming in laying hens; teeth clipping in pigs, and dehorning and disbudding in cattle. The pain associated with these procedures normally lasts a few days, but in some cases chronic pain may also result. Though these management procedures are often carried out on young animals, they too can feel pain. Some of the procedures are carried out to prevent other, potentially more severe, welfare problems. For example, beak trimming in laying hens prevents feather loss and injury during feather pecking. On the other hand, some procedures are intended to improve the quality of the product, e.g. castration eliminates the strong and often unpleasant odours and flavours of intact male pig meat. Usually, castration of male pigs is performed surgically without anaesthesia or post-operative analgesia despite evidence that castration at any age is painful and may have a detrimental influence on health (Prunier *et al.*, 2006). Some procedures are difficult to justify and seem merely to be a consequence of misconceptions or traditions. Tail docking in dairy cows, for example, is still done in some countries to reduce mastitis even though there is no scientific evidence for such an effect (Stull *et al.*, 2002).

When moving animals at slaughter, the combination of higher speeds and poorly designed handling systems is detrimental to animal welfare because handling the animals at this rate requires considerable coercion and triggers the use of goads and sticks. Shocking animals with electric goads results in lesions and pain, and significantly raises heart rate, open mouth breathing and many other physiological indicators of distress. The routine use of electric goads is also a poor reflection on the attitude of the stockperson towards the animals in his/her care.

### 2.4 Appropriate behaviour

The principle 'appropriate behaviour' includes the expression of social and other behaviours, good human-animal relationship and the absence of general fear. The latter criterion may also be labelled 'positive emotional state'. All farm species are social animals and as such are strongly motivated to have contact with conspecifics. Positive social interactions such as social licking have a desirable effect on welfare for at least two reasons. First, they have been shown to elicit physiological responses known to be pleasant. Second, they reduce the negative effects of stressful events; this is known as 'social buffering' of the stress response (Kikusui *et al.*, 2006).

Negative social interactions, such as aggression, impair animal welfare. Aggression may result in injuries, pain and, in extreme cases, the death of the animal. Secondly, aggression leads to fear and stress within the whole group (Fraser and Rushen, 1987). Fear is an aversive emotional state (see section 1 this chapter) and, although fear behaviour can be adaptive in ideal circumstances, its sudden, intense or prolonged elicitation (and the consequences thereof) is a major welfare problem (Jones, 1997; Jones, 1998; Faure and Jones, 2004). Stress may harm body functioning by impairing immune function and reproductive performance, and decreasing food intake. Also, negative social interactions may interfere with the expression of normal behaviour, particularly in low ranking animals, and thereby reduce food intake and resting time which may in turn lead to debilitation and health problems, such as lameness.

The expression of normal social behaviour may be prevented when animals are reared in isolation. For example, although sows kept in stalls may be aggressive towards their neighbours, their behaviour differs from that of group housed sows due to the impossibility of interaction and the reduced space. It does not involve injuries but fear and frustration and the duration of aggression can be high, because the aggressive interaction cannot be satisfactorily resolved (Velarde, 2007). Also, disruption of social groups (through mixing of unacquainted animals, for example) may lead to an increase in aggressive behaviour and a reduction in positive social interactions. Mixing of unfamiliar animals (often with a change of physical environment) is a common practice in pig husbandry and often happens at weaning, at the beginning of the growing-finishing period and during transport to slaughter. Mixing of unacquainted pigs has adverse effects on welfare and production, mainly because pigs fight in order to establish dominance relationships, most aggressive interactions being typically shown during the first few hours after grouping, it then steadily decreases to a very low level by 24-48 h post-grouping, when the hierarchy becomes fairly stable (Meese and Ewank, 1972).

Housing conditions that result in increased competition for resources may heighten the number of negative social interactions. This may happen when stocking density is too high or when access to resources is insufficient, for example when feeding space is limited. Ewbank and Bryant (1969) concluded that high stocking rates cause the dominance hierarchy to be less successful in controlling aggression within the group, increasing the incidence of agonistic behaviour. High stocking densities increase aggression because the easy escape of attacked individuals is thwarted. In cattle, for example, reduction of feeding space increases competition for feed and this in turn increases aggression. Animals respond by reducing the time they spend feeding per day and by eating faster (Nielsen, 1999). Subordinate animals show the greatest effects on feeding pattern and they spend longer waiting around feeders to access the feed compared with dominant animals (McPhee *et al.*, 1964; Harb *et al.*, 1985; Ketelaar-de Lauwere *et al.*, 1996; Olofsson, 1999). Competition for access to the feeder can also damage performance as well as welfare (González *et al.*, 2008).

Animals are strongly motivated to perform particular behaviour patterns. This is the case, for example, with rooting in pigs, nest building in sows and hens, ground pecking and scratching in poultry, and exploration in all species. In some circumstances, the inability to perform such behaviour patterns may cause distress and lead to the development of damaging behaviours. For example, tail biting in pigs and feather pecking in poultry may reflect the lack of opportunity to perform rooting and ground pecking/scratching, respectively (see above).

A poor human-animal relationship results in the animals being fearful of the stockpersons and other humans. Poor stockmanship is considered the main cause of bad human-animal relationships. The term 'stockmanship' covers the way that animals are handled, the quality of their daily management and health care, and how well problems other than disease are recognised and solved (Waibingler and Spoolder, 2007). At least three factors underlie individual differences in the quality of stockmanship: personality, attitude and behaviour (Hemsworth and Coleman, 1998; Jones, 1996). Personality, which can be defined as a person's unique combination of traits that affects how he/she interacts with the environment, is relatively stable over time. Attitudes (including those towards animals) are learnt and can be modified through experience; they are often seen as the most important factor explaining how a person interacts with social objects, including animals (Waibingler and Spoolder, 2007). Clearly, personality and attitudinal factors influence the way that stockpersons behave towards the animals in their care.

The quality of stockmanship has a profound effect on the animals' welfare and productivity (Boivin *et al.*, 2003). For instance, despite centuries of domestication exposure to human beings remains one of the most potentially alarming experiences for many farm animals. More specifically, unless they have become accustomed to human contact of either a neutral or positive nature the predominant reaction to people is one of fear (Duncan, 1990; Jones, 1997). Not surprisingly, the problem is exacerbated by exposure to rough, aversive and/or unpredictable handling. Indeed, many human-animal interactions in current farm practice are frightening; these include restraint, depopulation, veterinary treatment, etc., while few, other than feeding, are positively reinforcing. It is worth bearing in mind that contact with humans could become even more distressing if increasing automation results in reduced opportunities for animals to habituate to people. The stockpersons' behaviour, which can vary from calm, gentle, frequent and 'friendly' to infrequent, rough and rushed, is a major variable determining animals fear of or confidence in humans and, hence, the quality of the human-animal relationship. Chronic fear of humans is a major welfare problem that can lead to handling difficulties, injury, and stress as well as impaired growth, reproductive performance and product quality (Hemsworth and Coleman, 1998; Jones, 1997). For example, a series of studies found negative (and probably causal) correlations between fear of humans and productivity in the dairy, egg, broiler and pig industry (Hemsworth, 2003). Conversely, the regular experience of positive human-animal interactions can decrease the animals' general level of stress (Seabrook and Bartle, 1992) and enhance reproductive performance (Waiblinger *et al.*, 2006), and the presence of a familiar person can calm the animal in potentially aversive situations (Waiblinger *et al.*, 2006). Regular gentle handling reduces stress and fear of humans in chickens, cattle, sheep and pigs (Hemsworth and Coleman, 1998; Jones, 1996) but is not feasible in modern farming. However, in chickens at least, the handling phenomenon seems quite flexible. For example, simply letting chickens see people regularly effectively reduced their fear of humans (Jones, 1996; Hemsworth, 2003). This suggests that more frequent close examination of the flock by the stockperson would not only provide a better check that the birds are healthy and that the system is working properly but it could also help to reduce the birds' fear of humans.

Fear and anxiety are two emotional states induced by the perception of a danger or a potential danger, respectively, that threaten the integrity of the animal (Jones, 1987; Boissy, 1995). Fear and anxiety both involve physiological and behavioural changes that prepare the animal to cope with the danger. Although fear and anxiety have not always been clearly differentiated, fear can be operationally defined as states of apprehension focusing on isolated and recognisable dangers while anxieties are diffuse states of tension that magnify the illusion of unseen dangers (Rowan, 1988). General fear becomes a problem particularly when animals encounter new or unexpected stimuli, (e.g. a sudden noise or movement, an unfamiliar animal), or situations, e.g. a new housing facility, transportation. This has important implications for animal housing and management. For example, inappropriate handling, corridors/races and pen design, discontinuities in floor texture and colour, drafts and (poor) lighting may all induce fear and its undesirable consequences.

Many farm animals can hear over a greater range of frequencies than humans and are more sensitive to higher frequencies. For example, high sound levels in lairage (>100 dB) increase the levels of lactate and CPK in blood in pigs. The noise produced by the machinery, pressure hoses and human vocalization represents a source of stress.

When exposed to a threatening stimulus or situation, animals adopt generally adaptive behaviours such as escape, freeze, back off, shake, defensive or offensive aggression, or various other behaviours including lying down (Dalmau *et al.*, 2009). Social species like pigs can collaborate in defence against predators, and they tend to vocalize when caught or hurt. Species which are less able to defend themselves successfully, such as sheep, vocalize far less, perhaps because such an extreme response would simply 'inform' the predator that the animal attacked is severely injured and hence unlikely to be able to escape.

Fear has a strong genetic component and some breeds or individuals within breeds are likely to be more easily frightened than others. This means that breeding programmes intended to reduce fear should be possible. Fear is also sensitive to the animals' environment and previous experience. For instance, animals reared in barren environments are generally more fearful of novel situations than those that have been reared in enriched and varied environments.

## 3. Conclusions

### 3.1 What has been achieved?

Two of the main aims of this chapter were to emphasise the multi-factorial nature of animal welfare and to describe some of its many components within a framework of four principles and twelve criteria. While we do not claim to have provided an exhaustive list of influential factors we believe that the approach taken here will help to guide the ongoing consideration of the critical variables that determine an animal's welfare and the ways in which they might interact.

### 3.2 What has been neglected?

We cannot overemphasize the importance of increasing our understanding of the following crucial issues:

- What does the concept of animal welfare incorporate in terms of causal factors, biological mechanisms and outcomes?
- How can we objectively measure the above factors and outcomes?
- What risk factors should we be particularly aware of?
- How might the welfare problems associated with these risk factors arise?
- What are the likely consequences of these problems if they are not resolved?
- What can be done to eliminate or at least alleviate these risk factors and welfare problems?

### 3.3 What needs to be done?

Some of the answers to these questions are included in the present chapter although much work still remains to be done. Collectively however, the sort of information provided here can contribute significantly to the attainment of higher welfare levels in farm animals. For example, the development of standardised, practical and reliable systems for assessing animal welfare on farm and at slaughter in the WelfareQuality® project focuses to a large extent on the animal- or outcome-based measures mentioned here but in a truly integrative fashion it also includes several resource- and management-based ones, such as housing, husbandry, feeding, handling etc. In turn, the results of this and other assessment systems can facilitate the development and validation of practical and effective welfare improvement strategies. This is good news for the animals, the farmers and the public.

## References


Baxter, E.M, Jarvis, S., D'Eath, R.B., Ross, D.W., Robson, S.K., Farish, M., Nevison, I.M., Lawrence, A.B. and Edwards, S.A., 2008. Investigating the behavioural and physiological indicators of neonatal survival in pigs. Theriogenology 69, 773-783.

Bilcík, B. and Keeling, L.J., 2000. Relationship between feather pecking and ground pecking in laying hens and the effect of group size. Applied Animal Behaviour Science 68, 55-66.

Boissy, A., 1995. Fear and fearfulness in animals. Quarterly Review of Biology 70, 165-191.

Boivin, X., Lensink, B.J., Tallet, C. and Veissier, I., 2003. Stockmanship and farm animal welfare. Animal Welfare 12, 479-492.

Botreau, R., Veissier, I., Butterworth, A., Bracke, M.B.M. and Keeling, L.J., 2007. Definition of criteria for overall assessment of animal welfare. Animal Welfare 16, 225-228.

Bradshaw, R.H., Kirkden, R.D. and Broom, D.M. 2002. A review of the aetiology and pathology of leg weakness in broilers in relation to welfare. Avian and Poultry Biology Reviews 13, 45-103.

Brake, J., Baker, M., Morgan, G.W. and Thaxton, P., 1982. Physiological changes in caged layers during a forced moult. 4. Leucocytes and packed cell volume. Poultry Science 61, 790-795.

Brambell Committee, 1965. Report of the technical committee to enquire into the welfare of animals kept under intensive livestock husbandry systems. Her Majesty's Stationery Office, London, UK.

Broom, D.M., 1986. Indicators of poor welfare. British Veterinary Journal 142, 524-526.

Broom, D.M., 2001. Coping, stress and welfare? In: Broom, D.M. (Ed.). Coping with challenge. Welfare in animals including humans. Dahlem University Press, Berlin, Germany.

Clarkson, M.J., Downham, D.Y., Faull, W.B., Hughes, J.W., Manson, F.J., Merrit, J.D., Murray, R.D., Russell, W.B., Sutherst, J.E. and Ward, W.R., 1993. An epidemiological study to determine the risk factors of lameness in dairy cows. CSA 1379, Univ. Liverpool, Liverpool, UK.

Cannon, W.B., 1929. Organization for physiological homeostasis. Physiological Review 9, 399-431.

Dalmau, A., Fàbrega, E. and Velarde, A., 2009. Fear assessment in pigs exposed to a novel object test. Applied Animal Behaviour Science 117, 173-180.

Dawkins, M.S., 1990. From an animal's point of view: motivation, fitness and animal welfare. Behavioral and Brain Sciences 13, 1-16.

Dawkins, M.S., Donnelly, C.A. and Jones, T.A., 2004. Chicken welfare is influenced more by housing conditions than by stocking density. Nature 427, 342-344.

Duncan, I.J.H., 1990. Reactions of poultry to human beings. In: Zayan, R. and Dantzer, R. (Eds.). Social stress in domestic animals. Kluwer, Dordrecht, the Netherlands, pp. 121-131.

Duncan, I.J.H. and Fraser, D., 1997. Understanding animal welfare. In: Appleby, M.C. and Hughes, B.O. (Eds.). Animal welfare. CAB International, Wallingford, UK, pp. 19-31.

Edwards, S.A., 2002. Perinatal mortality in the pig: environmental or physiological solutions? Livestock Production Science 78, 3-12.

Edwards, S.A. and Fraser, D., 1997. Housing systems for farrowing and lactation. Pig Journal 39, 77-89.

Ewbank, R. and Bryant, M.J., 1969. The effects of population density upon the behaviour and economic performance of fattening pigs. Farm Building Progress 18, 14-15.

Faucitano, L. and Schaefer, A. (Eds.), 2008. Welfare of pigs from birth to slaughter. Wageningen Academic Publishers, Wageningen, the Netherlands, 316 pp.

Faure, J.M. and Jones, R.B., 2004. Genetic influences on resource use, fear and sociality. In: Perry, G.C. (Ed.). Welfare of the laying hen. 27th Poultry Science Symposium. CAB International, Wallingford, UK, pp. 99-108.

Fraser, D. and Rushen, J., 1987. Aggressive behavior. Veterinary Clinics of North America: Food Animal Practice, 3 (2).

Gentle, M.J. and Hunter, L.N., 1990. Physiological and behavioural responses associated with feather removal in *Gallus gallus var domesticus*. Research in Veterinary Science 50, 95-101.

González, L.A., Ferret, A., Manteca, X., Ruiz de la Torre, J.L., Calsamiglia, S., Devant, M. and Bach, A., 2008. Performance, behavior, and welfare of Friesian heifers housed in pens with two, four and eight individuals per concentrate feeding place. Journal of Animal Science 86, 1146-1458.

Green, L.E., Hedges, V.J., Schukken, Y.H., Blowey, R.W. and Packington, A.J., 2003. The impact of clinical lameness on the milk yield of dairy cows. Journal of Dairy Science 85, 2250-2256.

Gregory, N.G., 1998. Animal welfare and meat science. CAB International, London, UK.

Harb, M.Y., Reynolds, V.S. and Campling, R.C., 1985. Eating behavior, social dominance and voluntary intake of silage in group-fed milking cattle. Grass and Forage Science, 40, 113-118.

Hemsworth, P.H., 2003. Human-animal interactions in livestock production. Applied Animal Behaviour Science 81, 185-198.

Hemsworth, P.H. and Coleman, G.J., 1998. Human-livestock interactions: the stockperson and the productivity of intensively farmed animals. CAB International, Wallingford, UK.

Hörning, B., 2007. Lying down and standing up behavior. In: Geers, R. and Velarde, A. (Eds.). On farm monitoring of pig welfare. Wageningen Academic Publishers, Wageningen, the Netherlands, pp. 25-34.

Jensen, P., 1986. Observations on the maternal behaviour of free-ranging domestic pigs. Applied Animal Behaviour Science 16, 131-142.

Jones, R.B., 1987. The assessment of fear in the domestic fowl. In: Zayan, R. and Duncan, I.J.H. (Eds.). Cognitive aspects of social behaviour in the domestic fowl. Elsevier, Amsterdam, the Netherlands, pp. 40-81.

Jones, R.B., 1996. Fear and adaptability in poultry: insights, implications and imperatives. Worlds' Poultry Science Journal 52, 131-174.

Jones, R.B., 1997. Fear and distress In: Appleby, M.C. and Hughes, B.O. (Eds.). Animal welfare. CAB International, Wallingford, UK, pp 75-87.

Jones, R.B., 1998. Alleviating fear in poultry. In: Greenberg, G. and Haraway, M. (Eds.). Comparative psychology: a handbook, Garland Press, New York, NY, USA, pp. 339-347.

Jones, R.B., Blokhuis, H.J., de Jong, I.C., Keeling, L.J., McAdie, T.M. and Preisinger, R., 2004. Feather pecking in poultry: the application of science in a search for practical solutions. Animal Welfare 13, 215-219.

Keeling, L.J., 1994. Feather pecking – who in the group does it, how often and under what circumstances? Proceedings of the 9th European Poultry Conference, Glasgow, UK, pp. 288-289.

Kestin, S.C., Knowles, T.G., Tinch, A.E. and Gregory, N.G., 1992. Prevalence of leg weakness in broiler chickens and its relationship with genotype. Veterinary Record 131, 190-194.

Kestin, S.C., Su, G. and Sorensen, P., 1999. Different commercial broiler crosses have different susceptibilities to leg weakness. Poultry Science 78, 1085-1090.

Ketelaar-de Lauwere, C.C., Devir, S. and Metz. J.H.M., 1996. The influence of social hierarchy on the time budget of cows and their visits to an automatic milking system. Applied Animal Behaviour Science 49, 199-211.

Kikusui, T., Winslow, J.T. and Mori, Y., 2006. Social buffering: relief from stress and anxiety Philosophical Transactions of the Royal Society B 361, 2215-2228.

Kyriazakis, I. and Savory, C.J., 1997. Hunger and thirst. In: Appleby, M.C. and Hughes, B.O. (Eds.). Animal welfare. CAB International, Wallingford, UK, pp. 49-62.

Lawrence, A.B., Terlouw, E.M.C. and Kyriazakis, I., 1993. The behavioural effects of undernutrition in farm animals. Proceedings of the Nutrition Society 52, 219-229.

Manteca, X., Villalba, J.J., Atwood, S.B., Dziba, L. and Provenza, F.D., 2008. Is dietary choice important to animal welfare? Journal of Veterinary Behaviour: Clinical Applications and Research 3, 229-239.

Marchant, J.N. and Broom, D.M., 1996. Factors affecting posture-changing in loose housed and confined gestating sows. Animal Science 63, 477-485.

Martín-Peláez, S., Peralta, B., Creus, E., Dalmau, A., Velarde, A., Pérez, J., Mateu, E., Martín-Orúe, S., 2008. Different feed withdrawal times before slaughter influence caecal fermentation. The Veterinary Journal, doi:10.1016/j.tvjl.2008.08.002.

Mason, J.W., 1971. A re-evaluation of the concept of 'non-specificity' in stress theory. Journal of Psychological Research 8, 323-333.

Mason, G.J., Cooper, J. and Clarebrough, G., 2001. Frustrations of fur-farmed mink. Nature 40, 35-36.

McAdie, T.M. and Keeling, L.J., 2000. The social transmission of feather pecking in laying hens: effects of environment and age. Applied Animal Behaviour Science 75, 147-160.

McPhee, C.P., McBride, G. and James, J.W., 1964. Social behavior of domestic animals. III. Steers in small yards. Animal Production 6, 9-15.

Meat and Livestock Commission, 2006. The pig yearbook, 2006.

Meese, G.B. and Ewank, R., 1972. A note on instability of the dominance hierarchy and variations in level of aggression within groups of fattening pigs. Animal Production 14, 359-362.

Mellor, D.J. and Stafford, K.J., 2004. Animal welfare implications of neonatal mortality and morbidity in farm animals. The Veterinary Journal 168, 118-133.

Mendl, M., 2001. Animal husbandry: Assessing the welfare state. Nature 410, 31-32.

Moinard, C., Mendl, M., Nicol, C.J., Green, L.E., 2003. A case control study of on-farm risk factors for tail biting in pigs. Applied Animal Behaviour Science 81, 333-355.

National Research Council (NRC), 1981. Effect of environment on nutrient requirements of domestic animals. National Academy Press, Washington, DC, USA, pp. 75-84.

Nielsen, B.L., 1999. On the interpretation of feeding behaviour measures and the use of feeding rate as an indicator of social constraint. Applied Animal Behaviour Science 63, 79-91.

Olofsson, J., 1999. Competition for total mixed diets fed for *ad libitum* intake using one or four cows per feeding station. Journal of Dairy Science 82, 69-79.

Petherick J.C., 1983. A biological basis for the design of space in livestock housing. In: Baxter, S.H., Baxter, M.R. and MacCormak, J.A.D. (Eds.). Farm animal housing and welfare. Martinus Nijhoff Publishers, The Hague, the Netherlands, pp. 103-120.

Petherick, J.C. and Rushen, J., 1997. Behavioural restriction. In: Appleby, M.C. and Hughes B.O. (Eds.). Animal welfare. CAB International, Wallingford, UK, pp. 89-105.

Pickup, H.E., Cassidy, A.M., Danbury, T.C., Weeks, C.A., Waterman, A.E. and Kestin, S.C., 1997. Self selection of an analgesic by broiler chickens. British Poultry Science 38, 12-13.

Prunier, A., Bonneau, M., Von Borell, E.H., Cinotti, S., Gunn, M., Fredriksen, B., Giersing, M., Morton, D.B., Tuyttens, F.A.M. and Velarde, A., 2006. A review of the welfare consequences of surgical castration in piglets and the evaluation of non-surgical methods. Animal Welfare 15, 277-289.

Rowan, A.N., 1988. Animal anxiety and animal suffering. Applied Animal Behaviour Science 20, 135-142.

Sanotra, G.S., Lund, J.D., Ersboll, A.K., Petersen, J.S. and Vestergaard, K.S., 2001. Monitoring leg problems in broilers: a survey of commercial broiler production in Denmark. World's Poultry Science Journal 57, 55-69.

Schrøder-Petersen, D., Schrøder-Petersen, D.L., Simonsen, H.B. and Lawson, L.G. 2003. Tail-in-mouth behaviour among weaner pigs in relation to age, gender and group composition regarding gender. Acta Agriculturae Scandinavica. Section A. Animal Science 53, 29-34.

Schrøder-Petersen, D.L., Heiskanen, T. and Ersboll, A.K. 2004. Tail-in-mouth behaviour in slaughter pigs, in relation to internal factors such as: age, size, gender, and motivational background. Acta Agriculturae Scandinavica. Section A. Animal Science 54, 159-166.

Schrøder-Petersen, D.L. and Simonsen, H.B., 2001. Tail biting in pigs. The Veterinary Journal 162, 196-210.

Seabrook, M.F. and Bartle, N.C., 1992. Human factors. In: Phillips, C. and Piggins, D. (Eds.). Farm animals and the environment. CAB International, Wallingford, UK, pp. 111-125.

Selye, H., 1936. A syndrome produced by diverse noxious agents. Nature 138, 32-36.

Stull, C.L., Payne, M.A., Berry, S.L. and Hullinger, P.J., 2002. Evaluation of the scientific justification for tail docking in dairy cattle. Journal of the American Veterinary Medical Association 9, 1298-1303.

Terlouw, E.M.C., Schouten, W.G.P. and Ladewig, J., 1997. Physiology. In: Appleby, M.C. and Hughes, B.O. (Eds.). Animal welfare. CAB International, Wallingford, UK, pp. 143-158.

Van Putten, G., 1982. Welfare in veal calf units. Veterinary Record 111, 437.

Velarde, A., 2007. Agonistic behavior. In Geers, R. and Velarde, A. (Eds.). On farm monitoring of pig welfare. Wageningen Academic Publishers, Wageningen, the Netherlands, pp. 53-56.

Velarde, A., 2008. How can lairage conditions be improved? Proceedings Animal welfare at slaughter and killing for disease control, pp. 31-36.

Waiblinger, S. and Spoolder, H., 2007. Quality of stockpersonship. In: Geers, R. and Velarde, A. (Eds.). On farm monitoring of pig welfare. Wageningen Academic Publishers, Wageningen, the Netherlands, pp. 159-166.

Waiblinger, S., Boivin, X., Pedersen, V., Tosi, M., Janczak, A.M., Visser, E.K., Jones, R.B., 2006. Assessing the human-animal relationship in farmed species: a critical review. Applied Animal Behaviour Science 101, 185-242.

Warriss, P.D., Brown, S.J. and Adams, M., 1994. Relationship between subjective and objective assessment of stress at slaughter and meat quality in pigs. Meat Science 38, 329-340.

Weaver, A.D., 2000. Lameness. In: Andrews, A.H. (Ed.). The health of dairy cattle. Blackwell Science, Oxford, UK, pp 142-202

Webster, A.J.F., 1995. Animal welfare – a cool eye towards Eden. Blackwell Science, Oxford, UK.

Weeks, C.A., Danbury, T.D., Davies, H.C., Hunt, P. and Kestin, S.C., 2000. The behaviour of broiler chickens and its modification by lameness. Applied Animal Behaviour Science 67, 111-125.

Weiss J.M., 1970. Somatic effect of predictable and unpredictable shock. Psychosomatic Medicine 32, 397-408.

Weiss, J.M., 1971. Effects of coping behaviour with and without a feedback signal on stress pathology in rats. Journal of Comparative and Physiological Psychology 77, 22-30.

WelfareQuality®, 2004. Integration of animal welfare in the food quality chain: from public concern to improved welfare and transparent quality. EU Project FOOD-CT-2004-506508.

West, J.W., 2003. Effects of heat-stress on production in dairy cattle. Journal of Dairy Science 86, 2131-2144.

# Animal Welfare and housing

**Susanne Waiblinger**
Institute of Animal Husbandry and Animal Welfare, Department of Farm Animals and Veterinary Public Health, University of Veterinary Medicine Vienna, Veterinärplatz 1, 1210 Wien, Austria; susanne.waiblinger@vu-wien.ac.at

## Summary

The welfare of animals on a given farm is determined by the interactions of a range of environmental conditions, among which housing plays a central role for the following reasons: (1) it is difficult and costly to change existing housing conditions and thus these generally have impact on farm animals for a long time, (2) housing conditions often set the limits regarding the level of welfare that can be achieved, even when optimising all other factors, and (3) animals are subject to their housing environment for the duration of their lives. Poor welfare as a result of animals having difficulties to cope with their environment prevails if their species-specific control systems developed during evolution are disregarded. Intensive, barren housing may lead to problems, because animals cannot perform behaviour, for the expression of which they have an innate motivation. Furthermore, they may not be able to meet environmental challenges decidedly different than those they face in their natural habitat, so that adaptive behaviour could not evolve. Different aspects of housing are reviewed, both where they represent hazards to- and promoters of animal welfare. The following aspects are discussed in detail: confinement, space allowance, number and distributions of resources, structural elements, flooring (for activity and resting), enrichment, access to outdoor yards or pasture, design of equipment and climate. Although many welfare hazards are known, the implementation of such knowledge in the farming industry is not wide-spread, so many farm animals are still subjected to an inappropriate environment and poor welfare. Consequently, further efforts need to be made to develop more welfare-friendly housing systems and to ensure these are implemented.

**Keywords**: behaviour, health, stress, prevention, housing design

## 1. Introduction

The welfare of farm animals depends on the combination of different aspects of husbandry, mainly housing, feeding, management and daily care. Of these, housing is the least flexible due to costly investments associated with changing housing conditions. Thus, decisions regarding housing type (and details of its design features) influence the welfare of the animals on a given farm for many years or even for decades. Therefore, it is difficult to improve housing design when new knowledge becomes available, and the implementation in practice of such changes is time-consuming. For instance, although in dairy cow housing cubicle design has markedly improved, old cubicles imposing many welfare risks are still frequently found on farms. Moreover – as animals are in contact with their housing environment 24 hours every day – housing design represents an extremely important factor for the welfare situation of the farm animal population, notwithstanding the undisputed significance of the other factors mentioned. Would poor welfare be only due to deficiencies in management routines and health care, it could, in principle, be improved by remedying these relatively quickly. However, if poor welfare mainly results from housing features that can only be corrected by changes in construction, this is often more difficult to achieve and would require high investments and sometimes even rebuilding.

Housing provides the framework in which the other factors act and to which these have to be adapted. Thus, housing conditions can limit the level of welfare that can be achieved, even when the best practices for management, feeding and daily care are adhered to. This applies on a system level, where some systems have inherent deficiencies regarding animal needs, but also within systems, where several specifications allow predictions about welfare risks. On the other hand, some housing conditions have, in principle, the potential for ensuring a high quality of life for the animals, although the achievements of welfare can vary significantly, depending on the management and other aspects of welfare. An example is keeping fattening bulls on a fully slatted concrete floor or on deep litter. Whilst in both systems the welfare of the animals may be poor for reasons of management deficiencies (e.g. not enough straw in deep litter, inappropriate feeding), only the deep litter system has the potential for good welfare, as bulls held on fully slatted floors will – independent of the quality of management – suffer from pain and joint lesions caused by the inappropriate lying surface, which situation can only gradually be improved by the use of rubber mats (Schulze Westerath *et al.*, 2007; Gygax *et al.*, 2007a; Absmanner *et al.*, 2009).

This contribution aims to give some insight in the welfare risks and benefits associated with housing conditions. Whilst important housing-management interactions are mentioned where appropriate, other aspects of husbandry, management, health care and handling are discussed elsewhere in this Volume. It will first be discussed how housing conditions impact on the level of animal welfare and how these two variables are related with animal characteristics. Subsequently, the most important (relevant) welfare risks and benefits of indoor and outdoor housing are reviewed more specifically. Finally, it is discussed how a high standard of animal welfare may be achieved in future animal production.

## 2. Animal characteristics, housing design and animal welfare – how are they interrelated?

### 2.1 Species-specific characteristics

The wild ancestors of our domestic animals adapted to their environment by evolution in their body structures, their physiology and their behaviour (Darwin, 1859). This resulted in species-specific characteristics and behaviours that are still present in our domestic animals (regarding behaviour, (see e.g. Wood-Gush *et al.*, 1978; Reinhardt, 1980; Kiley-Worthington and Plain, 1983; Jensen, 1986; Stolba and Wood-Gush, 1989; McBride *et al.*, 2009) and which constitute their physical, physiological and behavioural needs. The differences in behaviour between farm animals and their wild counterparts are primarily quantitative rather than qualitative (Price, 1999), i.e. thresholds and frequencies change. The same is true for physical and physiological characteristics. Thus, generally spoken, our farm animals are still best adapted to the environment they have evolved in.

Housing conditions often differ substantially from those in nature. This is not necessarily a problem. If the relevant key stimuli and characteristics exist so that the species-specific, natural behaviour can be performed in a normal way (e.g. Stolba and Wood-Gush, 1984; Wechsler *et al.*, 1991; Wechsler, 1996) and physical and physiological limits of adaptation are taken into account, these housing systems have the potential for a high quality of life, i.e. they would provide the animals with positive experiences and emotions and would minimize negative ones. In this context, one should realize that even weakly motivated behaviour may be important to the animal, because it is associated with positive feelings of comfort, satisfaction or pleasure or because it contributes to a sense of control, social support or engagement (Mench, 1998).

In contrast, if the evolved species-specific control systems are disregarded during rearing and husbandry, animals will have more difficulties or are unable to cope with the conditions, so welfare problems will arise (Wechsler, 1993; Fraser *et al.*, 1997; Broom, 2007). The inability to cope can emerge when (1) animals are motivated to perform a behaviour but are not able to do so, or (2) the environment poses a challenge the animals are not able to meet (Buchholtz, 1993; Fraser *et al.*, 1997; Broom, 2007; Wechsler, 2007).

An example of a situation in which animals are unable to perform behaviour due to confinement or to a lack of environmental key features is a sow being prevented from nest-building behaviour e.g. when she is housed in a farrowing pen with confinement (i.e. in a crate) or is not provided with material for nest-building. When animals are prevented from performing a behaviour for which they are highly motivated, they may not only experience negative emotions or fail to experience positive ones but they may also suffer physical health problems through activation of the stress axis (Broom and Johnson, 2000) or simply due to a lack of naturally protecting behaviour. Hence, sows being prevented from nest building not only show elevated levels of cortisol and behavioural signs of stress, but also prolonged parturition, which might result in a higher stillbirth rate, less vital piglets and higher risk of MMA [mastitis, metritis, agalactia] (Weber and Troxler, 1988; Bertschinger *et al.*, 1994; Jarvis *et al.*, 2002; Wechsler and Weber, 2007). In addition, crated sows cannot separate the dunging- from the nest area, as they would naturally do, and are thus forced to lie down where they urinate and defecate, equally increasing the risk for infection (MMA). Other important examples are the prevention of behaviours related to searching for food, i.e. rooting in and chewing appropriate substrate in pigs or scratching in and pecking at appropriate substrate in laying hens. Pigs have a strong urge to root and explore, which in nature is necessary to search for food. Preventing pigs from doing so by keeping them in strawless systems without access to rooting material, increases the risk of behavioural problems such as tail biting and associated health problems (Moinard *et al.*, 2003; Zonderland *et al.*, 2008). Also, rearing laying hens without straw on wired floors (Blokhuis and Arkes, 1984; Johnsen *et al.*, 1998) or only giving them access to inappropriate foraging material such as sand (Huber-Eicher and Wechsler, 1997) increases the risk of feather-pecking in laying hens. Long-term effects on their behaviour are also likely to occur in piglets when they are reared in the barren environment of a farrowing crate: higher aggression and lower learning abilities were found compared to piglets raised in a more enriched environment such as a loose farrowing pen with straw-bedding (Wechsler and Weber, 2007). This emphasizes the importance of appropriate housing throughout the lifetime of an animal. All these behaviours served important functions in nature for survival and successful reproduction (i.e. foraging, providing a protected, warm area for piglets). This may explain the strong internally driven motivation to perform these behaviours, although the necessity no longer exists in animal husbandry where food and protection for piglets is provided by the human or housing system.

An example of animals being unable to meet an environmental challenge because an adaptive behaviour did not evolve in the natural habitat where this was unnecessary is the following. Many pig breeds have hairless, unpigmented skin. They can get severely sun-burned if offered access to an outside yard without shade and a wallowing site. Pigs did not develop behaviour to protect them against this hazard, because wild boars mainly live in forests and have pigmented brown skin and hair. Other limits to cope with a challenge lie in the animals' anatomy or physiology. For instance, cattle naturally graze while walking. In this 'pasture pace', with one front leg in front of the other, the head is closer to the ground compared to standing upright with both forelegs parallel. When eating from a feeding rack while standing, their forelegs must be parallel and this makes it harder for them to reach the ground. If the feeding table is not raised sufficiently, the load on the shoulder and forelegs is greater and the animals may become lame and develop shoulder lesions (Metzner, 1976; Molz, 1989; Waiblinger *et al.*, 2001b). A final example is a hard and synthetic lying surface for cattle. Cattle develop skin and joint lesions (hairless and wound patches, swellings) when they are forced to rest on rubber or other synthetic lying surfaces, especially when these surfaces are not soft enough (Friedl *et al.*, 1999; Buchwalder *et al.*, 2000; Wechsler *et al.*, 2000; Schulze Westerath *et al.*, 2007) as the load on their carpal joints during lying down and standing up as well as abrasion forces during lying are too high and too different from natural conditions. It is important to note that cattle show behavioural signs of stress and pain during lying behaviour on hard floors, but not when provided with soft synthetic floors [which nevertheless cause hock lesions] (Wechsler *et al.*, 2000).

In summary, housing systems enabling good welfare are designed according to the natural behaviour and characteristics of the animals. Behaviour can be categorized into the animal's functional systems, for instance social behaviour, feeding behaviour, locomotion (Tembrock, 1980). Housing systems can be subdivided in functional areas, for example a feeding area, walking area, lying area in cattle loose housing, or a feeding, dunging, activity and lying area in pig housing systems. Optimised housing design means that restrictions to animal behaviour are minimized by ensuring that the functional areas of the housing system are designed with regard to dimensions, number and composition according to the functional systems of animal behaviour.

## 2.2 General characteristics

Besides allowing animals to perform species-specific behaviour, some features of the environment are important for all species and will contribute to the quality of the animals' life. These include the chance to make choices and decisions, to be able to predict and have some control over the environment (Dickinson, 1980; Toates, 1987; Wiepkema, 1987; Veissier and Boissy, 2007). Animals have expectations (Dudink *et al.*, 2006) and become frustrated and stressed if these are not fulfilled (Carlstead, 1986). The implication for housing is that one must give the animals various options and not to decide for them what is 'better' (e.g. by giving give them permanent access to outside runs, even during bad weather). Furthermore, this means offering them a rich environment with diverse stimuli.

The hygienic situation of the environment is important regarding the risk of infectious diseases. Management has a huge impact here but it also interacts with housing design. For example, instalment of an automatic dung scrapper enables a frequent removal of dung and thus allows for a better hygienic situation in alleys.

It is important to note that best welfare can only be achieved by providing both a good hygienic situation and accounting for behavioural needs, thus supporting good immune function and health due to psychological well-being and reduction of stress (Henry and Stephens, 1977; Broom, 2006). Some of the most important welfare problems – be it behavioural problems such as feather pecking or tail-biting, or diseases such as mastitis or lameness – are multi-factorial. If several environmental challenges with which the animal has difficulties are occurring concertedly, the risk for poor welfare increases. Other aspects besides housing features often also contribute to the development of these complex welfare problems, such as genetics, production level and management. These are discussed elsewhere in this Volume.

## 3. Effects of housing conditions on animal welfare

After having discussed the principles of how features of the housing environment influence welfare, the different aspects of housing affecting welfare both in the sense of imposing risks as well as providing benefits need to be considered. Within the framework of this contribution it would be impossible to give an exhaustive review of all risk factors for the different farm animal species or to give comparisons of systems. Instead, the single aspects of housing that are most important to ensure welfare are discussed. For a more detailed discussion of risk factors and different housing systems the reader is referred to further books or reviews on the respective topic (Anonymous, 2000; Waiblinger *et al.*, 2003; Anonymous, 2005; Rushen *et al.*, 2007; Marchant-Forde, 2009).

### 3.1 Confinement

Keeping animals individually (for example cows in tie stalls, sows in crates, calves in individual hutches or pens) has several general implications for the welfare of the animals. In these housing systems, behaviour of most of the animal's functional systems is restricted due to lack of space and/or because animals are restrained by tethering. The possibilities to interact with the environment are extremely limited and most confinement housing systems are characterised by a very barren environment providing few environmental stimuli. The behavioural restrictions of confinement and lack of environmental stimuli cause psychological and physical challenges that are associated with occurrence of behavioural signs of reduced welfare (increased levels of abnormal behaviours, apathy) and with negative consequences for health. Additionally, all behaviours have to be performed at one place and thus conflicting requirements of different functional systems can hardly be solved, whereas in a loose housing system for groups or with more space different functional areas can be better adapted to the behavioural and physical requirements.

Most obviously, social behaviour and locomotion are compromised. Cattle in tie stalls are generally able to lick the neighbouring cow, but cannot select their partner for doing so. Also, subdominant animals are forced to stay close to a dominant animal which they would avoid otherwise. In a study comparing farms with tie stalls or loose housing systems, tied cows licked more than those kept in loose housing systems (Laister, 2009). This might be a sign of high social tension, in that social licking might be used for appeasement and tension reduction (Waiblinger *et al.*, 2002). Although, in dry sows, the amount of aggressive interactions was lower in animals housed in crates as compared to those housed in groups, the interactions escalated much more frequently in the confined situation where sows had higher overall free corticosteroid concentrations (for review Marchant-Forde, 2009). The inability to resolve aggressive interactions with neighbouring animals may be a considerable and represent a chronic psychological stressor (Marchant-Forde, 2009). This fits well with the concept that welfare is impaired in case the animals' lacks control over the environment.

Sows in crates more frequently show abnormal behaviour, such as stereotypies [mainly bar-biting, sham chewing] than do sows in loose housing (Wechsler and Weber, 2007; Marchant-Forde, 2009). Oral stereotypies are largely linked to feeding motivation and can thus be reduced when a possibility to perform food searching behaviour (rooting) and to manipulate and eat straw is given. However, this is mostly not possible in crates. Apathy develops when the animal has no way to alter negative events and thus stops trying it ['learned helplessness'] (Boissy *et al.*, 2007a). Therefore, apathy often develops in a confined or otherwise deprived situation (Ruiterkamp, 1987, cited in O'Conell, 2009). Also in tethered cattle, the occurrence of abnormal behaviour [tongue-rolling, nosing] is higher as compared with loose housing, especially immediately after the animal is first tethered (Redbo, 1992, 1993).

Pre-parturient sows are highly motivated to walk around during nest-building behaviour and sows confined in a farrowing crate exhibit stress responses [increase in glucocorticoid levels, increased heart rate] (for review Wechsler and Weber, 2007; Johnson and Marchant-Forde, 2009). But even in loose farrowing pens the sows are still restricted in their ability to perform locomotor activity during nest-building behaviour. The degree to which these restrictions impact the welfare of the sow is not known (Johnson and Marchant-Forde, 2009).

In tied dairy cows locomotion [walking, trotting] and exploration when being released to an exercise area increases with decreased access to such an area, suggesting a motivation to perform these behaviours (Loberg *et al.*, 2004). In calves, some elements of play behaviour cannot- or hardly be performed in single pens and thus are not, or very rarely, shown: calves cannot gallop and hardly perform play behaviour such as elevation of the hind leg (Jensen *et al.*, 1998). As a consequence, the motivation to perform locomotor play and trotting increased with increased length of confinement, which suggests that calves feel a need to perform these behaviours during long periods of confinement (Jensen, 2001). Calves in large single pens are also less active than in group pens. Thus both the lack of sufficient space for performing locomotor behaviour as well as lack of conspecifics contribute to the behavioural deprivation.

In laying hens the confined environment of cages leads (both in conventional and furnished cages) to very low frequencies of certain behaviours. For instance, wing flapping could not at all be observed and comfort behaviour is generally rarely performed even at low stocking densities in furnished cages (Albentosa and Cooper, 2004).

With regard to health of dairy cows, hock lesions, callosities at the carpal joints, teat injuries, tramped teats, mastitis [especially after the first lactation period], ketosis and retained placenta were all found to be more prevalent in tied than in loose housed cows (Blom, 1983; Bendixen *et al.*, 1987a,b, 1988a,b; Regula *et al.*, 2004). Regular exercise of sufficient duration can alleviate the problems (Gustafson, 1993; Gustafson and Lund-Magnussen, 1995; Keil *et al.*, 2006) as does access to pasture during summer. This points at the importance of locomotion for health. However, other factors such as an enriched sensory input may also contribute.

In calves, group housing often is refused with arguments of increased risk of diseases. Yet, when comparing individual and group housing of calves, the results are controversial, which indicates that the management and housing design play an important role. Group size seems to be important: whereas group housing as such does not represent a welfare hazard, keeping calves in large groups seems to negatively impact on health. Altogether, recent epidemiological and experimental studies show that unweaned calves can be kept in small groups without increased health problems provided housing, feeding and management is appropriate (Svensson and Liberg, 2006; Rushen *et al.*, 2007). Together with the effects on behaviour (see above) this indicates that housing calves in small groups will benefit their welfare.

Sows in crates have been shown to have (1) more traumatic injuries caused by flooring and metalwork (including severe teat lesions), (2) more claw lesions with increasing incidence over parity, (3) higher incidence of lameness, especially when compared with sows on straw, and (4) a reduced immune function and higher disease incidence, while physiological data are not clear-cut (Marchant-Forde, 2009).

### 3.2 Space allowance

According to Petherick (1983) space requirements can be subdivided into three main categories. Each of these contributes to the total area the animal needs. Firstly, it relates to the individual's physical space requirement, defined as the space an animal needs purely on the basis of its physical dimensions. The second category is behavioural space, i.e. the space needed for behaviours such as feeding, locomotion and dunging. The third is social space, i.e. the space needed to have appropriate social interactions and avoid social stress. Of course, in practice, physical space, behavioural space and social space often influence the same aspect, overlap or act concertedly, for example regarding space at the feeding place, cubicle- or alley width. Furthermore, space allowance is associated with the amount of manure per $m^2$ and thus can contribute to cleanliness. For example, in fattening bulls kept on a fully slatted floor with rubber cover, a higher space allowance was associated with greater cleanliness (Gygax *et al.*, 2007b). Given the same height of building, the space allowance also contributes to air quality (and thus welfare) by its association with air volume.

Distinguishing physical and behavioural space is hard in practice, as the necessary physical space depends on the behaviour performed, e.g. an animal lying in a sternal position needs less space than an animal lying on the side. Also, the space needed for dynamic behaviours such as lying down or separating dunging and nesting site is dependent of the size of the animal. Therefore, physical and behavioural space can be referred to as individual space as opposed to social space.

### *3.2.1 Individual space*

The individual space for performing appropriate 'individual' behaviour (i.e. not including social interactions) is particularly important for dimensions of individually used equipment within groups such as cubicles or feeding stalls, but also for the overall space allowance in group housing or the size of individual pens. Dimensions in confinement housing often only the static space requirements are considered, but the requirements during dynamic behaviours are ignored. Marchant-Forde (2009) states that 'the average sow stall currently in use on commercial farms does not come even close to meeting the dynamic space requirements, and with the increasing size of the commercial sow, many now are coming close to the static space requirements' (McGlone *et al.*, 2004). As the sow becomes larger in relation to the stall in which she is kept, she has increasing difficulty in standing up and lying down and she becomes more prone to injury (for review see Marchant-Forde, 2009). Also in cattle, the dynamic space as well as static space required for lying down and rising or comfortable lying positions are often neither provided in tie stall nor in cubicle systems, so that difficulties performing these behaviours or abnormal behaviour and bumping into parturitions are frequently observed (Hörning, 2003; Keil *et al.*, 2004; Mülleder and Waiblinger, 2004). If cubicles for cows are too small, too short or have a neck rail positioned too low or too much in the rear not only is normal lying down and rising hampered, but this situation is also associated with longer standing in the cubicles with the cows' hind legs in the alleys, shorter lying durations and, consequently, with higher occurrence of lameness (Hörning, 2003; Mülleder and Waiblinger, 2004; Tucker *et al.*, 2004, 2005; Bernardi *et al.*, 2009; Fregonesi *et al.*, 2009). In group housing with a free lying area, the individual behaviour is also hampered by insufficient space allowance (Gygax *et al.*, 2007b). In laying hens in a perchery system, moving, foraging and dust-bathing are decreased with increasing stocking density, while standing is increased (Carmichael *et al.*, 1999). Natural play behaviour of calves is not possible when housed singly (Jensen *et al.*, 1998), but it is also hampered by insufficient space allowance in group pens (Jensen and Kyhn, 2000). To enable separation of the dunging and lying area for pigs requires enough space. Also, loose farrowing pens require sufficient space to avoid high piglet mortality: in pens measuring 5 $m^2$ or less, pre-weaning piglet mortality was markedly higher compared to crate systems, whereas piglet mortality did not differ significantly between pens and crates in several studies with larger pens (Wechsler and Weber, 2007). This may be explained by the sow's inability to perform, in small pens, behavioural patterns which are typically demonstrated by sows prior to lying down. To allow undisturbed behaviour the minimum space required is 7.5 $m^2$ (Schmid, 1992, 2009).

The individual space requirements depend on the animal's size [see for example recommendations by CIGR (1994), which are based on animal's dimensions]. This makes it difficult to optimize dimensions for all individual animals on the farm and also over time. For example, with the increasing dimensions of cows resulting from breeding efforts over the last decades, the space requirements increased, while housing dimensions [e.g. cubicles, alley width, tie-stall dimensions] are hardly flexible, which has resulted in older buildings often leading to severe welfare problems. The same applies to pigs (see above).

Another problem of size-dependent equipment is the sometimes large variation in the size of animals within one farm (CIGR Working Group, 1994). Consequently, whereas equipment may be appropriate for some animals, it may have a negative impact on the welfare of the larger individuals. For instance, in dairy cows, even in quite large cubicles (125 cm inner width) larger cows (height at withers >1,45 cm) more rarely show comfortable lying positions with outstretched hind legs than do smaller cows (Keil *et al.*, 2004). Similarly, it is difficult to offer optimal conditions to growing animals.

*3.2.2 Social space*

The need for social space arises because animals attempt to keep an individual distance to conspecifics (Wilson, 2000). This individual distance depends on the affiliative and dominance relationship between the animals – bonded animals keeping and accepting closer distances (Bouissou *et al.*, 2001; Aschwanden *et al.*, 2008) – but also on the total space available, more available space leading to animals choosing to keep a greater distance (DeVries *et al.*, 2004). If possible, subordinate animals avoid to intrude the individual distance of the dominant and withdraw from approaching dominants that may occasionally show overt aggression (threats) (Bouissou *et al.*, 2001) as such avoidance of proximity reduces aggression. Accordingly, the number of agonistic social interactions and injuries increases with reduced space allowance (Wierenga, 1984; Andreae *et al.*, 1985; Kondo *et al.*, 1998; Weng *et al.*, 1998, Menke *et al.*, 1999; Fregonesi and Leaver, 2002; Boe *et al.*, 2006).

Besides effects on aggression, insufficient space affects many behaviours, health and production. In pigs, insufficient space reduces resting time (Ewbank and Bryant, 1969), mating rate (Hemsworth *et al.*, 1986), and weight gain and feed conversion efficiency (Brumm, 1996). Existence of an outside run, providing more space allowance, as well as higher total space allowance (≥8.6 $m^2$/cow) was associated with lower prevalence of lameness in dairy cows (Mülleder and Waiblinger, 2004). In dairy heifers, insufficient space causes reduced lying, hindered synchronisation of lying behaviour, higher variance in lying time, lower live-weight gain, increased net energy intake and higher prevalence of heel horn erosion (Hindhede *et al.*, 1996; Nielsen *et al.*, 1997; Mogensen *et al.*, 1997a,b). Comparing two groups with low (total space 6.7 $m^2$, lying area 4.5 $m^2$/ cow) and high (13.5 and 9 $m^2$/cow) space allowance in straw yards, Fregonesi and Leaver (2002) did not find a significant difference in lying time, synchronicity of lying, feeding or agonistic interactions in four groups of 6 Holstein Friesian cows (approximately 660 kg live weight). However, cows were dirtier in the low space condition and milk lactose and somatic cell count pointed at a higher incidence of mastitis in high yielding cows. Goats spent less time resting and rested less synchronized when the resting area was 0.5 $m^2$ compared to 0.75 or 1.0 $m^2$ per goat (Andersen and Boe,

2007). In broilers, a lower space allowance (40, 34 or 28 kg/m$^2$ terminal stocking density were compared) led to a decline in feed intake and in longer deep panting in the two higher stocking densities indicative of problems with thermal comfort (McLean *et al.*, 2002).

Space allowance is very often linked to the number of available resources such as lying- and feeding places (for more discussion on these aspects see below). Reduced space allowance reduces also the availability of resources, especially for low-ranking individuals, because their way to the resource is more easily blocked by higher-ranking individuals. Bottlenecks can cause jams of animals. This is especially the case in small alleys such as prevail in cubicle houses, and may be associated with reduced production (Konggaard, 1983b; Ostergaard *et al.*, 1986).

As regards alley width of cubicle houses, the interplay of individual space, social space and animal size is of special importance, the requirement being that at least two cows should be able to pass each other without (major) social disturbance. For example, the alley behind the feeding place in dairy cow cubicle housing would need to be at least as wide as a cow length plus 2 times cow width to theoretically enable cows during feeding time (when many cows are standing in the feeding place) to pass each other. Recommendations for alley width were based on such calculations (Zeeb, 1987; Bartussek *et al.*, 1995; McLean *et al.*, 2002), albeit that the cows' need for social individual distance is not yet considered in this calculation.

With a given amount of total space the arrangement of this space allowance may be very important. Very small alleys behind the feeding place or in passageways can only be compensated to a very limited extent by a larger space allowance elsewhere, e.g. an outside run. Furthermore, the frequency of crossways, i.e. the alley length in between crossways, influences the effects of alley width, as crossways may offer alternative ways for an individual (low ranking) animal to reach specific resources.

### 3.3 Number and distribution of resources

Animals are highly motivated to rest, feed and drink. A limited number of places for these activities causes increased competition and thus aggression, and can cause stress and injuries. Some animals, especially lower ranking individuals, may not be able to fulfil their nutritional needs or their need for resting as a result of restrictions in access to the resource by higher ranking individuals and thus experience an increased risk of diseases. Also, longer waiting times to get access may lead to frustration and increased standing which may also enhance the risk of leg disorders (for a more detailed discussion of feeding and lying places: see below). Although other limited resources (for example brushes or a salt for licking) can enhance competition, the decrease in welfare appears to be limited.

Besides the mere amount of resources their distribution or placement is also important. When several resources are clumped together [e.g. drinkers and concentrate feeders in close proximity], then encounters and aggression increase while spatial separation can relax the situation (Estevez *et al.*, 2007).

Despite these clear welfare benefits in general, pasture can also bear some welfare risks, especially related to climatic conditions (heat stress during hot sunny days when no access to sufficient shelter or showers is provided), parasitic load, under-nutrition, intoxication, poor walking tracks or, in chickens, even predation. Access to shelter or otherwise protection from adverse weather conditions is required. The other factors are largely management-dependent.

## 3.8 Climatic conditions

Climatic conditions (e.g. temperature, occurrence of draft or noxious gases, air exchange rate) also depend on housing design, including technical solutions for ventilation and dung or slurry removal, and may have a large impact on welfare.

To care for appropriate climatic conditions the different aspects such as light, air quality, air humidity and rainfall, air flow, temperature and solar radiation need to be considered. These are to be managed by natural or artificial light sources and appropriate ventilation in indoor housing, and by shelter or other adequate measures to protect the animal from adverse weather conditions when they are outdoors.

### *3.8.1 Light*

Light is necessary for the animals to be able to see and react properly to their environment. For example, with light intensities lower than 32 lux, cows walk with shorter step length and higher step rate, indicative of lower motion confidence (Phillips *et al.*, 2000). Laying hens or birds in general are able to perceive UV-light and the perception of the environment may be impaired when light does not contain UV (Prescott *et al.*, 2003). Poultry may also perceive the flicker of fluorescent luminaries (100 Hz) and it is advisable to consider using high frequency fluorescent lighting.

Furthermore, day-night rhythm and the duration and intensity of the light, are important for endocrine regulation. A longer light phase (16 light/8 hours dark) was shown to positively affect production and reproduction (e.g. milk yield and age of puberty in dairy animals (Dahl *et al.*, 2000; Biewenga and Winkel, 2003), and is a common approach to stimulate egg production in laying hens (Appleby *et al.*, 2004). A shorter day (8 light/16 hours dark) during the dry period increases the milk production of dairy cows in the subsequent lactation (Biewenga and Winkel, 2003). Electric light with a sufficient intensity allows for photoperiodicity and, at least in mammals, enables good vision of the environment. However, natural (especially sun-) light may have beneficial effects due to higher intensity.

### *3.8.2 Air quality*

The concentrations of different contaminants, mainly noxious gases, dust and bacteria as well as humidity constitute air quality. Carbon dioxide ($CO_2$) is a useful indicator for ventilation intensity. Insufficient ventilation can lead to high $CO_2$ concentration indoors. There is some scientific evidence for adverse effects on farm animals when the concentration of carbon dioxide is 3,000 ppm or higher (Scientific Veterinary Committee, 1997). The most important noxious gases in livestock housing are ammonia ($NH_3$) and hydrogen sulphide ($H_2S$). The ammonia concentration largely depends on the system of dung or slurry removal (fully slatted, solid with scrapers, deep-litter) and its management. The combination of dust, bacteria and ammonia in the air compromises lung health (Appleby *et al.*, 2004). The individual role of ammonia in the development of respiratory disease remains unclear, though it seems to act synergistically with other pollutants (especially with dust).

Regarding humidity both dry and very moist air ease respiratory diseases. High moisture content of air and surfaces favours the production and release of noxious gases from manure and soiled surfaces and supports bacterial growth

*3.8.3 Temperature and solar radiation*

If temperature gets too high animals need possibilities to cool down. They may lie on surfaces that allow cooling even to the expense of lying comfort (softness). For example, in pigs this may lead to higher ammonia levels in the air, soiling of animals and the lying area, so that the separation of dunging area and lying area no longer works. This also negatively impacts on labour required. Showers can be used for cooling the animals, to replace the natural wallowing behaviour in pigs. Also in cattle sprinklers are effectively used for cooling the animals (Tucker *et al.*, 2008). In addition, in buildings with natural ventilation additional ventilators may be installed to provide higher wind speeds.

For thermal comfort different variables such as temperature, humidity, wind speed and solar radiation interact. To allow for the inclusion of all these the heat load index was developed recently (Tucker *et al.*, 2008), i.e. an extension to the temperature-humidity index. Provision of shade is important during high solar radiation and high temperatures, and also can cool down the animals (Tucker *et al.*, 2008). For animals on pasture during summer this means they need to have sufficient access to a building or a shady area. In pigs the frequency of playing decreased with increasing temperature. When housed on the south side of the building pigs with access to shelter played more than those without, suggesting improved welfare due to the opportunity to regulate the body temperature by use of shade (Olsen *et al.*, 2002)

During periods with low temperatures animals need protection. During the winter in temperate and cold areas, free-range cattle need at least a dry lying area with protection against wind (Wallbaum *et al.*, 1997). In wet areas this can only be offered by a roofed area. Cattle on pasture use a shelter after two hours of rain (Vandenheede *et al.*, 1994).

### 3.9 Design of equipment

Housing equipment, such as drinkers, feeders, feeding place partitions or cubicles, can be designed quite differently, even if the basic shape is the same. In the following some important aspects are briefly mentioned. For all equipment a construction avoiding injuries is essential but not self-evident everywhere. For example, in feed racks jutting screws on the side of the animals can cause severe lesions at the shoulder.

### *3.9.1 Feeding place design*

For cattle, to avoid skin and claw lesions from a heavy load and to allow a relaxed feeding posture, the feeding area should have the following three features: (1) a raised feeding table or crib (20-25 cm in loose housing), (2) an inclined feeding rack, and (3) food always being within reach (Boxberger, 1983; Konggaard, 1983a; Hansen *et al.*, 1998; Waiblinger *et al.*, 2001b, see also section 2.1). Drinkers should not be too high – at most 80 cm for adults – so that the cattle can use their normal drinking posture (Metzner, 1976) and sufficient water flow is essential to avoid increased competition in groups and to allow species-typical drinking behaviour. Feeding racks/feeding place partitions can have quite different shapes which may influence the sight of the animal and the level of protection. For horned cattle or goats, head locks having a rail above the head are not suitable as animals have difficulties to leave the feeding place quickly and can thus be injured by other animals, whereas in head locks with open side above the head they can leave as quickly as hornless animals (Menke, 1996, Waiblinger *et al.*, 2001a).

In pigs, aspects of feeder design that can influence feed intake of weaned pigs includes depth of trough, size of feeder gap, lip height and side panels (O'Conell, 2009): The presence of drinkers in the trough seems not to be recommendable as weaned pigs may have difficulty learning to operate these drinkers, which leads to the accumulation of wet, unpalatable feed (O'Connell, 2009). The positive effects of parturitions at the trough were mentioned already in section 3.4.

High competition feeders, e.g. automated milk feeder for calves, concentrate feeders for dairy cows or electronic sow feeders should offer protection of the animals in the feeding station from displacement and aggression by others by closing gates (Menke 1996; Weber and Wechsler, 2001). The best solution may be a separate exit gate so that the animal is not subjected to aggression from the hind when leaving the feeder.

In poultry as feeding and drinking are closely associated (Fölsch, 1981), drinkers should be close to the feeders. Given a choice, hens generally prefer open water surfaces, as in bell or cup drinkers, rather than nipple drinkers. Drinking from nipples is not a natural behaviour and must be learned, but switching from nipple to bell drinkers also may cause problems (Appleby *et al.*, 2004). Rearing facilities should therefore be equipped with drinkers of various types, or at least the type that the hens will later find in the hen house.

## 4. Conclusions

### 4.1 What has been achieved?

Research in applied animal ethology and animal welfare started about 50 years ago. Since then, our knowledge about the behavioural needs of the animals and the interactions between environment and animal welfare has increased significantly and risk factors for poor welfare have been identified. This knowledge in combination with other societal developments has raised the awareness and sensitivity regarding farm animal husbandry as related to welfare issues and this has lead to some improvements in animal production practice, notably: (1) the realisation of the agrifood businesses that a proper consideration of animal welfare issues is not only beneficial for animal health and production efficiency, but equally contributes to improving the reputation of the farm sector which has allowed for higher pricing for quality products; (2) the formulation of national and European animal welfare legislation, which – based on research, that demonstrated the behavioural needs of animals, the extreme behavioural restrictions and associated welfare problems they have to face – provides the necessary tool to introduce improved minimum animal welfare standards. Probably the most far-reaching changes enforced by law, are a ban on the use of conventional cages (although enriched cages are still allowed except in few European countries), the ban of crates for housing sows during the dry period, and the decision to make group-housing of calves older than 9 weeks mandatory (with exception of small farms); and (3) the partly successful development of housing systems or elements thereof, that are based on animal needs but still allow a modern, quite intensive production. Only recently, more attention is paid to the concept of 'positive welfare' or 'life of quality', where, rather than merely avoiding negative emotions in animals, the main aim is to reach a quite high welfare standard by enabling animals to experience positive emotional states.

### 4.2 What has been neglected?

Despite this modest progress, the development towards more welfare-friendly systems in practice is slow. It would be naïve to think that with the afore-mentioned bans of confinement housing systems the situation for farm animals has greatly improved: still many – if not most – of them have to face an inappropriate environment and poor welfare. The risk factors and problems described in this chapter are widespread. Intensification and low labour costs are still often the main driving force and thus improvements apply to the minority of production units, i.e. are limited to the high quality label production, or to farmers who, through their own positive attitude towards animals, want to offer them a good environment. Thus knowledge about improved systems is implemented only to a limited degree. Even newly built and newly developed systems often neglect the needs of the animals instead of taking them as a starting point and trying to develop a system where all relevant key features are incorporated. An example in the latter direction would be the family pen for pigs (Stolba *et al.*, 1984; Wechsler *et al.*, 1991), where, however, still some questions need to be answered.

The important role of an appropriate welfare-friendly environment which also serves animal and public health and disease prevention goals has been largely neglected, not only by farmers, but – sadly – also by many veterinarians, who seem to often overemphasize the role of hygiene and disease and under-emphasize behavioural or brain function indicators of poor or good welfare (Broom, 2006).

### 4.3 What needs to be done?

For future improvements in animal welfare related to housing both research and implementation is needed. Although, in general, our knowledge regarding single welfare risks and animal needs is good, our knowledge about their interactions and relative importance in the farm setting is still limited. Furthermore, there are some unsolved problems also in alternative systems, e.g. feather pecking and cannibalism in non-caged laying hens, or tail biting in pigs on straw. While other factors than housing (management, hygiene, genetics) may contribute to a large degree to the success or failure of a system the role of specific housing factors need to be further highlighted.

In some areas research does not really allow to set limits where poor welfare is expected to cease. For example, we know little if anything about the effects of different alley width on dairy cow behaviour, health and welfare, except for effects of extremely small alleys which no longer prevail in farm practice. However, to set minimum standards the effects of different limits should be known. In recent years more groups focus on investigating such aspects (e.g. Jensen *et al.*, 1998; Huzzey *et al.*, 2006; Gygax *et al.*, 2007b).

An important aspect that deserves further research relates to the interactions of housing, psychological well-being (i.e. being in a positive emotional state) and health. This also should be included in economical models providing a larger perspective (e.g. environmental and societal costs of poor welfare) with the aim to reconcile economic and animal welfare requirements.

Another important point is implementation into practice. It is one thing to establish minimum standards for housing in legislation, preferably on a European level, to implement such standards, however, is no small feat and to allow for this to happen sufficient control measures and associated penalties are essential (Waiblinger and Wechsler, 2007). The sampling scheme of the EU (2% per year) seems insufficient to assess the true problems prevailing and to ensure that enough single farms are controlled. In contrast, in Switzerland 30% of farms obtaining subsidies are controlled annually. Obviously, other welfare indicators addressed elsewhere in this Volume need to equally be taken into account.

Besides setting minimum standards and enforcing legislation through penalties in case of non-compliance, incentives for improvement of housing conditions and animal welfare may offer an important alternative. The welfare of animals can range from very poor to very good. It is good to realize that minimum standards do not serve to guarantee a very high level of welfare but rather 'to ensure improved protection and respect for the welfare of animals as sentient beings' (Treaty of Amsterdam). Granting subsidies for incentives aiming at reaching high levels of animal welfare on the farm and particularly by supporting 'welfare labels' on products that have been generated in such a way [as proposed by the EU commission (Anonymous, 2006)] would enable consumers to make an informed choice of products. In the long term such would probably be the most effective way to ensure that animal welfare standards are properly considered.

## Acknowledgements

The author gratefully acknowledges Dr. Hans Spoolder and Dr. Knut Niebuhr for their valuable support during the conception of this contribution.

Jensen, M.B., 2001. A note on the effect of isolation during testing and length of previous confinement on locomotor behaviour during open-field test in dairy calves. Applied Animal Behaviour Science 70, 309-315.

Jensen, M.B., 2004. Computer-controlled milk feeding of dairy calves: the effects of number of calves per feeder and number of milk portions on use of feeder and social behavior. Journal of Dairy Science 87, 3428-3438.

Jensen, M.B. and Budde, M., 2006. The effects of milk feeding method and group size on feeding behavior and cross-sucking in group-housed dairy calves. Journal of Dairy Science 89, 4778-4783.

Jensen, M.B. and Kyhn, R., 2000. Play behaviour in group-housed dairy calves, the effect of space allowance. Applied Animal Behaviour Science 67, 35-46.

Jensen, M.B., Vestergaard, K.S. and Krohn, C.C., 1998. Play behaviour in dairy calves kept in pens: the effect of social contact and space allowance. Applied Animal Behaviour Science 56, 97-108.

Jensen, P., 1986. Observations on the maternal behaviour of free-ranging domestic pigs. Applied Animal Behaviour Science 16, 131-142.

Johnsen, P.F., Vestergaard, K.S., and Nørgaard-Nielsen, G., 1998. Influence of early rearing conditions on the development of feather pecking and cannibalism in domestic fowl. Applied Animal Behaviour Science 60, 25-41.

Johnson, A.K. and Marchant-Forde, J.N., 2009. Welfare of pigs in the farrowing environment. In: Marchant-Forde, J.N. (Ed.). The welfare of pigs. Springer, the Netherlands, pp. 141-181.

Jørgensen, B., 2003. Influence of floor type and stocking density on leg weakness, osteochondrosis and claw disorders in slaughter pigs. Animal Science 77, 439-449.

Jorgensen, G.H.M., Andersen, I.L. and Boe, K.E., 2007. Feed intake and social interactions in dairy goats – The effects of feeding space and type of roughage. Applied Animal Behaviour Science 107, 239-251.

Keil, N.M., Audigq, L. and Langhans, W., 2000. Factors associated with intersucking in Swiss dairy heifers. Preventive Veterinary Medicine 45, 305-323.

Keil, N.M., Gisiger, E. and Stauffacher, M., 2004. Evaluation von Liegeboxenabmessungen für Rindvieh aufgrund des Liegeverhaltens unterschiedlich großer Milchkühe. KBTL-Schrift 143. KBTL, Darmstadt, Germany, pp. 115-121.

Keil, N.M., Wiederkehr, T.U., Friedli, K. and Wechsler, B., 2006. Effects of frequency and duration of outdoor exercise on the prevalence of hock lesions in tied Swiss dairy cows. Preventive Veterinary Medicine 74, 142-153.

Keil, N.M., Zwicky, U. and Schrader, L., 2002. Einfluss der Umweltkomplexität auf Verhalten und gegenseitiges Besaugen von Aufzuchtkälbern in Gruppenhaltung. KTBL, Darmstadt, Germany, pp. 76-83.

KilBride, A.L., Gillman, C.E., Ossent, P. and Green, L.E., 2008. A cross-sectional study of the prevalence and associated risk factors for capped hock and the associations with bursitis in weaner, grower and finisher pigs from 93 commercial farms in England. Preventive Veterinary Medicine 83, 272-284.

Kiley-Worthington, M. and Plain, S.d.l., 1983. The behaviour of beef suckler cattle. Birkhäuser Verlag TH, Basel, Switzerland.

Kondo, S., Sekine, J., Okubo, M. and Asahida, Y., 1989. The effect of group size and space allowance on the agonistic and spacing behaviour of cattle. Applied Animal Behaviour Science 24, 127-135.

Konggaard, S.P., 1983a. Feeding conditions in relation to welfare for dairy cows in loose-housing systems. Martinus Nijhoff, The Hague, the Netherlands.

Konggaard, S.P., 1983b. Feeding conditions in relation to welfare for cows in loose- housing systems. In: Baxter, S.H., Baxter, M.R. and MacCormack, J.A.D. (Eds.). Farm animal housing and welfare. M. Nijhoff publishers, Boston, USA, pp. 272-282.

Kremer, P.V., Nueske, S., Scholz, A.M., Foerster, M., 2007. Comparison of claw health and milk yield in dairy cows on elastic or concrete flooring. Journal of Dairy Science 90, 4603-4611.

Krötzl, H. and Hauser, R., 1997. Ethologische Grundlagen zum Platzbedarf, zur Gestaltung und zum Betrieb von Laufhöfen bei Kühen im Laufstall. Agrartechnische Forschung 3, 141.

Laister, S., 2009. Suitability of selected behavioural paramters for on-farm welfare assessment in dairy and beef cattle. University of Natural Resources and Applied Life Sciences, Vienna, Austria. 71 p.

Leonard, F.C., O'Connell, J.M., and O'Farell, K.J., 1996. Effect of overcrowding on claw health in first-calved Friesian heifers. British Veterinary Journal, 152: 459-472.

Loberg, J., Telezhenko, E., Bergsten, C. and Lidfors, L., 2004. Behaviour and claw health in tied dairy cows with varying access to exercise in an outdoor paddock. Applied Animal Behaviour Science 89, 1-16.

Manninen, E., De Passillé, A.M., Rushen, J., Norring, M. and Saloniemi, H., 2002. Preferences of dairy cows kept in unheated buildings for different kind of cubicle flooring. Applied Animal Behaviour Science 75, 281-292.

Marchant-Forde, J.N., 2009. Welfare of dry sows. In: Marchant-Forde, J.N. (Ed.). The welfare of pigs. Springer, the Netherlands, pp. 94-139.

Marchant-Forde, J. N. (Ed.) 2009. The welfare of pigs, Springer, the Netherlands, pp 350.

McBride, G., Parer, L.G. and Foenander, F., 2009. The social organization and behaviour of the feral domestic fowl. Animal Behaviour Monographs 2, 126-182.

McGlone, J.J. and Curtis, S.E., 1985. Behavior and performance of weanling pigs in pens equipped with hide areas. Journal of Animal Science 60, 20-24.

McGlone, J.J., Vines, B., Rudine, A.C. and DuBois, P., 2004. The physical size of gestating sows. Journal of Animal Science 82, 2421-2427.

McLean, J.A., Savory, C.J. and Sparks, N.H.C., 2002. Welfare of male and female broiler chickens in relation to stocking density, as indicated by performance, health and behaviour. Animal Welfare 11, 55-73.

Mench, J.A., 1998. Thirty years after Brambell: whither animal welfare science? Journal of Applied Animal Welfare Science 1, 91-102.

Menke, C., 1996. Laufstallhaltung mit behornten Milchkühen. Dissertation, Institut für Nutztierwissenschaften, ETH-Zürich, Zurich, Switzerland.

Menke, C., Waiblinger, S. and Fölsch, D.W., 2000. Die Bedeutung von Managementmaßnahmen im Laufstall für das Sozialverhalten von Milchkühen. Dtsch. Tierärztliche Wschr. 107, 262-268.

Menke, C., Waiblinger, S., Fölsch, D.W. and Wiepkema, P.R., 1999. Social behaviour and injuries of horned cows in loose housing systems. Animal Welfare 8, 243-258.

Metzner, R., 1976. Kennwerte tiergerechter Versorgungseinrichtungen des Kurzstandes für Fleckviehkühe. TU München, Munchen, Germany.

Miller, K. and Wood-Gush, D.G.M., 1991. Some effects of housing on the social behaviour of dairy cows. Animal Production 53, 271-278.

Mogensen, L., Krohn, C.C., Sørensen, J.T., Hindhede, J. and Nielsen, L.H., 1997a. Association between resting behaviour and live weight gain in dairy heifers housed in pens with different space allowance and floor type. Applied Animal Behaviour Science 55, 11-19.

Mogensen, L., Nielsen, L.H., Hindhede, J., Sørensen, J.T. and Krohn, C.C., 1997b. Effect of space allowance in deep bedding systems on resting behaviour, production, and health of dairy heifers. Acta Agriculturae Scandinavica Section A, Animal Science 47, 178-186.

Moinard, C., Mendl, M., Nicol, C.J. and Green, L.E., 2003. A case control study of on-farm risk factors for tail biting in pigs. Applied Animal Behaviour Science 81, 333-355.

Molz, C., 1989. Beziehung zwischen haltungstechnischen Faktoren und Schäden beim Milchvieh in Boxenlaufställen. Dissertation LMU München, Munchen, Germany.

Mülleder, C. and Waiblinger, S., 2004. Analyse der Einflussfaktoren auf Tiergerechtheit, Tiergesundheit und Leistung von Milchkühen im Boxenlaufstall auf konventionellen und biologischen Betriebe unter besonderer Berücksichtigung der Mensch-Tier-Beziehung. Endbericht FP 1264. VUW, Wien, Austria. 165 p.

Mülleder, C., Absmanner, E., Kahrer, E., Zeiner, H., Stanek, C. and Troxler, J., 2008. Alternative Haltungssysteme für die Rindermast unter österreichischen Verhältnissen unter besonderer Berücksichtigung von Betonspaltenböden mit Gummiauflage. Institut für Tierhaltung und Tierschutz, Projektnr 1447, 174 p.

Mülleder, C., Palme, R., Menke, C. and Waiblinger, S., 2003. Individual differences in behaviour and in adrenocortical activity in beef-suckler cows. Applied Animal Behaviour Science 84, 167-183.

Nielsen, L.H., Mogensen, L., Krohn, C., Hindhede, J. and Sorensen, J.T., 1997. Resting and social behaviour of dairy heifers housed in slatted floor pens with different sized bedded lying areas. Applied Animal Behaviour Science 54, 307-316.

O'Conell, N.E., 2009. Housing the fattening pig. In: Marchant-Forde, J.N. (Ed.). The welfare of pigs. Springer, the Netherlands, pp. 189-210.

Olofsson, J., 1999. Competition for total mixed diets fed *ad libitum* intake using one or four cows per feeding station. Journal of Dairy Science 82, 69-79.

Olofsson, J., 2000. Feed availability and its effects on intake, production and behaviour in dairy cows. Acta-Universitatis-Agriculturae-Sueciae-Agraria.

Olsen, A.W., Simonsen, H.B. and Dybkjraer, L., 2002. Effect of access to roughage and shelter on selected behavioural indicators of welfare in pigs housed in a complex environment. Animal Welfare 11, 75-87.

Ostergaard, V., Munksgaard, L. and Henneberg, U., 1986. Sengestaldes belaeguing og dens betydning for malkekoens velfaerd og produktionssystemets okonomi. (Housing density in cubicle housing and its importance for the welfare of dairy cows and for the economics of production). Beretning-fra-Statens-Husdyrbrugsforsog 615, 52-58.

Petherick, J.C., 1983. A biological basis for the design of space in livestock housing. In: Baxter, S.H., Baxter, M.R. and MacCormack, J.A.D. (Eds.), Farm animal housing and welfare., Martin Nijhoff publisher, Boston, USA, pp. 103-120.

Phillips, C.J.C. and Morris, I.D., 2000. The locomotion of dairy cows on concrete floors that are dry, wet, or covered with a slurry of excreta. Journal of Dairy Science 83, 1767-1772.

Phillips, C.J.C. and Morris, I.D., 2001. The locomotion of dairy cows on floor surfaces with different frictional properties. Journal of Dairy Science 84, 623-628.

Phillips, C.J.C. and Morris, I.D., 2002. The ability of cattle to distinguish between, and their preference for, floors with different levels of friction, and their avoidance of floors contaminated with excreta. Animal Welfare 11, 21-29.

Phillips, C.J.C., Morris, I.D., Lomas, C.A. and Lockwood, S.J., 2000. The locomotion of dairy cows in passageways with different light intensities. Animal Welfare 9, 421-431.

Potter, M.J. and Broom, D.M., 1987. The behaviour and welfare of cows in relation to cubicle house design. In: Wierenga, H.K. and Peterse, D.J. (Eds.) Cattle housing systems, lameness and behaviour. Martinus Nijhoff, Dordrecht, the Netherlands, pp. 129-147.

Prescott, N.B., Wathes, C.M., and Jarvis, J.R., 2003. Light, vision and the welfare of poultry. Animal Welfare 12, 269-288.

Price, E.O., 1999. Behavioral development in animals undergoing domestication. Applied Animal Behaviour Science 65, 245-271.

Putz, K., 2002. Haltungsbedingte Verletzungen bei Sauen und Ferkeln in strohlosen Abferkelstallungen. Veterinärmedizinische Universität, Wien, Austria, 120 p.

Rasmussen, D.K., Weber, R., and Wechsler, B., 2006. Effects of animal/feeding-place ratio on the behaviour and performance of fattening pigs fed via sensor-controlled liquid feeding. Applied Animal Behaviour Science 98, 45-53.

Redbo, I., 1992. The influence of restraint on the occurrence of oral stereotypies in dairy cows. Applied Animal Behaviour Science 35, 115-123.

Redbo, I., 1993. Stereotypies and cortisol secretion in heifers subjected to tethering. Applied Animal Behaviour Science 38, 213-225.

Regula, G., Danuser, J., Spycher, B., and Wechsler, B., 2004. Health and welfare of dairy cows in different husbandry systems in Switzerland. Preventive Veterinary Medicine 66, 247-264.

Reinhardt, V., 1980. Untersuchungen zum Sozialverhalten des Rindes. Birkhäuser Verlag TH, Basel, Switzerland. 90 p.

Rushen, J., De Passillé, A.M., Keyserlingk, M.A.G. and Weary, D.M., 2007. The welfare of cattle. Springer, Berlin, Germany.

Schmid, H., 1992. Arttypische Strukturierung der Abferkelbucht. KTBL, Darmstadt, Germany, pp. 27-36.

Schmid, H., 2009. Ethological design of a practicable farrowing pen. KTBL, Darmstadt, Berlin, Germany, pp. 238-241.

Schrader, L., Keil, N., Rölli, D., and Nydegger, F., 2002. Einfluss eines erhöhten Tier-Fressplatzverhältnisses auf das Verhalten von Milchkühen unterschiedlichen Ranges im Laufstall. KTBL, Darmstadt, Germany, pp. 17-22.

Schrader, L., Roth, H.R., Winterling, C., Brodmann, N., Langhans, W., Geyer, H., and Graf, B., 2001. The occurrence of tail tip alterations in fattening bulls kept under different husbandry conditions. Animal Welfare 10, 119-130.

Schulze Westerath, H., Gygax, L., Mayer, C., and Wechsler, B., 2007. Leg lesions and cleanliness of finishing bulls kept in housing systems with different lying area surfaces. The Veterinary Journal 174, 77-85.

Schütz, K., Davison, D., and Matthews, L.R., 2006. Do different levels of moderate feed deprivation in dairy cows affect feeding motivation? Applied Animal Behaviour Science 101, 253-263.

Scientific Veterinary Committee, 1997. The welfare of intensively kept pigs. Report to the Directorate General XXIV of the European Commission. Adopted 30th September 1997. Doc XXIV/ScVc/0005/97. Scientific Veterinary Committee, Animal Welfare Section, Brussels, Belgium.

Scott, K., Chennells, D.J., Armstrong, D., Taylor, L., Gill, B.P., and Edwards, S.A., 2007. The welfare of finishing pigs under different housing and feeding systems: liquid versus dry feeding in fully-slatted and straw-based housing. Animal Welfare 16, 53-62.

Smulders, D., Hautekiet, V., Verbeke, G., and Geers, R., 2008. Tail and ear biting lesions in pigs: an epidemiological study. Animal Welfare 17, 61-69.

Somers, J.G.C.J., Metz, J.H.M., Noordhuizen-Stassen, E.N., and Frankena, K., 2001. Epidemiological study on claw disorders in dairy cattle: impact of floor systems. In: Shearer, J.K. (Ed.) 12th International symposium on lameness in ruminants. Orlando, FL, USA, pp. 350-352.

Sommer, T. and Troxler, J., 1986. Ethologische und veterinärmedizinische Beurteilungskriterien in Bezug auf die Tiergerechtheit von Loch- und Spaltenböden für Milchvieh. KTBL, Darmstadt, Germany, pp. 73-85.

Stolba, A. and Wood-Gush, D.G.M., 1989. The behaviour of pigs in a semi-natural environment. Animal Production 48, 419-425.

Stolba, A. and Wood-Gush, D.G.M., 1984. The identification of behavioural key features ad their incorporation into a housing design for pigs. Annual Research Veterinary 15, 287-299.

Stumpf, S., Nydegger, F., Wechsler, B., and Beyer, S., 2000. Untersuchungen zum Tier-Fressplatzverhältnis und zur Fressplatzgestaltung bei der Slebstfütterung von Milchkühen am Fahrsilo. KTBL, Darmstadt, Germany, pp. 103-110.

Svensson, C. and Liberg, P., 2006. The effect of group size on health and growth rate of Swedish dairy calves housed in pens with automatic milk-feeders. Preventive Veterinary Medicine 73, 43-53.

Telezhenko, E. and Bergsten, C., 2005. Influence of floor type on the locomotion of dairy cows. Applied Animal Behaviour Science 93, 183-197.

Tembrock, G., 1980. Grundriß der Verhaltenswissenschaften. Fischer, Jena, Germany.

Thomsen, P.T., Kjeldsen, A.M., Sørensen, J.T., Houe, H., and Ersboll, A.K., 2006. Herd-level risk factors for the mortality of cows in Danish dairy herds. Veterinary Record 158, 622-626.

Toates, F., 1987. The relevance of models of motivation and learning to animal welfare. In: Wiepkema, P.R. and Van Adrichem (Eds.), Biology of stress in farm animals: an integrative approache, M. Nijhoff Publishers, Dordrecht, the Netherlands, pp. 153-186.

Tucker, C.B., Rogers, A.R., and Schütz, K., 2008. Effect of solar radiation on dairy cattle behaviour, use of shade and body temperature in a pasture-based system. Applied Animal Behaviour Science 109, 141-154.

Tucker, C.B., Weary, D.M., and Fraser, D., 2004. Free-stall dimensions: effects on preference and stall usage. Journal of Dairy Science 87, 1208-1216.

Tucker, C.B., Weary, D.M., and Fraser, D., 2005. Influence of neck-rail placement on free-stall preference, use, and cleanliness. Journal of Dairy Science 88, 2730-2737.

Tucker, C.B., Weary, D.M., De Passillé, A.M., Campbell, B., and Rushen, J., 2006. Flooring in front of the feed bunk affects feeding behavior and use of freestalls by dairy cows. Journal of Dairy Science 89, 2065-2071.

Tucker, C.B., Weary, D.M., Von Keyserlingk, M.A.G., and Beauchemin, K.A., 2009. Cow comfort in tie-stalls: increased depth of shavings or straw bedding increases lying time. Journal of Dairy Science 92, 2684-2690.

Van der Mheen, H. and Spoolder, H.A.M., 2004. Effect of environmental enrichment and limited access on pasture damage by sows. In: Hänninen, L. and Valros, A. (Eds.) Proceedings of the annual meeting of the International Society for Applied Ethology, Helsinki, Finland, 2004, p. 135.

Vandenheede, M., Shehi, R., Nicks, B., Canart, B., Dufrasne, I, Biston, R. and Lecomte, P., 1994. Influence de la pluie sur l'utilisation d'un abri par des taurillons au paturage. Annales de Médicine Vétérinaire, 138: 91-94.

Veissier, I. and Boissy, A., 2007. Stress and welfare: Two complementary concepts that are intrinsically related to the animal's point of view. Physiology & Behavior 92, 429-433.

Von Keyserlingk, M.A.G., Brusius, L., and Weary, D.M., 2004. Competition for teats and feeding behavior by group-housed dairy calves. Journal of Dairy Science 87, 4190-4194.

Waiblinger, S. and Wechsler, B., 2007. Handlungsbedarf betreffend Mindestanforderungen für eine tiergerechte Rinderhaltung. KTBL, Darmstadt, Germany, pp. 11-22.

Waiblinger, S., Baars, T., and Menke, C., 2001a. Understanding the cow – the central role of human-animal relationship in keeping horned dairy cows in loose housing. In: Hovi, M. and Bouilhol, M. (Eds.), Network for animal health and welfare in organic agriculture (NAHWOA). University of Reading, Reading, UK.

Waiblinger, S., Baumgartner, J., Kiley-Worthington, M., and Niebuhr, K., 2003. Applied ethology – the basis for improved animal welfare in organic farming. In: Vaarst, M., Roderick, S., Lund, V., and Lockeretz, W. (Eds.), Animal health and welfare in organic agriculture. CABI Publishing, Wallingford, UK.

Waiblinger, S., Fresdorf, A., and Spitzer, G., 2002. The role of social licking in cattle for conflict resolution. Münster, Germany. 22 p.

Waiblinger, S., Reichmann, V., Troxler, J., Dreiseitl, H., Haller, J., and Windischbauer, G., 2001b. Einfluß eines Vorrückfressgitters auf Druckbelastungen und Schäden an den Schultern von Milchkühen. Institut für Agrartechnik der Universität Hohenheim, Giessen, Germany, pp. 462-465.

Wallbaum, F., Wassmuth, R., and Langholz, H.-J., 1997. Outdoor wintering of suckler cows in low mountain range. Arendonk J.A.M. (Ed.) Book of abstracts of the 48th annual meeting of the European Association for Animal Production, Vienna, Austria 25-28.8.97, pp. 226-226.

Washburn, S.P., White S.L., Green Jr., J.T. and Benson, G.A., 2002. Reproduction, mastitis, and body condition of seasonally calved Holstein and jersey cows in confinement of pasture systems. Journal of Dairy Science 85, 105-111.

Weber, R. and Troxler, J., 1988. Die Bedeutung der Zeitdauer der Geburt in verschiedenen Abferkelbuchten zur Beurteilung auf Tiergerechtigkeit. KTBL-Schrift, 323, KBTL, Darmstadt, Germany, pp. 172-184.

Weber, R. and Wechsler, B., 2001. Reduction in cross-sucking in calves by the use of a modified automatic teat feeder. Applied Animal Behaviour Science 72, 215-223.

Wechsler, B., 1993. Verhaltensstörungen und Wohlbefinden – ethologische Überlegungen. In: Martin, G. (Ed.) Leiden und Verhaltensstörungen bei Tieren. Birkhäuser, Basel, Switzerland, pp. 50-64.

Wechsler, B., 1996. Rearing pigs in species-specific family groups. Animal Welfare 5, 25-35.

Wechsler, B., 2007. Normal behaviour as a basis for animal welfare assessment. Animal Welfare 16, 107-110.

Wechsler, B. and Weber, R., 2007. Loose farrowing systems: challenges and solutions. Animal Welfare 16, 295-307.

Wechsler, B., Schaub, J., Friedl, K., and Hauser, R., 2000. Behaviour and leg injuries in dairy cows kept in cubicle systems with straw bedding or soft lying mats. Applied Animal Behaviour Science 69, 189-197.

Wechsler, B., Schmid, H., and Moser, H., 1991. Der Stolba-Familienstall für Hausschweine. Birkhäuser, Basel, Switzerland.

Weng, R.C., Edwards, S.A., and English, P.R., 1998. Behaviour, social interactions and lesion scores of group-housed sows in relation to floor space allowance. Applied Animal Behaviour Science 59, 307-316.

Wiepkema, P.R., 1987. Behavioural aspects of stress. In: Wiepkema, P.R. and Van Adrichem, (Eds.) Biology of stress in farm animals: an integrative approach, Martinus Nijhoff Publishers, Dordrecht, the Netherlands, pp. 113-133.

Wierenga, H.K., 1983. The influence of the space for walking and lying in a cubicle system on the behaviour of dairy cattle. In: Baxter, S.H., Baxter, M.R., and MacCormack, J.A.D. (Eds.) Farm animal housing and welfare. M. Nijhoff Publishers, Boston, USA, pp. 171-180.

Wierenga, H.K., 1984. The social behaviour of dairy cows: some differences between pasture and cubicle system. In: Unshelm, J., Van Putten, G., and Zeeb, K. (Eds.) Proc. Int. Congr. Appl. Ethol. KTBL, Darmstadt, Germany, pp. 135-138.

Wierenga, H.K., 1990. Social dominance in dairy cattle and the influences of housing and management. Applied Animal Behaviour Science 27, 201-229.

Wierenga, H.K., Metz, J.H.M., and Hopster, H., 1985. The effect of extra space on the behaviour of dairy cows kept in a cubicle house. In: Zayan, R, (Ed.) Social space for domestic animals. Springer, Berlin, Germany, pp. 160-170.

Wilson, E.O., 2000. Sociobiology: the new synthesis. Harvard University Press, Cambridge, MA, USA.

Wood-Gush, D.G.M., Duncan, I.J.H., and Savory, C.J., 1978. Observations on the social behaviour of domestic fowl in the wild. Biology of Behaviour 3, 193-205.

Zähner, M., Schrade, S., Schaeren, W., and Schmidtko, J., 2008. Neue Materialien als Einstreu in Liegeboxen von Milchviehställen. In: Freiland-Verband, Reinhard Geßl, Freiland-Verband, pp. 43-46.

Zeeb, K., 1987. Wieviel Lauffläche brauchen Milchkühe? Tierzüchter, 39: 169-170.

Zonderland, J.J., Wolthuis-Fillerup, M., van Reenen, C.G., Bracke, M.B.M., Kemp, B., Hartog, L.A. and Spoolder, H.A.M., 2008. Prevention and treatment of tail biting in weaned piglets. Applied Animal Behaviour Science, 110: 269-281.

# Animal welfare and nutrition

**John Webster**
University of Bristol, Old Sock Cottage, Mudford Sock, Yeovil, Somerset BA22 8EA, United Kingdom; john.webster@bris.ac.uk

## Summary

This chapter sets out the foundations necessary for the identification and assessment of welfare problems that may arise from inadequacies in feeding. The perceptions of hunger and appetite, the factors that motivate animals to seek food, their ability to recognise and select a metabolically 'sensible' diet and their potential for suffering if these needs are not met, are critically reviewed. Major criteria necessary to ensure 'good feeding' include (1) adequate provision of all essential nutrients, (2) feed whose chemical composition and physical form is consistent with stable digestion, (3) feed and opportunities for foraging that permit animals to achieve oral satisfaction, (4) feed that is free from harmful substances. It then reviews the most important practical problems that can arise within the context of modern agricultural production systems within the light of these basic principles and offers suggestions for new research where this is necessary. This contribution concludes with an introduction to the application of HACCP principles to the identification and classification of hazards and adverse outcomes within the context of nutrition and farm animal welfare and outlines a strategy for risk management.

**Keywords**: appetite, digestion, foraging, good feeding practice, metabolic hunger, oral satisfaction, stereotypy

## 1. Introduction

Food, for any animal, is the most basic of its physiological and behavioural needs. Thus hunger is one of the most primitive and unremitting of all motivating forces. When we recognise farm animals as sentient creatures we acknowledge that they associate the presence and absence of food with good or bad feelings. They experience pleasure both when they eat food and when they confidently anticipate that food will arrive. They can suffer both directly from the absence of food and indirectly when they are prevented from carrying out behaviours such as foraging, intended to lead them to food.

The behavioural need for food is more than just the desire to eat. It is a complex process that reflects the physiological need of the animal to obtain all the nutrients essential to maintain good health, reproductive fitness and peace of mind. Wild animals living in their ancestral habitat are usually able to select from a range of plants (and animals) suited to their digestive capacity and metabolic needs. Even when food is short, they can obtain some behavioural satisfaction from constructive foraging. When we domesticated farm animals we assumed responsibility to meet their nutritional needs. Animals such as beef cattle and sheep that graze extensive pastures still retain an element of choice as to what they eat and how much. Pigs and poultry in intensive housing have no choice but to eat the feed they are given and at a time that suits the farmer. Moreover, farm animals have been selected to produce food for humans (milk, meat and eggs) at rates that far exceed those which are consistent with fitness under conditions of natural selection in the wild. It is self evident therefore that the proper feeding of highly productive farm animals under intensive conditions requires a profound understanding of the science of nutrition, itself underpinned by an equally profound knowledge of the physiology of digestion and metabolism. However this chapter is not primarily concerned with the nutrient requirements of farm animals but with their perceptions of hunger and appetite, the factors that motivate them to seek food, their ability to recognise and select a metabolically 'sensible' diet and their potential for suffering if these needs are not met. Having outlined the principles that define nutrient supply and demand in farm animals I shall review how they determine an animal's perception of its nutritional state. On this basis I shall set out criteria for good feeding practice and highlight some of the most important problems likely to occur in practice. Finally I shall consider the assessment and management of nutritional risks and point to areas where good practice in the context of 'nutrition and welfare' does (and does not) call for new research.

## 2. Nutrient supply and metabolic demand

The basic steps involved in nutrient supply and demand are illustrated in Figure 1. An animal eats a diet made up of several ingredients (here given as A, B, and C). Digestion of these ingredients produces a supply of digestible nutrients with loss of some organic matter in faeces. In a simple-stomached animal like the pig, the digestibility of the diet is inversely proportional to the concentration of dietary fibre. Grazing animals are able to obtain digestible nutrients from fermentation of fibre (e.g. cellulose). Fermentation in the rumen also results in some loss of food energy as methane. Digestible nutrients absorbed from the gut are available for metabolism. Four broad categories of requirement for available nutrients are energy (conventionally described as metabolizable energy, ME) protein (or more specifically essential amino acids such as lysine and methionine) minerals (e.g. calcium, magnesium, copper, selenium) and vitamins. Quantitatively the greatest nutrient requirement by far is for ME to support the work of maintenance and the energy costs of synthetic processes such as growth and lactation. All this ME is dissipated as heat. The only energy available for subsequent human consumption is that retained in the body as muscle protein and fat or exported as milk or eggs. The demand for food is driven by feedback mechanisms (hatched lines in Figure 1) operating within the digestive tract, hepatic portal blood carrying nutrients absorbed from the digestive tract and within the central nervous system (CNS).

ME, consumed in excess of requirement for maintenance and the production of vital (and saleable) substances such as meat, milk and eggs, can be stored in the body as fat. In circumstances when ME supply is insufficient for the needs of maintenance and production, these fat reserves can help to fuel the work of metabolism and contribute to the synthesis of milk and eggs. Genetic selection for high-producing lines of dairy cow and laying hen has greatly increased ME demand for milk and egg production and thus increased the probability that these farm animals will have to draw on their energy reserves even when food supply

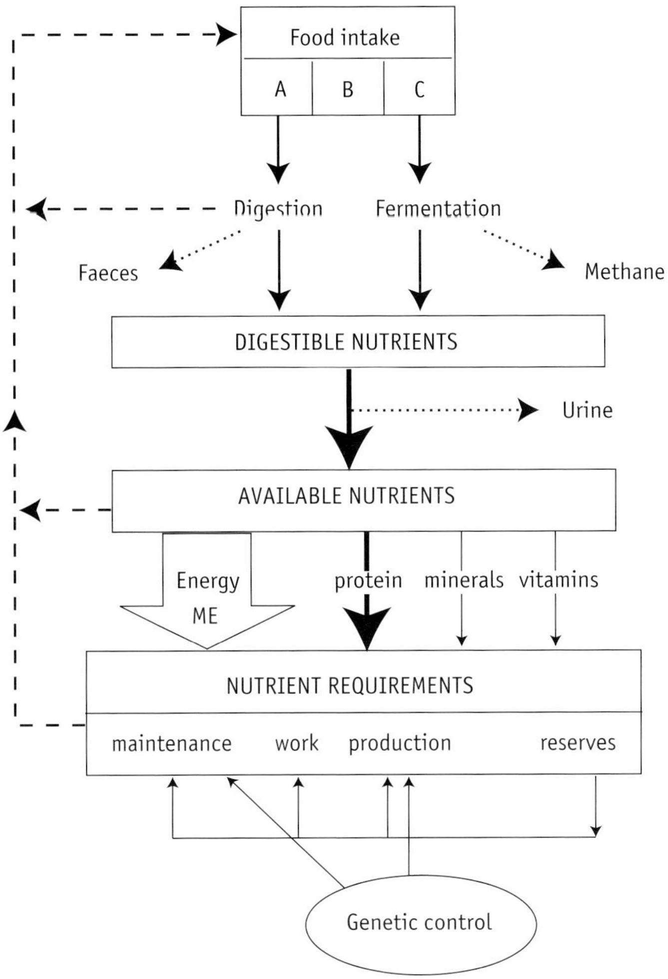


*Figure 1. Available nutrients: supply and requirement.*

is plentiful. Indeed, it is considered entirely normal for dairy cows to 'milk off their backs' in early lactation.

Requirements for protein, minerals and vitamins are relatively low at maintenance but increase with respect to ME with increasing level of production. Protein consumed in excess of requirement is catabolised as an energy source, with some loss of nitrogenous organic matter mostly as urea (or uric acid in poultry). Minerals and water-soluble vitamins consumed in excess of requirement are excreted in urine and faeces. Fat-soluble vitamins may be stored (unsurprisingly) in fat.

## 3. Perception of nutrition and welfare

The attitude of a sentient animal with regard to food, or the absence of food, can not be defined simply as hunger for food in general or even for specific nutrients. It is necessary to make a distinction between hunger, appetite and oral satisfaction. Hunger and appetite are similar, though not identical emotions that motivate actions designed to seek and ingest food; i.e. foraging and eating (Figure 2). Hunger, or more correctly, physiological state within a spectrum from starvation to satiety, represents the emotional integration of many symptoms from the internal environment, digestive state (e.g. gut fill), acute metabolic state in relation to current feeding and long-term metabolic state as defined by overall body condition. Both acute and chronic metabolic state are monitored largely (though not entirely) from concentrations of circulating metabolites (e.g. glucose and amino acids). Integration of these internal stimuli defines the animal's state of *metabolic hunger* within the spectrum from suffering to satiety and obviously has a major impact on its strength of motivation to eat. *Appetite* is determined by stimuli from both the internal and external environment; metabolic hunger for nutrients and the presence, or expectation, of attractive food. The reward for actions designed to assuage appetite and relieve hunger are twofold, the physiological satisfaction gained from acquiring nutrients and the behavioural satisfaction gained from discovering and consuming food. Both these needs are important to a sentient animal and can, to some extent, compensate for one another. An animal that is experiencing satiety is less motivated to forage, obviously. Equally, grazing animals such as out-wintered cattle and sheep that have access to forage can satisfy this behavioural need for *oral satisfaction* even though the quantity and quality of food may not meet their nutritional requirements. On the other hand, intensively housed pigs and poultry provided with high quality feed that meets nutrient requirement in very small, rapidly consumed meals may experience poor welfare through failure to obtain oral satisfaction.

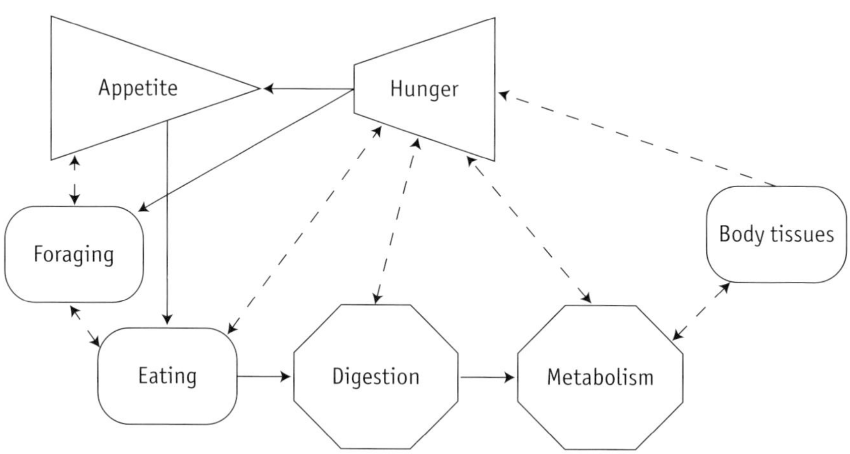


*Figure 2. Hunger and appetite as motivating forces: stimulus and response.*

### 3.1 Metabolic hunger

The physiological mechanisms whereby metabolic hunger motivates conscious appetite are too complex to review here in detail. The following reviews are offered as an introduction for those wishing to explore this subject in more depth (Forbes, 1988; Weigle, 1994). In brief, short-term sensations of hunger and satiety, leading to decisions to start and stop eating arise both from events occurring within the digestive tract, e.g. impulses from nerves sensing gastric and duodenal filling and emptying, and from nerves sensing concentrations of nutrients such as glucose and specific amino acids. Nutrients that cross the blood/brain barrier are sensed at specific sites within the CNS. These are commonly referred to as the appetite and satiety centres. However, peripheral sensory nerves are also able to sense concentrations of these chemicals at strategic locations within the circulation, e.g. at the entrance and exit of the liver.

Animals are also able to monitor their overall body condition against some innate model of what it should be in order to meet genetically programmed targets during pregnancy and growth or during recovery from illness or malnutrition. A good illustration of this is the phenomenon of compensatory growth (Saintz *et al.*, 1995). If young cattle, for example, are housed and fed over winter only on a relatively low quality diet such as grass silage, they will eat to appetite but grow at a slower rate than their genetic potential. At the end of the winter they will be larger but leaner than they were in the autumn. Specifically they will contain both less fat and muscle relative to their skeletal size. When they are given access to good quality grass in the spring (or cereal-based concentrates) they will eat more and grow lean tissue faster than an animal that does not sense that it is underweight for its age. In this illustration, as during pregnancy or recovery from illness, the motivation to eat is driven not by the need to maintain homeostasis but to achieve *allostasis,* the process of achieving stability through physiological and/or behavioural change: in this case the motivation to recover their target body condition.

### 3.2 Hunger for specific nutrients

Animals eat food primarily as a source of energy. At maintenance, when an adult animal is neither gaining nor losing weight, pregnant nor exporting nutrients as eggs or milk, approximately 90% of metabolizable nutrients are used as a source of energy to drive the work of metabolism and are dissipated as heat. It follows therefore that for many domestic animals and most wild animals, the primary source of metabolic hunger is a simple lack of ME in the diet, exacerbated by a perceived lack of energy reserves in body fat. This can arise in one or more of three ways (1) absolute lack of feed, (2) low digestibility of feed, (3) excessive metabolic demand. The first of these causes is self-evident and a potential threat to the welfare of all animals. The second can arise in ruminants housed and fed poor quality conserved forage or outside on dormant pastures during the winter in temperate latitudes or the dry season in the tropics. In this case the rate and extent to which the rumen microorganisms can extract ME (as volatile fatty acids) from potentially fermentable plant material is insufficient to meet ME needs within the constraints of gut fill. In physiological terms, intake is constrained by the amount of indigestible fibre retained in the rumen. In welfare terms, the animal is presented by two problems: it feels both hungry and full up (Webster, 1993).

The third situation in which an animal may experience a sense of hunger is when it is faced by an unsustainable metabolic demand. This experience is common for high yielding dairy cows in early lactation. The metabolic demand of the mammary gland for nutrients to support milk synthesis, exceeds the capacity of the digestive tract to supply nutrients from the processes of ruminant digestion (Webster, 1993). Once again the animal is hungry but full up. This problem is exacerbated when dairy cows are fed high levels of dietary protein. Amino acid supply is a first limiting factor for milk synthesis *in the short term*. Thus increasing protein intake in early lactation can increase milk production, but increase the rate at which the cow must draw on energy from body reserves. The increases the economic risk of infertility and the welfare risks of excessive hunger, loss of body condition and immune suppression.

The balance between amino acid supply and demand has a major influence on appetite. When young pigs (and children) are fed protein-deficient diets, they eat more and get fatter as they seek to meet their physiological need to acquire sufficient amino acids to meet their targets for lean tissue growth (Kyriazakis and Emmans, 1991). Selection for extreme leanness in pigs and Belgian Blue cattle is associated with reduced feed intake and growth rate. The appetite of caged laying hens in intensive units is restricted by feeding high-protein diets and keeping the birds at a high temperature to restrict ME demand to maintain homeothermy. Grazing sheep infested with parasitic roundworms that increase protein losses from the gastrointestinal epithelium adjust their grazing habits to select grasses and legumes higher in protein concentration (Hutchins *et al.*, 2002). Excessively high concentrations of circulating amino acids and/or end-products of protein catabolism (e.g. urea) are known to be a cause of reduced appetite. This is seen in humans, arising either from disease (e.g. cachexia in terminal cancer patients) or from 'abnormal' dietary habits (e.g. the high protein/low carbohydrate 'Atkins diet'). Feeding protein to excess is unlikely to occur in farm animals for economic reasons. However appetite in ruminants can be depressed by excessive absorption of ammonia from diets containing an excess of rumen-degradable N relative to fermentable energy (Tolkamp *et al.*, 2002). This is relatively common when the feeding of high-N grass silages is not balanced by a suitable energy source. The success of 'Total Mixed Rations' (TMR) for cattle can be attributed in major part to achieving this balance.

All this evidence strongly supports the conclusion that appetite and satiety in animals are powerfully motivated by the physiological need to sustain vital mass in essential (high protein) tissues like brain, liver, the epithelium and immunological cells of the intestine. Muscle and skin are of secondary importance. Fat content and fat synthesis are of little physiological significance to animals, at least in the short term. It is not therefore surprising to discover that intake of dietary fat and the quantity of body fat reserves have little influence on the perception of hunger and satiety.

There is good clinical and experimental evidence to show that farm animals, especially grazing animals, are motivated to seek and eat specific minerals to combat specific deficiencies (Underwood, 1980). Cattle and other ruminants can smell sodium salt at vanishingly small concentrations and in salt-deficient areas will seek our natural sodium (Na) sources at a distance of several kilometres. Salt moreover is highly palatable. Animals will overeat salt and this can constitute a welfare problem when water supply is limited (e.g. for weaner pigs).

The evidence that farm animals have a 'nutritional wisdom' with respect to other minerals is less convincing. The balance of evidence would suggest that cattle can sense a phosphorus (P) deficiency and will seek sources of P in soil, bones etc. This behaviour has been described as 'pica' or depraved appetite; which reflects a serious misunderstanding, since this behaviour represents an adaptation essential to the survival of grazing animals on the P-deficient continental plains of Africa and the Americas. It is almost certain that cattle have neither the ability to sense a magnesium (Mg) deficiency nor to recognise the beneficial effects of Mg supplements. Indeed addition of Mg salts to a trough of drinking water render the water less palatable and encourages cattle to drink from alternative sources such as natural streams. This increases the risk of acute hypomagnesaemic tetany in cattle at pasture.

The ability of grazing animals to recognise a physiological need for Na and P but neither Mg nor (probably) trace metals such as copper, may be seen as an evolutionary adaptation to the fact that Na and P are marginal-to-deficient in many natural pastures. Mg, on the other hand, is seldom deficient in natural and permanent pastures. The problem of acute hypomagnesaemia in dairy cattle may be considered a production disease that has arisen very recently (in evolutionary terms) through a combination of increased production of grass from monocultures and the intensive use of N and K fertilisers that reduce Mg availability.

### 3.3 Digestion, indigestion and feelings

Farm animals are obviously not motivated to eat by the need to sustain rates of growth or milk production that are economically rewarding for the farmer. Less obviously, a cow is not *directly* motivated to eat by the physiological need to produce enough milk production to ensure the survival of her calf, nor the calf motivated by the need to build up sufficient energy reserves to carry it through its first winter. Appetite, satiety and food aversion in sentient farm animals are driven by how they *feel* with respect to the feeds on offer both before and after the meal. For most sentient animals, most of the time, food is a source of satisfaction, indeed pleasure. Offered a wholesome, nutritionally balanced, easily digestible diet, most farm animals will eat to excess. To achieve optimum intakes it is necessary therefore either to restrict the ration or to dilute it with indigestible fibre to the point where satiety is achieved at an intake that matches the production targets of the farmer. The latter approach is likely to be better for animal welfare, though it may prove uneconomic. Restricting the ration fed to a growing animal will induce some degree of hunger (though not to the point of inducing suffering). More seriously it can create practical problems in (e.g.) batches of growing pigs. Competition for feed can lead to variable growth rates and inconsistent 'finish' at the time for slaughter. Aggression at the feed trough can result in serious injury.

Diets that contain too high a concentration of starchy cereals relative to digestible fibre present a welfare problem for ruminants in consequence of excessively rapid fermentation in the rumen, leading to ruminal and metabolic acidosis (Enemark *et al.*, 1998; Gabel, 1990). In extreme cases of grain overload (e.g. cattle breaking into a grain store) this can cause death within hours. Such cases are rare and can usually be ascribed to negligence. However less acute welfare problems arising from feeding amounts of cereals that are inconsistent with healthy digestion are common both in high-yielding dairy cows and in beef cattle being finished fast on feedlots. Cattle experiencing mild ruminal acidosis (e.g. after eating over 4 kg concentrate feed in the milking parlour) show a depressed appetite for forage. As mentioned earlier, the success of Total Mixed Rations that contain at least 40% dietary fibre and encourage rumen fermentation at a stable pH, can be attributed mainly to the fact that they encourage dry-matter intakes that may be 20% greater than those observed in animals with the same nutrient demand but fed forage/concentration mixtures in the wrong proportions or at the wrong times. In many high production dairy herds, dietary supplements in the form of sodium bicarbonate, yeast products or mineral clays are marketed as aids to increased production through increased food intake. The mode of action of these supplements usually involves the stabilisation of rumen pH. Thus they are not appetite stimulants as such but medicines designed to offset appetite depression arising from the discomfort of indigestion.

Indigestion is not the only potential cause of appetite depression. Farm animals can sense how they feel (better or worse) in consequence of absorbing the products of digestion and make the association with the meal (or specific plant) they ate some hours earlier. Experimental studies of food aversions have involved dosing test feeds with Lithium chloride (LiCl) which induces a sense of nausea (Thorhallsdottir *et al.*, 1987; Burritt and Provenza, 1990). Hens offered grains dosed with LiCl and coloured with an innocuous blue dye, will remember the colour and avoid blue grains after intervals of more than one year. They will also work to discourage their chicks from eating blue grains ever when these do not contain LiCl and are perfectly wholesome (Nicol and Pope, 1996). This learnt behaviour probably explains the success of animals raised with their mothers in extensive circumstances in avoiding poisoning by toxic plants relative to animals reared in more artificial circumstances that do not provide them with the opportunities for a proper education.

Animals also learn to favour food sources that make them feel better even when this may start to occur many hours after the meal. This explains the preference of parasitized sheep for high protein grasses and clovers. A more dramatic experimental demonstration has been provided by the work of Danbury *et al.* (1999) with broiler chickens offered a choice between conventional feed and feed dosed with the anti-inflammatory, analgesic drug carprofen. Birds that showed no signs of lameness preferred the conventional feed. However birds that showed any visual evidence of lameness preferentially selected feed containing the analgesic. They chose this feed because it made them feel better: it eased their pain.

### 3.4 Oral satisfaction and oral stereotypies

A sentient animal is motivated to eat and to seek food in order to improve its emotional state. A pig does not have to be stressed by hunger in order to welcome the appearance of a palatable meal and consume it with every appearance of pleasure. Moreover all the main farm species are powerfully motivated to spend hours foraging or scavenging for feed. Cattle, sheep and horses graze and browse grasses and shrubs. Pigs root in the earth for nuts and worms, hens peck for grains and insects. These behaviours have a powerful instinctive basis, being amongst the most essential of evolutionary adaptations. There is clear evidence that welfare problems can arise when farm animals are provided with diets that meet their nutritional requirements but fail to satisfy their instinctive need to acquire oral satisfaction through behaviour appropriate to their species. The most visible evidence of failure to achieve oral satisfaction is the performance of stereotypic behaviour (Mason, 1991; Lawrence and Rushen, 1993). Examples include bar chewing in sows without access to foraging substrate, tongue rolling and cross sucking in veal calves fed liquid diets. Feather pecking in hens is also more likely to be an expression of redirected foraging behaviour than a manifestation of aggression.

The assumption that stereotypic behaviour is due to lack of oral satisfaction prompts the conclusion that signs of stereotypic behaviour reflect a failure of the farming system to meet the behavioural needs of the animals. One should however reflect with care before reaching such a conclusion, particularly on the basis of brief observation, e.g. during a farm inspection for quality assurance purposes. Stereotypic behaviour may be defined as 'fixed patterns of behaviour performed repetitively and having no obvious function'. The sow that chews the bars of its stall when it hears or smells the arrival of food is performing an anticipatory behaviour associated with the positive emotion of hope. Incessant, repetitive bar-chewing is more likely to be an expression of hopelessness. We do not know for certain how animals feel when engaged in stereotypic behaviour. There is much debate as to whether specific stereotypies are an expression of distress or a coping mechanism designed to achieve some degree of satisfaction when the more natural sources are absent. Whatever may be the answer to this question, we may conclude with confidence that true stereotypies (prolonged, repetitive and emancipated from the original stimulus) reflect a failure in the provision of good husbandry.

Although oral stereotypies are conventionally described as an expression of failure to obtain oral satisfaction, this is not necessarily the case. Recent evidence indicates that crib-biting and wind-sucking in horses (a behaviour in which the animal grasps a solid object with its teeth and swallows air) is more likely to originate in foals fed high starch diets and is probably a mechanism designed to relieve abdominal discomfort by increasing salivation and gastrointestinal motility (Nicol *et al.*, 2002). From first principles it is possible that bar chewing in sows, and highly probable that tongue rolling and cross sucking in veal calves may have the effect of reducing physical discomfort as well as mental disquiet.

## 4. Good feeding practice

It should now be clear that the proper feeding of farm animals to promote efficient production, good health and welfare involves more than the provision of a nutritionally balanced diet. The four essentials for good feeding practice may be summarised as follows:

- Meet nutrient requirements: through provision of a diet that meets all nutrient requirements for maintenance, growth, work and reproduction.
- Promote healthy digestion: through provision of food that promotes normal physical, biochemical and immunological function in the digestive tract.
- Promote oral satisfaction: through provision of feed and feeding systems consistent with behavioural needs.
- Do no harm: through provision of feed that is free from harmful substances.

## 5. Nutrition and welfare problems in practice

To recapitulate in brief the story so far: feeding for good welfare involves provision of a wholesome diet, nutritionally balanced with respect to physiological state (e.g. growth, lactation), in composition and physical form consistent with healthy digestion, and presented in such a way as to promote oral satisfaction. Left to themselves, in a suitable habitat and with sufficient resources, farm animals can meet these needs through the performance of appropriate behaviours such as foraging and diet selection. Even when nutrient supply is inadequate, free-ranging animals can obtain some satisfaction from constructive foraging. This section will consider practical farming situations where one or more of these criteria may be lacking (Table 1), identify the welfare problems that may arise, describe the associated clinical and behavioural signs, and suggest solutions.

### 5.1 Dairy cattle

In the last 30 years annual lactation yields of the most productive herds of dairy cattle have increased from less than 5,000 to over 10,000 litres. These total yields correspond to peak daily yields of approximately 25 and 50 litres/day. This has been achieved by parallel processes of genetic selection for increased yield of milk solids accompanied by changes in diet and feeding practice designed to increase nutrient supply. Table 2 estimates the maximum food intake (expressed as dry matter, DM) that high genetic merit Holstein cows may reasonably be expected to achieve at pasture, when fed silage plus concentrates twice daily or fed a well-balanced 'Total Mixed Ration'. It then relates these to the level of milk production (l/day) that this amount of food can sustain. A cow on excellent pasture with an ME concentration of 11.9 MJ ME/kg DM is restricted by time and availability of grass to an intake of about 16kg DM per day. This will provide sufficient nutrition to sustain a yield of

*Table 1. Nutrition and welfare: problems and solutions for farm animals.*

| | Problems | Solutions |
|---|---|---|
| Dairy cows | Metabolic hunger | Increase nutritive value of diet |
| | | Reduce milk yield (diet and/or genotype) |
| | Ruminal indigestion | Encourage stable fermentation |
| | | Control composition, physical form and presentation of diet |
| Beef cows | Hunger, mineral deficiencies | Supplements, provision for foraging |
| Finishing cattle | Ruminal acidosis, laminitis | Increase dietary fibre, gradual diet changes |
| Breeding ewes | Hunger, mineral deficiencies | Supplements, provision for foraging |
| Red deer | Winter starvation/hypothermia | Good condition in autumn |
| | | Winter shelter and supplementary feeding |
| Breeding sows | Frustrated foraging | Create foraging opportunities |
| Growing pigs | Aggression, irregular growth | Diet dilution, create foraging opportunities |
| Broiler chickens | Lameness | Feed restriction, strain selection |
| Broiler breeders | Unnatural hunger | Diet dilution |
| Laying hens | Osteoporosis, bone fractures | Dietary improvements to bone metabolism |

*Table 2. Milk yields and dry matter intakes of Holstein/Friesian cows in different circumstances.*

| | DM intake (kg /d) | Milk yield (l/d) | ME concentration MJ ME/kg DM |
|---|---|---|---|
| Pasture | 16 | 25 | 11.9 |
| Grass silage + concentrates (×2/d) | 20 | 36 | 12.3 |
| Total mixed ration | 27 | 55 | 12.6 |

25 l/day. The housed cow with continuous access to grass silage but given concentrate feeds only during twice-daily milking may eat 20 kg DM and this can sustain 36 l/day. A hungry, healthy cow producing 55 l/day may eat 26 kg DM/day or more if she has continuous access to a balanced diet, presented in suitable physical and chemical form to promote rapid, safe fermentation of food energy and digestion of protein in the rumen. The formulation and presentation of balanced Total Mixed Rations (TMR) has permitted high genetic merit dairy cows to express their physiological potential to produce milk at rates of 50-70 l/day. Thus milk yields can be two to three times greater than at pasture, through provision of feed that is more available, is of higher nutritive value (ca. 12.6 MJ ME/kg DM) and promotes rapid but stable digestion in the rumen. It follows however that cows can only sustain yields over 50 l/day if they are kept off grass throughout most or all of their lactation.

However well formulated the diet for high yielding dairy cows ME demand in early lactation exceeds that which can be achieved by the cow within the constraints of appetite. This problem of energy imbalance is exacerbated by feeding diets high in protein (Webster 1993; Collard *et al.*, 2000; Ingvartsen *et al.*, 2003). The physiological demand for nutrients to support lactation creates a condition of 'metabolic hunger' and the intensity of this hunger increases as body condition falls. This is evident from the delayed effect on appetite of injecting bovine growth hormone (BST) to stimulate increased milk yield (Chalupa and Galligan, 1989). It is reasonable to conclude that motivation to start and stop eating in the high yielding dairy cow is driven not by the need to promote satisfaction but by the need to minimise *discomfort*. Metabolic hunger, the primary stimulus to eat, conflicts with powerful stimuli to *stop* eating. When housed and given *ad libitum* access to conserved forage, intake is constrained physically by gut fill, itself determined by the rate at which the diet can be fermented and reduced to a size that will pass through the reticulo-omasal orifice. When forage is supplemented by concentrates in one or two daily meals asynchrony between the digestion of crude protein and carbohydrate in the rumen can increase the absorption of ammonia, giving rise to an increased concentration of urea in blood and this can impair appetite. The aim of the TMR is to promote rapid, stable fermentation, so reduce the appetite constraints presented both by rumen fill and high blood urea concentrations. This problem can be particularly acute at the beginning of lactation before the rumen has adapted fully to the twin challenges of increased feed intake and increased starch : fibre ratio in the diet. This problem can be overcome by careful attention to the feeding and diet formulation in the 'transition period' prior to calving (Grant and Albright, 1995; McNamara *et al.*, 2003).

To summarise this physiology in terms of how it feels to the high yielding cow: she is motivated to eat by metabolic hunger (a function of both milk yield and body condition). She is motivated to *stop* eating by sensations (conscious or unconscious) associated with gut fill, unbalanced absorption of certain end products of rumen digestion (especially ammonia), and the conflicting desire to do something other than eat, such as rest. At pasture, the first constraint on food intake is the rate at which the cow can physically consume the grass. Thus the motivation for the cow to stop eating is most likely to arise from the desire to rest than from the sensation of gut fill. On very high nitrogen pastures food intake may be constrained by ammonia absorption. A high yielding cow offered poorly-made, high ammonia grass silage plus concentrate in parlour only, faces an impossible compromise between metabolic hunger, gut fill, chemically induced inappetence and the need to rest.

This summary of the conflicts between metabolic need, digestive capacity and the need for rest provokes a number of difficult practical questions concerning dairy cow welfare.

- 'Is it more stressful to a dairy cow to produce 55 l milk/day than 25 l/day?'
- 'Is it stressful to restrict a cow with the genetic potential to produce 55 l/day to a diet that can only sustain 25 l/day?'
- 'Is it stressful to restrict a cow to a zero-grazing system for most or all of her adult life?'

There is no simple answer to any of these questions. Sustained production of 55 l milk/day is hard work. Indeed the dairy cow is by far the hardest worked of all the farm animals. Scaling energy demand by metabolic size (kg weight$^{0.75}$) reveals that the metabolic demand for 55 l/day is 65% greater than that for a hen laying one egg/day and 2.2 times that required for maximum growth in a pig. Moreover the secondary consequences of loss of condition, of which the greatest is infertility (Beam and Butler, 1999), are a major cause of premature forced culling. Nevertheless, the overall welfare of dairy cows in such high yielding herds can be as good as, or better than that on many traditional and extensive farms. It is just more difficult to achieve.

The use of high genetic merit Holsteins (60 l/day potential) has presented welfare problems manifested most obviously by infertility and early forced culling (Pryce *et al.*, 1997). This is particularly evident on pastoral systems (e.g. in New Zealand) where ME intake can only sustain 25 l/day (Harris and Winckleman, 2000). The best dairy farmers operating this system are now working towards a better balance between supply and demand by (1) improving the quality of the pasture and (2) selection within the Holstein breed, or through cross-breeding to produce a more robust cow with a lower potential yield but an improved lifetime performance.

The strategic decision to maximise milk yield per cow inevitably imposes the need to provide lactating cows with continuous access to a highly nutritious, premixed TMR. This is incompatible with a system that allows cows to have daily access to pasture. The welfare problems of permanent confinement are outside the scope of this chapter. Nevertheless it is appropriate to consider what benefits, other than feed, that cows may derive from access to pasture. These may include improved comfort, opportunities for exercise, and social behaviour. It is possible to address these needs within intensive, zero-grazing systems, but seldom achieved to the cows' satisfaction.

## 5.2 Beef cattle, sheep and deer

### *5.2.1 Breeding animals*

Adult female suckler beef cattle and sheep are expected to subsist on just enough feed to support maintenance, pregnancy and milk production sufficient to raise their offspring. In most circumstances they will be expected to fend for themselves on pasture with little or no supplementary feeding. During the winter in temperate latitudes and the dry season in the tropics, the availability and nutritive value of the grazing land will, in many cases, be inadequate to sustain even maintenance and the animals will have to recruit body reserves of fat and protein. Thus the primary welfare problem for these animals is simple hunger for food energy. However this can be offset if the animals are presented with sufficient dormant grasses to provide the behavioural satisfaction of foraging and the physiological satisfaction of rumen fill. Two problems may arise. If the dormant forage is very highly lignified, fermentation rate will be very slow and all ruminants can approach starvation conditions while their rumen is full of indigestible lignin. The grazing behaviour of cattle makes them unable to graze very short pastures. This can make them especially subject to under-nutrition when grazed with other species, sheep, deer and/or horses.

The problem of under-nutrition in winter can be exacerbated by cold stress. Acute cold stress increases energy requirement so increases the rate of loss of body reserves when ME is lacking. Loss of body fat decreases thermal insulation so increases susceptibility to cold. In general, adult cattle and sheep are extremely tolerant to acute cold stress and can tolerate temperatures of -20 °C in still air conditions without undue distress (Webster, 1983). The most severe cold stresses occur during conditions of driving wind and rain or snow, which radically reduce the thermal insulation provided by the coat of hair or wool. Thus the most effective remedy is shelter, which need not involve a building. A windbreak of trees is very good. A forest is excellent. The main welfare problem for out-wintered ruminants is one of starvation and metabolic exhaustion induced by prolonged exposure to cold and under-nutrition. Deaths occur in late winter when body reserves are exhausted. It follows therefore that the animals most likely to survive are those that have acquired sufficient fat reserves by the end of the grazing season. Thus genetic predisposition to deposit relatively large amounts of fat is critical for survival of breeding females expected to winter outside. This conflicts with the need to select for lean tissue growth in the slaughter generation and stresses the importance of divergent selection and cross breeding in suckler beef cattle and sheep systems that are required to produce relatively large lean offspring from relatively small, fat, low maintenance mothers. This problem is critical for lean red deer which are particularly susceptible to starvation if over-wintered without supplementary feed (Simpson *et al.*, 1978).

Dormant winter pastures are also likely to be deficient in nutrients other than energy. These include protein, phosphorus, magnesium and trace elements such as copper and selenium. A useful strategy to meet the essential but quantitatively small demand for these nutrients is to provide them in the form of feed blocks set out on the pastures. Protein requirement can be met through the provision of urea and other sources of non-protein nitrogen that will be converted into true protein by microbial action in the rumen.

#### 5.4.2 Broiler breeders

The broiler breeder is a bird that has developed a pathologically abnormal appetite as a result of selection for rapid growth. Birds destined for breeding must be fed on a restricted diet from about 4 weeks of age and only given a maintenance ration once they achieve mature weight. This is essential to prevent loss of libido and premature death from obesity. It follows that it would be an abuse of physical welfare to allow these birds full expression of their abnormal behavioural need. On the other hand simple provision of a high quality diet designed to sustain health through provision of correct amounts of nutrients will lead to expressions of distress, manifested by aggression or self-mutilation (Farm Animal Welfare Council, 1998). In this case the need is once again to provide some behavioural satisfaction without significantly increasing nutrient supply. The aim should be to provide a foraging substrate just sufficient to produce a lasting reward without leading to obesity.

#### 5.4.3 Laying hens

Laying hens in nearly all commercial units, whether caged, in aviaries or with access to free range, are given scientifically formulated diets calculated to meet all their nutritional needs. Despite this, the prevalence of osteoporosis leading to broken bones is a major welfare problem (Whitehead, 1992). Inactivity, especially within the confinement of the battery cage, has been cited as a major predisposing factor. However, the prevalence of broken bones can be even greater in aviary and free range systems that allow the birds greater freedom of movement (Norgard-Nielsen, 1990). In these systems most healed fractures are seen in the furculum and keel bone (Wilkins *et al.*, 2004). Risk factors include injuries resulting from hard landings on perches and possibly vigorous wing flapping. In any event, the risk of fractures increases after about 40 weeks in lay, which implies a failure to maintain bone homeostasis during prolonged egg-laying. Attempts to reduce this problem through adjustments to mineral nutrition have been disappointing. A more promising approach is to study the nutrition required to sustain the integrity of the connective tissue in the organic matrix of bone. Diets rich in omega-3 fatty acids may offer a solution but this conclusion must be tentative pending further research.

## 6. Assessment and management of nutritional risks to welfare

The first objective of this chapter was to describe the principles that characterise possible causes of poor farm animal welfare in terms of their physiological and behavioural needs for proper nutrition, wholesome, healthy food and oral satisfaction. It then described in brief the most common problems that may occur in practice. Improved management of these conditions will, in a few cases, require new research. However the greatest priority must be for improvements to the identification, characterisation and management of risks on farm in the light of existing knowledge. HACCP (Hazard Analysis and Critical Control Point) principles may be applied to the management of welfare risks at farm level (Noordhuizen and Welpelo, 1996; Bell *et al.*, 2009). The first steps in this process are to describe the hazards and relate these to adverse effects (Whay *et al.*, 2003). Table 3 presents an illustration of this approach as applied to nutrition and welfare problems in dairy cattle and growing pigs. This table

*Table 3. Hazard description and adverse effects: examples for dairy cows and growing pigs.*

| | Hazard description | Adverse effect |
|---|---|---|
| Dairy cows | Metabolic | |
| | [ME] too low | loss of body condition, infertility, failure to sustain milk yield |
| | [Protein]/[ME] too high | |
| | Digestive | |
| | poor quality fibre (forage) | loss of body condition, cud spitting, infertility, failure to sustain milk yield |
| | insufficient fibre | ruminal acidosis, inappetence, laminitis |
| | improper feeding schedule | loss of body condition, infertility, failure to sustain milk yield |
| | Behavioural | |
| | Inadequate feeding time at pasture | loss of body condition, infertility, failure to sustain milk yield |
| Growing pigs | Inadequate feeder design | inconsistent growth rates, aggression, tail biting |
| | Restricted feeding | |
| | Lack of suitable foraging substrate | |

should be largely self-explanatory in the light of my earlier exposition of these problems. Hazards for the dairy cow have been listed within three categories, metabolic, digestive and behavioural. Thus poor quality forage is characterised as a digestive hazard leading to adverse effects such as loss of body condition, infertility and failure to sustain milk yield. It also lists 'cud spitting' as an occasional sign that the forage (silage) is inconsistent with normal rumination. In some high-energy rations for high yielding dairy cows the absolute quantity of digestible forage may be insufficient to promote stable rumen fermentation leading to adverse effects such as ruminal acidosis, inappetence and possibly laminitis.

Problems for growing pigs may arise from restricted feeding or inadequate feeding design leading to inconsistent growth and aggression at mealtimes, or to failure to provide a satisfactory foraging substrate leading to chronic problems of frustration with potentially harmful consequences such as injuries from tail-biting.

Having described the hazards one can proceed through the sequence of actions called for in the implementation of HACCP procedures. The magnitude of risks can be assessed both from measures of provision (e.g. diet formulation) and outcome (e.g. ruminal acidosis, loss of body condition). The next stage is to identify critical control points and create a prioritized action plan. The effectiveness of the action plan will be assessed and modified as necessary after an appropriate interval by further monitoring of risks attached to the identified hazards and changes in outcome.

# 7. Conclusions

## 7.1 What has been achieved?

The first principle of provision for good welfare, whether set out in the 'Five Freedoms' or the standards of the pan-European WelfareQuality® programme (Boitreau *et al.*, 2007) is freedom from hunger. It is now commonly accepted that, to meet this provision, it is necessary to provide: (1) an adequate quantity of feed, defined primarily by ME supply, (2) an adequate supply of all essential nutrients, achieved either by provision of scientifically formulated diets, or by offering animals sufficient choice to permit them to acquire all essential nutrients through diet selection, (3) feed whose chemical composition and physical form is consistent with stable digestion and the health of the gastro-intestinal tract, and, finally, (4) feed and opportunities for foraging that permit animals to achieve oral satisfaction and minimise the risks of disturbed behaviour; e.g. oral stereotypies. More specifically, some of the major achievements in relation to nutrition and welfare include: the recently acquired scientific information that has allowed the improvement of nutrition in the high-yielding dairy cow in early lactation and in the transition period, the better understanding of the role of dietary fibre in promoting healthy digestion and oral satisfaction in ruminants, veal calves and growing pigs, and, finally, the improvements in feeding practice that have succeeded in reducing aggression in growing pigs.

## 7.2 What has been neglected?

To date, there is a lack of scientific information that could help to solve the following major problems: (1) Metabolic hunger for specific metabolites: e.g. minerals in grazing ruminants, specific amino acids in growing pigs and poultry, (2) Diet selection: study of how animals (especially poultry; broiler chickens, turkeys, ducks) may optimise welfare and performance through selection of nutrients to meet their own perception of physiological need, and (3) The incorporation of nutritional needs (nutrient balance, healthy digestion, oral satisfaction) into HACCP approaches to managing the complex hazards faced by farmers seeking to reconcile efficient production with good welfare.

## 7.3 What needs to be done?

The greatest benefits for the industry and for the animals will accrue through improved methods for the identification, classification and management of nutritional risks to welfare at farm level. Although, obviously, for all the problems already mentioned (be they unsolved or only partially solved) we need to conduct more research, the following areas need to be addressed with some degree of priority: (1) study of the possible pathophysiological origins of oral stereotypies in pigs and veal calves, (2) further study of the nutritional, genetic and environmental origins of osteoporosis in laying hens, and (3) studies of genotype/environment interactions with the aim of producing improved genotypes for less intensive, higher welfare, more sustainable production systems (e.g. pastoral systems for dairy cows, organic meat production).

## References

Beattie, V.E., Walker, N. and Sneddon, I.A., 1998. Preference testing of substrates by growing pigs. Animal Welfare 7, 27-34.

Beam, S.W. and Butler, W.R., 1999. Effects of energy balance on follicular development and first ovulation in postpartum dairy cows. Journal of Reproduction and Fertility. Supplement 54, 411-424.

Bell, N.J, Bell, M.J., Knowles, T.G., Whay, H.R., Main, D.C.J. and Webster, A.J.F., 2009. The development, implementation and testing of a lameness-control programme based on HACCP principles and designed for heifers on dairy farms. The Veterinary Journal 180, 178-188.

Boitreau, R., Veissier, I., Butterworth, A., Bracke, M.B.M., Keeling, L., 2007. Definition of criteria for overall assessment of animal welfare. Animal Welfare 16, 225-228.

Burritt, E.A. and Provenza, F.D., 1990. Food aversion learning in sheep persistence of conditioned taste aversions to palatable shrubs. Journal of Animal Science 68, 1003-1007.

Chalupa, W. and Galligan, D.T., 1989. Nutritional implications of somatotropin for lactating dairy cows. J. Dairy Sci. 2510-2524.

Collard, B.L., Boettcher, P.J., Dekkers, J.C., Petitclerc, D. and Schaeffer, L.R., 2000. Relationships between energy balance and health traits of dairy cattle in early lactation. Journal of Dairy Science 83, 2683-2690.

Danbury, T.C., Weeks, C.A., Chambers, J.P., Waterman-Pearson, A.E. and Kestin, S.C., 1999. Self-selection of the analgesic drug carprofen by lame broiler chickens. Veterinary Record 146, 307-311.

Enemark, J.M.D., Jorgensen, R.J. and Enemark, P.S., 1998. Ruminal acidosis, with special emphasis on diagnostic aspects of subclinical acidosis. A review. Dansk Veterinaertidsskrift 81, 886-897.

Farm Animal Welfare Council (FAWC), 1998. Report on the welfare of broiler breeders. Farm Animal Welfare Council Secretariat, London, UK. Available at: http://www.fawc.org.uk/reports/pb3907/broiltoc.htm.

Forbes, J.M., 1988. Metabolic aspects of the regulation of voluntary food intake and appetite. Nutrition Research Reviews 1, 145-168.

Gabel, G., 1990. Rumen acidosis: interactions between changes in the lumen and rumen wall. Übersichten zur Tierernährung 18, 11-38.

Grant, R.J. and Albright, J.L., 1995. Feeding behaviour and management factors during the transition period in dairy cattle. Journal of Animal Science 73, 2791-2803.

Harris, B.L. and Winkelman, A.M., 2000. Influence of North American Holstein genetics on dairy cattle performance in New Zealand. Proceedings New Zealand Large Herds Conference 6, pp. 122-136.

Hutchings, M.R., Gordon, I.J., Kyriazakis, I., Robertson, E. and Jackson, F., 2002. Grazing in heterogenous environments: infra- and supraparasite distributions determine herbivore grazing decisions. Oecologia 132, 453-460.

Ingvartsen, K.L., Dewhurst, R.J. and Friggens, N.C., 2003. On the relationship between lactational performance and health: is it yield or metabolic imbalance that causes diseases in dairy cattle? A position paper. Livestock Production Science 83, 277-308.

Kelly, D., Smyth, J.A. and McCracken, K.J., 1990. Effect of creep feeding on structural and functional changes of the gut of early weaned pigs. Research in Veterinary Science 48, 350-356.

Knowles, T.G., Kestin, S.C., Haslam, S.M., Brown, S.N., Green, L.E., Butterworth, A., Pope, S.J., Pfeiffer, D and Nicol, C.J., 2008. Leg disorders in broiler chickens: prevalence, risk factors and prevention. PLoSOne 3(2): e1545. doi: 10.1371/journal.pone.0001545

Kyriazakis, I and Emmans, G.C., 1991. Diet selection in pigs: dietary choices made by pigs following a period of underfeeding with protein. Animal Production 52, 337-346.

Lawrence, A.B. and Rushen, J., 1993. Stereotypic animal behaviour: fundamentals and applications to welfare. CAB International, Wallingford, UK.

Li., D.F., Nelssen, J.L., Reddy, P.G., Blecha, F., Klemm, R. and Goodband, R.D., 1991. Interelationships between hypersensitivity to soybean proteins and growth performance in early-weaned pigs. Journal of Animal Science 69, 4062-4069.

McNamara, S., Murphy, J.J., Rath, M. and O'Mara, F.P., 2003. Effects of different transition diets on energy balance, blood metabolites and reproductive performance in dairy cows. Livestock Production Science 84, 195-206.

Mason, G.J. 1991. Stereotypies: a critical review. Animal Behaviour 41, 1015-1037.

Moinard, C., Mendl, M., Nicol C.J. and Green, L.E., 2003. A case control study of on-farm risk factors for tail-biting in pigs. Applied Animal Behaviour Science 81, 333-355.

Nicol, C.J. and Pope, S.J., 1996. The maternal feeding display of domestic hens is sensitive to perceived chick error. Animal Behaviour 52, 767-774.

Nicol, C.J., Davidson, H.P.B., Harris, P.A., Waters, A.J. and Wilson, A.D., 2002. Study of crib-biting and gastric inflammation and ulceration in young horses. Veterinary Record 151, 658-660.

Noordhuizen, J.P. and Welpelo, H.J., 1996. Sustainable improvement of animal health care by systematic quality risk management according to the HACCP concept. Veterinary Quarterly 18, 121-126.

Norgard-Nielsen, G.J., 1990. Bone strength of laying hens kept in an alternative system compared with hens in cages or on deep litter. British Poultry Science 31, 81-89.

Pedersen, L.J, Holm, L., Jensen, M.B. and Jorgensen, E., 2005. The strength of pigs' preferences for different rooting materials using concurrent schedules of reinforcement. Applied Animal Behaviour Science 94, 31-48.

Pryce, J.E., Veerkamp, R.F., Thompson, R., Hill, W.G. and Simm, G., 1997. Genetic aspects of common health disorders and measures of fertility in Holstein Friesian dairy cattle. Animal Science 65, 353-360.

Saintz, R.D., Torre, F. de la and Oltjen, J.W., 1995. Compensatory growth and carcass quality in growth-restricted and refed beef steers. Journal of Animal Science 73, 2971-2979.

Simpson, A., Webster, A.J.F., Smith, J.S. and Simpson, C.A., 1978. Energy and nitrogen metabolism of Red Deer (*Cervus Elaphus*) in cold environments; a comparison with cattle and sheep. Comparative Biochemistry and Physiology 60, 251-256.

Stacey, K.F., Parsons, D.J., Frost, A.R., Fisher C., Filmer, D. and Fothergill, A., 2004. An automatic growth and nutrition control system for broiler production. Biosystems Engineering 89, 363-371.

Thorhallsdottir, A.G., Provenza, F.D. and Balph, D.F., 1987. Food aversion learning in lambs without a mother: discrimination, novelty and persistence. Applied Animal Behaviour Science 18, 327-340.

Tolkamp, B.J., Friggens, N.C., Emmans, G.C., Kyriazakis, I and Oldham, J.D., 2002. Meal patterns of dairy cows consuming mixed foods with a high or a low ration of concentrate to grass silage. Animal Science 74, 369-382.

Underwood, E.J., 1980. Mineral nutrition of livestock. 2$^{nd}$ edition. CAB Slough, UK.

Webster, A.J.F., 1983. Nutrition and the thermal environment. In: Rook, J.A.F. and Thomas, P.C. (Eds.). Nutritional physiology of farm animals. Longman, London, UK, pp. 639-669.

Webster, J., 1993. Understanding the dairy cow. 2$^{nd}$ Edition. Blackwell, Oxford, UK, 374 pp.

Weigle, D.S., 1994. Appetite and the regulation of body composition. FASEB Journal 8, 302-301.

Welchman, D.B. and Baust, G.N., 1987. A survey of abomasal ulceration in veal calves. Veterinary Record 121, 586-590.

Whay, H.R., Main, D.C.J., Green, L.E. and Webster, A.J.F., 2003. Animal-based measures for the assessment of welfare state of dairy cattle, pigs and laying hens: consensus of expert opinion. Animal Welfare 12, 205-219.

Wilkins, L.J., Brown, S.N., Zimmerman, P.H., Leeb, C. and Nicol, C.J., 2004. Prevalence of keel and furculum damage in laying hens: validation of a palpation method. Veterinary Record 155, 547-549.

Whitehead, C.C., 1992. Bone biology and skeletal disorders in poultry. Butterworth Heineman Ltd., London, UK.

# Animal welfare and management

**Paul H. Hemsworth[1] and Grahame J. Coleman[2]**
[1]The Melbourne School of Land and Environment, The University of Melbourne, and the Department of Primary Industries (Victoria), Parkville, Victoria 3010, Australia; phh@unimelb.edu.au
[2]School of Psychology, Psychiatry and Psychological Medicine, Monash University, Victoria 3800, Australia

## Summary

The principle that management, including supervising and managing animals, affects farm animal welfare is widely recognised within the livestock industries. However, the manner in which management affects animal welfare, both directly and indirectly, is probably not fully appreciated. At the level of farm management, human resource management practices, including employee selection and training, and animal management practices, such as best practice in housing and husbandry, and implementation of welfare protocols and audits, all impact on farm animal welfare. At the stockperson level, together with the opportunity to perform their tasks well, stockpeople require a range of well developed husbandry skills and knowledge to effectively care for and manage farm animals. There are three main classes of stockperson characteristics that can be considered to contribute to a stockperson's work performance: capacity, willingness and opportunity. Capacity includes variables such as skills, health, ability and knowledge while willingness includes motivation, job satisfaction, attitude to the animals and work attitude and opportunity includes working conditions, actions of co-workers and organisational policies and rules. This chapter considers how management, at the level of the farm and the stockperson, affects animal welfare and the opportunities for the livestock industries through auditing, training and selecting stockpeople to improve animal welfare under the headings of (1) stockperson capacity, (2) stockperson willingness and (3) stockperson opportunity.

**Keywords**: stockpeople, management, animal welfare, attitudes, training, selection

## 1. Introduction

Close relationships between humans and animals exist in intensive livestock production systems. Studies focused on the human-farm animal relationship commenced in the 1970s, initially because of the implications of this relationship for farm animal productivity but more recently because of its critical effects on farm animal welfare, an increasingly prominent contemporary social issue. In a study of dairy herds in the United Kingdom, Seabrook (1972) reported a significant inter-farm relationship between the personality of the stockperson and the productivity of the herd; stockpeople who were introverted and confident managed herds with higher milk yields. Seabrook also commented that herds in which cows appeared to readily approach the stockperson had higher productivity than herds in which cows appeared to less readily approach the stockperson. Hemsworth *et al.* (1981) reported a significant negative inter-farm relationship between the fear response of breeding sows to humans and their reproductive performance, indicating that productivity tended to be lower at farms in which breeding pigs showed less approach to humans. Neither of these early studies identified the variables that account for the observed relationships, in particular, the possible causal relationships between stockperson variables, animal variables and welfare and production outcomes.

Subsequent research in both in experimental and commercial situations over the last 25 years has shown that interactions between stockpeople and their animals can markedly limit both the productivity and welfare of livestock (Hemsworth and Coleman, 1998; Hemsworth, 2004). While stockpeople utilise a range of behaviours to inspect and handle their animals, the frequent use of some of these routine behaviours has been shown to result in these animals becoming fearful of humans as reflected in increased avoidance of humans. It has also been shown that these fear levels, through stress, limit animal welfare and productivity. While domestic animals, such as farm animals, may be intensively managed, the profound effects of human interactions, which are often fairly infrequent and of short duration, on the behaviour and physiology of these animals are surprising. This is particularly so when it is considered that many of these interactions are routinely but briefly used by stockpeople and many of these human interactions intuitively appear to be innocuous. A major finding from these studies was that the attitude of the stockperson towards interacting with the animal was a key variable affecting the way the stockperson behaved and thus the animal's fear of humans. Furthermore, the attitude of the stockperson towards working with the animal may affect animal welfare by influencing the level of inspection and promptness of intervention when welfare problems arise.

There are other job-related characteristics of the stockperson that are likely to affect animal welfare and these include technical knowledge and skills, job motivation, job satisfaction and attitudes. Blumberg and Pringle (1982) recognized three classes of contributing factors to a person's work performance, capacity, willingness and opportunity, and these can be adapted to account for the work performance of the stockperson in relation to effectively caring and managing the farm animal (Figure 1).

This chapter will review how, at the level of the stockperson, these three classes of stockperson characteristics, stockperson capacity, stockperson willingness and stockperson opportunity, contribute to animal welfare. Furthermore, this chapter will also review, at the farm level, how farm management practices, including employee selection and training, can affect these stockperson characteristics. An understanding of the impact of these stockperson characteristics and how farm management practices affect these characteristics can be utilised to develop strategies to improve stockperson performance as well as monitoring stockperson performance.

## 2. Stockperson characteristics affecting animal welfare

At the individual level, stockpeople require a range of well-developed husbandry skills and knowledge as well as access to the appropriate facilities and opportunity to use these

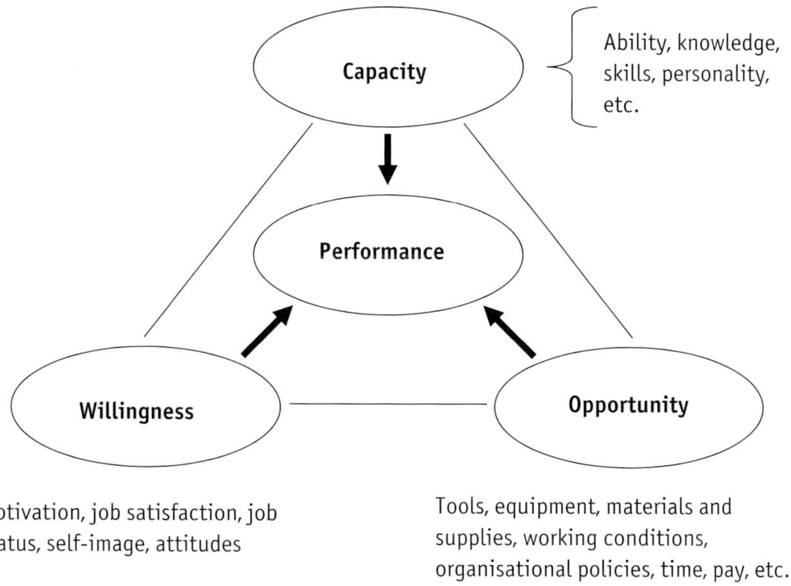


*Figure 1. The main factors that affect the performance of the stockperson in relation to animal welfare. Adapted from Blumberg and Pringle (1982).*

skills and knowledge, to effectively care for and manage farm animals. These factors, which contribute to a stockperson's work performance, will be considered in this section under three categories, stockperson capacity, willingness and opportunity.

## 2.1 Stockperson capacity

### *2.1.1 Technical skills and knowledge*

Knowing and being skilled at the techniques that must be used to accomplish a task are clearly prerequisites to being able to perform that task. Thus these job-related characteristics will limit job performance in situations where specific technical skills and knowledge are required to perform the tasks. There are little data on this for the agricultural industries, however this basic premise is widely accepted (Hemsworth and Coleman, 1998).

Some of the key characteristics that stockpeople require in successfully caring for and maintaining their animals include a good general knowledge of the nutritional, climatic, social and health requirements of the animal; practical experience in the care and maintenance of the animal; and ability to identify quickly any departures in the behaviour, health or performance of the animal and promptly provide or seek appropriate support to address these departures. Indeed, Waiblinger and Spoolder (2007) propose that 'stockmanship' summarises the different aspects of taking care of animals, particularly the quality of handling animals, daily care (e.g. feeding, cleaning, etc.), health care (e.g. how quickly sick animals are recognized and effectively treated) and problem solving management (e.g. how management problems are recognized and effectively solved).

## 2.2 Stockperson willingness

### *2.2.1 Attitudes and behaviour*

Consistent findings of negative inter-farm correlations between fear of humans, assessed on the basis of the behavioural response to humans, and the productivity of dairy cattle, pigs and poultry have stimulated research to identify behaviours used by stockpeople associated with these fear responses. Coleman *et al.* (1998) and Hemsworth *et al.* (1989) found that the use of negative tactile behaviours by stockpeople, such as slaps and hits, was positively correlated with avoidance of humans by breeding sows. In studies on dairy cows housed outdoors all year round on pastures, Breuer *et al.* (2000) and Hemsworth *et al.* (2000) found that the use of negative tactile interactions, such as slaps, pushes and hits, was positively associated with avoidance of humans by cows. Similarly, in a study of dairy cows housed indoors, Waiblinger *et al.* (2002, 2003) found that the use of positive behaviours, such as touching and talking, was negatively associated with avoidance of humans by cows, while the use of negative behaviours, such as forceful slaps and hits and shouting, was positively associated with avoidance of humans. Lensink *et al.* (2001) found that the use of positive behaviours, such as touching, petting and allowing calves to suck the stockperson's fingers, was negatively associated with avoidance of humans by veal calves. Hemsworth *et al.* (1994b) and Cransberg *et al.* (2000) found that the speed of movement by the stockperson was positively correlated with avoidance of humans by meat chickens. Edwards (2009) found that the incidence of noise made by stockpeople, such as shouting and cleaning with an air hose or leaf-blower, was positively associated with avoidance of humans by laying hens, while the times stockpeople spent standing stationary and spent close to the birds' cages were negatively associated with avoidance of humans by hens.

These field studies that demonstrate relationships between stockperson behaviour and fear of humans in farm animals are supported by experimental studies that show that negative or aversive handling of farm animals, imposed briefly but regularly, will increase their fear of humans (Hemsworth and Coleman, 1998). Furthermore, these handling studies show that negative handling, imposed briefly but regularly, will reduce growth, feed conversion efficiency, reproduction and health of pigs and poultry (Hemsworth and Coleman, 1998; Waiblinger *et al.*, 2006; Hemsworth *et al.*, 2009). A chronic stress response has been implicated in these effects of fear on productivity since in some of the handling studies it was shown that handling treatments which resulted in high fear levels, also produced either a sustained elevation in the basal free cortisol concentrations or an enlargement of the adrenal glands (Hemsworth and Coleman, 1998). Fear is one of the primary emotions that governs the way animals respond to their social and physical environment and fear is now widely regarded as an undesirable state of suffering and as a powerful and damaging stressor (Jones, 1997). One of the key recommendations proposed to the United Kingdom Parliament by the Brambell Committee in 1965 (Brambell *et al.*, 1965) was that intensive-housed livestock should be free from fear. Thus there are likely to be concerns for the welfare of farm animals that are fearful of humans and the studies reviewed here on human-animal relationships in the livestock industries indicate that the attitudes and behaviour of stockpeople have important implications on animal fear and thus animal welfare.

A number of the field studies on stockperson and animal behaviour have shown that the attitudes of stockpeople towards interacting with their animals are related to the behaviour of the stockpeople towards their animals. Questionnaires have been used to assess the attitudes of the stockpeople on the basis of the stockpeople's beliefs about their behaviour and the behaviour of their animals. In general positive attitudes to the use of petting and the use of verbal and physical effort to handle animals were negatively correlated with the use of negative tactile interactions, such as slaps, pushes and hits, in the dairy (Breuer *et al.*, 2000; Hemsworth *et al.*, 2000; Waiblinger *et al.*, 2002) and pig (Coleman *et al.*, 1998; Hemsworth *et al.*, 1989) industries. Lensink *et al.* (2000) also found that a positive attitude to the sensitivity of calves to human contact was positively correlated with the frequency of positive behaviour used by stockpeople towards veal calves, while Edwards (2009) found that negative attitudes to the sensitivity of hens to human contact as well as negative general beliefs about hens were positively correlated with more noise produced by the stockperson, faster speed of movement and negatively correlated with less time spent stationary near caged hens.

*2.2.2 Job motivation and commitment*

It is convenient to consider group job motivation and commitment together for the purpose of this review. These variables generally refer to the extent to which a person applies his or her skills and knowledge to the management of the animals under his or her care (e.g. how reliable, thorough, conscientious, etc. a person is). Factors such as job satisfaction, meaningfulness of work and utilisation of skills will affect work motivation and commitment. High job performance in any industry relies on a combination of motivation, technical knowledge and skills and an opportunity to perform the job and low motivation will limit job performance regardless of technical skills and knowledge of the individual.

There are various theories on how work conditions may affect job satisfaction and work motivation (Hemsworth and Coleman, 1998). For example, motivation to perform the tasks will improve with rewards such as feelings of pride or accomplishment. A hierarchy of needs may influence motivation, with esteem needs (recognition by others) and self-actualisation (self-fulfillment) motivating workers once physical, safety and social needs of the individual are met (Maslow, 1970).

Work motivation is important because of its relationship to other job-related characteristics such as skill variety, task significance (i.e. responsibility for outcomes) and autonomy (Hackman and Oldham, 1976) as well as motivation to learn new skills and knowledge and thus in turn, technical skills and knowledge. In this way, work motivation exerts indirect effects on job performance (Hemsworth and Coleman, 1998).

### 2.3.1 Working conditions

Apart from the physical working conditions that may make the workplace more or less uncomfortable or dangerous, the nature of the job is related to worker motivation, general satisfaction, absenteeism and work effectiveness (Hackman and Oldham, 1976). Aspects of the job that have been found to be relevant include skill variety, task significance, autonomy and feedback. There appears to be no published data on management of these factors for stockpeople employed in the livestock industries. Whether this implies that such practices do not exist or that they do exist but have not been systematically evaluated in uncertain. At the very least, a review of current industry practices should be undertaken.

### 2.3.2 Actions of co-workers

There is a large literature comparing individuals working alone with those who work in groups. In general, the group environment affects productivity, decision making, turnover, and absenteeism. Such research has shown that the interdependence inherent in work groups does predict group output (Sundstrom *et al.*, 2000). In addition, groups have been found to have effects on the behaviour and affective states of their members (Hackman, 1987). Individuals belonging to the same group have been found to display similar moods (Bartel and Saavedra, 2000), similar work culture (Levine and Moreland, 1991), and similar absenteeism levels (Gellatly and Luchak, 1998; Rumery, 2004). All of these factors contribute to the individual within a work group being pressured to behave in a way that is consistent with group norms. Clearly if a group culture does not favour good animal welfare practices, any individual joining that group will tend to succumb to the group norm no matter what predispositions he or she may have had previously.

### 2.3.3 Organisational policies

Management practices can affect levels of stress and job satisfaction in workers. For example, when management techniques involving employee participation with an emphasis on the three components of setting of goals, participation, and feedback about task fulfillment, are employed, workers experience lower levels of stress and higher levels of job satisfaction (Konradt *et al.*, 2003). Also, when they believe that managers identify clear roles for them and set high standards of performance, workers tended to show higher job satisfaction. When senior managers show ethical behaviour and integrity, work group performance and spirit tend to be higher (Church, 1995).

As indicated before, there is no systematic research on the application of current management practices to stockperson work groups on-farm, despite the high turnover experienced in the livestock industries and the need for a strong work group ethic in relation to animal welfare. There is every reason to suppose that practices that are effective in a wide range of other contexts would also apply to the livestock industries. Some developments in this area are discussed in the next section.

## 3. Opportunities to improve animal welfare through farm management practices

This review so far has shown that at the level of the stockperson, several classes of stockperson characteristics, stockperson capacity, stockperson willingness and stockperson opportunity, either have been demonstrated, or have the capacity, to contribute to farm animal welfare. It is clear from this understanding of the impact of these stockperson characteristics, that at the farm level, farm management practices including employee selection and training, can indirectly affect the welfare of farm animals by affecting these stockperson characteristics.

### 3.1 Stockperson training

The sequential relationships between stockperson attitudes and behaviour and animal fear and productivity that have been found in the dairy and pork industries (Hemsworth *et al.*, 1989, 2000; Coleman *et al.*, 1998; Breuer *et al.*, 2000; Lensink *et al.*, 2000, 2001; Waiblinger *et al.*, 2002) demonstrate the opportunities that exist to improve animal productivity and welfare by appropriate training of stockpeople. Subsequent studies in the dairy and pork industries (Hemsworth *et al.*, 1994a, 2002; Coleman *et al.*, 2000) have shown that cognitive-behavioural training, in which the key attitudes and behaviour of stockpeople are targeted, can successfully improve the attitudes and behaviour of stockpeople towards their animals, with consequent beneficial effects on animal fear and productivity.

Cognitive-behavioural techniques basically involve retraining people in terms of their behaviour by firstly targeting both the beliefs that underlie the behaviour (attitude) and the behaviour in question and secondly, maintaining these changed beliefs and behaviours (Hemsworth and Coleman, 1998). This process of inducing behavioural change is a comprehensive procedure in which all of the personal and external factors that are relevant to the behavioural situation are explicitly targeted. Therefore to improve the stockperson's beliefs about their animals and particularly their beliefs about handling and working with their animals, stockpeople should be provided with key information on their livestock, such as the ease with which they can and should be handled, their sensitivity to the range of negative behaviours used by stockpeople (and their sensitivity to stressors in general), and the adverse effects of these negative behaviours on their fear of humans, which in turn can have negative consequences on their welfare, productivity and ease of handling. The training should also provide stockpeople with information on the positive behaviours which can be used to reduce fear in their animals. To address the behavioural aspects of the intervention, stockpeople should have the opportunity to rehearse either directly or vicariously the relevant behaviours. Where it is not practical to engage in direct animal handling during training, video footage of the behaviour of stockpeople in their industry can be used with emphasis on those patterns of behaviour that increase the animals' fear of humans. Video footage of the behavioural responses of animals to stockpeople can also be presented to assist stockpeople in recognising and assessing fear responses in their animals. To reinforce the information targeting improvements in both beliefs and behaviours, stockpeople should be provided with on-going follow-up support, perhaps in the form of written material including a booklet, posters and newsletters.

Cognitive-behavioural intervention studies in the dairy (Hemsworth *et al.*, 2002) and pig industries (Hemsworth *et al.*, 1994a; Coleman *et al.*, 2000), targeting the key attitudes and behaviour of stockpeople that were previously found to be correlated with fear responses of cows and pigs to humans resulted in reductions in these fear responses. Furthermore, concurrent improvements in animal productivity were also observed; there were improvements in the milk yield of dairy cows and a marked tendency for an improvement in the reproductive performance of sows. These intervention studies in commercial settings, in conjunction with earlier research on handling in experimental settings and human-animal relationships in commercial settings, provide evidence of causal relationships between these stockperson and animal variables and establish a strong case for stockperson training courses that target the attitudes and behaviour of the stockperson.

These intervention studies by Hemsworth *et al.* (1994a, 2002) and Coleman *et al.* (2000), demonstrate that this approach to training is practical and effective on a wide range of stockpeople working in a variety of situations. Therefore, there is a strong case for introducing this type of training in the livestock industries. A commercial multimedia training program called 'ProHand Pigs' (Animal Welfare Science Centre, 2005), based on the cognitive-behavioural intervention techniques studied in the dairy and pig industries, has been developed and validated and is currently being used in Australia, New Zealand and the USA. A similar training program for the dairy industry (ProHand Dairy) has been developed within the 6$^{th}$ Framework WelfareQuality® program (Ruis *et al.*, 2009).

Stockpeople clearly require a basic knowledge of both the behaviour of the farm animal and its requirements together with a range of well developed husbandry and management skills to effectively care for and manage the farm animal. Therefore, while cognitive-behavioural training addressing the key attitudes and behaviour of stockpeople that affect animal fear is important in improving animal welfare, knowledge and skills training are also fundamental to improving the welfare of commercial livestock.

### 3.2 Stockperson selection

Coleman (2000) reported the results of a study that identified the stockperson characteristics that best predicted stockperson performance. The aim of this study was to determine the contribution of stockperson characteristics to the prediction of performance and turnover of stockpeople in the pig industry. The design of the study was in two phases: Phase One of the project involved stockpeople, at commencement of employment, completing a set of computerised questionnaires which included measures of personality, motivation, turnover potential, performance potential, and attitudes and empathy towards pigs. Phase Two of the project involved those stockpeople who remained at the piggery six months following employment as well as a group of stockpeople who had been at the piggery for at least one year, completing a second set of computerised questionnaires, which included measures of job satisfaction, internal work motivation, organisational commitment and work-group functioning. In addition, direct supervisors were asked to assess job performance in terms of animal handling, conscientiousness and other skills. Behavioural observations of job performance in the piggery were also obtained at this time.

Females were rated as more conscientious by their supervisors, and were independently rated as having greater technical knowledge and a better work ethic compared to males. The self-reported PDI-EI Performance measure (Berman, 1993) was significantly correlated with all three measures of independently observed performance. A person scoring high on this measure is likely to adhere to rules, show stability of behaviour, take care while performing tasks, and take responsibility. On the other hand, the PDI-EI Tenure measure correlated with intention to remain in employment over the next year. A positive attitude toward the characteristics of pigs correlated significantly with conscientiousness and intention to remain in the job. Finally, empathy toward animals correlated with technical knowledge, behaviour toward pigs, and intention to remain in the job. Another important finding was that 50% of new stockpeople left their jobs within six months of employment.

These tests have the potential to assist in screening potential stockpeople both to aid selection and, perhaps more importantly, to identify areas where targeted induction training is indicated. Given the difficulty that many employers face in recruiting stockpeople, a tool that may widen that range of acceptable candidates by identifying relevant on-the-job training is of considerable benefit to the livestock industries.

### 3.3 Welfare protocols and audits

The authors have recently proposed that monitoring the key characteristics of the stockperson, particularly attitudes to animals and working with these animals, empathy, work motivation and technical knowledge and skills, provides the opportunity to detect deficits in stockmanship and the necessity for further targeted training (Hemsworth *et al.*, 2009).

Monitoring stockperson and animal behaviour provides the opportunity to directly examine human-animal interactions (De Passillé and Rushen, 2005; Spoolder, 2007). However, attitudes to animals and working with these animals, empathy, work motivation and technical knowledge and skills of stockpeople underpin the influence of the stockperson on animal welfare. While there is support for measuring stockperson behaviour and fear responses of animals to humans as measures of stockmanship, several authors have identified constraints to standardising such tests (De Passillé and Rushen, 2005; Spoolder, 2007; Hemsworth *et al.*, 2009). In contrast, attitude and technical knowledge and skills questionnaires are easier to both standardise and to deliver in a standardised manner. However it is essential that standardised, validated questionnaires be identified for the various livestock species, normative data sets be established and standardised modes of delivery in an on-farm context be developed. Because of the strong relationships between stockperson attitudes to their behaviours and, in turn, to animal fear responses, as well as the relationships between attitudes and other job-related characteristics, Hemsworth *et al.* (2009) argued that attitudes, together with empathy, work motivation and technical knowledge and skills, should be the principal focus of measuring stockmanship in on-farm welfare monitoring schemes. Identification of deficits in stockmanship can be immediately targeted to address limitations to animal welfare.

# 4. Conclusions

## 4.1 What has been achieved?

While the welfare of farm animals is affected by a diverse range of factors such as their physical and social environment and their nutrition and health status, there is considerable research that shows that management practices at the farm level and at the level of the stockperson play a pivotal role in determining the welfare of farm animals. Stockpeople have direct effects on animal welfare, via their behaviour towards animals, their husbandry skills and knowledge, and their motivation to apply these skills and knowledge. The capacity and willingness of stockpeople to supervise and manage farm animals are important attributes determining animal welfare since these factors influence the likelihood of best practice by stockpeople in terms of their handling behaviour, their husbandry skills, their technical knowledge and their willingness to implement these. Furthermore, the human resource management practices of the farm, including employee selection and training, management style and organisational policies and rules that affect working conditions, can affect these important attributes of stockpeople.

## 4.2 What has been neglected?

The role of the stockperson in determining animal welfare has generally been neglected in the animal farming industries. Recent research has shown that the role and impact of the stockperson on the animal should not underestimated and that to do so will seriously risk the welfare and productivity of livestock. It is likely that in the near future both the livestock industries and the general community will place an increasing emphasis on ensuring the competency of stockpeople that manage livestock. In the future, welfare audits will necessarily include stockmanship.

## 4.3 What needs to be done?

Farm management approaches that maximize stockperson performance are an essential part of farm animal production and there is limited relevant research in agriculture and little evidence of systematic utilisation of these approaches. Therefore an understanding of these influential characteristics of stockpeople that affect farm animal welfare and the effects of local factors, such as management style and working conditions, on these influential stockperson characteristics is important in developing human resource and organizational policies to ensure that the stockperson is equipped to effectively care for and manage farm animals. Such policies would include employee selection and training targeting these influential stockperson characteristics, as well as a working environment that reduces employee levels of stress, enhances a positive group culture to the animal and the animal management tasks, and provides jobs that offer skill variety, task significance, autonomy and feedback. Welfare audits that include stockperson performance will provide a mechanism for delivering feedback to stockpeople, management and the wider community.

## References

Ajzen, I. and Fishbein, M., 1969. The prediction of behavioral intentions in a choice situation. Journal of Experimental Social Psychology 5, 400-416.

Animal Welfare Science Centre, 2005. ProHand Pigs. Available at: www.animal-welfare.org.au. Accessed 10 November 2008.

Barrick, M.P. and Mount, M., 1991. The big five personality dimensions and job performance: a meta-analysis. Personnel Psychology 44, 1-26.

Bartel, C.A. and Saavedra, R., 2000. The collective construction of work group moods. Administrative Science Quarterly 45, 197-231.

Berman, J.A., 1993. Validation of the P.D.I. employment inventory in a retail chain. Journal of Business and Psychology 7, 413-419.

Beveridge, L.M., 1996. Studies on the influence of human characteristics and training on stockperson work performance and farm animal behaviour. PhD thesis. University of Aberdeen, UK.

Blumberg, M. and Pringle, C.D., 1982. The missing opportunity in organisational research: some implications for a theory of work performance. Academy of Management Review 7 (4), 560-569.

Brambell, F.W.R., Barbour, D.S. Barnett, M.B., Ewer, T.K., Hobson, A., Pitchforth, H., Smith, W.R., Thorpe, W.H. and Winship, F.J.W., 1965. Report of the technical committee to enquire into the welfare of animals kept under intensive husbandry systems. Her Majesty's Stationery Office, London, UK.

Breuer, K., Hemsworth, P.H., Barnett, J.L., Matthews, L.R. and Coleman, G.J., 2000. Behavioural response to humans and the productivity of commercial dairy cows. Applied Animal Behaviour Science 66, 273-288.

Church, A.H., 1995. Managerial behaviors and work group climate as predictors of employee outcomes. Human Resource Development Quarterly 6, 173-205.

Coleman, G.J., 2000. Selection and training of stockpeople to improve sow productivity. In: Proceedings of the Congreso Mercosur de produccion porcina. 22 October 2000. Buenos Aires, Argentina, pp. 73-81.

Coleman, G.J., 2001. Selection of stockpeople to improve productivity. In: The 4th Industrial and Organisational Psychology Conference. 21-24 June 2001. Sydney, Australia, p. 30.

Coleman, G.J., 2004. Personnel management in agricultural systems. In: Rollin, B.E. and Benson, J. (Eds.). Maximizing well-being and minimizing suffering in farm animals. Iowa State University Press, IA, USA, pp. 167-181.

Coleman, G.J., Hemsworth, P.H., Hay, M. and Cox, M., 1998. Predicting stockperson behaviour towards pigs from attitudinal and job-related variables and empathy. Applied Animal Behaviour Science 58, 63-75.

Coleman, G.J., Hemsworth, P.H., Hay, M. and Cox, M., 2000. Modifying stockperson attitudes and behaviour towards pigs at a large commercial farm. Applied Animal Behaviour Science 66, 11-20.

Cransberg, P.H., Hemsworth, P.H. and Coleman, G.J., 2000. Human factors affecting the behaviour and productivity of commercial broiler chickens. British Poultry Science 41, 272-279.

De Passillé, A.M.B. and Rushen, J., 2005. Can we measure human-animal interactions in on-farm welfare assessment? Some unresolved issues. Applied Animal Behaviour Science 92, 193-209.

Edwards, L.E., 2009. The human-animal relationship in the laying hen. PhD. thesis, University of Melbourne, Australia.

English, P., Burgess, G., Segundo, R. and Dunne, J., 1992. Stockmanship: improving the care of the pig and other livestock. Farming Press Books, Ipswich, UK.

Gellatly, I.R. and Luchak, A.A., 1998. Personal and organizational determinants of perceived absence norms. Human Relations 51, 1085-1102.

Hackman, J.R., 1987. The design of work teams. In: Lorsch, J. (Ed.). Handbook of organizational behavior. Prentice-Hall, Englewood Cliffs, NJ, USA, pp. 315-342.

Hackman, J.R. and Oldham, G.R., 1976. Motivation through the design of work: Test of a theory. Organizational Behavior & Human Performance 16, 250-279.

Hemsworth, P.H., 2004. Human-livestock interaction. In: Benson, G.J. and Rollin, B.E. (Eds.). The well-being of farm animals, challenges and solutions. Blackwell Publishing, IA, USA, pp. 21-38.

Hemsworth, P.H and Coleman, G.J., 1998. Human-livestock interactions: the stockperson and the productivity and welfare of intensively-farmed animals. CAB International, Wallingford, UK.

Hemsworth, P.H., Brand, A. and Willems, P.J., 1981. The behavioural response of sows to the presence of human beings and their productivity. Livestock Production Science 8, 67-74.

Hemsworth, P.H., Barnett, J.L., Coleman. G.J. and Hansen, C., 1989. A study of the relationships between the attitudinal and behavioural profiles of stockpersons and the level of fear of humans and reproductive performance of commercial pigs. Applied Animal Behaviour Science 23, 301-314.

Hemsworth, P.H., Coleman, G.J. and Barnett, J.L., 1994a. Improving the attitude and behaviour of stockpersons towards pigs and the consequences on the behaviour and reproductive performance of commercial pigs. Applied Animal Behaviour Science 39, 349-362.

Hemsworth, P.H., Coleman, G.J., Barnett, J.L. and Jones, R.B., 1994b. Fear of humans and the productivity of commercial broiler chickens. Applied Animal Behaviour Science 41, 101-114.

Hemsworth, P.H., Coleman, G.J., Barnett, J.L. and Borg, S., 2000. Relationships between human-animal interactions and productivity of commercial dairy cows. Journal of Animal Science 78, 2821-2831.

Hemsworth, P.H., Coleman, G.J., Barnett, J.L. Borg, S. and Dowling, S., 2002. The effects of cognitive behavioral intervention on the attitude and behavior of stockpersons and the behavior and productivity of commercial dairy cows. Journal of Animal Science 80, 68-78.

Hemsworth, P.H., Barnett, J.L. and Coleman, G.J., 2009. The integration of human-animal relations into animal welfare monitoring schemes. Animal Welfare (in press).

Jones, R.B., 1997. Fear and distress. In: Appleby, M.C. and Hughes, B.O. (Eds.). Animal welfare. CAB International, Wallingford, UK, pp. 75-87.

Konradt U. Hertel, G. and Schmook, R., 2003. Quality of management by objectives, task-related stressors, and non-task-related stressors as predictors of stress and job satisfaction among teleworkers. European Journal of Work & Organizational Psychology 12, 61-79.

Lensink, B.J., Boissym, A. and Veissier, I., 2000. The relationship between farmers' attitude and behaviour towards calves, and productivity of veal units. Ann. Zootech. 49, 313-327.

Lensink, B.J., Veissier, I. and Florland, L., 2001. The farmers' influence on calves' behaviour, health and production of a veal unit. Animal Science 72, 105-116.

Levine, J.M and Moreland, R.L., 1991. Culture and socialization in work groups. In: Resnick, L.B., Levine, J.M. and Teasley, S.D. (Eds.). Perspectives on socially shared cognition. American Psychological Association, Washington, DC, USA, pp. 257-279.

Maslow, A.H., 1970. Motivation and personality. Second edition. Harper and Row, New York, NY, USA.

Ravel, A., D'Allaire, S.D., Bigras-Poulin, M. and Ward, R., 1996. Personality traits of stockpeople working in farrowing units on two types of farms in Quebec. In: Proceedings of 14$^{th}$ Congress of the International Pig Veterinary Society, 7-10 July 2000. Bologna, Italy.

Ruis, M., Coleman, G.J., Waibliger, S. and Boivin, X., 2009. A multimedia-based cognitive-behavioural intervention programme improves the attitude of stockpeople to handling pigs. In: Proceedings of 43$^{rd}$ Congress of the International Society of Applied Ethology. 6-10 July 2009. Cairns, Australia (in press).

Rumery, S.M., 2004.The influence of work group culture on employee turnover. Dissertation Abstracts International: Section B: The Sciences and Engineering 65, 470.

Rusbult, C.E, Farrell, D., Rogers, G. and Mainous, A.G., III., 1988. Impact of exchange variables on exit, voice, loyalty, and neglect: an integrative model of responses to declining job satisfaction. Academy of Management Journal 31, 599-627.

Seabrook, M.F., 1972. A study to determine the influence of the herdsman's personality on milk yield. J. Agr. Labour Sci. 1, 45-59.

Seabrook, M.F., 1991. The influence of the personality of the stockperson on the behaviour of pigs. Applied Animal Behaviour Science 30, 187 (Abstract).

Spoolder, H.A.M., 2007. Fear of humans. In: Verlarde, A. and Geers, R. (Eds.). On farm monitoring of pig welfare. Wageningen Academic Publishers, Wageningen, the Netherlands, pp. 35-39.

Sundstrom, E., McIntyre, M., Halfhill, T. and Richards, H., 2000. Work groups: from the Hawthorne studies to work teams of the 1990s and beyond. Group Dynamics: Therapy, Research, and Practice 4, 44-67.

Waiblinger, S. and Spoolder, H.A.M., 2007. Quality of stockmanship. In: Verlarde, A. and Geers, R. (Eds.). On farm monitoring of pig welfare. Wageningen Academic Publishers, Wageningen, the Netherlands, pp. 156-166.

Waiblinger, S., Menke, C. and Coleman, G., 2002. The relationship between attitudes, personal characteristics and behavior of stockpeople and subsequent behavior and production of dairy cows. Applied Animal Behaviour Science 79, 195-219.

Waiblinger, S., Menke, C. and Fölsch, D.W., 2003. Influences on the avoidance and approach behaviour of dairy cows towards 35 farms. Applied Animal Behaviour Science 84, 23-39

Waiblinger, S., Boivin, X., Pedersen, V., Tosi, M-V., Janczak, A.M., Visser, E.K. and Jones, R.B., 2006. Assessing the human-animal relationship in farmed species: A critical review. Applied Animal Behaviour Science 101, 185-242.

# Animal welfare and transport

**Jörg Hartung and Annette C. Springorum**
Institute for Animal Hygiene, Animal Welfare and Farm Animal Behaviour, University of Veterinary Medicine Hannover, Foundation, Bünteweg 17p, 30559 Hannover, Germany; itt@tiho-hannover.de

## Summary

Transportation can be stressful for untrained animals. Therefore animal transport is one of the most controversially discussed welfare topics during the last decades in Europe. About 340 million red meat farm animals and billions of poultry are transported by road every year within, from and to the European Union. With the introduction of the Council Regulation 1/2005 on the protection of animals during transport several improvements were introduced in order to protect health and well-being of the animals. However, the welfare of farm animals in transit is still a cause for considerable concern, especially for long transports. The most important risk factors for health and welfare of animals during transport are seen in loading and unloading, bad handling, inappropriate driving performance (often in combination with poor road conditions), insufficient ventilation (micro-climate), high stocking densities, mixing of unfamiliar animals in groups, insufficient supply of water and food, vehicle vibration, journey length and the frequency and length of break times. In the future the organisation of transports must more closely observe the needs of the animals. This includes a preparation of the animals for the transport in order to reduce stress. The animals should be familiar with the type of feed and the drinking water technique on the lorry. Mixing of animals from groups of different origin should be avoided. Education of the personal is crucial for a proper handling of the animals. Loading and unloading is one of the most stressful situations and should be reduced to a minimum. The welfare of animals is limited by their needs not by a fixed maximum transport time, provided vehicle, animal density and other transport conditions such as micro-climate and feed and water supply are appropriate.

**Keywords**: well-being, stressors, EU legislation, risk factors

## 1. Introduction

The transport of farm animals belongs to the most controversially discussed animal welfare topics of the last three decades in Europe. Since pictures and television programmes revealed mal-treatment of slaughter cattle especially during long transports on extra-EU routes public opinion has put pressure on politicians to introduce legislation for the protection of animals during transport. There is a variety of reasons which give cause for concern for both short and long animal transports: i.e. (1) transport can cause severe stress in animals entailing poor welfare, (2) stressful transports can cause high mortality among transported animals, (3) stressful transports may have negative effects on meat quality, (4) stressful transports can reduce immunity, cause excretion of latent pathogens and enhance infectiousness which can promote development of diseases during and after transport, (5) there is the risk of spread of infectious diseases among the animals on a transport vehicle, an assembly centre or a place of rest or transfer, (6) animal welfare can be impaired by poor pre-transport and transport conditions, mal-handling, insufficient supply with water and food (if appropriate) and injury, (7) There is the risk of transmission of infectious diseases over large distances, e.g. FMD, and (8) animal transport contributes also to the pollution of the environment by emissions from the vehicle engines (e.g. $CO_2$, dust particles-PM2.5 and 10).

Among these concerns animal welfare receives highest public attention and has led to legal regulations such as the early Council Directive (1991) and the recent comprehensive EU Transport Regulation of 2005 (EC, 2005).

Transport and trade of farm animals have a long tradition in Europe and are driven by market forces. Van Reenen *et al.*, (2008) report that at the beginning of the $17^{th}$ century ten thousands of oxen travelled by road or sea from Denmark and Schleswig-Holstein to the Netherlands, the total trade volume of live cattle was estimated to be more than half a million per year in Europe at that time. Today the numbers of animals which are transported for commercial reasons are very much higher. Two recent scientific reports summarize welfare, health and bio-security aspects of the transport of horses, pigs, sheep and cattle (SCAHAW, 2002) and of broilers, laying hens, turkeys, ducks, geese, pigeons, quails, ostriches and other rare species, deer, reindeer, rabbits, dogs, cats, rodents, primates, fish and exotic animals (EFSA, 2004a).

The number of transported animals per year is large but exact figures on many species are difficult to obtain. Table 1 provides an assumption of the numbers of animals transported in the European Union (EU) based on slaughter figures and on the assumption that farm animals are moved usually twice during their lifetime, e.g. from the place of birth or hatching to another unit, e.g. for fattening, and later to the slaughter house (Eurostat, 2009), which results in about 700 million individual journeys of farm animals, poultry not included. These figures seem to be realistic because in 2001 (i.e. before introduction of the single market and omission of the inner EU border controls in 2002), more than 315 million farm animals, incl. horses, were transported within the EU-MS (Eurostat, 2009), and about 13 million animals were traded across borders of EU member states or imported or exported from or to the EU (Nagel, 2001). The total figures for the EU27 may be even higher taking into account the numbers of farm animals actually kept for farming purposes (Eurostat, 2009, in million places): pigs (160), cattle (89 incl. cows and calves), sheep (99, figures available only for 2002), although sheep may not be transported twice in their life. Real figures are probably much higher.

Figures from year 2000 suggest that the number of broiler chicks produced worldwide was approximately 40 billion (FAOSTAT, 2002) and 4.3 billion in the European Union (EU-14).

*Table 1. Number of slaughtered animals in millions in the EU 27 in 2008 (Eurostat, 2009).*

| Species | Pig | Cattle (total) | Cow (incl.) | Calf (incl.) | Sheep |
|---|---|---|---|---|---|
| Numbers in millions | 255 | 29 | 8 | 6 | 65 |

The corresponding figure for layer chicks was 290 million (EU-15) and for turkey 245 million (EU-14). The placement of broiler breeder chicks in the EU during the same period was approximately 30 million (Poultry Bulletin, 2002). All these birds are transported at least twice during their life time, firstly from hatcheries to the production or rearing barns within 1-3 days after hatching and secondly to the slaughter and processing plants.

The formation of the European Single Market enhanced animal transports by an increasing specialization in the meat producing sectors. The fattening units are settled mainly in northern central Europe, leading to overproduction of animals in some of these states. Denmark for example produces 400% of its own requirement for pigs and Ireland even the tenth fold of its beef and veal consumption. Similar statistics of overproduction in some finishing sectors can be found for the Netherlands and Belgium. In contrast, southern European countries offer low labour costs and slaughter house capacities resulting in more long journeys to the farms or slaughter houses (Van Reenen *et al.*, 2008). Taking into account the figures of Nagel (2001) it is assumed that about 20 million animals are transported longer than 8 hours on special transport vehicles, pass through markets, sampling points or places of rest or transfer where they may be unloaded and loaded again before travelling to their final destination.

A special, at times important but heavily criticised market is the export of live slaughter cattle from the EU to the near east region. Although transport duration is long in this scenario, it is driven by additional criteria: (1) the market demand is for live animals, because there are not sufficient cooling chains for fresh meat in the importing countries, and the simplest way to keep the meat fresh over longer periods is the living animal, (2) the consumers pay higher prices for living animals because they want to see the animals before slaughter, (3) low slaughter costs in the importing countries and the use of the 'fifth quarter' (Nearly all parts of the animals are used for consumption), and finally (4) religious reasons: slaughter without stunning.

The stressors livestock are facing during transportation, and which may possibly lead to poor health and welfare, can derive from a large range of hazards in their environment such as poorly designed and equipped vehicles, inappropriate handling, inadequate micro-climate (insufficient ventilation, air velocity, air quality), insufficient space allowance, inappropriate floors and bedding materials, mixing unfamiliar animals, separating familiar animals, rough or abusive handling and inappropriate driving. In addition transport duration, lack of (appropriate) food and water on longer journeys, poor fitness and health status prior to transport, insufficient fasting or inappropriate food supply prior to transport, introduction of pathogens before and during transport, inappropriate application of resting periods during transport (EFSA, 2004a) may pose risks for animal welfare. The most aversive factors when transporting animals are seen in loading and unloading, bad handling, inadequate micro-climate and journey time. However, the reactions of animals to transport situations depend to a high degree on species, gender and temperament (Marahrens *et al.*, 2003).

This contribution gives some information on health and welfare associated risks during transport and shows examples of reactions of animals in typical critical situations such as loading, unloading, resting periods and the implications of ventilation. Also an attempt is made to rank the various risk factors in a particular order of their influence on health and welfare of animals in transit. Finally, recommendations are made to minimize stress inducing factors for future consideration in legislation and practice.

## 2. The EU regulation on animal transport

The EU Transport Regulation No. 1/2005 of December 22$^{nd}$ 2004 on the protection of animals during transport and related operations, which came into force on 5 January 2007 aims at 'limiting the transport of animals over long journeys as far as possible' and at 'improving the protection and welfare of animals as well as preventing the occurrence and spread of infectious animal diseases' and will 'safeguard the welfare and health of animals during and after transport'. The creation of the regulation was accompanied by intensive debates of all stakeholders trying to bring in their specific views and interests. The most disputed topics were the transport times and the unloading rule for 24 hour breaks in resting places. Proposals such as a general limitation of the transport time to 9 h followed by a break of 12 h for rest (which would have fit with the social regulations for one-driver transport journeys) and feeding on the lorry without unloading and without any limitation of the total transport time did not find acceptance and were postponed for future consultations.

One of the most important rules is the limitation of the maximum transport time to 8 hours on normal transport vehicles. If the transport has not reached the destination the animals have to be unloaded in a place of rest for 24 hours. After that time they can continue their journey for another 8 hours, etc. Transports planned for longer than 8 hours have to be carried out on special vehicles (type 2) which provide litter, drinking water and forced ventilation. For cattle journeys up to 29 hours are allowed, with a 1-hour break after 14 hrs, before the animals have to be unloaded in a place of rest for 24 hours where they get feed and water. Pigs heavier than 30 kg can be transported for 24 hours allowing free access to water on the vehicle. Feeding of pigs is unusual because they are prone to nausea and vomiting in transit when fed. The maximum transport time for adult poultry of 12 hours seems rather long because transport losses can dramatically increase with journey length (e.g. Warriss *et al.*, 1993).

Although the regulation is recognised to have a great potential for safeguarding health and welfare of transported animals, some serious shortcomings can be identified such as the journey times which are not based on scientific evidence and which do not always meet the needs of the respective (modern) farm animal species, the strict unloading rule after the allowed maximum transport times and the micro-climate recommendations which are to some extent useful in terms of ventilation rules but which do not meet the climatic requirements in all of Europe.

Some of the most important improvements and the persisting drawbacks of the regulation compared to the former Directive of 1991 are given in Table 2.

*Table 2. Improvements and drawbacks of the EU Transport Regulation No 1/5 (EC, 2005) as compared with former EU Directive (EC, 1991).*

Improvements:
1. It is a regulation which becomes effective directly in the Member States.
2. Owners, keepers and attendants have to take more responsibility for the fitness of the animals.
3. Journey log books are compulsory and have to be signed by the sending and receiving person. The documents have to be kept for control purposes for 3 years.
4. All drivers and personnel on markets have to hold certificates of appropriate training courses.
5. Markets and assembly centres have to provide written instructions and training courses for staff to reduce the risk of transmission of infectious diseases.
6. A link to the TRACES system is established. Usefulness has still to be proven.
7. Harmonised formats for certificates in all Member States.
8. Air temperature sensors, water meters and data recorders on vehicles for long transports.
9. Contact points in all Member States improve exchange of information between competent authorities of Member States in case of problems.

Drawbacks:
1. The transport time schedules do not always meet the needs of the animal species or gender.
2. The strict off loading rule for the 24 h resting period remains in place for all animals.
3. No harmonisation of travel periods and driving times for vehicle drivers.
4. Increase in administration, electronic data and paper work have to be kept in parallel.

# 3. Risks for animal health and welfare typically associated with transport

Transport to 'finishing' sites or to slaughterhouses is an event which can be stressful particularly for untrained animals. Long journeys, high animal density, lack of mobility, exposure to temperature extremes, crowding with unfamiliar animals and also bad driving behaviour are factors that cause stress and can result in injuries such as bruising, skin lesions and even bone fractures and can also reduce immune competence. Insufficient ventilation and lack of water can lead to dehydration and heat stroke. Some animals develop motion sickness while the vehicle is moving. Immune suppression can trigger the excretion of pathogens from latent carriers favouring the spread among the animals and, in case of slaughter animals excreting pathogens with a zoonotic potential, along the food chain to the consumer (Nowak *et al.*, 2007).

## 3.1 Mortality, injury and muscle damage

The term welfare is relevant only when an animal is alive but death during handling and transport is usually preceded by a period of poor welfare. Mortality records during journeys and on arrival at slaughterhouses (Dead on Arrival, DoA) as well as rates of carcass degradation are often the only data which give information about the possible welfare status of the animals during a journey and the severity of the problems encountered.

The measurement of oxytocin has not been of particular value in animal transport studies (e.g. Hall *et al.*, 1998). However, elevated oxytocin concentration in blood is associated with pleasurable sensations in several different types of circumstances. Plasma ß-endorphin levels were found to increase during the loading of pigs (Bradshaw *et al.*, 1996b) and measurement of beta-endorphin levels in blood is useful to complement ACTH or cortisol measurement.

When sheep were loaded on to a vehicle for the first time, all showed elevated plasma and saliva cortisol for at least the first hour (Broom *et al.*, 1996; Parrott *et al.*, 1998). Cortisol levels in transported pigs were affected by being on a rough rather than a smooth journey (Bradshaw *et al.*, 1996b), being mixed with pigs from different origin and being on a moving vehicle rather than being stationary (Bradshaw *et al.*, 1996c).

However, also the periods in resting stations can be disturbing and exhausting for animals. Figure 4 shows the creatin kinase (CK) concentrations in a group of 12 bulls, steers and heifers during a journey of 29 hrs and at the end of the resting period. While the CK concentrations remained on usual levels, the group of bulls showed an enormous increase indicating high muscular activity.

When high heart frequencies, catecholamine levels and creatine kinase concentrations indicate a status of reduced welfare during loading, unloading and even in the resting pen (bulls only) it is to consider whether the uniform transport schemes are appropriate for all animals of the same species. Frequent loading and unloading also increases the risk of injuries. Therefore, the repeated unloading and loading during long distance transports should be avoided to reduce stress and the risk of injuries. Resting, feeding and watering on the vehicle in properly equipped supply stations should be considered as a possible alternative.

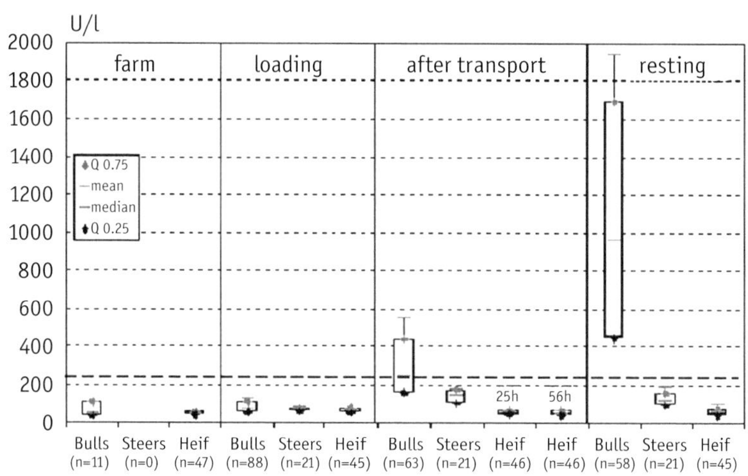


*Figure 4. Plasma activity of creatine kinase (50% confidence interval, median, mean) in blood plasma of bulls, steers and heifers before and after transport and after resting (Marahrens* et al., *2003).*

## 4.2 Journey length, feeding and lying behaviour

There is a debate whether animals lie down during transport or stand all the time. Once journeys start, some species of farm animals explore the compartment in which they are placed and try to find a suitable place to sit or lie down. Unfortunately for the animals, many journeys involve so many lateral movements or sudden braking or accelerations that the animals cannot lie down. The number of pigs, or other animals, which remain standing during transport is a relevant measure of welfare in relation to the roughness of the journey. For example, Bradshaw *et al.* (1996a) found that more pigs remain standing during a rough journey, measured in terms of accelerations in three possible planes, than during a smooth journey. In journeys with certain vibration characteristics, pigs show behavioural evidence of motion sickness in that they retch and vomit (Bradshaw *et al.*, 1996b; Randall and Bradshaw, 1998).

Observations of Marahrens *et al.* (2003) showed that only about 20% of bulls lie down during long transports independent of whether the vehicle stood during a break or was driving. Own observations on long transports revealed the opposite. About 5 hours after smooth transport nearly all of heifers were lying and resting. The different behaviour may be caused by different temperament and activity of heifers and bulls, but also because pregnant heifers may experience fatigue earlier. During transport lipomobilization is activated in all types of cattle indicated by an increased level of NEFA (non-esterified fatty acids) in the blood plasma. The process starts in heifers already during loading and continues to increase during transport, and even in the resting place. It seems that because heifers have less carbohydrate reserves more fat reserves have to be mobilised. The high levels of NEFA and 3-hydroxybutyrate (not shown) indicate the beginning of a ketotic metabolism (Marahrens *et al.*, 2003). Figure 5 shows the considerable increase in NEFA in heifers. Consequently, the travel times for heifers should be reconsidered offering more breaks for feeding and resting compared to bull or steers. Observation on 199 cattle showed that after a transport time of 12 hours about 35% of the animal was foraging on the vehicle during a break of nearly two hours while during the journey less than 5% was seen eating (Marahrens *et al.*, 2003).

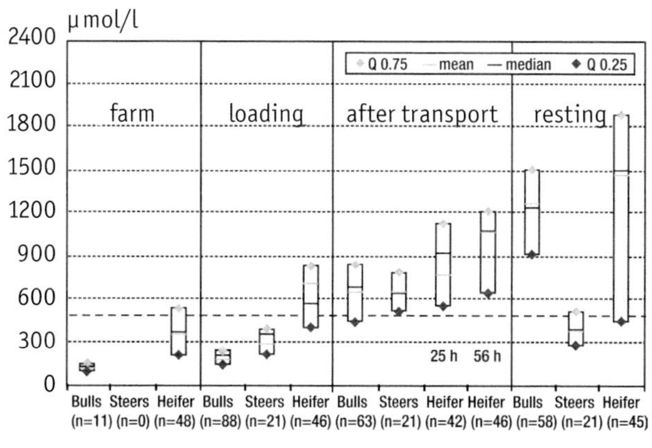


*Figure 5. NEFAs (non-esterified fatty acids) in blood plasma of bulls, steers, and heifers before and after transport and after resting (Marahrens* et al., *2003).*

Physiological indicators such as heart frequency do not reach any clinical thresholds of ill-health during long transport of cattle. Welfare can be poor when water and food supply or ventilation are insufficient. Research shows that after 6 hours of transport pregnant heifers can develop an energy deficit (Marahrens *et al.*, 2003). The energy deficit is obviously caused by the additional efforts of the animals to keep the balance when the lorry is driving. It also depends on the feeding of the animals prior to transport. Steers which are kept on pastures with the opportunity to move freely are usually in a better physical condition than bulls kept in confined buildings. More breaks and time to take up feed seems helpful to avoid metabolic deficits in heifers. This example stresses the necessity of a careful preparation of the animals for transport and that the transport times given for cattle in the legal regulation do not seem to meet the needs of all types of cattle. There are significant differences between cattle gender (Hartung *et al.*, 2003).

### 4.3 Microclimate on transport vehicles

One of the most important stress factors for animals during transport is thermal stress, being a major reason for suffering and losses of lives both in the standing as well as in the moving vehicle. Therefore, sufficient ventilation of vehicles is important (EFSA, 2004b). Ventilation in livestock transport vehicles takes place usually through openings along the sides of the vehicle and by electricity powered mechanical ventilation. The incoming air reduces temperature, relative humidity, gas levels and levels of other contaminants. Little is known about the quality and emissions which are deriving from animal transporters. This may be of interest with respect to environmental pollution by gases such as ammonia and methane and by particles including infectious agents which can contribute to the spread of diseases. Up to 18 ppm ammonia were measured on a Swedish cattle transport vehicle (Wikner *et al.*, 2003).

The temperature inside transport vehicles increases with reduced ventilation and increased animal density. Stops to load new animals lower the temperature and relative humidity in winter time when the doors are opened. In summer, temperature and relative humidity increase within minutes dramatically when the vehicle is standing (Sällvik *et al.*, 2004).

Figure 6 shows typical temperature curves from inside a loaded vehicle on a transport from the north of Germany to Italy during three different seasons. The temperature increases quickly during loading. When driving air exchange increases and provide fresh air through the openings on the vehicles.

There is a clear relationship between the outside temperature, ventilation openings and the inside temperature. During loading the temperature increases to 30 °C in the hot summer situation. The temperature curve in the moderate season follows the same pattern. The ventilation openings were not reduced during the crossing of the Alps at night and the temperatures dropped to 6 °C in the early morning. The temperature curve of the winter journey is different because ventilation intensity was reduced to keep the temperature inside the vehicle between 10 and 15 °C. Temperatures are falling in the evening and during the night. Breaks and tunnel passages increase temperatures again because of reduced ventilation or increased temperatures inside the tunnel.

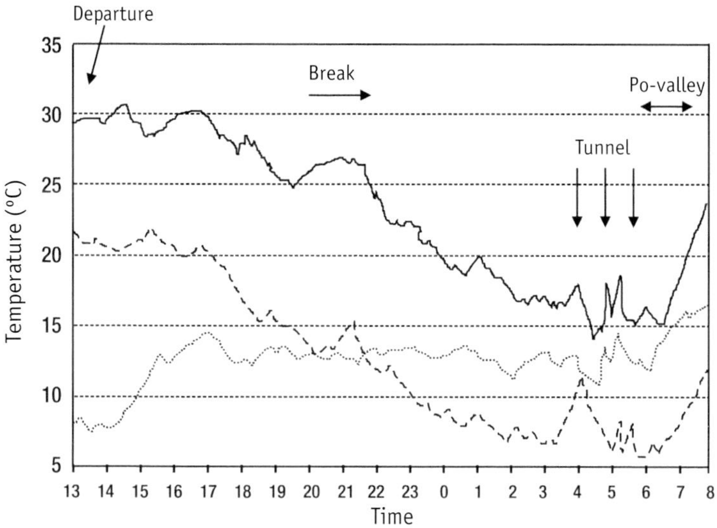


*Figure 6. Temperature curves in commercial cattle transport vehicles in hot (—), moderate (---) and cold (···) weather conditions during journeys from northern Germany to Italy (after Brüser-Pieper, 2006).*

Ventilation of the standing vehicle can also be supported by movable roofs which can be lifted allowing the warm air to escape. By mechanical ventilation temperature and humidity can be better controlled in the standing vehicle. Modern vehicles can be equipped with temperature sensors inside the loading decks giving information to the driver when upper or lower temperature limits are exceeded. Such systems can considerably help to avoid heat stress accidents.

If temperatures continue to rise during hot summer days water spraying is an option to cool the animals particularly in pig transport. Figures 7 and 8 compare the efficiency of natural ventilation with forced ventilation and water spraying.

To avoid heat and cold stress control systems for the microclimate should be further improved measuring not only air temperature but also air humidity. Through careful planning and mindful executing of transports, with adequately trained personnel, it is possible to minimise the stress on the animals before critical temperatures are reached. The future records of the position of the vehicles (e.g. by global positioning systems, GPS) may also help to avoid heat stress by early advisory models.

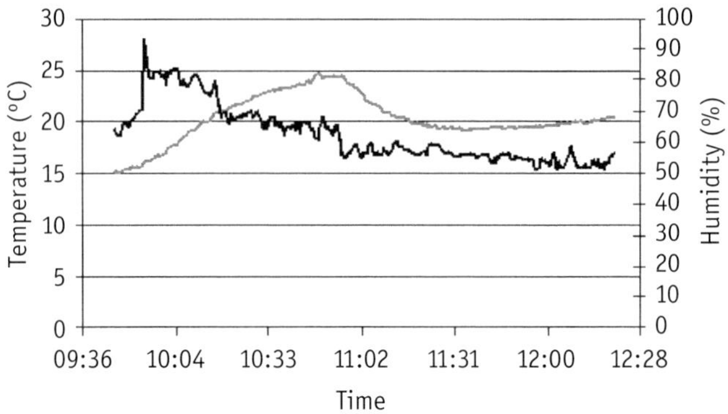


*Figure 7. Air temperature and relative humidity in a vehicle with mechanical ventilation (/ = temperature, / = rel. humidity) (Christensen and Jonsson, 2007; pers. comm.).*

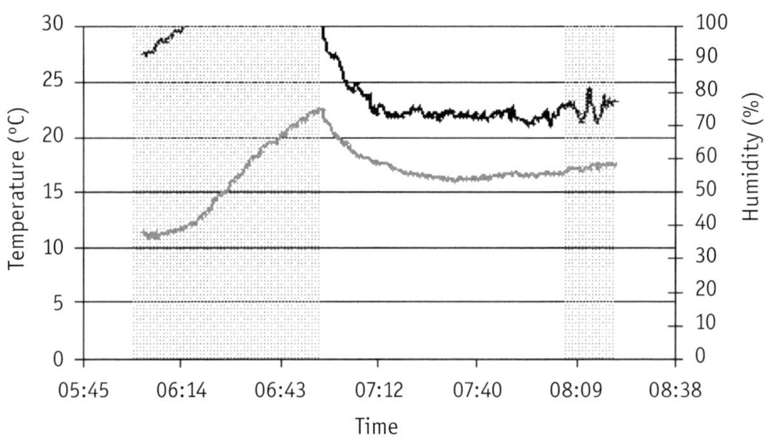


*Figure 8. Air temperature and relative humidity in a vehicle with natural ventilation (/ = temperature, / = rel. humidity, xx = standing) (Christensen and Jonsson, 2007; pers. comm.).*

# 5. Conclusions

## 5.1 What has been achieved?

The quality of farm animal transport has improved during the last decades, in particular in long transports. This is documented by decreasing numbers of dead on arrival in red meat species. The EU transport regulation of 2005 harmonises transport conditions throughout Europe.

Owners, keepers and attendants have to take more responsibility for the fitness of the animals, to which end a journey log book has shown to be useful for helping to make animal transports more transparent.

Compulsory training courses and certificates for qualification have improved knowledge and skills of drivers and personnel on markets and have reduced the risk of poor welfare and of transmission of infectious diseases. With newly designed transport vehicles having become available loading systems have become easier. In addition, these vehicles have less slippery floors, forced ventilation, and water spraying systems allowing avoiding heat stress in the summer.

### 5.2 What has been neglected?

In spite of these achievements the duration of transports still represent a major point of concern. Short transports up to 8 hours should be given more attention. For some species of broilers should be adapted to suit the specifics of modern breeds. For instance, it has been shown that 12 hour transports for turkeys and broilers increase mortality.

In 2004, the European Commission introduced the TRACES system, allowing the integration within a central server of all veterinary information relating to trade of animals and products thereof within the EU or with third countries and the linking up with the required animal health certification by keeping up-to-date all regulatory bases which apply. Unfortunately, the system is not yet fully working in long animal transports.

The introduction of air temperature sensors and water meters needs further attention, as does the systematic recording of these measurements. Also, the usefulness of the presently practised transport times, number and duration of breaks, as well as loading densities and feed supply on longer journeys need to be thoroughly discussed. An improved exchange of information would appear to be lacking and particularly the role the various contacts points in Member States could play for this purpose is not yet enough visible.

### 5.3 What needs to be done?

The following represent a number of actions that would appear to deserve being pursued with some priority: (1) adapt transport time schedules to the needs of the animal species and gender, (2) harmonise transport times for the animals and driving times of vehicle drivers, (3) investigate usefulness of loading and unloading in places of rest in regard to stress, injuries and transmission of infectious diseases, (4) investigate options of resting, feeding and watering on the vehicles in properly equipped resting places in comparison to repeated loading procedures, (5) avoid additional administrative work by developing electronic data sheets, (6) carry out proper risk assessments on all aspects of animal transport of all relevant species, and (7) More species-related technical and organisational solutions in animal transport should be developed in an interdisciplinary approach between animal scientists, practitioners, engineers and legislators.

## References


Bradshaw, R.H., Hall, S.J.G. and Broom, D.M., 1996a. Behavioural and cortisol responses of pigs and sheep during transport. Veterinary Record 138, 233-234.

Bradshaw, R.H., Parrott, R.F., Forsling, M.L., Goode, J.A., Lloyd, D.M., Rodway, R.G. and Broom, D.M., 1996b. Stress and travel sickness in pigs: effects of road transport on plasma concentrations of cortisol, beta-endorphin and lysine vasopressin. Animal Science 63, 507-516.

Bradshaw, R.H., Parrott, R.F., Goode, J.A., Lloyd, D.M., Rodway, R.G. and Broom, D.M., 1996c. Behavioural and hormonal responses of pigs during transport: effect of mixing and duration of journey. Animal Science 62, 547-554.

Brogden, K.A., Lehmkuhl, H.D. and Cutlip, R.C., 1998. *Pasteurella haemolytica* complicated respiratory infections in sheep and goats. Veterinary Research 29, 233-254.

Broom, D.M., 2000. Welfare assessment and welfare problem areas during handling and transport. In: Grandin, T. (Ed.). Livestock handling and transport. 2nd Edition. CABI Publishing, Wallingford, UK, pp. 42-62.

Broom, D.M., 2003. Transport stress in cattle and sheep with details of physiological, ethological and other indicators. Dtsch. tierärztl. Wschr. 110, 83-89.

Broom, D.M. and Johnson, K.G., 1993. Stress and animal welfare. Kluwer, Dordrecht, the Netherlands.

Broom, D.M. and Kirkden, R.D., 2003. Welfare, stress, behaviour and pathophysiology. In: Dunlop, R.H. (Ed.). Veterinary pathophysiology. Iowa State University Press, Ames, IA, USA.

Broom, D.M., Goode, J.A., Hall, S.J.G., Lloyd, D.M. and Parrott, R.F., 1996. Hormonal and physiological effects of a 15 hour road journey in sheep: comparison with the responses to loading, handling and penning in the absence of transport. British Veterinary Journal 152, 593-604.

Brüser-Pieper, C., 2006. Untersuchungen zur thermischen Belastung von Schlachtrindern beim Straßentransport auf Langstrecken. (Thermal load of slaughter bulls during long distance road transport.) PhD thesis. University of Veterinary Medicine Hannover, Foundation, Germany.

European Commission (EC), 1991. Council Directive 91/628/EEC of 19 November 1991 on the protection of animals during transport (Official Journal L 340, 17-27), amended by Council Directive 95/29 of 29 June 1995. Official Journal L 148, 52-63.

European Commission (EC), 2005. EU Transport Regulation No. 1/2005 of 22 December 2004 on the protection of animals during transport and related operations. Official Journal L 3, 1-43.

Dunnington, E.A. and Siegel, P.B., 1984. Thermoregulation in newly hatched chicks. Poultry Science 63, 1303-1313.

Ebedes, H., Van Rooyen, J. and Du Toit, J.G., 2002. Capturing wild animals. In: Game Ranch Management, Ed. J. du P. Bothma, Van Scheik, Pretoria, South Africa, 382-440.

European Food Safety Authority (EFSA), 2004a. The welfare of animals during transport. Available at: http://www.efsa.eu.int.

European Food Safety Authority (EFSA), 2004b. Opinion of the Scientific Panel on Animal Health and Welfare on a request from the Commission related to the Standards for the microclimate inside animal road transport vehicles, The EFSA Journal 122, 1-25.

EUROSTAT, 2009. Table on slaughtered animals for meat production. Available at: http://nui.epp.eurostat.ec.europa.eu. Accessed on 23 June 2009.

FAOSTAT, 2002. Available at: http://apps.fao.org. Accessed on 05 Mai 2009.

Freeman, B.M., 1964. The relationship between oxygen consumption, body temperature and surface area in the hatching and young chick. British Poultry Science 6, 67-72.

Freeman, B.M., 1984. Transportation of poultry. World's Poultry Science Journal 40, 19-30.

Gregory, N.G. and Wilkins, L.J., 1989. Broken bones in domestic fowl: handling and processing damage in end-of-lay battery hens. British Poultry Science 30, 555-562.

Guise, J. and Penny, R.H.C., 1989. Factors affecting the welfare, carcass and meat quality of pigs. Animal Production 49, 517-521.

Hackel, G. and Kaleta, E.F., 1997. Methods for the estimation of the age of broilers and effects of water and feed deprivation within the first days of life. Arch. Geflügelk. 61, 280-286.

Hall, S.J.G., Broom, D.M and Kiddy, G.N.S., 1998. Effect of transportation on plasma cortisol and packed cell volume in different genotypes of sheep. Small Ruminant Research 29, 233-237.

Hartung, J., Marahrens, M., Parvizi, N., Schmeiduch, S., Hiegemann, H., Aussel, M., Zerbe, F. and Ulbrich, H., 2000. Investigations on stress response of heifers during long distance road transport (Zur Belastung von Rindern beim Straßentransport auf Langstrecken). In: Proceedings Congress Bundesverband der beamteten Tierarzte (BbT), Staffelstein, 3-4 Mai 2000, pp. 33-41.

Hartung, J., Marahrens, M. and Von Holleben, K., 2003. Recommendations for future development in cattle transport in Europe. Dtsch. tierärztl. Wschr. 110, 128-130.

Higgs, A.R.B., Norris, R.T. and Richards, R.B., 1993. Epidemiology of salmonellosis in the live sheep export industry. Australian Veterinary Journal 70, 330-335.

Honkavaara, M., Rintasalo, E., Ylonen, J. and Pudas, T., 2003. Meat quality and transport stress of cattle. Dtsch. Tierärztl. Wochenschr. 110, 125-128.

Knowles, T.G. and Broom, D.M., 1990b. Limb bone strength and movement in laying hens from different housing systems. Veterinary Record 126, 354-356.

Knowles, T.G. and Broom, D.M., 1990a. The handling and transport of broilers and spent hens. Applied Animal Behaviour Science 28, 75-91.

Löliger, H.-Ch. and Torges, H.-G., 1977. Tierschutzgerechter Transport beim Geflügel. Du-und-das-Tier 7(1), 27-30.

Marahrens, M., Von Richthofen, I., Schmeiduch, S. and Hartung, J., 2003: Special problems of long-distance road transports of cattle. Dtsch. tierärztl. Wschr. 110, 120-125.

Nagel, R., 2001 Tiertransporte in Europa – Umfang und Tendenzen. (Animal transports in Europe – scope and trends). Dtsch. tierärztl. Wschr. 108, 324- 325.

Nowak, B., Mueffling, T.v., Chaunchom, S., Hartung, J., 2007. *Salmonella* contamination in pigs at slaughter and on the farm: A field study using an antibody ELISA test and a PCR technique. International Journal of Food Microbiology 115, 259-267.

Owen, R.A., Fullerton, J. and Barnum, D.A., 1983. Effects of transportation, surgery, and antibiotic therapy in ponies infected with *Salmonella*. American Journal of Veterinary Research 44, 46-50.

Parrott, R.F., Hall, S.J.G., Lloyd, D.M., Goode, J.A. and Broom, D.M., 1998. Effects of a maximum permissible journey time, 31 h on physiological responses of fleeced and shorn sheep to transport, with observations on behaviour during a short, 1 h rest-stop. Animal Science 66, 197-207.

Parrott, R.F., Lloyd, D.M. and Brown, D.M., 1999. Transport stress and exercise hyperthermia recorded in sheep by radiotelemetry. Animal Welfare 8, 27-34.

Poultry Bulletin, 2002 Chicken market quarterly, January 2002, D. Taylor (Ed). National Farmers Union, Newmarket, Suffolk, p. 15.

Quinn, P.J, Markey, B.K., Carter, M.E., Donnelley, W.J. and Leonard, F.C., 2000. Veterinary microbiology and microbial diseases, Blackwell Science, Oxford, UK.

Radostits, O.M., Gay, C.C., Blood, D.C. and Hinchcliff, K.W., 2000. Veterinary medicine: a textbook of the diseases of cattle, sheep, pigs, goats and horses. 9th Edition. W.B. Saunders, London, UK.

Randall, J.M. and Bradshaw, R.H., 1998. Vehicle motion and motion sickness in pigs. Animal Science 66, 239-245.

Sällvik, K., Gebresenbet, G., Gustafsson, G., Jeppsson K-H., Lundgren, A-K., Rantzer D. and Svendsen, J., 2004. Transport of fattening pigs from farm to abattoir – Studies of the climate, pig behaviour and injuries, and vibrations in the truck; proposals for improvements. Swedish University of Agricultural Sciences, Dpt. of Agricultural Biosýstems and Technology, Report 1.

Scientific Committee on Animal Health and Animal Welfare (SCAHAW), 2002: The welfare of animals during transport (details for horses, pigs, sheep and cattle). Available at: http://ec.europa.eu/food/fs/sc/scah/out71_en.pdf. Accessed on 20 June 2009.

Schlüter, H. and Kramer, M., 2001. Epidemiologische Beispiele zur Seuchenausbreitung. Dtsch. Tierärztl. Wschr. 108, 338-343.

Schneider, V., 2000. Einfluss der Fang- und Transportbedingungen auf die Transportverluste bei Masthühnern. Influence of catching and transport conditions on transport losses of broiler chickens. PhD thesis. University of Veterinary Medicine Hannover, Foundation, Germany.

Schütte, A., Mergens, A., Pott, U. and Venthien, S., 1996. Effect of transport conditions (straw, stoppage) and unloading procedures on physiological and meat quality. Proceedings of a seminar 'New information on welfare and meat quality of pigs as related to handling, transport and lairage conditions'. Bundesforschungsanstalt für Landwirtschaft (FAL), Institut für Tierzucht und Tierverhalten, Mariensee 29-30 June 1996, pp. 117-132.

Steffens, C., 1999. Untersuchungen zur Belastung von Schlachtschweinen beim LKW-Transport und im Schlachttierstall. PhD thesis. University of Veterinary Medicine Hannover, Foundation, Germany.

Tarrant, P.V., 1989. The effects of handling, transport, slaughter and killing on meat quality and yield in pigs – a review. Irish Journal of Food Science and Technology 13, 79-107.

Thiry, E., Saliki, J., Bublot, M. and Pastoret, P.-P., 1987. Reactivation of infectious bovine rhinotracheitis virus by transport. Comparative Immunology, Microbiology & Infectious Diseases 10, 59-63.

Trunkfield, H.R., Broom, D.M., Maatje, K., Wierenga, H.K., Lambooy, E. and Kooijman, J., 1991. Effects of housing on responses of veal calves to handling and transport. In: Metz, J.H.M. and Groenestein, C.M. (Eds.). New trends in veal calf production, Pudoc, Wageningen, the Netherlands, pp. 40-43.

Van Putten, G., 1982. Handling of slaughter pigs prior to loading and during loading on a lorry. In: Moss, R. (Ed.). Transport of animals intended for breeding, production and slaughter. Current topics in veterinary medicine animal science 18. Martinus Nijhoff, The Hague, the Netherlands, pp. 15-25.

Van Putten, G. and Elshof, W.J., 1978. Observations on the effect of transport on the well being and lean quality of slaughter pigs. Animal Regulation Studies 1, 247-271.

Van Reenen, K., Reimert, H., Gerritzen, M., Leenstra, F. and Lambooij, B., 2008. Hazard identification and characterization of welfare aspects during transport of farm animals. Report nr. 152. Animal Sciences Group Lelystad, the Netherlands, pp. 1-26.

Warriss, P.D., Kestin, S.C., Brown, S.N., Knowles, T.G., Wilkins, L.J., Edwards, J.E., Austin, S.D. and Nicol, C.J., 1993. The depletion of glycogen stores and indices of dehydration in transported broilers. British Veterinary Journal 149, 391-398.

Wierup, M.,1994. Control and prevention of *Salmonella* in livestock farms. In: $16^{th}$ Conf. OIE Regional Commission for Europe, Stockholm (Sweden) 28 June-1 July 1994, pp. 249-269.

Wikner, I., Gebresenbet, G. and Nilsson, C. 2003. Air quality in a cattle transport vehicle in Scandinavian summer and winter conditions. Dtsch. tierärztl. Wschr. 110, 100-104.

# Stunning and slaughter methods

**Haluk Anil[1] and Bert Lambooij[2]**
[1]University of Bristol, Department of Clinical Veterinary Science, Langford BS40 5DU, United Kingdom; haluk_anil@hotmail.com
[2]Wageningen UR, Life stock Research, P.O. Box 65, 8200 AB Lelystad, the Netherlands

## Summary

Stunning and slaughter methods mark the end of animal production and have implications for animal welfare as well as product quality, operator safety and public health. Animal welfare at the time of slaughter must be maintained so that no unnecessary suffering occurs. Product quality can be influenced and these include undesirable defects such as haemorrhages, broken bones, pale soft exudative (PSE) and dark firm dry (DFD) meat, short shelf life and lead to downgrading. Therefore it is imperative to employ optimum methods during slaughter. Preslaughter handling and slaughter techniques used must also be developed to protect operator safety. Lastly, public safety for the consumers needs to be considered at all stages of slaughter and product preparation. Methods employed are usually regulated by national as well as international legislation and codes of practice and recommendations for different species, conditions and economic considerations exist. Different methods may be suitable for certain animals and can have disadvantages in certain situations. This chapter deals with aspects of conventional methods, their variations and effects. In addition, novel methods as well as religious slaughter are also considered.

**Keywords**: welfare, unconsciousness, exsanguination, captive bolt, electro-narcosis, $CO_2$, gases, halal, shechita

## 1. Introduction

According to the EU Council Directive (EC, 1993) on the protection of animals at the time of slaughter it is stated that horses, ruminants, pigs, rabbits and poultry brought into abattoirs for slaughter shall be (a) moved and if necessary lairaged, (b) restrained and (c) stunned before slaughter. Animals must be restrained in an appropriate manner, so as to spare them any avoidable pain, suffering, agitation, injury or contusions. Animals must not be suspended before stunning or killing. However, poultry and rabbits may be suspended for slaughter provided that appropriate measures are taken to ensure that they are in a sufficiently relaxed state for stunning. Permitted methods for stunning are (1) captive bolt pistol, (2) concussion, (3) electro-narcosis and (4) exposure to special gas mixtures.

## 2. Electrical stunning methods

Electrical methods are by far the most commonly used stunning and slaughter applications in meat animals (Anil *et al.*, 1997; Gregory and Wotton, 1985; Lambooij *et al.*, 1997) Rendering

post stun reflexes and unconsciousness (Bodnar, 1984; Cook, 1999; Tume and Shaw, 1992). Combining head-only stunning with exsanguination has a synergistic affect on the release of glutamate and aspartate, which increases the duration of unconsciousness (Cook, 1999; Cook *et al.*, 1995, 1996). Sticking following a stun should be carried out as promptly as possible when using head-only stunning as it takes time depending on the species before brain responsiveness is lost following sticking (Anil *et al.*, 2000a; Hoenderken, 1978). It is widely recognised that inducing a cardiac arrest at stunning has distinct welfare advantages: (1) it results in a rapid loss of brain function, (2) it ensures that the animal will not regain consciousness and (3) it does not depend on the operator performing an accurate stick (Anil, 1991; Cook and Devine, 2002; Gregory, 1994; Wotton *et al.*, 1992).

A major point that deserves particular attention is whether stunning actually renders the animal unconscious and insensible. Although this is usually believed to be the case, recent experimental findings cast some doubts upon this assumption (Lambooij, 2004).

## 3. Mechanical stunning methods

Mechanical methods, can be divided into penetrating and non-penetrating applications and principally operate by causing concussion. Mechanical methods that principally refer to captive bolt guns also include use of free bullet and rifles. In addition to conventional stunning cattle, guns are routinely used for emergency killing casualties, on-farm culling and for disease control. Captive bolt guns are usually required to be present as back up in case of failure of the main stunning equipment at abattoirs.

Missiles used for stunning and killing of animals include free bullet, a bolt, and water jet and air pressure. Immediately after stunning animals express a tonic spasm for approximately 10 s prior to relaxation, however, excessive convulsions often follow (Lambooij and Spanjaard, 1981). Immediately after shooting major changes (delta and theta waves tending to an isoelectric line) are seen on the EEG (electroencephalogram) and it is assumed that the animal is unconscious by analogy due to similar EEG changes described in man (Daly, 1987; Daly and Whittington, 1989; Daly *et al.*, 1986, 1987; Lambooij, 1982b; Lopes da Silva, 1983). In general, penetration of a missile into the brain can cause injury in the following three ways, depending on it 1986s velocity and shape: by laceration and crushing (<100 m/s), by shock waves [about 100 to 300 m/s] (Hopkinson and Marshal, 1967) and by temporary cavitation effect (>300 m/s). In fact, using the formula of $e = m \times v_2$, where e = energy, m = mass, and v = velocity, it has been shown that the delivered energy required for effective stunning is determined by the velocity of the missile (Crockard *et al.*, 1987; Daly *et al.*, 1987; Lambooij *et al.*, 2007). However, secondary tissue damage by penetration also prevents any chance of recovery.

### 3.1 Penetrating captive bolt stunning

The aim of captive bolt stunning methods is to cause concussion by transmitting the energy from the missile (bolt) into the cranium and brain. Captive bolt stunning is based on energy transfer via an agent (a cylindrical steel bolt) which retains mass and shape, and operates at low speed (<100 m/s). The velocity of a bolt of a captive bolt can be about 100 m/s in the air. This relative low velocity and shape of the bolt should crush the cortex and deeper parts of the brain and cause haemorrhages either by the bolt itself or by forward shock waves. Concussion and shearing forces also result in haemorrhages and lacerations. Captive bolt stunning is widely used for red meat farm animals. Cartridges filled with gunpowder, compressed air or springs under tension are used to drive bolts (missiles) against and through the skull of farm animals. The ideal shooting position is frontally on the head. 33 Captive bolt stunning is a relatively safe technique, but requires that deer is adequately restrained (DEFRA, undated). Positions for placing the captive bolt pistol are similar to those recommended for cattle, see Figures 2 and 3.

Figure 2. Position and placement of the captive bolt on the head of cattle.

Figure 3. Positions for placing the captive bolt pistol in deer (from Australian Best Practice Programme for Deer Farming, June 2001).

### 3.2 Non-penetrating captive bolt stunning

Cerebral concussion is generally agreed to be a traumatically induced derangement of the nervous system, resulting in an instantaneous diminution or loss of consciousness without gross anatomical changes in the brain (Ommaya and Gennarelli, 1974; Ommaya *et al.*, 1964, 1971). Irrespective of the type of force which produces the traumatic depolarisation of the cell membrane there is now evidence that powerful pressure waves are provoked within the cranial cavity by a blow on the head and that the frequency and force of the waves vary in different parts of the brain (Ommaya *et al.*, 1971). It has been suggested that it is not the pressure as such developed by these waves that is the important factor but the rapid oscillations in this pressure (Lambooij *et al.*, 1981). It should be noted that many investigators (EFSA) consider blood flow impairment as being primarily responsible for the electrical changes in the brain, although the immediate changes in the brain cannot be explained by this theory.

### 3.3 Free bullet and rifle

Killing by free bullets is applied on free (hunted) and farmed game, and occasionally in cases of emergency slaughter or euthanasia. Guidelines for effective killing of farmed deer by free bullets have been issued in many countries (e.g. UK: DEFRA, undated; HSE Information, 2004; Taylor 1996; Germany: Bavarian Research Center for Agriculture, undated). They basically address:

- appropriate bullet diameter and energy;
- placement of the shot (frontal head, high neck or thorax);
- precautions that bystanders are not injured and that remaining animals are not injured or unduly disturbed.

Free bullets have a lower mass than bolts of captive bolt stunners, and travel with higher velocity (typically >300 m/s for rifles). To improve energy transfer, bullets are constructed to fragment and/or deform when hitting the target. The energy of the bullet and the flatness of the bullet trajectory is mainly determined by its velocity (up to 1000 m/s), while the bullet mass (ca. 7-15 g for wild deer) has some relevance for the stability of the trajectory and for the mode of the energy transfer within the target tissues. Detailed requirements have been formulated by DEFRA (Table 1). The Bavarian State Research Center for Agriculture recommends bullets of min. 6.5 mm diameter with initial energy of 2,000 J (or 5.6 mm with 300 J for fallow deer <25 m; head shots).

For killing of wild deer, it is recommended that the animal is standing broadside and the target areas are the head (neurocranium)/neck or the cranial part of the thorax (see also Winkelmayer, 2009). Shots in the head will do no damage to the carcass, and are, thus preferred by professional hunters (Urquhart and McKendrick, 2006), both for wild as well as farmed game. While advantageous for 'meat-getters', and effecting immediate death of the animal, the head should be considered an ideal target only under optimum conditions (distance, weather). Its relatively small size, the fact that deer as well as other wild ruminants tend to move the head suddenly for various reasons (grazing, vigilance, etc.), and finally, that shooting distances for deer may be up to 200 m, make it a relatively small and motile target.

*Table 1. DEFRA recommendations on bullet diameter, shooting distance and target organ.*

| Location (target organ)/distance | <1 m | >1-20 m | 20 – 40 m | >40 m |
|---|---|---|---|---|
| Head (brain)[1] | yes, 0.22′ long rifle sufficient | yes, min. 0.243′ bullet (6.2 mm) diameter | | nr[2] |
| High neck (spinal cord) | ns[3] | yes, min. 0.243′ bullet (6.2 mm) diameter | yes, min. 0.243′ bullet (6.2 mm) diameter | only by proven marksmen |
| Thorax (heart-lung) | ns | yes, min. 0.243′ bullet (6.2 mm) diameter | yes, min. 0.243′ bullet (6.2 mm) diameter | yes, min. 0.243′ bullet (6.2 mm) diameter |

[1] Through brain; frontal or poll position, or halfway between eye and ear.

[2] nr, not recommended.

[3] ns, not explicitly specified, but possible for bullet diameter min. 0.243′.

Also, some wild ruminant species carry antlers which are highly valued trophies in some countries and should not be destroyed during killing of the animal. The other target area (thorax) is an area delineated by the caudal contour of the shoulder blade, the humerus and a line drawn from the elbow to the caudodorsal end of the shoulder blade. A hit in this area will cause destruction to the basis of the heart, and also rupture the large blood vessels. Also, lungs will be affected (Winkelmayer *et al.*, 2005). Depending on the exact point of impact, secondary lesions can be affected by bone fragments, etc. This target area also considers some possible deviations in the point of impact due to bad weather conditions (wind), sudden movement of the animal etc. Implications of killing method on the microbiological quality of game meat have been recently reviewed (Gill, 2007).

DEFRA (undated) recommends the following precautions to prevent injuries of people or animals: 'A safe backstop for the bullet is needed and care must be taken in shooting one deer not to injure others. Shooting from an elevated position such as a high seat or trailer is often helpful in these respects. Sensible precautions for public safety include shooting in the early morning when few people are around; walking the perimeter fence of small farms or paddocks on large farms to ensure all is clear; shooting away from roads, houses and gardens.' Also, DEFRA recommends that 'shooting should be undertaken, preferably by the regular stockmen, when deer are quiet, as will occur at a selected regular feeding site when they are being hand fed. Under such circumstances it may be possible to shoot 10 or more deer from a large group before the remainder become unduly disturbed. Factors to take into accounts in assessing the effect shooting will have on the rest of the herd and deciding the number to shoot include the size of the original group, stocking density and the amount of cover. Care should be taken not to leave too few since small numbers become unsettled and try to escape. There is also a risk of panic if too small a paddock is used'. Handling and use of firearms is subject to national legislation, which in most cases requires some familiarity in use of the guns and also should secure sufficient marksmanship.

### 3.4 Water jet and air pressure

High water jets developed for cutting and drilling in solid materials are available for use as stunners. Experiments to explore the suitability of water jets for stunning and killing purposes were conducted under laboratory conditions using post mortem materials (e.g. pig heads) and also on live slaughter pigs (Schatzmann *et al.*, 1990). Immediate unconsciousness as determined by EEG, was initiated by a rapid penetration of the skin and skull. In these studies destruction of the brain occurred within 0.2 to 0.4 s. The water jet, if employed, should be aimed frontally on the head and injected into the cranial cavity at the intersection of the imaginary lines from the ear to the opposite eye. A potential problem with water jets could be the excessive convulsions, that can appear after the use of this stunning method (Lambooij and Schatzmann, 1994).This is because – whenever an animal is decerebrated – convulsions (i.e. muscle contractions) of the carcass, caused by stimuli evoked in the medulla oblongata, mainly occur in the hind limbs.

Development of captive bolt stunning has been negated in most species due to lack of means to prevent post-stun convulsions. Recently, a captive needle stunning device for broilers has been developed, by which air pressure is injected into the brains and partly directed towards the spinal cord (Hillebrand *et al.*, 1996a,b). The latter extension is thought to prevent the convulsions. In broilers the air pressure stunning reduced post stun convulsions to less than 13% of the level of convulsions (Hillebrand *et al.*, 1996b). Additionally, a captive bolt stunning method for broilers has been modified so that air pressure is used to block post stun convulsions (Hillebrand *et al.*, 1996). In order to improve the method of practical application a commercial air tacker was modified (Lambooij *et al.*, 1999a). The plunger of the original design was replaced by two needles, which penetrate the skin and skull at an angle of 15o in caudal direction. Both needles were provided with small holes allowing air through in different directions. The stunning position was at the intersection of two imaginary lines drawn from the ear on one side to the inner corner of the eye on the other side. A trigger starts the injection of compressed atmospheric air when the needles penetrate the skull, and the duration of air injection was electronically controlled. The duration of injection as well as the air pressure was adjusted to a shooting pressure of 8 bars and an air injection of 3 bars for 1.5 s. It is hypothesized that the compressed atmospheric air administered through the needle in the captive pistol, placed more anteriorly on the animals' head, damages the higher brain regions to provide unconsciousness, while the other needle damages the upper spinal cord to prevent post stun convulsions (Lambooij *et al.*, 1999a,b). The captive needle pistol was adapted to guinea pigs, eels and cat fish regarding the length and shape of the needle (Lambooij *et al.*, 2002a). Only one cone shaped needle of 16 mm was used, which pressed the air in 3 directions radial 120°, where one direction was caudally towards the spinal cord. For correct positioning on the head an adapter was placed at the barrel of the pistol.

### 3.5 Neck dislocation and neck cutting

It is well known that in nature some predators use the method of cervical dislocation to mmobilize and kill their victims. In this method the head is turned in opposite direction to the body while stretching the neck and concomitantly turning and stretching, blood vessel are crushed and bleeding occurs. For neck dislocation in practice, the spinal cord is destroyed by thrusting a knife into the intervertebral space between the head and the 1$^{st}$ or 2$^{nd}$ vertebra. After dislocation or thrusting a knife a tonic cramp occurs resulting in paralysis after 5 to 10 s (Gregory and Wotton, 1990). Removal or inhibition of the contact between brain and spinal cord causes apnoea and loss of (pain) sensory perception from the body and spinal shock, with the exception of the face innervated by the 5$^{th}$ cranial nerve (Eichbaum *et al.*, 1975).

### 3.6 Decapitation and bleeding

In this method the head is separated from the body of the animal using a knife. An alternative method for animals with a long neck involves stretching by hand. And as a result bleeding occurs. A guillotine has also been developed for use in rats (Eichbaum *et al.*, 1975). By this method direct bleeding occurs by severing the neck including the main blood vessels.

## 4. Other physical methods

### 4.1 Heating

Since the end of the 19$^{th}$ century high frequency electric currents have been used to heat tissues. Long wave diathermy, using frequencies in the order 1 MHz required the use of electrodes which were in direct contact with the skin and consequently the risk of burning was high. Later frequencies known as short wave diathermy were introduced with the advantage that it was not necessary for the electrodes and the skin to be in contact being air between it (Lambooij *et al.*, 1990). In a reported procedure (Guy and Chou, 1982) heads of rats were irradiated with micro waves of 2,450 MHz for 1 s the temperature in the brain increased up to 75-90 °C within the next 1 s. Consequently it was shown the brain enzymes are inactivated very rapidly, that they can be used in neuro-chemical investigations. It was observed that an increase of about 10 °C in the brain resulted in a clinical state of unconsciousness using 2,450 MHz (6kW) for 1,5 to 2 sec. A change of 6.4 °C at a depth of 3 mm could cause a stunning effect using 915 MHz. After seizure the rats lay in an unconscious state for a period of 4 to 5 min.

### 4.2 Cooling down

The current pre-slaughter process used for fish consists of live chilling to immobilize them prior to evisceration. Assessment of live chilling revealed that this method is stressful as vigorous activity of the animals and irregular heart rates were observed (Lambooij *et al.*, 2002b). Responses to pain stimuli disappeared at a body temperature of approximately 8 to 10 °C, that occurred after 10 to 15 min, suggesting that consciousness is lost by this time. A patented alternative method of stunning and killing eels (Lambooij *et al.*, 2002b; Rorvik *et al.*,

2001) involves cooling them down gradually until death. According to the patent description the eels should ideally remain at least for 10 min in a medium with a temperature below -20 °C. A saturated brine solution at -15 °C may also be used. In addition, the eels should be stunned prior to killing by cooling down the body temperature to between 0 and 5 °C. Placing eels in brine at -18 °C is an effective method to kill the eels. However, it cannot be recommended to place conscious eels in cold brine water, because it takes more than 27 s before unconsciousness may be induced.

### 4.3 Fragmentation

Instantaneous fragmentation in a high-speed grinder may kill a small animal within a second. Grinders with rotating blades with a speed of 2,800 turns per min and a power of 4 KW may be useful for small birds (Kettlewell, 1986). For a correct result the grinder may not be overloaded. A capacity can not be recommended. However, it is argued that for welfare reasons may be animals should be unconscious first (e.g. chick placed in $CO_2$ atmosphere first).

### 4.4 Magnetic stimulation

All stunning methods have disadvantages relating to quality, public health as well as possible misstuns. There is a need for research to develop alternative, ideally non invasive, stunning methods (Knight and Anil, 2001). A non-invasive method that does not result in tissue damage before death could also be acceptable by Jewish and Muslim communities. In practice, an intense magnetic field is generated by passing a large current through a copper coil. The coil is positioned close to the head so that the brain lies within this magnetic field. Transcranial magnetic stimulation (TMS) has been used in humans for years. The technique also reliably initiates seizures in humans as an alternative to ECT for the treatment of depression (Lisbany, 2002). Bristol research has provided evidence for insensibility during the TMS application (Anil *et al.*, 2000b). Using similar technology, studies aimed at producing seizure activity and prolonged insensibility without a painful induction are being conducted using new equipment and special coils in animals. If fully developed, magnetic stimulation, a potential technique for stunning animals, can be used in future.

## 5. Gas stunning methods

### 5.1 Carbon dioxide

Because conventional electrical stunning methods can have adverse effects on carcass and meat quality gas stunning methods have been introduced and used in the last 3 decades. Carbon dioxide, the principle agent, and inert gases such as argon and nitrogen can be used and pigs and poultry are the chosen species. Carbon dioxide is also used for depopulation and disease eradication in poultry houses (Raj, 1996, 1997, 1999, 2003). Gasses used for euthanasia can be divided into narcotic and anoxic gasses. Narcotic gasses include for example ether, chloroform, halothane and methoxy-flurane. During induction by a narcotic gas excitation occurs. The severity of this excitation depends on the gas used (Abraini *et al.*, 1988). $CO_2$ in concentrations higher than 12% depresses the central nervous system directly

(Eisle *et al.*, 1967). It is well recognized that $CO_2$ is an anaesthetic gas which produces rapid unconsciousness when inhaled at high concentrations. However, signs of asphyxia and behavioural excitation were observed due to occurrence of both hypercapnia and hypoxia (Ernsting, 1963). Moreover, $CO_2$ is an acidic gas and has been found to be painful, causing unpleasant sensations on the nasal mucosa, lips and forehead in human subjects, when gas puffed stimuli were administered in concentrations 41 over 65% (Gregory *et al.*, 1990). The right hemisphere of the human SII cortex is dominantly involved in this response, which may suggest emotional/motivational aspects of trigeminal pain, and is in agreement with the role of the trigeminal pathways as a general warning system (Erlichman and Leiter, 1997). It has been shown that broilers can detect $CO_2$ in air at concentrations greater than 10% (Raj *et al.*, 2006a,b; Sandiland *et al.*, 2006). Increased head shaking and the elicitation of withdrawal from feeding at 55% $CO_2$ and above suggest that mixtures of $CO_2$ in air at above concentration may have an aversive effect. In rats a low concentration of $CO_2$ and addition of $O_2$ and use of humidified gases could ameliorate these negative effects. In the latter case almost no signs of asphyxia and excitation were observed. The main action of $CO_2$ is not its suffocating activity, but its anaesthetic activity. However, pigs have been shown to aversive to high concentrations of $CO_2$ (Raj, 1999).

## 5.2 Other gases and combinations

There is a wide range of anoxic gasses available such as carbon monoxide (CO), carbon dioxide ($CO_2$), HCN, dinitrogenoxide or laughing gas ($N_2O$), nitrogen ($N_2$) and argon (Ar). The most popularly used gases are CO and $CO_2$ or a combination of both. Narcotic gasses cause a depression of the Central Neural System with induction of unconsciousness and muscle flaccidity (Eisle *et al.*, 1967). An overdose leads to paralysis of the breathing muscles and the animal is killed by apnoea.

Presence of anoxic gasses reduces the flow of oxygen ($O_2$) to the tissues in one way or another. HCN prevents the flow of $O_2$ to the tissues and causes an extreme excitation phase (Kennedy *et al.*, 1992). CO has a stronger affinity for haemoglobin than $O_2$, hence reduces the $O_2$ concentration in the blood. The CO also causes a vascular dysfunction which in turn leads to haemorrhages by ruptures and diapedesis (Lambooij *et al.*, 1985). Other gasses, such as $N_2$, Ar or $CO_2$ work by replacing $O_2$ from the air to be inhaled A problem with O2 – replacing gasses is their lower efficacy in younger animals (Lambooij and Spanjaard, 1980). However, Ar can be easily administered in gas stunning, because it is heavier than air (as is $CO_2$), tasteless and odourless. Another option is a low concentration of $CO_2$ in Ar. Research has shown that both last mentioned gas mixtures caused a rapid loss of brain function in chickens, turkeys and pigs for stunning purposes before slaughter (Danneman *et al.*, 1997; Raj,1999; Raj and Gregory, 1990, 1994; Raj *et al.*, 1992a,b). However, anoxic mixtures induced vigorous wing flapping. EEG analysis suggested that anoxic wing flapping occurred during periods in which a form of consciousness could not be excluded (McKeegan and Smith, 2005; McKeegan *et al.*, 2003).

## 6. Religious slaughter

Religious slaughter of food animals is a controversial and often contentious issue that has received much attention in recent decades. Increases in Muslim populations in European countries, meat exports into the Middle and Far East have also intensified the debate and stimulated reviews of practices and legislation. Most religious slaughter in Europe and the Western countries, where allowed by law, is carried out either by the Jewish (Shechita) or the Muslim/Halal methods. Although legislation in most European countries requires preslaughter stunning, there can be exemptions for animals slaughtered by religious methods decided by individual countries. Several countries in Europe have banned slaughter without stunning (e.g. Sweden, Denmark, Switzerland). An EC funded project, DIALREL, is currently in progress collecting information and organising consultations on religious slaughter (http://www.dialrel.eu).

The Farm Animal Welfare Council in the UK published a report on religious slaughter methods in 2003 (FAWC, 2003) on the welfare of livestock when slaughtered by religious methods. This report recommended that UK government should repeal the exemption, however this was declined.

The following three questions have frequently been addressed and are still being debated:
1. The possibility of undue stress during handling prior to slaughter (Dunn 1990; Grandin, 1994; Grandin and Regenstein, 1994);
2. The possibility of the neck incision being painful during the cut and/or immediately afterwards (Gibson *et al.*, 2009a,b,c,d);

Whether sensibility is lost quickly enough following exsanguination ['sticking'] (Anil *et al.*, 1995; Daly *et al.*, 1988; Grandin and Regenstein, 1994; Kalweit *et al.*, 1989; Rosen, 2004). Regarding the stress of handling, as there are no specific religious requirements, the first question also applies to all other methods of slaughter. However, there are some traditional methods of handling animals before religious slaughter which are of concern. These practices include the use of a casting pen (no longer permitted in the UK), tying the legs of and blindfolding sheep, and hoisting cattle by one of the hind legs prior to slaughter. The above practices may be regarded as unduly stressful where inimized.

The second and third questions are related. Unfortunately, scientific method to attempting to measure 'pain' have limitations and not provided any 'proof' to answer the second question conclusively. However, irrefutable loss of sensibility can be assessed by measuring evoked responses and brain activity. Studies on this have reported early loss of sensibility (10-20 seconds) following incision although others have shown delays lasting up to 2 minutes. There is some evidence to suggest problems may occur in calves and adult cattle if the cut carotid ends develop clots restricting blood flow following slaughter (Anil *et al.*, 1995). The development of these occlusions is usually attributed to inadequate sharpness of the knife. However, this problem has also been reported following Shechita slaughter employing a razor-sharp knife. (Anil, personal observations; Levinger, 1976). The most recent series of publications by Gibson *et al.* have examined EEG patterns in calves following neck cutting [Gibson *et al.*, 2009a,b]. Their comparative analysis conclude that ventral neck cutting results in responses to noxious stimuli, in particular blood vessels are severed. These reports provide scientific evidence to suggest pain after slaughter without stunning and the debate continues. In contrast Rosen (2004) claim that Shechita cut is painless and no pain can be possible because brain function is lost immediately.

The potential problem of sensibility during and after neck cutting could be inimized by the use of preslaughter electrical stunning provided that this is correctly applied and is acceptable. In regard to Shechita stunning is precluded. In contrats, stunning that does not stop the heart before exsanguination for Halal slaughter is used in some countries in Europe and invariably employed in New Zealand and Australia for export to countries in the Middle and Far East with the approval of the appropriate religious authorities. However, religious Muslim slaughter rules are currently being examined in Dialrel project.

### 6.1 Halal slaughter

As far as Muslim slaughter is concerned there seems to be variations in the way slaughter is racticed. The variations are possibly due to differences in the interpretation of the Koran and the Hadith (the sayings of the prophet Mohammed). Pig meat is forbidden.

The act of slaughter (Al-Dhabh) is allowed in the name of God; therefore pronouncing the name of Allah is the usual practice. This is to remind the slaughterer that he is taking the life of a living creature. Animals are restrained but there are no specific religious regulations as to how this should be done. Following restraint, slaughter is achieved by severing both carotid arteries and jugular veins using a sharp knife. The usual type of incision is severance of the vessels in the retrograde fashion following an initial stab incision in the neck.

Exsanguination after Halal slaughter has been a source of concern in that in some cases occlusions can impede bleed out rate and delay loss of consciousness (Anil *et al.*, 1995). Another claim was that stunning methods could impede blood loss during halal slaughter. Comparative studies in sheep and cattle have shown, however, that there is no significant difference between stunned and non-stunned sheep (Anil *et al.*, 2004) and cattle (Anil *et al.*, 2006).

### 6.2 Shechita

Jews consume beef, veal, mutton, lamb and poultry, but not pork. These meats must be slaughtered and prepared in accordance with the rabbinical laws (Levinger, 1976).

Slaughter may only be carried out by an approved and trained slaughterman of the Jewish faith, called a Shocet. He must be a person of recognised high moral character and consistent religious practice, often a Rabbi. He must obtain a licence from the Rabbinical Commission in addition to the usual local authority licence.

The Jewish method of slaughter, Shechita, is preceded by positioning the animal, though this is not subject to regulation by the religious authorities. Sheep are placed on their backs in a cradle; cattle are placed in a restraint apparatus in an upright position and their neck is extended by a mechanically operated 'chin lift' or turmed over in a rotary pen.

The slaughter is carried out by a Shocet. A single, transverse cut is made across the neck using a very sharp, special knife (chalaf). The knife has to be examined for its sharpness between each cut. It is usually 16 inches long for cattle. In the UK, after the neck cut, captive bolt stunning is used at some abattoirs.

Once an animal is dead, an incision is made through the abdominal wall and a Jewish Inspector feels at arm's length into the thorax to check for pleural adhesions or any other signs of abnormality. If any abnormality is found, the entire carcass is rejected for Jewish consumption on the ground that the animal was not healthy at the time of slaughter.

## 7. Effects on product quality and welfare

Various stunning methods and electrical parameters have been reported to have a different effect on *post-mortem* rigor development (Bilgili, 1992, 1999; Devine *et al.*, 1984; Hillebrand *et al.*, 1996a,b; Roth *et al.*, 2002). The *post-mortem* metabolism is largely a consequence of indirect stimulation through nervous pathways. Broken vertebrae can occur when stunned with head-to-back electrode positioning if the voltage and the current is too high (Troeger and Woltersdorf, 1990; Wotton *et al.*, 1992). Sinusoidal alternating currents of 50 Hz have a large stimulatory effect on skeletal muscles which can be reduced by increasing the current frequency to an extent that prevents the occurrence of broken backs. The prevalence of broken vertebrae and pelvises could be reduced to zero by increasing the frequency from 50 to 1,500 Hz. The drawback of this approach is that the effect on fibrillating the heart is also reduced (Anil and McKinstry, 1991; Wotton *et al.*, 1992) Haemorrhages can be induced by stunning and killing, however, the underlying mechanism is considered to be multi-factorial (Kranen *et al.*, 2000a; Troeger and Woltersdorf, 1990). Investigations revealed that the morphology of haemorrhages was dependent on the tissue in which they occurred. In the pectoral muscles extravasating blood was found to follow the direction of the muscle fibres. In fat tissue, the majority of haemorrhages had a petechial appearance. More diffuse haemorrhages were found in loose connective tissue (Hillebrand *et al.*, 1996a,b; Kranen *et al.*, 2000b). The histological study of haemorrhages in different types of muscles showed that the morphological appearance of the blood extravasation is determined by the structure of the tissue as well as by the amount of blood leaving the circulation. Some haemorrhages were associated with hyper contracted and disrupted muscle fibres, indicating that they were caused by severe muscular strain. Many haemorrhages were found near venules or veins, and were packed with erythrocytes, surrounded by intact adipocytes and connective tissue. Rupture was observed only in venous structures, such as post-capillary venules and small veins, not in arterial vessels. This strongly indicates that a local rise in venous blood pressure can cause rupture of venules and small veins (Kranen *et al.*, 2000a).

## 8. Stunning methods and public health implications

Although, the focus of this review is on animal welfare, in some instances, public health measures and concerns, especially as a result of the BSE threat, have inevitable welfare consequences too. To this end, a detailed EFSA opinion on stunning methods and public health implications has been prepared (EFSA, 2004a,b). Potential public health concerns from TSE infected animals have been considered and reviewed (Anil and Austin, 2001; Knight and Anil, 2003). In cattle stunned with a penetrating captive bolt (PCB) method a frequency of 4%, CNS embolism in jugular blood of with a 95% has been reported (Coore *et al.*, 2004, 2005). In sheep, higher frequencies (23% and 14% respectively for cartridge activated and pneumatically activated guns) of CNS embolism in jugular blood have been reported (Anil and Harbour 2001; Coore *et al.*, 2004). As the heart continues pumping for several minutes between the stunning and the end of exsanguinations, some of the embolic CNS material dislodged by the penetrating captive bolt gun might enter venous blood vessels draining the head and consequently be disseminated to other organs/tissues. This can happen not only with use of a penetrating gun that injects air into the brain (Anil *et al.*, 1999; Schmidt *et al.*, 1999) but also when stunning is performed without air injection (Anil *et al.*, 2002; Coore *et al.*, 2004, 2005) In NPCB stunned cattle, CNS material was detected in jugular blood of 2% animals (Coore *et al.*, 2004, 2005). In addition to haematogenous contamination of edible tissues with CNS material, other public health concerns may also be associated with PCB methods. For example, cross- or airborne contamination of the stunning gun operator, the environment such as the stun-box and/or the animals consecutively stunned with the same gun could occur, based on studies using experimental contamination with marker bacteria (Daly *et al.*, 2001; Prendergast *et al.*, 2004).

## 9. Conclusions

### 9.1 What has been achieved?

In the last 3 decades there has been much advance made in examining stunning and slaughter methods, identifying welfare and quality problems as well as addressing them and finding solutions (EFSA, 2004a,b, 2006). In comparison to earlier survey results showing shortcomings more recent practices indicate improvements in animal welfare due to better understanding, training, inspection and improved application of methods.

Each stunning technique referred to is assumed to be capable of being applied correctly under normal circumstances. Optimum application parameters for each species such as tong and gun positions, voltage and amperage levels, cartridge strength/size (as colour codes), gas concentrations and duration of application exist in various documents. These include national legislative documents, good practice guidelines and codes of practice as well as manufacturers' recommendations contained in manuals.

### 9.2 What has been neglected?

The insight into the stunning process that has come from neuro-physiological studies is of important significance. Assessment of more parameters than general epileptiform insult and analgesia may support the humaneness of the stunning and killing system. Most studies investigating the mechanism of concussion were performed with laboratory animals (i.e. rats, cats, monkeys). It is evident from these investigations that concussion does not always cause loss of consciousness. In man retrograde amnesia that often occur after the blow thereby making interpretation of the effects less certain.

In general, exposure of animals to gas mixtures and inhalation anaesthetics, unlike some other methods of euthanasia, does not produce immediate loss of consciousness in animals. Therefore, it is important to focus on further research for gas mixtures that are non-aversive and do not induce distress or pain prior to loss of consciousness. Others include problems relating to religious slaughter methods, insufficient support for developing improved methods to protect public health (e.g. BSE outbreak) and novel methods.

### 9.3 What needs to be done?

All stunning methods have disadvantages relating to quality, public health as well as possible misstuns. There is a need for research to develop alternative, ideally non-invasive, stunning methods. A non-invasive method that does not result in tissue damage before death could also be acceptable by Jewish and Muslim communities.

## References

Abraini, J.H., Rostain, J.C. and Kreim, B., 1998. Sigmoidal compression retaedependence of inert gas narcotic potency in rats: implication for lipid vs protein theories of iner gas action in the central nervous system. Brain Research, 808: 300-304.

Anil, M.H., 1991. Studies on the return of physical reflexes in pigs following electrical stunning. Meat Science 30, 13-21.

Anil, M.H. and Austin, A., 2003. Bovine spongiform ecephalopathy: a review of some factors that influence meat safety. Agrippa. Available at: http://www.fao.org//DOCREP/ARTICLE/AGRIPPA/590_en.htm#P11_1213. Accessed 19 December, 2008.

Anil, M.H. and Harbour, D., 2001. Current stunning and slaughter methods in cattle and sheep: potential for carcass contamination with central nervous tissue and microorganisms. Fleischwirtschaft International 3, 26-27.

Anil, M.H. and McKinstry, J.L., 1991. Reflexes and loss of sensibility following Head-to-back electrical stunning in sheep. Veterinary Record 128, 106-107.

Anil, M.H. and McKinstry, J.L., 1992. The effectiveness of high frequency electrical stunning of pigs. Meat Science 31, 481-491.

Anil, M.H. and McKinstry, J.L., 1998. Variations in electrical stunning tong placements and relative consequences in slaughter pigs. The Veterinary Journal 155, 85-90.

Anil, M.H., McKinstry, J.L., Gregory, N.G., Wotton, S.B. and Symonds, H., 1995. Welfare of calves. 2. Increase in vertebral artery blood flow following exsanguination by neck sticking and evaluation of chest sticking as an alternative slaughter method. Meat Science 41, 113-123.

Anil, M.H., McKinstry, J.L. and Wotton, S.B., 1997. Electrical stunning and slaughter of pigs. Guidelines for good welfare assurance. Fleischwirtschaft 77, 632-635.

Anil, M.H., Raj, A.B. and McKinstry, J.L., 1998. Electrical stunning in commercial rabbits: effective currents, spontaneous physical activity and reflex behaviour. Meat Science 48, 21-28.

Anil, M.H., Love, S., Williams, S., Shand, A., McKinstry, I.L., Helps, C.R., Waterman- Pearson, A., Seghatchian, J. and Harbour, D., 1999. Potential contamination of beef carcasses with brain tissue at slaughter. Veterinary Record 145, 460-462.

Anil, M.H., Whittington, P.E. and Mckinstry, J.L. 2000a. The effect of sticking method on the welfare of slaughter pigs. Meat Science 55, 315-319.

Anil, M.H., Butler, S.R., Johnson, C.B. and McKinstry, J.L., 2000b. Suppression of somatosensory evoked potentials by transcranial magnetic stimulation in the rabbit. Proc. Physiological Society, 526 p.

Anil, M.H., Love, S., Helps, C.R. and Harbour, D., 2002. Potential for carcass contamination with brain tissue following stunning and slaughter in cattle and sheep. Food Control 13, 431-436.

Anil, M.H., Yesildere, T., Aksu, H., Matur, E., McKinstry, J.L., Erdogan, O., Hughes, S. and Mason, C., 2004. Comparison of religious slaughter of sheep with methods that include preslaughter stunning and the lack of differences in exsanguination, packed cell volume and quality parameters. Animal Welfare 13, 387-392.

Anil, M.H., Yesildere, T., Aksu, H., Matur, E., McKinstry, J.L., Erdogan, O., Hughes, S. and Mason, C., 2006. Comparison of Halal slaughter with captive bolt stunning and neck cutting in cattle: exsanguination and quality parameters. Animal Welfare 15, 325-330.

Bavarian State Research Center for Agriculture (undated): 'Landwirtschaftliche Wildhaltung – Ein Leitfaden' 12/004; ISSN 1611-4159: Available at: http://www.lflneu.bayern.de/ publikationen/daten/schriftenreihe_url_1_21.pdf. Accessed 19 December, 2008.

Bilgili, S.F., 1992. Electrical stunning of broilers – basic concepts and carcass quality implications: a review. Journal of Applied Poultry Science 1, 135-146.

Bilgili, S.F., 1999. Recent advances in electrical stunning. Poultry Science 78, 282-286.

Bodnar, R.J. 1984. Types of stress which induce analgesia. In: Tricklebark, M.D. and Curzon, G. (Eds.) Stress-induced analgesia, Chichester: John Wiley & Sons. 19-32.

Cook, C.J., 1999. Neurological measures to quantify welfare aspects of stunning. In: Proceedings of the International Workshop on Stunning Systems for Pigs and Animal Welfare, held during 25-27 August 1999, Billund, Denmark.

Cook, C.J. and Devine, C.E., 2002. Electrical stunning of cattle: aspects of animal welfare and meat quality. WBC, Hannover, Germany.

Cook, C.J., Devine, C.E., Gilbert, K.V., Tavener, A. and Day, A.M., 1991. Electroencephalograms and electrocardiograms in young bulls following upper cervical vertebrae-to-brisket stunning. New Zealand Veterinary Journal 39, 121-125.

Cook, C.J., Devine, C.E., Tavener, A. and Gilbert, K.V., 1992. Contribution of amino acid transmitters to epileptiform activity and reflex suppression in electrically head stunned sheep. Research in Veterinary Science 52, 48-56.

Cook, C.J., Devine, C.E., Gilbert, K.V., Smith, D.D. and Maasland, S.A., 1995. The effect of electrical head-only stun duration on electroencephalographic-measured seizure and brain amino acid neurotransmitter release. Meat Science 40, 137-147.

Cook, C.J., Maasland, S.A., Devine, C.E., Gilbert, K.V. and Blackmore, D.K., 1996. Changes in the release of amino acid neurotransmitters in the brains of calves and sheep after headonly electrical stunning and throatcutting. Research in Veterinary Science 60, 225-261.

Coore, R.R., Love, S., McKinstry, J.L., Weaver, H.R., Phillips, A., Hillman, T., Hiles, M.J., Shand, A., Helps, C.R. and Anil, M.H., 2004. Research note: Dissemination of brain emboli following captive bolt stunning of sheep: Capacity for entry into the systemic arterial system. Journal of Food Protection 67, 1050-1052.

Coore, R.R., Love, S., McKinstry, J.L., Weaver, H.R., Philips, A., Hillman, T., Hiles, M., Helps, C.R. and Anil, M.H., 2005. Brain tissue fragments in jugular-vein blood of cattle stunned by use of penetrating or non-penetrating captive bolt guns. Journal of Food Protection 68, 882-884.

Crockard, H.A., Brown, F.D., Johns, L.M. and Mullan, S., 1977. An experimental cerebral missile injury model in primates. Journal of Neurosurgery 46, 776-783.

Daly, C.C., 1987. Recent developments in captive bolt stunning. In: Humane slaughter of animals for food. Universities Federation for Animal Welfare, 15-19.

Daly, C.C. and Whittington, P.E., 1989. Investigation into the principal determinants of effective captive bolt stunning sheep. Research in Veterinary Science 46, 406-408.

Daly, C.C., Gregory, N.G., Wotton, S.B. and Whittington, P.E., 1986. Concussive methods for preslaughter stunning in sheep: assessment of brain function using cortical evoked responses. Research in Veterinary-Science 41, 349-352.

Daly, C.C., Gregory, N.G. and Wotton, S.B., 1987. Captive bolt stunning of cattle: effects on brain function and role of bolt velocity. British Veterinary Journal 143, 574-580.

Daly, C.C., Kallweit, E. and Ellendorf, F., 1988. Cortical function in cattle during slaughter: conventional captive bolt stunning followed by exsanguination compared with shechita slaughter. The Veterinary Record 122, 325-329.

Daly, D.J., Prendergast, D.M., Sheridan, J.J., Blair, I.S. and McDowell, D.A., 2001. Use of a marker organism to model the spread of central nervous system tissue in cattle and the abattoir environment during commercial stunning and carcass dressing. Applied and Environmental Microbiology 68, 791-798.

Danneman, P.J., Stein, S. and Walshaw, S.O., 1997. Humane and practical implications of using carbon dioxide mixed with oxygen for anesthesia or euthanasia of rats. Laboratory Animal Science 47, 376-385.

DEFRA. Available at: http://www.defra.gov.uk/animalh/welfare/farmed/othersps/deer/pb0055/deercode.htm#13.

Devine, C.E., Ellery, S., Wade, L. and Chrystall, B.B., 1984. Differential effects of electrical stunning on the early *post mortem* glycolysis in sheep. Meat Science 11, 301-309.

Dunn, CS., 1990. Stress reactions of cattle undergoing ritual slaughter using two methods of restraint. Veterinary Record 26, 522-525.

Eichbaum, F.W., Slewer, O. and Yasaka, W.J., 1975. Postdecapitation convulsions and their inhibition by drugs. Experimental Neurology 49, 802-812.

Erlichman, J.S. and Leiter, J.C., 1997. Comparative aspects of central $CO_2$ $CO_2$ chemoreception. Respiration Physiology 110, 177-185.

Eisle, J.H., Eger, E.I. and Muallem, M., 1967. Narcotic properties of carbon dioxide in the dog. Aneasthesiology 28, 856-865.

European Community (EC), 1993. Directive 93/119/EC on the protection of animals at the time of slaughter or killing. Official Journal L 340, 21-34.

European Food Safety Authority (EFSA), 2004a. Welfare aspects of animal stunning and killing methods. Scientific Report of the Scientific Panel for Animal Health and Welfare on a request from the Commission related to welfare aspects of animal stunning and killing methods (Question N° EFSA-Q-2003-093). Report AHAW/04 027, 241 pp. Available at: http://www.efsa.europa.eu/en/science/ahaw/ahaw_opinions/495.html. Accessed 19 December 2008.

European Food Safety Authority (EFSA), 2004b. Welfare aspects of animal stunning and killing methods. Available at: http://www.efsa.eu.int/science/ahaw/ahaw_opinions/495_en.html. Accessed 19 December 2008.

European Food Safety Authority (EFSA), 2006. The welfare aspects of the main systems of stunning and killing applied to commercially farmed deer, goats, rabbits, ostriches, ducks, geese and quail (Question N° EFSA-Q-2005-005). Annex to the EFSA Journal (2006) 326, 1-18. Report, 71 pp.

FAWC, 2003. Report on the welfare of farmed animals at slaughter or killing. Part 1. Red Meat Animals. Farm Animal Welfare Council, London, UK.

Gibson, T.J., Johnson, C.B., Murrel, J.C., Hull, C.M., Mitchinson, S.L., Stafford, K.J., Johnstone, A.C. and Mellor, D.J., 2009a. Electroencephalographic responses of halothane-anaesthetised calves to slaughter by ventral-neck incision without prior stunning. New Zealand Veterinary Journal 57(2), 77-83.

Gibson, T.J., Johnson C.B., Murrell, J.C., Chambers, J.P., Stafford, K.J. and Mellor, D.J., 2009b. Components of electroencephalographic responses to slaughter in halothane-anaesthetised calves: Effects of cutting neck tissues compared with major blood vessels. New Zealand Veterinary Journal 57(2), 84-89.

Gibson, T.J., Johnson, C.B., Murrel, J.C., Hull, C.M., Mitchinson, S.L., Stafford, K.J., Johnstone, A.C. and Mellor, D.J., 2009c. Electroencephalographic responses to concussive non-penetrative captive-bolt stunning in halothane-anaesthetised calves. New Zealand Veterinary Journal 57(2), 90-95.

Gibson, T.J., Johnson, C.B., Murrel, J.C., Mitchinson, S.L., Stafford, K.J., Johnstone, A.C. and Mellor, D.J., 2009d. Amelioration of electroencephalographic responses to slaughter by non-penetrative captive-bolt stunning after ventral-neck incision in halothane-anaesthetised calves. New Zealand Veterinary Journal 57(2), 96-101.

Gill, C.O., 2007. Microbiological conditions of meats from large game animals and birds. Meat Science 77, 149-160.

Grandin, T. and Regenstein, J.M., 1994. Religious slaughter and animal welfare: a discussion for meat scientists. Meat Focus International, March 1994: 115-123. Available at: http://www.grandin.com/ritual/kosher.slaugh.html. Accessed 19 December 2008.

Gregory, N.G., 1994. Preslaughter handling, stunning and slaughter. Meat Science 36, 45-56.

Gregory, N.B. and Wotton, S.B., 1985. Sheep slaughtering procedures IV. Responsiveness of the brain following electrical stunning. British Veterinary Journal 141, 74-81.

Gregory, N.G. and Wotton, S.B., 1990. Comparison of ne ck dislocation and percussion of the hand on visual evoked responses in the chicken's brain. Veterinary Record 126, 370-572.

Gregory, N.G., Raj, A.B.M., Audsley, A.R.S. and Daly, C.C., 1990. Effects of carbon dioxide on man. In: The use of carbon dioxide for the stunning of slaughter pigs. Reportof a meeting of experts held in Heeze from the 26-27th January 1990. Flieschwirtschaft 70, 1173-1174.

Gregory, N.G., Anil, M.H., McKinstry, J.L. and Daly, C.C. 1996. Prevalence and duration of insensibility following electrical stunning in calves. New Zealand Veterinary Journal 44, 1-3.

Guy, A.S.W. and Chou, C.K., 1982. Effects of high intensity micro wave pulse exposure of rat brain. Radio Science 17, 169S-178S.

Hillebrand, S.J.W., Lambooy, E. and Veerkamp, C.H., 1996a. The effects of alternative electrical and mechanical stunning methods on hemorrhaging and meat quality of broiler breast and thigh muscles. Poultry Science 75, 664-671.

Hillebrand, S.J.W., Pieterse, C. and Lambooy, E., 1996b. Air pressure stunning of broilers – changes in the spontaneous electroencephalogram, immobilisation and haemorrhaging aspects. In: Proceedings of the 42nd International Congress of Meat Science and Technology, Lillehammer, Norway, pp. 456-457.

Hoenderken, R, 1978. Elektrische bedwelming van slachtvarkens. Ph.D. thesis, State University of Utrecht, Utrecht, the Netherlands.

Hopkinson, D.A.W. and Marshall, T.K., 1967. Firearm injuries. British Journal for Surgery 54, 344-353.

HSE Information Sheet: Deer Farming, AIS7, revised, 2004) available at: http://www.bdfa.co.uk/deerfarming/JApXvCVdN.pdf. http://www.hsa.org.uk/Resources/Publications/Technical%20Notes/TN25.pdf.

Kalweit, E., Ellendorf, F., Daly, C. and Smidt, D., 1989. 1 Physiological reactions during slaughter of cattle and sheep with and without stunning. Deutsch. Tierärztl. Wochenschr. 96, 89-92.

Kennedy, R.R., Stokes, J.W. and Downing, P., 1992. Anaesthesia and inert gases with special reference to xenon. Anaesthesia and Intensive Care 20, 66-70.

Kestin, S.C., 2002. Protocol for assessing brain function in fish and the effectiveness of methods used to stun and kill them. Veterinary Record 150, 302-307.

Kettlewell, P.J., 1986. Engineering aspects of humane killing of poultry. In: Cotract Report (no. CR/173/86/8333) of the National Institute of Agricultural Engineering, Wrest Park, Silsoe, Bedford, MK45 4HZ, United Kingdom.

Knight, A.C. and Anil., M.H., 2003. EC Accompanying Measure QLK2-CT-2002 30531, UK.

Kranen, RW., Lambooy, E., Veerkamp, C.H., van Kuppevelt, T.H. and Veerkamp, J.H., 2000a. Histological characterization of hemorrhages in muscles of broiler chickens. Poultry Science 79, 110-116.

Kranen, R.W., Lambooij, E., Veerkamp, C.H., van Kuppevelt, T.H. and Veerkamp, J.H., 2000b. Haemorrhages in muscles of broiler chickens. World's Poultry Sci. J. 56, 94-126.

Lambooij, E., 1982a. Electrical stunning of sheep. Meat Science 6, 123-135.

Lambooy, E., 1982b. Some aspects of the effectiveness of stunning in sheep by captive bolt. Meat Science 7, 51-57.

Lambooij, E., 2004. Electrical stunning. Encyclopedia of meat sciences. In: Jensen, W.K., Devine, C. and Dikeman, M. (Eds.). Elsevier, Oxford, UK, pp. 1342-1348.

Lambooij, E. and Schatzmann, U., 1994. The use of high pressure waterjet combined with electroimmobilization for the stunning of slaughter pigs: some aspects of meat quality. Meat Science 37, 381-389.

Lambooij, E. and Spanjaard, W., 1980. Euthanasia of young pigs with carbon monoxide. Veterinary Record 107, 59-61.

Lambooij, E. and Spanjaard, W.J., 1981. Effect of the shooting position on the stunning of veal calves by captive bolt. The Veterinary Record 109, 359-361.

Lambooij, E., Spanjaard, W.J. and Eikelenboom, G., 1981. Concussion stunning of veal calves. Fleischwirtschaft 61, 98-100.

Lambooy, E., Roelofs, J.A. and Van Voorst, N., 1985. Euthanasia of mink with carbon monoxide. Veterinary Record 116, 416.

Lambooy, E., Lagendijk, J.J.W. and van Rhoon, G.C., 1990. Feasibility of stunning slaughter pigs with microwaves at 434 MHz. Fleischwirtschaft International 2, 3-5.

Lambooij, B., Merkus, S.M., Van Voorst, N. and Pieterse, C., 1997. Effect of low voltage with a high frequency electrical stunning on unconsciousness in slaughter pigs. Fleischwirtschaft International 2, 13-14.

Lambooij, E., Pieterse, C., Hillebrand, S.J.W. and Dijksterhuis, G.B., 1999a. The effects of captive bolt and electrical stunning, and restraining methods on broiler meat quality. Poultry Science 78, 600-607.

Lambooij, E., Pieterse, C., Potgieter, C.M., Snyman, J.D. and Nortjé, G.L., 1999b. Some neural and behavioural aspects of electrical and mechanical stunning in ostriches. Meat Science 52, 339-345.

Lambooij, E., van de Vis, J.W., Kloosterboer, R.J. and Pieterse, C., 2002a. Evaluation of captive needle stunning of farmed eel (*Anguilla anguilla, L.*): suitability for humane slaughter. Aquaculture 212, 141-148.

Lambooij, E., van de Vis, J.W., Kloosterboer, R.J. and Pieterse, C., 2002b. Welfare aspects of live chilling and freezing of farmed eel (*Anguilla anguilla, L.*): neurological and behavioural assessment. Aquaculture 210, 159-169.

Lambooij, E., van de Vis, J.W., Kuhlmann, H., Münkner, W., Oehlenschläger, J., Kloosterboer, R.J. and Pieterse, C., 2002c. A feasible method for humane slaughter of eel (*Anguilla anguilla, L.*); electrical stunning in fresh water prior to gutting. Aquaculture Research 33, 643-652.

Lambooij, E., Pilarczyk, M., Bialowas, H., Van den Boogaart, J.G.M., and Van de Vis, J.W., 2007. Electrical and percussive stunning of the common carp (*Cyprinus carpio, L.*): neurological and bahavioural assessment. Agricultural Engineering 37, 171-179.

Levinger, I.M., 1976. Physiological and general medical aspects of Shechita. In: Munk, M.L. and Munk, E. (Eds.). Jerusalem Gur Aryeh Publications, Jerusalem, Israel, pp. 147-149.

Lisanby, S.H., 2002. Update on magnetic seizure therapy: a novel form of convulsive therapy. Journal of ECT 18(4), 182-188.

Lopes da Silva, H.F., 1983. The assessment of consciousness: General principles and practical aspects. In: Eikelenboom, G. (Ed.). Stunning of animals for slaughter, Martinus Nijhoff, The Hague, the Netherlands, pp. 3-12.

McKeegan, D.E.F. and Smith, F.S., 2005. Behavioural correlation of olfactory and trigeminal gaseous stimulation in chickens, *Gallus domesticus*. Physiology and Behaviour 84, 761-768.

McKeegan, D.E.F., Demmers, T.G.M., Wathers, C.M. and Jones, R.B., 2003. Chemosensitivity responses to gaseous pollutants and carbon dioxide: implications for poultry welfare. Poultry Science 82 (supplement 1), 16.

Ommaya, A.K. and Gennarelli, T.A., 1974. Cerebral concussion and traumatic uncousciousness: correlation of the experimental and clinical observations of blunt head injuries. Brain 97, 633-654.

Ommaya, A.K., Rockoff, SD. and Baldwin, M., 1964. Experimental concussion: a first report. Journal of Neurosurgery 21, 249-264.

Ommaya, A.K., Grub, R.L. and Naumann, R.A., 1971. Coup and contre-coup injury: observations on the mechanics of visible brain injuries in the rhesus monkey. Journal of Neurosurgery 35, 503-516.

Prendergast, D.M., Sheridan, J.J., Daly, D.J., McDowell, D.A. and Blair, I.S., 2004. The use of a marked strain of Pseudomonas fluorescens to model the spread of brain tissue to the musculature of cattle after shooting with a captive bolt gun. Journal of Applied Microbiology 96, 437-446.

Raj, A.B.M., 1996. Aversive reactions of turkeys to argon, carbon dioxide, and a mixture of carbon dioxide and argon. The Veterinary Record 138, 592-593.

Raj, A.B.M., 1997. Novel on-farm killing system. Poultry International, August issue, 48-49.

Raj, A.B.M., 1999. Behaviour of pigs exposed to mixtures of gases and the time required to stun and kill them: welfare implications. Veterinary Record 144, 165-168.

Raj, A.B.M., 2003. A critical appraisal of electrical stunning in chickens. World's Poultry Science Journal 59, 89-98.

Raj, A.B.M. and Gregory, N.G., 1990. Investigation into the batch stunning/killing of chickens using carbon dioxide or argon-induced hypoxia. Research in Veterinary Science 49, 364-366.

Raj, A.B.M. and Gregory, N.G., 1994. An evaluation of humane gas stunning methods for turkeys. The Veterinary Record 135, 222-223.

Raj, AB.M. and O'Callaghan, M., 2004a. Effects of amount and frequency of head-oly stunning currents on the electroencephalogram and somatosensory evoked potentials in broilers. Animal Welfare 13, 159-170.

Raj, A.B.M. and O'Callaghan, M., 2004b. Effects of electrical water bath stunning current frequencies on the spontaneous electroencephalogram and somatosensory evoked potentials in hens. British Poultry Science 45, 230-236.

Raj, A.B.M., Wotton, S.B. and Gregory, N.G., 1992a. Changes in the somatosensory evoked potentials and spontaneous electroencephalogram of hens during stunning with a carbon dioxide and argon mixture. British Veterinary Journal 148, 147-156.

Raj, A.B.M., Wotton, S.B. and Whittington, P.E., 1992b. Changes in the spontaneous and evoked electrical activity in the brain of hens during stunning with 30% carbon dioxide in argon with 5% residual oxygen. Research in Veterinary Science 53, 126-129.

Raj, A.B.M., Wotton, S.B., McKinstry, J.L. Hillebrand, S.J.W. and Pieterse, C., 1998a. Changes in the somatosensory evoked potentials and spontaneous electroencephalogram of broiler chickens during exposure to gas mixtures. British Poultry Science 39, 686-695.

Raj, A.B.M., Richardson, R.I., Wilkins, L.J. and Wotton, S.B., 1998b. Carcass and meat quality in ducks killed with either gas mixtures or an electric current and processed under commercial conditions. British Poultry Science 39, 404-407.

Raj, A.B.M., Sandilands, V. and Sparks, N.H.C., 2006a. Review of gaseous methods of killing poultry on farm for disease control purposes. The Veterinary Record 159, 229-235.

Raj, A.B.M., Smith, C. and Hickman, G., 2006b. Novel method of killing poultry in houses with dry foam created using nitrogen. EU Workshop, Brussels. American Veterinary Medical Association (AVMA) 2000. Report of the AVMA Panel on Euthanasia. Internet: http://www.avma.org/issues/animal_welfare/euthanasia.pdf. Accessed 19 December 2008.

Rorvik, K.A., Skervol,d P.O., Fjaera, S.O., Morkore, T., and Steien S.H., 2001. Body temperature and seawater adaptation in farmed Atlantic salmon and rainbow trout during prolonged chilling. Journal of Fisheries Biology 59, 330-337.

Rosen, S.D., 2004. Physiological insights into Shechita. The Veterinary Record 154, 759-765.

Roth, B., Moeller, D., Veland, J.O., Imsland, A. and Slinde, E., 2002. The effect of stunning methods on rigor mortis and texture properties of Atlantic salmon (*Salmo salar*). Journal of Food Science 67, 1462-1466.

Sandilands, V., Raj, A.B.M., Baker, L. and Sparks, N.C.H., 2006. Aversion of chickens to various gases: Methods for humane culling. In: Proceedings of the 40$^{th}$ International Congress of the International Society for Applied Ethology (ISAE) held in Bristol, United Kingdom during 8-12 August 2006, pp. 64.

Savenije, B., Lambooij, E., Gerritzen, M.A. and Korf, J., 2002. Development of brain damage as measured by brain impedance recordings, and changes in heart rate, and blood pressure induced by different stunning and killing methods. Poultry Science 81, 572-578.

Schatzmann, U., Leuenberger, Th., Fuchs, P., Howald, M. and Howard, J., 1990. Untersuchungen über die Anwendbarkeit eines Hochdruckwasserstrahls zur Betäubung von Schlachtschweinen. Fleischwirtschaft 70, 890-894.

Schmidt, G.R., Hossner, K.L., Yemm, R.S. and Gould, D.H., 1999. Potential for disruption of central nervous system tissue in beef cattle by different types of captive bolt stunners. Journal of Food Protection 62,390-393.

Taylor, G.B., 1986. Deer slaughter. Veterinary Record 118 (12), 344.

Troeger, K., and Woltersdorf, W., 1990. Electrical stunning and meat quality in the pig. Fleischwirtschaft International 4, 3-10.

Troeger, K. and Woltersdorf, W., 1991. Gas anaesthesia of slaughter pigs. Stunning experiments under laboratory conditions with fat pigs of known halothane reaction type: meat quality and animal protection. Fleischwirtschaft 71, 1063-1068.

Tume, R.K. and Shaw, F.D., 1992. Beta endorphin and cortisol concentration in plasma blood samples collected during exsanguination of cattle. Meat Science 31, 211-217.

Urquhart, K.A. and McKendrick, I.J., 2006. Prevalence of head shooting and the characteristics of the wounds in culled wild Scottish red deer. Veterinary Record 159, 75-79.

Winkelmayer, R., Malleczek, D., Paulsen, P. and Vodnansky, M., 2005. Mitteilung zu röntgenanatomischen Untersuchungen des Thorax des Rehwilds im Hinblick auf den optimalen Zielpunkt für den tierschutzgerechten und wildfleischhygienisch einwandfreien Schuss. Wiener Tierärztliche Monatsschrift 92, 40-45.

Winkelmayer, R., 2009. Animal welfare during hunting: the ethical perspective. In: Smulders, F.J.M. and Algers, B. (Eds.). Food safety assurance and veterinary public health, Vol.5, Welfare of production animals: assessment and management of risks. Wageningen Academic Publishers, Wageningen, the Netherlands, pp. 205-220.

Wotton, S.B. and O' Callaghan, M., 2002. Electrical stunning of pigs: the effect of applied voltage on impedance to current flow and the operation of a fail-safe device. Meat Science 60, 203-208.

Wotton, S.B., Anil, M.H., Whittington, P.E. and McKinstrey, J.L., 1992. Pig slaughtering procedures: head-to-back stunning. Meat Science 32, 245-255.

# Animal welfare and killing for disease control

**Marien A. Gerritzen[1] and Mohan A.B. Raj[2]**
[1]Animal Sciences Group of Wageningen UR, Department of Animal husbandry, Division Animal Welfare & Health, P.O. Box 65, 8200 AB Lelystad, the Netherlands; marien.gerritzen@wur.nl
[2]Division of Farm Animal Science, School of Clinical Veterinary Science, University of Bristol, Langford, Bristol BS40 5DU, United Kingdom

## Summary

Killing animals for disease control is a complex process that involves different interests and conflicts between various stakeholders. Main objective of killing animals from an epidemic view point is to prevent the disease from spreading. Second interest is human safety, minimal human-animal interaction is preferred. A third important interest is animal welfare. Animals are killed on farms, mostly with the minimum of facilities and resources required to completely protecting animal welfare during killing. When it concerns poultry, it is preferred to kill all birds in their housings without catching or handling. Gassing whole houses is the most common practice for whole house killing of poultry. Advantages of this approach include large capacity, minimal human-animal interaction and a minimal use of personnel. However, whole house killing is in many situations not possible and handling live animals can in these situations not be prevented. Killing methods for pigs and ruminants can be grouped into three categories: mechanical (free bullet, penetrating and non-penetrating captive bolts), electrical and gas mixtures. A potential animal welfare benefit of using gas mixtures, in comparison with the mechanical and electrical methods, is that they could be administered to small groups of animals contained in cages or crates without the need for restraint. Some of the animal welfare issues associated with the use of these methods on farms are highlighted in this paper.

**Keywords**: killing, disease control, animal welfare, poultry, pigs, ruminants

## 1. Introduction: general welfare problems during killing for disease control

The first definition of animal welfare as proposed by the Brambell committee (1965) states that 'welfare is a broad term that embraces both the physical and mental well-being of the animal'. This implies that to protect the welfare of animals we should prevent them from experiencing avoidable pain and suffering. The killing of animals always interferes with their welfare and there is no ideal way of killing animals. Important risk factors to animal welfare are catching, handling and restraining which in many situations are inevitable steps involved in the process of killing. The extent to which animals must be handled or restrained and the choice of the killing method depends mainly on species and housing conditions. Conditions differ from regular slaughter for human consumption because killing for disease control purposes is usually performed on farms with minimum or no handling and restraining facilities. This implies that the outcomes of deploying various killing methods on farms are not always as efficient as in the slaughterhouse. Additionally, when it concerns highly infectious or zoonotic diseases the required speed of action places extra pressure on personnel.

It is clear that killing for disease control is very complex and that different interests are involved which could often result in conflict with each other (Figure 1).

To stop the disease from spreading it is essential that control measures are efficient. This requires that all susceptible animals be killed preferably without additional interference in large numbers with a minimal use of labour and materials. However, Member States have a legal obligation to humanely kill animals and therefore 'minimal use of labour and material' can conflict with efficiency and with capacity.

Differences in farms size, location and housing types added to regional or cultural differences make it difficult to guarantee use of the most efficient control measure. On grounds of efficiency, it is important to choose the best applicable method available for the situation at hand. The use of an efficient killing method prevents the disease from spreading and therefore, reduces the number of animals that have to be killed, which benefits animal welfare. On the other hand, methods that could be applied to large numbers of animals, or requiring minimal labour and material use are not always applicable or the best choice from an animal welfare point of view.

From a human safety viewpoint, minimal human-animal interaction is desirable, especially during zoonotic disease outbreaks. Minimal handling of live animals is also desirable to benefit animal welfare by reducing fear and distress due to catching, restraint and separation from companion animals. On the other hand, minimizing human-animal interaction may reduce control and monitoring of procedures and thus increasing the risk of compromising animal welfare.

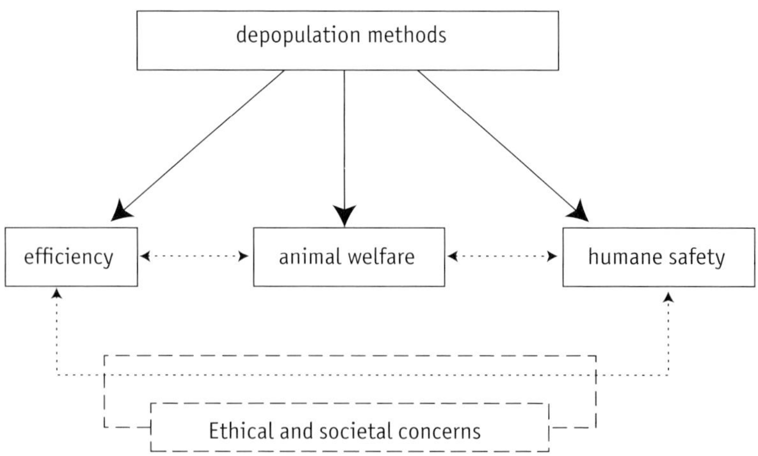


*Figure 1. Factors or criteria (⟶) of depopulation methods and relations or conflicts (◄---►) between the different criteria.*

Worldwide, there are many social and cultural differences regarding animal welfare and the killing of animals, which may have to be considered. Some of these social and cultural differences can be in conflict with animal welfare and/or efficient disease control measures. Furthermore, killing animals for disease control has considerable (socio) economic consequences. In many countries the financial cost of killing procedures can be a limitation.

In order to assure the best possible standard of animal welfare during emergency killing we should consider the risks to different animal species and different (housing) situations and with this in mind choose the best applicable species-specific methods. To enable us to make such a choice from an animal welfare point of view we first need to identify the specific risks involved when using the different killing methods.

In the following passages, a description is given of some of the most commonly used methods for on-farm killing of animals, including poultry, in the EU in terms of the risks to reduced animal welfare.

## 2. Specific animal welfare risk factors when killing poultry

Procedures applied for on-farm killing of poultry can be divided in two main categories; methods that involve handling live animals and those without handling. Both approaches possess specific risks factors concerning animal welfare.

### 2.1 Methods that include handling

It is generally recognized that catching and handling of poultry will cause fear and distress. The animals regard humans as predators and that separation from their companion animals' increases distress. It is also known that rough catching and handling or crating of poultry often results in painful bruises and/or fractured wings and legs, especially if performed by poorly trained personnel.

The capacity of mobile killing devices is limited due on the one hand to equipment or on the other hand to availability of skilled personnel. To enlarge capacity a lot of labour and equipment is needed. However, there is an evident increased risk of spreading the disease when more equipment and /or more personnel become involved.

*2.1.1 Containerized gassing*

Mobile containers, bins and bags of different sizes have been developed and used for on farm killing of poultry. All of these systems are filled with a minimum concentration of 60% carbon dioxide usually from liquid delivery cylinders or occasionally delivered as dry ice (solid $CO_2$). Birds are caught and carried with 3-5 animals in each hand from the poultry house to the gas filled containers usually kept outside. The birds are dropped into the containers in batches. As mentioned, catching and handling of live birds can be detrimental to their welfare. Throwing batches of live birds through small openings into the prefilled containers drastically reduces their welfare. If large numbers of birds are rapidly introduced into these systems, some birds may be suffocated under layers of birds. Introducing poultry into high concentrations of carbon dioxide has been described as painful and inhalation of high concentrations (40% or more by volume) has been found to be aversive to poultry. On the other hand, if carbon dioxide concentrations drop (due to mixing with air or after introducing large numbers of birds into small systems) onset of unconsciousness will be prolonged resulting in even more reduced welfare state.

#### *2.1.2 Mobile water bath electrocution lines*

In addition to the previously mentioned risks involved with handling of live birds there are several other welfare risks concerning electrocution in water baths. Electrocution using a water bath stunner involves painful shackling of live birds. Conscious birds could receive potentially painful electric shocks prior to electrocution due to the fact that leading wings make contact with the electrified water bath before the heads are immersed, especially in turkeys. If electrocution lines fail and as a result is stopped for a certain period birds will be hanging by their legs on shackles for a longer period. An important criterion with electrocution is that the current applied is high enough to ensure death *via* cardiac arrest or ventricular fibrillation. Cardiac arrest could be induced in some species of birds, such as ducks and geese, while they remain conscious due to the fact that their heads are not always immersed in the water bath, instead base of the neck or crop makes contact with the electrified water bath. The amount of current flowing through individual birds in a water bath is inversely proportional to the electrical impedance of individual birds and therefore some birds with high electrical impedance may not receive sufficient current required to inducing cardiac arrest. The height of the water bath should be adequate to ensure even the smallest bird in the flock (runt) is also immersed up to its base of the wings in the water bath. The identification and humane killing of surviving birds is therefore crucial and this would be difficult to achieve in electrocution systems that involve automatic removal of carcasses from the shackle.

#### *2.1.3 Individual bird procedures*

Killing methods that are performed on individual bird's, i.e. lethal injection, neck dislocation or decapitation and captive bolt stunning all involve handling and restraining of live birds, which are known to be distressing to birds. Although, these methods are relatively easy to apply there are still potential risks to animal welfare. For example, effective application of these methods depends on human skills. Mistakes such as non-intra venous injection of non-aqueous solutions, intra-peritoneal injection of irritants, poorly performed neck dislocation or a captive bolt fired at the wrong position on the head can be potentially painful and distressing to birds. Furthermore, due to the large numbers of birds that are killed during a disease outbreak the risk of poor performance will increase with time due to operator fatigue.

### 2.2 Whole house or flock treatment

In general, killing methods that could be applied simultaneously to large number of birds, such as whole house gassing, are less controllable than those applied to individual birds. For example, identification and humane killing of survivors after whole house gassing is in most situations only possible after the end of the gassing and only when the house is declared safe to enter on human health and safety grounds. Some stunning/killing methods do not induce immediate loss of consciousness or onset of death and thus it is possible that a period of reduced animal welfare may exist prior to the onset of death in birds. Additionally, given a free choice birds avoid exposure to some of the killing agents. On the other hand these methods do not involve catching or handling which is a major benefit to their welfare.

### *2.2.1 Whole house gassing*

Whole house gassing using carbon dioxide or carbon monoxide has in recent years been applied on a large scale especially for the control of Highly Pathogenic Avian Influenza (HPAI). The advantages of different gasses and gas mixtures as well as the different gassing methods have been reviewed previously (Gerritzen, 2006; Raj *et al.*, 2006). The advantages of whole house gassing are that is the method can be fast, has the potential to kill birds in very large numbers, and that there is almost no contact of humans with infected birds and materials, which improves biosecurity. From an animal welfare point of view, the main advantage is that animals are killed in their housings without being separated from their companion animals. The adverse effects of catching, handling and restraining are eliminated with whole house gassing methods. The disadvantage is that carbon dioxide is aversive to poultry and is potentially painful to inhale in high concentrations required to killing poultry. However, during most whole house gassing procedures carbon dioxide will increase gradually which will induce unconsciousness before animals will be exposed to high levels of carbon dioxide (Gerritzen *et al.*, 2006, 2007). Convulsions occur in unconscious birds and are not considered to be a welfare risk.

Carbon monoxide is generally considered as acceptable from an animal welfare point of view. However, exposure of birds to carbon monoxide introduces severe convulsions and even if convulsions start immediately after loss of consciousness they can reduce animal welfare in a flock of birds. If large numbers of birds have major convulsions unaffected companion birds will panic and move away from the convulsing birds (Gerritzen, personel observation). During whole house gassing with carbon monoxide increased vocalisation is measured (Gerritzen *et al.*, 2006). For human health and safety reasons, the use of carbon monoxide may not be suitable for whole house gassing (Gerritzen *et al.*, 2006).

Carbon dioxide is the most commonly used gas for whole house gassing. During whole house gassing the carbon dioxide concentration increases gradually from 0 to 45% in the breathing air. Loss of consciousness occurs gradually as the concentration of this gas increases in side the house. The time to loss of consciousness during exposure to carbon dioxide depends upon the rate of increase and final concentration of this gas in side the house, at bird levels. However, the way carbon dioxide is administrated to a poultry house strongly influences possible animal welfare risks. When carbon dioxide is injected into a poultry house in liquid form it will gradually vaporize due to the higher temperature in the house. When using a single injection point, gas concentration will not increase gradually through the whole building which implies that not all animals will be affected by the gas at the same time. Birds in the vicinity of gas entry point will be instantly exposed to high carbon dioxide concentrations, where as, those remaining away from the gas inlet will be exposed to rising concentrations of carbon dioxide. A negative aspect of this method is that the temperature around the birds can drop below freezing (e.g. -50 °C) and remain at that level for several minutes. Therefore, conscious birds can suffer from extreme cold or freezing. If carbon dioxide is injected under pressure at multiple injection points the temperature drop may not be more than 10-15 °C. High pressure multi injection point leads to a gradual increase of the carbon dioxide in the whole building which will decrease the risk on exposure of conscious birds to high carbon dioxide concentrations (Gerritzen, 2007).

Other potential risk factors associated with whole house gassing include distress caused by noise levels and turbulence and injury due to conscious birds inadvertently walking into the gas injection jet stream. The design, layout and construction of poultry houses and the fittings in side them vary widely and these factors affect the uniform distribution and maintenance of high carbon dioxide levels. In spite of all these risks, whole house gassing is still considered to be acceptable on humanitarian grounds.

Gasses such as nitrogen and argon replace oxygen in the breathing air and birds are rendered unconscious due to resultant hypoxia. Unconsciousness and death are induced in an acceptable way when concentrations of less than 2% oxygen in the breathing air are achieved, which will be difficult to achieve in commercial poultry houses due to the fact that poultry housing are never air tight. However, hypoxic conditions necessary to kill poultry may be achieved if the gas loss is compensated by increased gas administration flow rate. The feasibility of whole house gassing of poultry with nitrogen or argon has not been evaluated.

### *2.2.2 Air or gas filled foam*

In the USA air filled high density water based foam has been tested and conditionally approved by the United States Department of Agriculture (USDA) Animal and Plant Health Inspection Service (APHIS) for use with floor reared poultry (Benson *et al.*, 2009). A blanket of high density foam is created and spread to cover all the birds. The dense foam blocks the airways resulting in death by suffocation (Benson *et al.*, 2007). In general, suffocation is not accepted as a humane method for killing animals, including poultry. In order to investigate and solve the problem of suffocation different studies have been carried out (Gerritzen and Sparrey, 2008; Raj *et al.*, 2008). In these studies, different expansion ratios, bubble sizes and gasses have been investigated. The expansion ratio determines the amount of water in the foam. A low expansion will lead to wet foam with a small bubble size. A large expansion ratio will create a 'dry' foam with a larger bubble size. Foam with an expansion ratio between 1:250 and 1:350 seems to be the optimum compromise between foam stability, water content, bubble size and wetness. A higher expasion ratio will make the foam too dry and brittle and the foam then will not flow. Use of wet foam on the other hand will be equivalent to drowning of birds in water.

Large bubble sizes (e.g. >10 mm diameter) allow animals to remain breathing the air or gas that is trapped in the foam. When using a very low oxygen concentration in the foam created from a 100% nitrogen source animals will die very quickly. However, low oxygen levels induce strong convulsions by which the foam can be broken down. The nitrogen released from the bubbles could then mix with air resulting in some birds surviving the

# Genetic improvements for production and animal welfare

**Pascal A. Oltenacu**
Department of Animal Science, Oklahoma State University, Stillwater, OK 74074, USA; toni.oltenacu@okstate.edu

## Summary

As a result of selective breeding and intensification of the production, in many developed countries the production of farm animals has increased dramatically in the last 50 years. Milk production per cow has more than doubled and cows producing more than 20,000 kg of milk per lactation are now the norm. The increase in production has been accompanied by declining fertility, increasing incidence of health problems, and declining longevity in modern dairy cows. In broilers, due to increased growth rate and improved feed efficiency brought about by selection in the last 40 years, the time required to produce a 1,815 g broiler has decreased from 101 days to 32 days with a threefold decrease in the amount of feed consumed. Fast early growth rate was linked with increased mortality, high susceptibility to metabolic disorders and low locomotor activity. In egg layers, the production increased from 116 eggs per hen per year in 1930 to more than 300 eggs nowadays but this increase has not been without consequences, especially for the animals that are now more at risk for behavioural, physiological, and immunological disorders. The pig breeding programs resulted in high growth rate, high lean meat percent, increased feed efficiency and piglet production and these changes were associated with increased aggressiveness, leg weakness and susceptibility to stress.

Genetic selection for increased production is more and more viewed as increasing profit at the expense of reducing animal welfare. Sustainable breeding goals aimed at improving welfare in all farm animals should be developed and implemented. The welfare problems should be addressed before there is widespread condemnation of breeding and management practices.

**Keywords:** selective breeding, genetic improvement, health, dairy, poultry, swine

## 1. Introduction

Farm animals have been undergoing human managed selection ever since their original domestication. Initially, selection was probably limited to docility and manageability but in the last 60 years breeding programs have focused on the genetic improvement of production traits, such as milk yield, growth rate and number of eggs.

A major advance in selection practice occurred in the mid 20$^{th}$ century, with the advent of quantitative genetics based on principles of heredity and modern statistical theory (Hazel, 1943). For the first time, it was possible to impose selection systematically to improve specific target traits in production, permitting dramatic improvements in milk production and leading to development of faster-growing pigs and broilers. In practice, a phenotype is recorded on an individual and a substantial number of its relatives to compute the likelihood that the individual is transmitting a favorable set of alleles for the trait of interest. Although the method is still based on phenotypic selection, it more easily captures variation at loci having a relatively small effect and represents an important advance. Essentially, the statistical genetics method develops an average of all genetic loci contributing to a trait as transmitted by the individual, and reports it as an estimated breeding value [EBV] (Lynch and Walsh, 1998).

The timing of the development of quantitative theory was propitious, for the dairy, poultry and swine industry in the United States and Europe changed substantially in the last century (Ensminger and Perry, 1997).

In response to changes in dietary preference of consumers, the selection emphasis has changed from just increased output to increased production efficiency and product quality. For example, dairy production focus shifted from milk quantity, which was very successfully increased by intense selection, to milk quality and quantity (VanRaden, 2004) and in broiler and swine industry production efficiency and meat quality were added to growth and size as producers cannot afford to raise products that consume excessive amounts of energy or contain undesirable levels of fat.

However effective, selection for high production efficiency was often associated with undesired side effects on animal health and welfare.

One of the first attempts to explain the negative collateral consequences of selection for increased production were presented by Goddard and Beilharz (1977) who suggested the 'Resource Allocation Theory'. The resources an animal has are limited and as a result, if output is increased through one biological process, such as producing more milk, other functions such as fertility, maintenance, movement, immune defense, etc. will be affected. The resources that one process demands can be increased to a certain extent. Management factors, such as increasing access to feed and nutrients, could increase fitness of the animal until resources became limited again. Any further increase in fitness would imply a reallocation of resources and thus modify other outputs such as disease resistance or behaviour (Beilharz *et al.*, 1993). Reviewing the negative side effects of selection for high production, Rauw *et al.* (1998) concluded that 'when a population is genetically driven towards high production, [...] less resources will be left to respond adequately to other demands like coping with stressors; i.e. buffer capacity is [negatively] affected'.

In this chapter the effects of selection for increased production on health and other welfare indicators in dairy cattle, poultry and swine will be discussed.

## 2. Selection for increased yield and its consequences on welfare of dairy cattle

The North American Holstein Breed began with imports from Northern Europe in the late 1800's. The breed stayed in North America until the early 1970's when large scale exports began initially of live animals followed by semen and embryos. Over a 25-year span, US Holstein semen exports grew from about 400,000 units in 1973 to almost 8 million units in 1997. Early exports were mainly to EU member countries, particularly Italy, the Netherlands, Germany and France. In addition several countries, such as the Netherlands, France and others, implemented an aggressive embryo import program as a source of breeding stock for their internal genetic improvement programs. The major factors responsible for this trend were:

- it became known worldwide that Holsteins gave higher milk yields than most breeds;
- dairy farmers' breeding objectives worldwide became increasingly focused on income from sale of milk;
- technology existed to import Holsteins from USA into other countries.

The import of semen and embryos has resulted in the 'Holsteinization' of much of Europe. If we take UK as an example, in the last 30 years what was predominantly a Friesian dairy cattle population is now, on average, 90% North American Holstein (Brotherstone and Goddard, 2005).

The high input/high outputs are the most important dairy production systems and are used in most of the EU. These systems account for 83% of all EU dairy cows (with 95% of them being Holstein-Friesian breed) and 85% of all EU milk production (OECD, 1999). This chapter, unless otherwise specified, refers to Holstein dairy cows performing in a high input/high output dairy production system.

The dairy industry's goal has always been to produce quality milk for the consumer market. In many countries yield per cow has more than doubled in the last 40 years. The dramatic increase in yield per cow is due to rapid progress in genetics and management. The average energy corrected milk (ECM) yield for Swedish dairy cows increased from 4,200 kg to 9,000 kg between 1957 and 2003 (SHS Annual reports, 2003). Data from National Milk Records in the UK show an increase in average yields of dairy cows of about 200 kg/year from 1996 to 2002 and 50% of the progress in milk yield is attributed to genetics (Pryce and Veerkamp, 2001).

The picture is similar in the US where between 1957 and 2007 the average milk production/cow increased by 5,997 kg and 3,390 kg of this increase, or 56%, was due to genetics. Interesting to note that, until mid-1980s most of the increase in milk yield was the result of improved management, in particular better application of nutritional standards and improved quality of roughage. Since then, genetics has become the major factor due to effective use of artificial insemination (AI), intense selection based on progeny testing of bulls and worldwide distribution of semen from bulls with high genetic merit for production.

There are several practical reasons why one sided genetic selection for high production of milk, almost generally practiced since late 1990's, should be viewed with concern: (a) the increase in milk yield has generally been accompanied by declining ability to reproduce, increasing incidence of health problems, and declining longevity in modern dairy cows, all of which are indicative of reduced animal welfare; (b) substantial antagonistic genetic correlation exists between milk yield and fertility and between milk yield and several production diseases indicating that a proportion of the observed decrease in reproductive performance and of the increase in incidence rate of production diseases are correlated responses associated with past one sided selection for increase yield; (c) with increasing production cows need to spend more time eating and thus have less time available for other activities, and may not be able to allocate time enough to fulfil their need for important activities such as resting; (d) the selection for high milk production has produced a cow that is dependent on a high level of management in order to maintain its health, and which requires certain management practices to maintain its high milk output, which may themselves reduce animal welfare e.g. high-starch grain-based diets, and minimal grazing.

### 2.1 Metabolic stress

As the genetic ability to produce milk increases, more cows have sub-fertility or production diseases. As more cows are culled for health or fertility reasons, the productive life of modern cows is rapidly declining. The associations between increased production and increased risk of production disease as well as decreased fertility are well documented, but less is known about the biological mechanisms behind these relationships.

To address the growing perception that the pursuit of ever-increasing milk production is detrimental to cows' welfare, Ingvartsen *et al.* (2003) developed a framework for future research. The framework links the genotype, nutritional environment and management of the cow through its metabolic status to fertility and disease susceptibility and suggests that mobilization of body reserves has the potential to be the key factor.

High producing dairy cows need to mobilize body reserve to support their milk production. In the early 1/3 of the lactation period, until energy intake catches up with the requirements, high producing cows enter a state of negative energy balance, losing excessive amounts of body condition. The terms 'metabolic load' and 'metabolic stress' are used (Clarkson *et al.*, 1996) to describe the effects of high production on dairy cows. The metabolic load is defined as 'the burden imposed by the synthesis and secretion of milk' and metabolic stress as 'that amount of metabolic load which cannot be sustained, such that some energetic processes, including those that maintain good fertility and general health, must be down regulated.' The extent and type of down regulation would be indicative of the degree of metabolic stress. The term metabolic load is often used to describe only the part of the total energetic burden of lactation that is met by mobilization of body reserves, i.e. metabolic load is the same as negative energy balance.

Selection for production increases milk secretion and, therefore, the demand for energy. It also changes the partitioning of available energy by increasing the priority with which energy is allocated to support milk synthesis. Selection for production also increases feed intake but, with a genetic correlation between yield and feed intake ranging from 0.46 to 0.65 (Veerkamp, 1998), the gap between energy input and output during early lactation is increasing. There is little evidence for more efficient digestion or utilization of metabolizable energy in high-genetic-merit cows; so, the correlated response to selection for yield is increased body-tissue mobilization and increased metabolic load to bridge the gap between energy available through feed intake and energy needed to support increased milk production. Furthermore, an increasing demand for energy may lead to time constraints since cows like all other living creatures only have a total of 24 hours per day to allocate to different activities including eating behaviour.

Animal welfare ranges from poor to good and an objective way to assess it is in terms of directly measurable biological functions such as reproductive success, disease incidence, survival and behavioural changes (Anonymous, 2001). Duration, prevalence and severity are aspects that need to be considered to assess the importance of any welfare indicator. For dairy cattle, the major indicators of poor welfare that are associated with selection for high milk production are lameness, mastitis, other metabolic disorders, sub-fertility, and longevity. A similar study presented by Whay *et al.* (2003) revealed that lameness, limb lesions, mortality records, medical records, lung pathology, feeding behaviour, and body condition were considered to be the most important animal-based measures of dairy cattle welfare.

The impact of selection and breeding can be judged on the basis of its impacts on the welfare of individual offspring and future generations.

*2.1.1 Health*

Ingvartsen *et al.* (2003) reviewed 14 genetic studies on the relationship between milk performance and health in dairy cattle. These studies showed an unfavorable genetic correlation between milk yield and incidence of ketosis (0.26-0.65), ovarian cyst (0.23-0.42), mastitis (0.15-0.68) and lameness (0.24-0.48) indicating that continued selection for higher milk yield will increase lactational incidence rates (LIR) for these production diseases.

As pointed out earlier, some of the most serious welfare problems in dairy cattle are the consequence of a lack of balance in genetic selection (Rauw *et al.*, 1998; Sandøe *et al.*, 1999). It is also important to recognize that rectifying welfare problems in commercial dairy herds that may have resulted from past breeding strategy might be difficult to do in a short time. Considerable efforts may be needed to encourage definition and adoption of more balanced breeding goals if such welfare problems are to be addressed.

A total of 22 scientists participated anonymously (Delphi method) to develop a conceptual framework for assessment of farm animal welfare, to identify the major welfare performance indicators and rank them in order of importance. For dairy cattle the major welfare indicators in order of priority were: lameness, mastitis, sub-fertility, and longevity (Anonymous, 2001).

Lameness

The incidence of production-related diseases has increased greatly over the last decades. A case in point is lameness. Its incidence increased in UK dairy herds from lactational incidence rate (LIR) <10% reported prior to 1980 (Russell *et al.*, 1982) to >20% after 1990 (Clarkson *et al.*, 1996). Many more countries are now reporting a high incidence of lameness (Barkema *et al.*, 1994; Boettcher *et al.*, 1998; Philipot *et al.*, 1994). For the US, Guard (1999) reported a 38% LIR whilst Espejo *et al.* (2006) report a mean prevalence in the country of 25%. Lameness decreases milk yield and is an important cause of culling (Rajala-Schultz *et al.*, 1999). Guard (1999) estimated direct cost due to lameness in a 100-cow herd to be $7,600.

Mastitis

The genetic antagonism between mastitis resistance and production traits has been well established. In their review, Mrode and Swanson (1996) reported a weighted average genetic correlation between somatic cell count (SCC) and milk yield in first lactation of 0.14. Pryce and Brotherstone (1999) and Rupp and Boichard (1999) reported similar results.

The genetic antagonism between yields and clinical mastitis is more pronounced. Based on Scandinavian data, Emanuelson *et al.* (1988) indicated an average genetic correlation of 0.30 and, 12 years later, Heringstad *et al.* (2000) reported genetic correlations ranging from 0.24 to 0.55 with an average of 0.43. Pryce *et al.* (1998), Rupp and Boichard (1999), and Kadarmideen *et al.* (2000) reported genetic correlations in the same range, i.e. 0.29, 0.49 and 0.35, respectively.

Interesting to note that higher incidence of clinical mastitis is also associated with lower reproductive performance. Pryce *et al.* (1998) and Kadarmideen *et al.* (2000) reported genetic correlations with calving interval ranging from 0.16 to 0.41 and with conception rate to first service ranging from -0.21 to -0.58. Castillo-Juarez *et al.* (2000) also reported that high somatic cell counts were correlated to reduced conception rate at first service and longer calving interval with genetic correlation of -0.21 and 0.14, respectively.

As expected, mastitis resistance is found to be an important component of cow's longevity. Sander-Nielsen *et al.* (1999) reported genetic correlations between udder health and survival from first calving to end of second lactation ranging from -0.37 to -0.75 according to the breed. Pryce and Brotherstone (1999) reported genetic correlations of lifespan with somatic cell counts and clinical mastitis ranging from -0.11 to -0.32.

Up to the mid-90s, in most countries, the breeding objective included primarily production traits (mainly protein and fat yield), milk composition (protein and fat contents), and several morphological traits, particularly capacity and udder type. The Scandinavian countries were an exception with selection based on broader breeding objectives that included many functional traits and, particularly, mastitis resistance (Heringstad *et al.*, 2000). Including resistance to mastitis in selection objectives proved to be effective. In Norway incidence of clinical mastitis increased from 0.15 cows treated per cow-year in 1975 to 0.44 in 1994, and then decreased to 0.23 in 2002 (Osteras *et al.*, 2007).

In the last five years, the continuous and unfavorable trend for fertility and mastitis susceptibility led most European dairy populations to update their breeding objective and to increase the weight of non-production trait. Evaluating some of the recently defined breeding objectives put forth by Colleau and Le Bilan-Duval (1995) and by Pedersen *et al.* (2002) shows that mastitis resistance accounts for 10 to 30% of the total weight. This weight is large enough to substantially decrease SCC and to stop any deterioration of conventional milking (CM) frequency, even if SCC information only is available.

There are some concerns that continuous selection for low SCC may influence the cow's capacity for leukocyte recruitment and, therefore, her ability to respond to intramammary infection. Further research is needed to clarify if this concern is indeed justified.

### *2.1.2 Sub-fertility and longevity*

Decreased reproductive success may not in itself indicate reduced welfare. However, it is clear that many of the reproductive problems associated with highly productive dairy cows result from disease, such as uterine infections or other disorders (Bell and Roberts, 2007; Dobson *et al.*, 2007; Sheldon *et al.*, 2008) or from metabolic stress associated with milk production. For these reasons, reduced reproductive performance of modern dairy breeds is a welfare concern.

Calving interval increased from < 13.0 months to > 14.5 months and the number of inseminations per conception from 2.0 to > 3.5 from 1980 to 2000 in 143 US commercial herds (Lucy, 2001). A decline in pregnancy rate to first service of 0.5% per year between 1975 and 1997 was reported in the US (Beam and Butler, 1999). Poor reproductive performance often leads to premature culling of dairy cows.

Declining reproductive efficiency is not limited to the US. In the UK (Royal *et al.*, 2000) pregnancy rate to first service decreased from 56% in 1975-82 to about 40% in 1995-98, a decrease of about 1% per year. Similar decreases in conception rate and other reproductive measures have been reported in Sweden (Roxstrom, 2001) and many other countries.

Behaviour may also play a critical role in the declining reproductive performance of genetically high-producing cows. In a study of 17 commercial herds that used an electronic estrus-monitoring system, Dransfield *et al.* (1998) showed that a higher proportion of cows with production above herd average exhibited estrus with only low intensity and short duration relative to lower-producing cows (24% *vs.* 16%). Lopez *et al.* (2004) also reported an unfavorable association between milk production and estrus behaviour with shorter estrus periods (5.5 h *vs.* 11.1 h) in high relative to low producing cows (>40 kg/d *vs.* <30 kg/d). Emanuelson and Oltenacu (1998) found extended interval to first breeding and prolonged days open in herds with poorer estrus detection. The decline in fertility also has economic consequences and several studies reported increasing reproductive costs for dairy cattle (Lindhe and Philipsson, 1999; Lucy, 2001; Royal *et al.*, 2000).

## 2.2 Body form

Although selection for yield traits has received primary emphasis in the selection goals of dairy cattle, substantial emphasis has been placed on other traits, particularly in North America. Many of these non-yield traits are related to the outward appearance of cows, such as overall conformation or 'type', udder type traits, body size (including height, chest width and body depth), and angularity. Consequently, selection for increased milk production has also increased the body size and weight despite evidence that smaller cows have advantages for longevity, health, and welfare (Hansen, 2000). Emphasis on angularity (or dairy type) may have also contributed to increased metabolic stress, particularly in early lactation, resulting in cows that are more prone to metabolic problems. The changes in body size and weight for US Holstein cows are presented by Hansen (2000). Similar changes occurred in other Holstein-Friesian populations. In the Dutch Holstein-Friesian dairy population average hip height of heifers increased from 130 cm in 1981 to 144 cm in 2007 while the 305-days milk yield of the pedigree cows in the same population increased from 5,765 kg in 1985 to 8,720 kg in 2007 (NRS Statistics, 2007).

Changes in body form, size and weight affect the mechanics of movements of the cow in two major respects: a) the space that is needed to execute her movement freely and b) the scale of forces to be exerted for the movement. The changes in body size and weight are also changing the environmental requirements needed to maintain cows' good health and welfare.

In conclusion, the selection program to increase milk production has also changed the form, size and weight of dairy cows. These changes have a major effect on the physical health and welfare of the dairy cows when kept in a restrained, indoor environment prevailing in today's dairy farming. The spatial requirements of the dairy cow have increased as well as their vulnerability for mechanical impacts and wounds on the exterior parts of the body, the skin, limbs and claws.

In relation to cow welfare, udder shape and volume are of specific concern: on its own, with respect to normal locomotion and prevention of lameness, and for the resting comfort of the cows in the most common barn types.

### 2.3 Interaction of genetics and environment

It is clear that selection for production may lead to problems in health and fertility. As animals tend to adapt to the environment they are selected in, it is likely that selection for increased yield may also lead to environmental sensitivity expressed as genotype × environment (G × E) interaction. Castillo-Juarez *et al.* (2000) and Kearney *et al.* (2004) showed that the magnitude of the antagonistic genetic correlations between milk yield and somatic cell score and between milk yield and conception rate were significantly higher in a poor environment relative to a good environment. Dairy producers in several grazing countries have expressed concern regarding the declining fertility of cows with an increased proportion of Holstein genes. Harris and Winkelman (2000) and Verkerk *et al.* (2000) reported significant differences between cows of New Zealand origin and North American origin for conception rate, services per conception, and days to first service. These studies indicate that the negative genetic correlations between production, fertility and health in modern dairy cows, already large when producing in an intensive production environment, are even larger when cows are producing in a less intensive production environment. The increase in negative genetic correlation between production and fitness traits in less favorable environments is indicative of a decline in adaptability associated with selection for increased yield in the modern dairy cow.

## 2.4 Inbreeding

Inbreeding results from the mating of related individuals is also increasing. Inbreeding in the UK currently stands at around 3% and has been rising at 0.17% per year (Brotherstone and Goddard, 2005). The Holstein and Jersey breeds in USA have rates of inbreeding of 0.2% per year (Thompson *et al.*, 2000) corresponding to an 'effective' population size (or N*e*) of 50 and the picture is relatively similar in all Holstein populations across Europe. Movement of genes between countries and focus on similar breeding objectives has meant the selection of the same cattle everywhere. Low N*e* causes inbreeding and loss of genetic variation in a population. The current N*e* of 50 in the US Holstein is less than required to maximize long-term genetic improvement, but it is a recent phenomenon so little genetic variance has been lost to date.

Although inbreeding is not currently considered a problem, if it continues to rise it will become a real problem in the future. Inbreeding has three major undesirable effects. It causes inbreeding depression (including an increase in the incidence of abnormalities caused by recessive alleles), loss of genetic variance and random drift in the population mean. Inbreeding depression reduces the value of many traits, particularly those related to fitness, such as fertility, ability to remain healthy, and other traits indirectly affecting welfare.

Breeding organizations have the responsibility to maintain the genetic variability and prevent the increase of inbreeding in the dairy population. This can be accomplished by broadening the breeding objectives to include health, fertility and other fitness traits along with production traits, by considering the genotype by environment interactions, by implementing selection strategies that minimizes the average relationship of selected individuals with the rest of the breeding population and by taking advantage of the molecular genetics tools already available or in the developmental phase. It should be acknowledged that several of these strategies are already adopted by many breeding organizations in EU and other countries.

Other factors that would contribute to conservation of within and between breeds genetic variance, are a more realistic payment system for the meat value of cull dairy cows that would encourage more crossbreeding with beef breeds and better knowledge of breed differences that may decrease the dominance of the Holstein breed. For instance, breed differences in fertility, disease resistance and feed conversion have not always been known or considered when choosing Holstein over other breeds. Crossbreeding of dairy cattle should benefit from heterosis in many traits contributing to profit, and for two breeds of similar profit, it is likely that crosses between them are more profitable than either. Until now crossbreeding has not been used much in dairy cattle but recently, it is receiving some attention. In New Zealand, an across-breed genetic evaluation for profit indicates that, under their conditions, Jerseys are at least as profitable as Holsteins (Montgomerie, 2004). In the US, crossbreeding of Holstein cows with Montbeliard, Swedish and Norwegian Red as well as other dairy breeds is gaining in popularity and a three breeds crossing system is generally recommended. Use of crossbreeding automatically leads to a demand for cattle of more than one breed. Therefore, a rational use of crossbreeding should, in the long run, support the conservation of a variety of breeds.

such as the red and white breeds in the Nordic countries (Philipsson and Lindhe, 2003). It is also clear that, as long as the majority of the black and white dairy bulls tested in the Nordic countries originate from sires outside Scandinavia, where no estimates for genetic merit for health and fertility traits are available, the Nordic effort to improve these traits will have only limited impact (Christensen, 1998). An active international germplasm market makes breeding of dairy cattle truly global.

For the Holstein/Friesian breed, which is the dominant dairy breed in the world and accounts for about 80% of all dairy cows in Europe (Van Arendonk and Liinamo, 2003), increased emphasis on selection for fitness traits should occur in all countries, particularly those that dominate the international germplasm market. Since mid 1990's several breeding organizations in Europe as well as North America have included fertility and health (at least mastitis) in their breeding objectives. Recently several Nordic Countries included lameness in the breeding objectives and their lead should be followed by other breeding organizations. A multi-trait selection programme in which health, fertility and welfare traits are included in the breeding objective is suggested. For the multi-trait approach to be effective in improving welfare, higher weight should be given to fitness and welfare traits, a total merit should be calculated *via* selection index and selection should be based on total merit.

Europe should address the animal welfare issues in its breeding programs to avoid public condemnation of breeding and management practices for dairy cows.

In order to improve the welfare and adaptability of dairy cows through genetic selection long term, the cooperation of breeding experts, geneticists, epidemiologists, nutritionists, ethologists and others concerned with animal welfare problems is required. Sustainable breeding goals aimed at improving fitness and robustness is necessary to prevent the decrease in the quality of life of the animals and, perhaps, enhance it. The effectiveness of a selection program to improve welfare should be enhanced if selection acts directly on causes of poor welfare and not only on its symptoms. To implement such a program, research is needed to clarify the relationship between production, negative energy balance, metabolic stress and welfare indicators and to develop practical methods for measuring negative energy balance and metabolic stress. This research should identify traits directly related to welfare status, such as negative energy balance, body condition score, onset of cyclicity after calving etc., and, ultimately, provide better selection tools to improve welfare status in dairy cows.

## 3. Selection for increased yield and its consequences on welfare of poultry

There is only a limited number of major internationally operating poultry breeding companies that have to provide broilers (meat type) or egg laying hens worldwide.

Traditional breeding based on quantitative genetics principles has resulted in a large increase in production. Selection process is sequential and is very intense. An example of sequential selection and its intensity for male and female lines typically carried out by a poultry breeding organization is shown in Table 1.

*Table 1. Example of sequential selection likely to occur in males and females lines of a poultry breeding organization.*

| Age | Number males | Number females |
|---|---|---|
| Day old | 30,000 | 30,000 |
| 42 days | 1,800 | 7,500 |
| 20 weeks | 800 | 6,000 |
| 40 weeks | 160 | 1,600 |

A genetic pyramidal structure with intense selection and great accuracy combined with the application and efficiencies of vertical integration, led to the development of the modern poultry industry and are important contributing factors to genetic progress. A breeding company supplies one day old parents or grandparents to multipliers, which in turn supplies one-day old chicks to egg layers or broiler producers. A typical genetic pipeline is illustrated in Figure 2.

### 3.1 Broilers

The broiler industry has grown consistently since 1940s and now occupies the second place in volume in the world, just after pork. The growth has been based on strong consumer demand for poultry meat that is perceived as affordable, safe and healthy.

The genetic program has focused on several economically important broiler and breeder traits. The broiler traits receiving greatest emphasis are early growth rate, feed efficiency, livability and carcass yield, while the breeder traits are fertility, hatchability and livability.

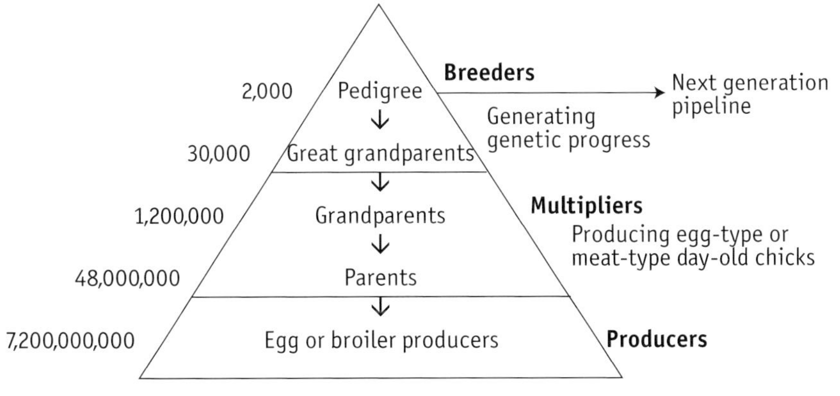


*Figure 2. Genetic pipeline illustrating the multiplication process occurring after one generation of selection in the poultry industry.*

remain healthy and are not wrongly selected as AS-resistant, an effective high-challenge system must be implemented. As soon as the major genes controlling the AS susceptibility are identified and sequenced and appropriate DNA markers are developed, a marker assisted selection program against AS should be implemented.

Addressing the leg disorders through selective breeding is more difficult as the leg problems are a direct consequence of selection for high growth rate. It has been shown that the incidence of twisted legs and tibial dischondroplasia can be reduced by genetic selection (Sorensen, 1989) but, because of the negative correlation between leg problems and growth rate, this option is not fully exploited in commercial breeding companies.

It is clear that many, if not all, skeletal disorders in broilers are consequences of rapid growth rate and a slower growth rate may alleviate these problems. Locomotor activity is important for the ossification of the bones in growing birds and there is a causal interrelationship between fast growth, low locomotor activity and leg problems.

### 3.2 Egg layers

The egg layer industry has a similar structure and traditional breeding has resulted in a rapid increase in egg production; in 1930 the average production was 116 eggs per hen per year, whereas nowadays the average production is increased to more than 300 eggs per hen per year. Furthermore, production became more efficient by intensification; farms increased in size and birds were kept at a higher density.

The combination of breeding for increased production and the intensification of housing conditions have not been without consequences, especially for the animals. Laying hens have become more at risk for behavioural, physiological, and immunological disorders (Rauw *et al.*, 1998) and, consequently, for reduced animal welfare. Behavioural disorders include cannibalism, feather pecking, and absence of broodiness behaviour (Newberry, 2004); physiological disorders include asymmetric growth (Yngvesson and Keeling, 2001; Tuyttens, 2003) and osteoporosis (Whitehead *et al.*, 2003); and immunological disorders include increased susceptibility against Marek's disease (Dalgaard *et al.*, 2003).

Selective breeding can and should be used to improve health and welfare related traits in laying hens. To effectively address these problems, behavioural, immunological and physiological traits with moderate to high heritabilities should be identified and incorporated in the breeding objectives.

Among behavioural traits, parameters describing fearful behaviour, feather pecking and cannibalism can be used in selection. Estimated heritabilities for fearful behaviour and feather pecking ranges between 0.35 and 0.60, and between 0.10 and 0.24, respectively (Rodenburg *et al.*, 2004).

Animals with a well developed immune system are more likely to stay healthy and immunological competence can be enhanced through selective breeding. Siwek *et al.* (2006) estimated heritabilities for natural antibodies determined in blood ranging between 0.11

and 0.42, whereas Bovenhuis *et al.* (2002) estimated heritabilities for specific antibodies ranging between 0.16 and 0.19.

Laying hens selected for high egg numbers can become prone to osteoporosis, particularly towards the end of the laying cycle, because of the high metabolism of calcium for egg shell formation. Bishop *et al.* (2000) found that traits describing bone strength are moderately to strongly inherited, estimated heritabilities ranging between 0.30 and 0.45.

It is clear that many welfare traits have a genetic component and selective breeding can be effective in reducing incidence of behavioural disorders, improve resistance to a wide range of pathogens, alleviate the problem of osteoporosis and, ultimately, improve the welfare in laying hens.

## 4. Selection for increased yield and its consequences on welfare of pigs

Traditionally, pig breeding programs have focused on the genetic improvement of production traits, such as high growth rate, high lean meat percent, feed efficiency, and piglet production. The focus on production traits was motivated by the perceived economic importance of these traits. However, there is increasing public concern about the undesired side effects of selection for high production efficiency on animal health and welfare (Harper and Henson, 2001).

Pig breeding organizations cannot neglect these public concerns and must adopt sustainable breeding goals that should include not only economic values of traits, but also traits related to health and welfare of animals.

Selective breeding can effectively address those concerns that can be measured on individuals and for which a proportion of the variability among individuals is genetic in origin. Kanis *et al.* (2003) gave an overview of various possible societal concerns in pork production and identified several traits related to pig welfare and health that are animal-related, can be considered for genetic selection and should receive high priority. These traits can be classified into (1) behavioural traits (aggressiveness, docility, aberrant behaviour such as ear or tail baiting); (2) physiological traits (leg problems mostly caused by osteochondrosis and other skeletal disorders, cardiovascular problems, etc.); and (3) immunological traits, including stress and disease resistance.

Many of these traits are sufficiently heritable to be included effectively in selection programs. Possibilities for selective breeding to control aggressiveness and reduce leg problems are discussed.

### 4.1 Social behaviour, aggressiveness and fearfulness

Intense selection to achieve genetic growth potential may be associated with changes in behaviour among grouped animals. Animal welfare concerns become an issue when some animals tend to be aggressive and cause injury under poor environmental conditions. Fighting mostly occurs between pigs after regrouping to establish a social hierarchy. The level of aggressiveness has been measured with the backtest or lesion score for individual pigs. The backtest, for which an individual pig is held on its back and the frequency of escape attempts (struggling bouts) is recorded for one minute, a genetic component is assumed (Cassady, 2007). Skin lesion score has also been used as an indicator of post-mixing aggression in commercially housed pigs (Turner *et al.*, 2006a) and has an estimate of heritability of 0.22 (Turner *et al.*, 2006b). It is clear that selective breeding can be used to control aggressiveness, but more studies of the genetic correlation between aggressive and growth traits are needed in order to improve welfare without decline in production efficiency traits.

### 4.2 Osteochondrosis and other leg problems

Ostechondrosis is an abnormal bone development represented by ossification on the growth plates and cartilages at the bone joints. It is regarded as contributory or main cause of leg weakness and it has a genetic component, with reported estimated heritabilities in the 0.1 to 0.3 range (Jorgensen and Nielsen, 2005). Given its moderate heritability, selection against osteochondrosis is possible and should be implemented. As the susceptibility to this bone abnormality seems to be genetically associated with growth and meat quality traits, continued selection for increased growth rate is likely to exacerbate this disease unless resistance to this disease is included in the breeding goals.

### 4.3 Porcine Stress Syndrome

The porcine stress syndrome (PSS) gene is found to cause malignant hyperthermia (heat shock) when pigs are exposed to environmental stressors. The PSS condition is a single locus inherited disorder in swine. This gene has two alleles, one dominant (N) and one recessive (n), and is commonly referred to as the halothane gene, because the PSS condition can be triggered by exposing pigs to the anesthetic, halothane gas.

The PSS gene causes animal suffering from the stress syndrome. Animals carrying the PSS gene lack the ability to adapt to stresses and once the PSS condition is triggered, they exhibit symptoms quickly. Typical symptoms range from rapid respiration and increased body temperature to sudden collapse and death. Although any severely stressed pig can produce pale, soft and exudative (PSE) pork, animals that carry the PSS gene are much more likely to develop PSE pork than PSS gene-free animals.

In the 1990s, researchers in Canada developed a quick, simple and accurate DNA test for PSS gene detection, which makes it possible to identify all three PSS genotypes. The DNA test provides the pork industry with a powerful tool to detect the PSS gene in live pigs and eradicate it from the entire pig population.

In conclusion, for some welfare traits, such as aggressiveness and susceptibility to osteochondrosis, knowledge regarding their genetic component is already available and they seem sufficiently heritable to be included in selection programs. However, other important welfare traits are not yet accurately defined and their heritabilities and genetic correlations with production traits are often not known precisely enough for application in breeding programs. Both heritabilities and genetic correlations can only be estimated accurately in large datasets with a sufficient pedigree structure. Multidisciplinary research programs should be developed to provide the necessary information for successful implementation of selection to improve welfare.

## 5. Conclusions

Traditionally, breeding programs for farm animals have focused on genetic improvement of economically important production traits. Consequently, productivity in farm animals rose dramatically during the second half of the $20^{th}$ century and effective selective breeding programs were a major factor. The milk yield per cow in dairy and growth rate and feed conversion efficiency of broiler chickens illustrate these changes vividly. However, the combination of selective breeding narrowly focused on production traits and the intensification of animal production systems have not been without consequences, especially for the animals which have become more at risk for behavioural, physiological, and immunological disorders, i.e., compromised animal welfare.

Farm animal welfare has become increasingly important and relevant from the societal point of view and its importance is now recognized by all stakeholders in the farm animal production chain. Breeding organizations working with farm animals cannot neglect these public concerns and must develop and implement breeding goal designed to restore or improve animal welfare.

### 5.1 What has been achieved?

There is evidence that several welfare aspects have a genetic basis and that selective breeding for a better animal welfare, be it in combination with selection for production or not, could be successful. Broader selection goals have been implemented in some dairy cattle breeding programs. Nordic dairy cattle breeding programs combine production, health, fertility and longevity traits optimally combined in a 'total merit index', or TMI. More generally, it is clear that many breeding companies are working towards developing more balanced breeding goals by incorporating functional traits.

In swine, selection for meat quality by means of the halothane test is a good example of breeding for an economic trait that resulted in better welfare (reduced stress susceptibility) as a side effect.

Many pig breeding programs are currently paying attention to some welfare-related traits, such as stress susceptibility, quality of legs, longevity of sows, vitality of piglets, and heritable defects, mainly because of their economic importance. These efforts are, of course, positive but to have a substantial effect on welfare, the weights given to these traits in the breeding goal should be much greater. In addition, some behavioural traits, such as aggressiveness and abnormal behaviour should be included.

The situation is very similar in poultry, with some welfare-related traits considered, such as ascites syndrome and skeletal disorders in broilers and bone strength and disease resistance in egg-layers, but the weights of these traits is probably lower than they need to be to improve welfare.

### 5.2 What has been neglected?

Conventional selection can be used to bring about positive change in both the health and behaviour of animals. Of course, the goal is to improve the welfare of farm animals, but a consensus of how to describe and measure welfare has not yet been reached.

One definition of animal welfare put forward by Broom (1996) states that 'the welfare of an individual is its state as regards its attempts to cope with its environment.' Animal welfare ranges from poor to good and an objective way to assess it is in terms of directly measurable biological functions such as reproductive success, disease incidence, survival, and behavioural changes (Anonymous, 2001). Duration, prevalence, and severity are aspects that need to be considered to assess the importance of any welfare indicator. Lately, efforts have been made to develop composite animal welfare scales (Scott *et al.*, 2001).

Aggregation of welfare indicators into an operational welfare system could be performed as an iterative process to assess relevance, marginal value, and suitability of each factor (Rousing *et al.*, 2001). In an attempt to assess welfare in loose-housed dairy cows at farm level, Capdeville and Vessier (2001) used the five freedoms (FAWC – Farm Animal Welfare Council, 1992) as needs for a starting point. They subdivided those freedoms into 16 basic needs. Forty-nine welfare indicators were identified, scaled, and pooled. Infertility and fattening score were amongst those. Logical rules to limit compensation between scores were set. Step by step data were pooled into a single score for each of the five freedoms, one of them expressing the freedom from hunger and thirst (Capdeville and Vessier, 2001).

Concerns are often expressed by welfare science that, by developing measurable indicators, welfare may come to be seen more in terms of measurable health parameters and aberrant behaviour while those aspects of welfare that belongs to the social life of the animals will be ignored. The point is that animal welfare is a complex concept and, therefore, difficult to define operationally (Rowan, 1997), though progress is being made and more ongoing research efforts are aimed at development of welfare indicators.

Christiansen and Sandøe (2000) mentioned that breeding for animals that are better suited for intensive farming instead of adapting the farming system may be considered violations of animal integrity. It is assumed in this chapter that use of selective breeding to improve welfare by improving the ability to cope with stress or by controlling aggressive behaviour is not a violation of animal integrity, but the approach is still open to various interpretations.

### 5.3 What needs to be done?

In order for breeding organizations to implement a selection for improvement of welfare in farm animals, the traits related to welfare need to be clearly defined, their heritabilities and genetic (co)variances need to be known, and recording of those traits on individual animals need to be feasible. The challenge is to define a limited number of welfare and health traits with sufficiently large genetic component and for which information on individual animals can be obtained routinely. Multidisciplinary research programs should be developed, therefore, to provide the necessary information for successful implementation of selection for these societal important traits.

## References

Adamec, V., Cassell, B.G., Smith, E.P., and Pearson, R.E., 2006. Effects of inbreeding in the dam on dystocia and stillbirths in US Holsteins. Journal of Dairy Science 89, 307-314.

Anonymous, 2001. Scientists' Assessment of the impact of housing and management on animal welfare. Journal of Applied Animal Welfare Science 4, 3-52.

Barkema, H.W., Westrik, J.D., van Keulen, K.A.S., Schukken, Y.H., and Brand, A., 1994. The effects of lameness on reproductive performance, milk production and culling in Dutch herds. Preventive Veterinary Medicine 20, 149-259.

Beam, S.W. and Butler, W.R., 1999. Effects of energy balance on follicular development and first ovulation in postpartum dairy cows. Journal of Reproduction and Fertility Supplement 54, 411-424.

Beilharz, R.G., Luxford, B.G. and Wilkinson, J.L., 1993. Quantitative genetics and evolution: Is our understanding of genetics sufficient to explain evolution? Journal of Animal Breeding and Genetics 110, 161-170.

Bell, M.J. and Roberts, D.J., 2007. The impact of uterine infection on a dairy cow's performance. Theriogenology 68, 1074-1079.

Bishop, S.C., Fleming, R.H., McCormack, H.A., Flock, D.K., and Whitehead, C.C., 2000. Inheritance of bone characteristics affecting osteoporosis in laying hens. British Poulty Science 41, 33-40.

Boettcher, P.J., Dekkers, J.C., Warnick, L.D., and Wells, S.J., 1998. Genetic analysis of clinical lameness in dairy cattle. Journal of Dairy Science 81, 1148-1156.

Bovenhuis, H., Bralten, H., Nieuwland, M.G. and Parmentier, H.K., 2002. Genetic parameters for antibody response of chickens to sheep red blood cells based on a selection experiment. Poultry Science 81, 309-315.

Broom, D.M., 1996. Animal welfare defined in terms of attempts to cope with the environment. Acta Veterinaria Scandinavica, Sect. A, Animal Science 27, 22-28.

Broom, D.M., 2002. Effects of dairy cattle breeding and production methods on animal welfare. Proc. 21st World Buiatrics Congres., Punta del Este, Uruguay, pp. 1-7.

Brotherstone, S. and Goddard, M., 2005. Artificial selection and maintenance of genetic variance in the global dairy cow population. Philosophical Transactions of the Royal Society B: Biological Sciences, 360, 1479-1488.

Capdeville, J. and Vessier, I., 2001. A method of assesing welfare in loose-housed dairy cows at farm level, focusing on animal observations. Acta Veterinaria Scandinavica., Sect.A, Animal Science Supplement 30, 62-68.

Cassady, J.P., 2007. Evidence of phenotypic relationships among behavioral characteristics of individual pigs and performance. Journal of Animal Science 85, 218-224.

Castillo-Juarez, H., Oltenacu, P.A., Blake, R.W., McCulloch, C.E., and Cienfuegos-Rivas, E.G., 2000. Effect of herd environment on the genetic and phenotypic relationships among milk yield, conception rate, and somatic cell score in Holstein cattle. Journal of Dairy Science. 83, 807-814.

Cheema, M.A., Qureshi, M.A., and Havenstein, G.B., 2003. A comparison of the immune response of a 2001 commercial broiler with a 1957 randombred broiler strain when fed representative 1957 and 2001 broiler diets. Poultry Science 82, 1519-1529.

Christensen, L.G., 1998. Possibilities for genetic improvement of disease resistance, functional traits and animal welfare. Acta Veterinaria Scandinavica Sect.A, Animal Science 29, 77-89.

Christiansen, S.B. and Sandøe, P., 2000. Bioethics: limits to the interference with life. Animal Reproduction Science 60-61, 15-29.

Clarkson, M.J., Downham, D.Y., Faull, W.B., Hughes, J.W., Manson, F.J., Merritt, J.B., Murray, R.D., Russell, W.B., Sutherst, J.E., and Ward, W.R., 1996. Incidence and prevalence of lameness in dairy cattle. Veterinary Record 138, 563-567.

Colleau, J.J. and Le Bihan-Duval, E., 1995. A simulation study of selection methods to improve mastitis resistance of dairy cows. Journal of Dairy Science 78, 659-671.

Dalgaard, T.S., Hojsgaard, S., Skjodt, K., and Juul-Madsen, H.R., 2003. Differences in chicken major histocompatibility complex (MHC) class Ialpha gene expression between Marek's disease-resistant and -susceptible MHC haplotypes. Scandinavian Journal of Immunology 57, 135-143.

Dobson, H., Smith, R., Royal, M., Knight, C. and Sheldon, I., 2007. The high-producing dairy cow and its reproductive performance. Reproduction in Domestic Animals 42 Suppl 2, 17-23.

Dransfield, M.B., Nebel, R.L., Pearson, R.E., and Warnick, L.D., 1998. Timing of insemination for dairy cows identified in estrus by a radiotelemetric estrus detection system. Journal of Dairy Science. 81, 1874-1882.

Druyan, S. and Cahaner, A., 2007. Segregation among test-cross progeny suggests that two complementary dominant genes explain the difference between ascites-resistant and ascites-susceptible broiler lines. Poultry Science 86, 2295-2300.

Druyan, S., Hadad, Y. and Cahaner, A., 2008. Growth rate of ascites-resistant versus ascites-susceptible broilers in commercial and experimental lines. Poultry Science 87, 904-911.

Emanuelson, U. and Oltenacu, P.A., 1998. Incidences and effects of diseases on the performance of Swedish dairy herds stratified by production. Journal of Dairy Science 81, 2376-2382.

Emanuelson, U., Danell, B. and Philipsson, J., 1988. Genetic parameters for clinical mastitis, somatic cell counts, and milk production estimated by multiple-trait restricted maximum likelihood. Journal of Dairy Science 71, 467-476.

Ensminger, M.E. and Perry, R.C., 1997. Beef cattle science. The interstate printers and publishers, Illinois.

Espejo, L.A., Endres, M.I., and Salfer, J.A., 2006. Prevalence of lameness in high-producing holstein cows housed in freestall barns in Minnesota. Journal of Dairy Science 89, 3052-3058.

FAWC – Farm Animal Welfare Council, 1992. FAWC updates the five freedoms. Veterinary Research 17, 357-360.

FAWC – Farm Animal Welfare Council, 1997. Report on the welfare of dairy cattle. Available at: http://www.fawc.co.uk/repors/dairycow/dcowrtoc.htm.

Goddard, M.E. and Beilharz, R.G., 1977. Natural selection and animal breeding. Proc. 3rd Int. Congr. SABRAO, Animal Breeding Papers pp. 4.19-4.21.

Grashorn, M., Bessei, W., Thiele, H.A., and Seeman, G., 1998. Inheritance of Troponin T levels in meat-type chicken. Archiv für Geflügelkunde 62, 283-286.

Guard, C., 1999. Control programs for digital dermatitis. In: Kenelly (Ed.). The tools for success in the new Millenium: Vol. II. Advances in dairy technology. University of Alberta, Edmonton, Canada, pp. 235-242.

Hansen, L.B., 2000. Consequences of selection for milk yield from a geneticist's viewpoint. Journal of Dairy Science 83, 1145-1150.

Harper, G.C. and Henson, S.J., 2001. Consumer concerns about animal welfare and the impact on food choice. EU FAIR CT98-3678.Centre for Food Economics Research, University of Reading, UK.

Harris, B.L. and Winkelman, A.M., 2000. Influence of North American Holstein genetics on dairy cattle performance in New Zealand. Proc. N.Z. Large Herds Conf. 6, 122-136.

Havenstein, G.B., Ferket, P.R., and Qureshi, M.A., 2003. Growth, livability, and feed conversion of 1957 versus 2001 broilers when fed representative 1957 and 2001 broiler diets. Poultry Science. 82, 1500-1508.

Hazel, L.N., 1943. The genetic basis for constructing selection indexes. Genetics 28, 476-490.

Heringstad, B., Klemetsdal, G. and Ruane, J., 2000. Selection for mastitis in dairy cattle: a review with focus on the situation of the Nordic countries. Livestock Production Science 64, 95-106.

Ingvartsen, K.L., Dewhurst, R.J., and Friggens, N.C., 2003. On the relationship between lactational performance and health: is it yield or metabolic imbalance that causes diseases in dairy cattle? A position paper. Livestock Production Science 83, 277-308.

Jorgensen, B. and Nielsen, B., 2005. Genetic parameters for osteochondrosis traits in elbow joints of crossbred pigs and relationships with production traits. Animal Scienc 81, 319-324.

Kadarmideen, H.N., Thompson, R., and Simm, G., 2000. Linear and threshold genetic parameters for disease, fertility and milk production in dairy cattle. Animal Scienc 71, 411-419.

Kanis, E., Groen, A.F., and De Greef, K.H., 2003. Societal concerns about pork and pork production and their relationships to the production systems. Journal of Agricultural and Environmental Ethics 16, 137-162.

Kearney, J.F., Schutz, M.M. and Boettcher, P.J., 2004. Genotype x environment interaction for grazing *vs.* confinement. II. Health and reproduction traits. Journal of Dairy Science 87, 510-516.

Lindhe, B. and Philipsson, J., 1999. The Scandinavian experience of including reproductive traits in breeding programs. In: Diskin M.G. (Ed.). Fertility in the high-producing dairy cow. Occasional Pub. No. 26, 251-261.

Lopez, H., Satter, L.D. and Wiltbank, M.C., 2004. Relationship between level of milk production and estrous behavior of lactating dairy cows. Animal Reproduction Science 81, 209-223.

Lucy, M.C., 2001. Reproductive loss in high-producing dairy cattle: where will it end? Journal of Dairy Science 84, 1277-1293.

Lynch, M. and Walsh, B., 1998. Genetic analysis of quantitative traits. In: Sinauer Associates, Sunderland, MA.

Montgomerie, W.A., 2004. Future genetic progress of dairy cattle in New Zealand. NZ Soc. Animimal Production 64, 96-100.

Mrode, R.A. and Swanson, G.J.T., 1996. Genetic and statistical properties of somatic cell count and its suitability as an indirect means of reducing the incidence of mastitis in dairy cattle. Animal Breeding Abstracts 64, 847-857.

Newberry, R.C., 2004. Cannibalism. In: Perry, G.C. (Ed.). Welfare of the laying hen. CAB International, Oxfordshire, UK, pp. 239-258.

NRS Statistics, 2007. The annual statistics for 2006/2007 calculated by The Royal Dutch Cattle Syndicate. See: http://park.org/Netherlands/pavilions/typical_dutch/cows/cattle/nrs/.

OECD, 1999. OECD Agricultural Outlook, 1999-2004. OECD, Paris.

Osteras, O., Solbu, H., Refsdal, A.O., Roalkvam, T., Filseth, O., and Minsaas, A., 2007. Results and evaluation of thirty years of health recordings in the Norwegian dairy cattle population. Journal of Dairy Science 90, 4483-4497.

Pedersen, J., Sander-Nielsen, U. and Aamand, G.P., 2002. Economic values in the Danish total merit index. Interbull, Bulletin 29, 150-154.

Philipot, J.M., Pluvinage, P., Cimarosti, I., Sulpice, P. and Bugnard, F., 1994. Risk factors of dairy cow lameness associated with housing conditions. Veterinary Research 25, 244-248.

Philipsson, J. and Lindhe, B., 2003. Experiences of including reproduction and health traits in Scandinavian dairy cattle breeding programs. Livestock Production Science 83, 99-112.

Pryce, J.E. and Brotherstone, S., 1999. Estimation of lifespan breeding values in the UK and their relationship with health and fertility traits. Interbull, Bulletin 21, 166-169.

Pryce, J.E. and Veerkamp, R.F., 2001. The incorporation of fertility indices in genetic improvement programs. In: Diskin, M.G. (Ed.). Fertility in the high-producing dairy cow. British Soc. Anim. Sci., Galway, Ireland, Occasional Pub. No. 26, pp. 237-249.

Pryce, J.E., Esslemont, R.J., Thompson, R., Veerkamp, R.F., Kossaibati, M.A., and Simm, G., 1998. Estimation of genetic parameters using health, fertility and production data from a management recording system for dairy cattle. Animal Scieince 66, 577-584.

Pryce, J.E., Simm, G., Amer, P.R., Coffey, M.P. and Stott, A.W., 2000. Returns from genetic improvement of dairy cattle. Proc. 2000 British Soc. Anim. Sci. Winter Mtg.

Rajala-Schultz, P.J., Grohn, Y.T. and McCulloch, C.E., 1999. Effects of milk fever, ketosis, and lameness on milk yield in dairy cows. Journal of Dairy Science 82, 288-294.

Rauw, W.M., Kanis, E., Noordhuizen-Stassen, E.N., and Grommers, F.J., 1998. Undesirable side effects of selection for high production efficiency in farm animals: a review. Livestock Production Science 56, 15-33.

Rodenburg, T.B., Buitenhuis, A.J., Ask, B., Uitdehaag, K.A., Koene, P., van der Poel, J.J., van Arendonk, J.A., and Bovenhuis, H., 2004. Genetic and phenotypic correlations between feather pecking and open-field response in laying hens at two different ages. Behavior Genetics 34, 407-415.

Rousing, T., Bonde, M. and Sorensen, J.T., 2001. Aggregating welfare indicators into an operational welfare-assesment system: a bottom-up approach. Acta Veterinaria Scandinavica., Sect.A, Animal Science Supplement. 30, 53-57.

Rowan, A.N., 1997. The concepts of animal welfare and animal suffering. In: Van Zutphen, L.M.F. and Balls, M. (Eds.). Animal alternatives, welfare and ethics. Elsevier, Amsterdam, the Netherlands 157-168.

Roxstrom, A., 2001. Genetic aspects of fertility and longevity in dairy cattle. Dissertation.

Royal, M.D., Darwash, A.O., Flint, A.P.F., Webb, R., Woolliams, J.E., and Lamming, G.E., 2000. Declining fertility in dairy cattle: changes in traditional and endocrine parameters of fertility. Animal Science 70, 487-501.

Rupp, R. and Boichard, D., 1999. Genetic parameters for clinical mastitis, somatic cell score, production, udder type traits, and milking ease in first lactation Holsteins. Journal of Dairy Science 82, 2198-2204.

Rushen, J. and De Passillé, A.M., 1998. Behaviour, welfare and productivity in dairy cattle. Canadian J. Animal Science 78 (Suppl), 3-21.

Russell, A.M., Rowlands, G.J., Shaw, S.R. and Weaver, A.D., 1982. Survey of lameness in British dairy cattle. Veterinary Record. 111, 155-160.

Sander-Nielsen, U., Pedersen, G.A., Pedersen, J., and Sorensen, P., 1999. Genetic variation in disease traits and their relationships with survival in Danish dairy cattle. Interbull, Bulletin 21, 170-177.

Sandøe, P., Nielsen, B.L., Christensen, L.G., and Sorensen, P., 1999. Staying good while playing god – the ethics of breeding farm animals. Animal Welfare 8, 313-328.

Scott, E.M., Nolan, A.M. and Fitzpatrick, J.L., 2001. Conceptual and methodological issues realted to welfare assesment: a framework for measurement. Acta Veterinaria Scandinavica, Sect.A, Animal Science Suppl. 30, 5-10.

Sheldon, I.M., Williams, E.J., Miller, A.N., Nash, D.M., and Herath, S., 2008. Uterine diseases in cattle after parturition. Veterinary Journal 176, 115-121.

SHS Annual reports., 2003. Swedish Association for Livestock Breeding and Production, Eskilstuna, Sweden, Reps.

Simm, G., 1998. Genetic improvement of cattle and sheep. In: Farming Press, Ipswich, UK.

Siwek, M., Buitenhuis, B., Cornelissen, S., Nieuwland, M., Knol, E.F., Crooijmans, R., Groenen, M., Parmentier, H., and van der Poel, J., 2006. Detection of QTL for innate: non-specific antibody levels binding LPS and LTA in two independent populations of laying hens. Developmental and Comparative Immunology 30, 659-666.

Smith, L.A., Cassell, B.G., and Pearson, R.E., 1998. The effects of inbreeding on the lifetime performance of dairy cattle. Journal of Dairy Science 81, 2729-2737.

Sorensen, P., 1989. Broiler selection and welfare. Proc. 3rd Eur.Symp.Poultry Welfare pp. 45-48

Thompson, J.R., Everett, R.W. and Hammerschmidt, N.L., 2000. Effects of inbreeding on production and survival in Holsteins. Journal of Dairy Science 83, 1856-1864.

Turner, S.P., Farnworth, M.J., White, I.M.S. and Brotherstone, S., 2006a. The accumulation of skin lessions and their use as a predictor of individual aggresiveness in pigs. Applied Animal Behaviour Science 96, 245-259.

Turner, S.P., White, I.M.S., Brotherstone, S., and Farnworth, M.J., 2006b. Heritability of post-mixing aggresiveness in grower-stage pigs and its relationship with production traits. Animal Science 82, 615-620.

Tuyttens, F.A.M., 2003. Measures of developmental instability as integrated, a posteriori indicators of farm animal welfare. Animal Welfare. 12, 535-540.

Van Arendonk, J.A. and Liinamo, A.E., 2003. Dairy cattle production in Europe. Theriogenology 59, 563-569.

VanRaden, P.M., 2004. Invited review: selection on net merit to improve lifetime profit. Journal of Dairy Science 87, 3125-3131.

Veerkamp, R.F., 1998. Selection for economic efficiency of dairy cattle using information on live weight and feed intake: a review. Journal of Dairy Science. 81, 1109-1119.

Verkerk, G.A., Morgan, S. and Kolver, S., 2000. Comparison of selected reproductive characteristics in Overseas and New Zealand Holstein-Friesian cows grazing pasture or fed a total mixed ration. Proc. N.Z. Soc. Animal Production 60, 270-274.

Whay, H.R., Main, D.C., Green, L.E. and Webster, A.J., 2003. Animal-based measures for the assesment of welfare state of dairy cattle, pigs and laying hens: consensus of expert opinion. Animal Welfare 12, 205-217.

Whitehead, C.C., Fleming, R.H., Julian, R.J. and Sorensen, P., 2003. Skeletal problems associated with selection for increased production. In: Muir, W.M. and Aggrey, S.E. (Eds.). Poultry Genetics, Breeding and Biotechnology. CAB International, Oxfordshire, UK, pp. 29-52.

Yngvesson, J. and Keeling, L.J., 2001. Body size and fluctuating asymmetry in relation to cannibalistic behaviour in laying hens. Animal Behaviour 61, 609-615.

# Monitoring animal welfare

**Terence Cassidy**
European Commission, Directorate General for Health and Consumers, Food and Veterinary Office, Grange, Dunsany, Co. Meath, Ireland; Terence.Cassidy@ec.europa.eu

## Summary

The competent authorities of the EU Member States have, for several decades, been monitoring animal welfare conditions on farms, during transport and at slaughterhouses to assess compliance with EU regulations. Considerable experience on how this work has been carried out has been obtained through the missions performed by the Food and Veterinary Office of the European Commission. The relevant legislation has increased over the years and recently there have been significant changes on how controls shall be organised. Such developments have helped animal welfare to get a higher priority and have improved the implementation of controls, as authorities need to ensure greater levels of co-ordination and co-operation both within and between the services involved. Competent authorities work with limited resources and it is therefore important to target inspections where there is the greatest likelihood of offences occurring. Apart from ensuring that the key welfare requirements are assessed at subsequent inspections, authorities also need a good system of reporting to obtain data which can be analysed and subsequently used to establish risk criteria, revise procedures, identify training needs and facilitate supervision. If models of risk assessment for animal welfare are to be useful to competent authorities, these need to take into account that authorities need to make comparisons between the different livestock sectors, and establishments within a sector, regarding the likelihood and severity of potential animal welfare problems occurring. Models which help identify areas for more frequent or intense controls would be invaluable to authorities who need to decide how best to deploy their limited inspection resources.

**Keywords:** animal welfare, competent authorities, inspections, monitoring, official controls, risk assessment

## 1. Introduction

The competent authorities (CAs) of the EU Member States have been required to monitor and enforce EU animal welfare standards since the first Directive on welfare at slaughter was adopted in 1974. Rules on animal welfare during transport followed in 1977 and detailed standards for intensive farming systems in the 1980s and 1990s. The controls carried out by the CAs have in turn been evaluated as part of the inspections of the Food and Veterinary Office (FVO) of the European Commission. The various approaches to monitoring animal welfare requirements presented here are based on the reports of FVO missions; however the views expressed in this paper do not necessarily represent the views of the Commission services.. The aspects which influence the effectiveness of controls are presented in the context of the requirements of Regulation (EC) No. 882/2004 (hereafter: 'the Official Control

Regulation'), which, since 1 January 2006, gives the framework in which controls in the food and feed sector, including checks of animal welfare, must be organised. This means that monitoring of animal welfare requirements, and subsequent enforcement, requires that a CA is designated, or as is often the case several CAs are responsible, to deal with animal welfare, that resources, training and procedures are provided for carrying out these tasks, and that the effectiveness of controls is subsequently verified and audited.

## 2. Resources for performance of controls

In order to monitor animal welfare Member States must first designate the CAs responsible for official controls. Article 4 of the Official Control Regulation requires the CA to have access to a sufficient number of suitably qualified and experienced staff, that have appropriate and properly maintained facilities and equipment and that have the legal powers to carry out official controls. There must be efficient and effective co-ordination and co-operation between the various organisations involved. Article 6 of the Official Control Regulation requires that staff receive appropriate training, and are kept up-to-date in their competencies.

### 2.1 Resources for checks on farms and at collection centres

In several Member States private veterinary practitioners are authorised to carry out certain official controls on holdings, at control posts and at places where animals are collected. Veterinarians working under such arrangements must be subject to guidance, training and supervision, equivalent to their full time colleagues, otherwise the quality and reliability of inspections is undermined.

To adequately monitor animal welfare conditions certain equipment needs to be provided, for measuring the dimensions of pens, light intensity, gas concentrations and temperature. A camera is also a basic piece of equipment for providing evidence which can be effectively used in follow-up enforcement action.

### 2.2 Resources for checks during transport

In relation to animal welfare during transport, when checks are carried out during the journey this invariably involves the police. Unless it is clearly specified that it regards animal welfare, the police may only give a cursory look to the official documentation accompanying the animals. In such situations it has been found that if a consignment is accompanied by a signed veterinary certificate the police may be reluctant to do any further inspection. Such checks have been better where the police participate in basic training with the veterinary authorities and inspections are subsequently coordinated with them. Monitoring animal welfare during transport frequently requires this type of multi-disciplinary approach, and joint training sessions provide an opportunity for the personnel involved to share experience and knowledge. Inspectors from the CA can, for example, learn from police services and gain a better understanding of available enforcement measures [DG(SANCO)/7980/2008]. Equally police officers obtain a basic knowledge of animal welfare requirements which allows them to carry out more meaningful checks with a greater degree of autonomy.

In order to evaluate certain equipment used on vehicles for long distance transport most CAs have involved the services of inspection bodies which have already been involved in vehicle testing, such as those involved in checks for road worthiness of vehicles. Again there have been benefits from this multi-disciplinary approach to vehicle approval, as such bodies learn from the veterinary services regarding the practical aspects of animal transport which may be relevant to their technical assessment, such as evaluation of ventilation capacity and siting of temperature monitoring equipment. Assessment of drivers' records [Regulation (EEC) No 3821/85] also requires co-operation between different control bodies. Most veterinary authorities claim not to have the legal powers to gain access to records, such as tachographs, which should be examined in certain cases to verify animal journey times. Obtaining such records is useful for retrospective checks of returned journey logs and is a requirement indicated in Regulation (EC) No. 1/2005; however, with the exception of controls at EU exit points and at road-side checks, where police or customs officials can obtain these records, limited use has been made of these to monitor journey times.

### 2.3 Resources for checks of welfare at slaughter

In certain Member States the Ministry designated for official controls in slaughterhouses is different from that responsible for enforcement of animal welfare. At least in these Member States this has resulted in Official Veterinarians (OVs) giving a low priority to checks of animal welfare or, even where animal welfare issues are adequately monitored, the co-ordination with the other Ministry has not been sufficient to ensure that problems are addressed.

A higher degree of technical specialisation is needed when dealing with certain sectors, such as the poultry sector, and the CA must provide sufficient training to ensure that those tasked with such inspections can competently assess the transport and slaughter procedures involved. Equally inspectors should make use of the expert bodies available to them to obtain advice on issues such as electrical or gas stunning.

## 3. Registration/authorisation of operators

In order to monitor the various operators involved in livestock production it is usually a first step to register them. Farms and slaughterhouses were already registered under legislation other than animal welfare legislation, but animal welfare Directives and Regulations have introduced additional requirements applicable to the registration of laying hen farms and the authorisation of livestock transporters.

### 3.1 Registration of farms

In order to register a laying hen establishment, certain data is required to comply with marketing requirements and to allow for tracing eggs in relation to the system of production, i.e. caged, barn, free range, organic. Co-ordination between the body responsible for registration and the service carrying out on the spot checks is needed to ensure that this registered data is accurate. Exchange of information between these two services also allows inspectors to better prepare for farm inspections by having access to data on the capacity of the establishment, flock records and building plans.

### 3.2 Authorisation of transporters

There are two types of transporter authorisations required by Regulation (EC) No 1/2005, authorisation for transport up to eight hours and authorisation to transport animals on journeys of more than eight hours. The CA needs to record authorisations of transporters operating on journeys of more than eight hours in an electronic database and to make the transporter's name and authorisation publicly available. This facilitates the inspection activities of those CAs in places where these transporters operate, which are distant from the place where the authorisation was granted.

A corollary to granting authorisations to livestock transporters is that an authorisation can be withdrawn as a sanction. However withdrawal of transporter authorisations for repeated infringements has been used with limited success due to legal challenges from transporters or the company opting to operate under a different authorised name.

### 3.3 Registration of slaughterhouses

Under Directive 93/119/EEC equipment and installations for restraint and stunning need only be assessed by the CA after the slaughterhouse is in operation. Approval of food business establishments, including slaughterhouses, is mainly focused on public health requirements. Article 31(2) of the Official Control Regulation lays down the framework for this approval and includes a limited number of animal welfare requirements. So before the CA grants approval, apart from the adequacy of the lairages, animal welfare aspects do not need to be considered. During missions several examples have been seen where recently installed equipment for stunning and restraining animals had been recently put into operation but had created major animal welfare problems in the way it was operated.

## 4. Inspections based on risk

Article 3 of the Official Control Regulation requires that animal welfare inspections should be based on risk. This Regulation was primarily drawn up with food safety in mind, where methodologies such as HACCP (Codex Alimentarius, 2003) were already well established as part of food safety controls. It is only relatively recently that the European Food Safety Agency (EFSA, 2007) has been asked to develop a similar risk based approach towards animal welfare.

### 4.1 Risk assessment for farm welfare inspections

A risk approach is already part of the system where farms are inspected in relation to cross-compliance [Regulation (EC) No 796/2004/EC], where the single farm payment is linked to respecting legal requirements including animal welfare. This general application of a risk based approach has led to identification of farms with animal welfare problems, as management problems giving rise to deficiencies in one sector have been found to be more likely to give rise to deficiencies in other sectors and animal welfare problems have been detected as a result [DG(SANCO)/7980/2008]. Farms that have had previous non-compliances detected during animal welfare inspections and farms with none are also weighted accordingly for future inspection programmes.

Certain CAs have identified risk criteria by analysing the results of previous farm inspections. In the Netherlands the CA identified that certain types of farms were more likely not to comply with the ban on tethering of sows and also that the timing of inspections in relation to the stage of the production cycle was important on farms keeping calves for veal. Inspections were then targeted on those categories of farms which were more likely to be non-compliant [DG(SANCO)/7512/2005].

Although checks should in principle be carried out without prior notification, it can be beneficial to make a general announcement that animal welfare inspections are to take place. In the Netherlands the CA sent a letter to all farmers fulfilling certain risk criteria to inform them that they were likely to be inspected. At the subsequent inspections it was found that farmers had made adjustments prior to inspections taking place. It is likely that a higher level of compliance was also achieved on farms which were not inspected, as even though these farmers were not subsequently inspected, it is likely that they would have made similar efforts to improve their level of compliance after being notified that they may be subject to an inspection.

### 4.2 Risk assessment in relation to welfare during transport

Regulation (EC) No 1/2005 establishes a certain risk basis for the organisation of inspections, requiring the CA to increase the number of inspections where it is found that transporters are not respecting the rules. This Regulation also sets out that where a CA establishes that a transporter has not observed, or a means of transport does not comply with this regulation, it may subject the transporter concerned to additional checks, in particular requiring the presence of a veterinarian at the time of loading of the animals.

Checks at places of departure provide an ideal occasion to ensure that animals are fit for the intended journey and that vehicles are appropriately equipped. Certain CAs have identified a higher risk of problems occurring during transport, due to the unusual nature of a particular consignment or because they have been notified by other CAs of infringements detected on previous journeys. The targeting of more rigorous controls on operators who repeatedly breach legal requirements has led to better levels of compliance. This requires the CA to ensure, not only the basic legal requirement for a veterinarian to be present to certify consignments for intra-community trade, but to be actually present at the time of loading and to take action appropriately.

A small number of CAs have used data available from slaughterhouses to identify risks in relation to the fitness of animals for transport. Following close monitoring of the data on the transport of end of lay hens, the CA in Denmark applied strict controls to their transport which led to a *de facto* ban on the transport of birds from caged systems [DG(SANCO)8039/2006].

### 4.3 Risk assessment in relation to welfare at slaughter

Records already kept by food business operators as a result of their own checks have been used to identify risks. Many food business operators will have data including mortality rates occurring during transport and records of the parameters applied in electrical or gas stunning of animals. Official controls can therefore be targeted on the weak points identified. A review of the operator's own checks also provides an opportunity to get access to information on issues such as training of slaughterhouse personnel and on the maintenance and functioning of equipment for handling and stunning animals. Again the intensity of official controls can be adjusted according to the risks identified.

## 5. Other criteria for determining the frequency of inspection

### 5.1 Other criteria for farm welfare inspections

In addition to requirements for inspections to be carried out on the basis of risk, there are specific rules on pig and calf welfare (Article 7 of Directives 91/629/EEC and Article 7 of 91/630/EEC) which require checks of a statistically representative sample of holdings. Such random inspections do have the benefit of ensuring that certain types of production are still included in inspection programmes, even though they have not been identified as a high risk, and provides a useful way of assessing if the risk analysis was correct.

### 5.2 Other criteria for transport inspections

A specific requirement of Regulation (EC) No 1/2005 for the CA is to carry out non-discriminatory inspections of animals, means of transport and accompanying documents on an adequate proportion of animals transported each year. Additionally this Regulation provides for mandatory checks at border inspection posts and exit points of the EU.

Determining what represents 'an adequate proportion of animals transported' is however not an easy task. The CAs have ready access to data on consignments transported for the purposes of intra-community trade, but information on the amount of domestic transport is not as easily accessible. Nevertheless several CAs have used data from slaughterhouses and transport associations to determine the volume of this transport and to set inspection targets accordingly.

The approval of journey logs is an administrative task which is required for all consignments going on long distance transport, but does not necessarily need direct veterinary involvement. CA administrative procedures to assess the feasibility of the times indicated incorporate a wide variety of internet and computer software packages. The results provided by some internet sites often underestimate the time needed, as such assessments should take into account loading and unloading times, mandatory stops, likely delays (e.g. at ferry ports) and be based on realistic average speeds. The requirement of Regulation (EC) No 1/2005 to verify such data by checking driver's records is another important tool for the CA to use in monitoring compliance with journey times and the introduction of data collection from treatment. However, nitrogen foam is considered a very promising method that is still under investigation, especially the robustness of the foam.

Creating gas-filled foam using 100% carbon dioxide leads to stable foam with a large bubble size. An atmosphere of >99% carbon dioxide can be created around the birds and they die rapidly. Muscle convulsions are seen when using carbon dioxide foam and this leads to break down of the foam structure. However, destroying carbon dioxide foam structure still leads to high carbon dioxide concentrations around the birds at floor level because the gas is denser than air. Argon is also denser than air. Therefore, the chance of birds surviving foaming is lower when using carbon dioxide or argon than when using nitrogen. The feasibility of using argon, nitrogen or carbon dioxide gas for practical application on large scale is still under evaluation.

## 3. Specific animal welfare risks when killing pigs and ruminants

### 3.1 General principles

Published standard operating procedures (SOPs) are not available for killing pigs and ruminants on farms during disease outbreaks, which is disconcerting on animal welfare as well as biosecurity grounds. Animal restraining facilities are required for proper application of certain stunning and killing methods but these are not always available on farms. Straw bales can be used to construct temporary animal handling and holding facilities.

Since the intention of humane slaughter or killing regulations is to avoid as much as possible anxiety, pain, distress or suffering at slaughter, stunning and stun/killing methods should ideally fulfil the following criteria:

- induce immediate (e.g. <1 sec) and unequivocal loss of consciousness and sensibility (e.g. captive bolts and electrical methods); or
- when loss of unconsciousness is not immediate, the induction of unconsciousness should be non-aversive and should not cause anxiety, pain, distress, or suffering in conscious animals (e.g. gas mixtures).

Effective stunning should be achieved with a single application and, in the event of a failure the animal should be re-stunned immediately using a backup method, which is usually a captive bolt gun. If a stunning device fails on two consecutive applications, the device should not be used again on animals until the fault has been investigated and rectified.

Humane slaughter or killing regulations also require that the duration of unconsciousness induced by a stunning method should be longer than the sum of the time interval between the end of stun and sticking (bleeding) and the time it takes for blood loss to cause death. Ideally, sticking should therefore be performed within 15 seconds following stunning (all species of animals stunned with all the stunning methods) and, in this process, the major blood vessels supplying oxygenated blood to the brain must be severed to ensure rapid onset of death (e.g. pigs, Wotton and Gregory, 1986). In cattle and pigs, sticking is performed by a procedure known as 'chest sticking', severing the all the major blood vessels by inserting a sharp knife at the thoracic inlet, and the size of wound through which bleeding occurs should be at least 5 cm long to facilitate rapid blood loss and prevent recovery following stunning. If carotid arteries in the neck are to be severed in cattle then attention should be paid to avoid blood clots from occurring at the thoracic end, which will lead to occlusion. Carotid occlusion will lead to poor bleeding and the possibility of recovery of consciousness in cattle. The possibility for recovery of consciousness following severance of carotids in the necks of cattle is also increased by the continuation of oxygenated blood supply to the brain *via* the vertebral arteries. In sheep and goat however carotid arteries and jugular veins in the neck are severed during sticking, which does not imposes any animal welfare risk.

Watching slaughter of conspecifics does not cause physiological stress responses in pigs (Anil *et al.*, 1997) and sheep (Anil *et al.*, 1996). However, since sticking or bleeding of animals on farm is not conducive for maintaining good biosecurity during outbreaks of infectious diseases, stunning methods that rely on bleeding to cause death are not preferred. On the other hand, use of destruction of the brain *via* pithing through hole made by penetrating captive bolt stunning or induction of cardiac arrest immediately following the application of any stunning method are practically feasible and impose no risk to animal welfare when properly applied.

Nevertheless, established stunning/killing methods can be grouped into three categories:
- mechanical – free bullet, penetrating and non-penetrating captive bolts;
- electrical – head-only or head-to-body;
- gas mixtures – carbon dioxide, inert gases or mixtures.

### 3.2 Mechanical methods

Free bullets are high velocity missiles (>300 m per second) that induce structural damage to the brain when fired into the cranium. Free bullets (including shotguns) are not routinely used for killing animals during disease outbreaks due to the personnel health and safety concerns. However, shotguns (12, 16, 20, 28 bore), rifles (0.22 rimfire) and handguns (calibres range from 0.32 to 0.45) can be used effectively by trained and licensed operators or marksmen to kill different species of animals. Shooting in the heads of animals is vital to ensuring their welfare.

Penetrating and non-penetrating captive bolts are low velocity missiles (typical bolt velocity 100 m per second) propelled by blank cartridges and they induce brain concussion leading to loss of consciousness and sensibility when fired frontally into the cranium. The diameter, penetration depth and velocity of the bolt are some of the factors that determine the effectiveness and outcome of shooting animals with captive bolts. Animals should be restrained appropriately to facilitate effective application of the captive bolt. The species specific shooting positions are described in various reports (e.g. see EFSA, 2004). Nevertheless, shooting of captive bolts away from the ideal position on the head can be a welfare risk, which could occur due to poorly trained operators or lack of appropriate animal restraint and facilities. Barrels of the captive bolt guns will heat up due to continuous use and eventually the gun would cease to function. Poor cleaning and maintenance of the captive bolt guns, and build up of dirt and carbon in the expansion chamber of the gun will lead to failures. Operator fatigue is another risk factor. The guns and cartridges or air line pressure used in pneumatically operated guns should be appropriate for the species of animals to be killed.

Destruction of the brain by pithing can be used for effectively ensuring death in animals that have been stunned using penetrating captive bolts. However, captive bolt stunning and pithing may cause bleeding through the bolt hole, which is not conducive for good biosecurity. As mentioned previously, induction of cardiac arrest as a killing method for captive bolt stunned animals needs to be evaluated.

### 3.3 Electrical methods

Electrical stunning is based on the scientific principle that stimulation of brain with an electric current of sufficient magnitude induces generalised epilepsy accompanied by unconsciousness and insensibility. It is known that spreading depression (SD) also occurs as a pathophysiological consequence of generalised epilepsy and can be recognised from the occurrence of a quiescent electroencephalogram (EEG). The duration of unconsciousness induced by the stun equates to the duration of these two pathological EEG manifestations.

There are three methods of electrical stunning/killing:
1. Head-only electrical stunning is applied by placing the electrodes on either side of the head such that the electrodes span the brain (Figure 2).
2. Head-to-body (known as head-to-back stunning) application of a 50 Hz sine wave alternating current (AC) in which the electrodes span the brain and the heart induces unconsciousness and cardiac ventricular fibrillation (leading to cardiac arrest).
3. Head-only stunning is immediately followed by a second application of a 50 Hz sine wave alternating current (AC) across the chest to induce cardiac ventricular fibrillation (leading to cardiac arrest). The interval between the termination of the first current cycle and start of the second current cycle should not exceed 15 seconds.

Note: Methods 2 and 3 are known as electrical stun/kill methods.

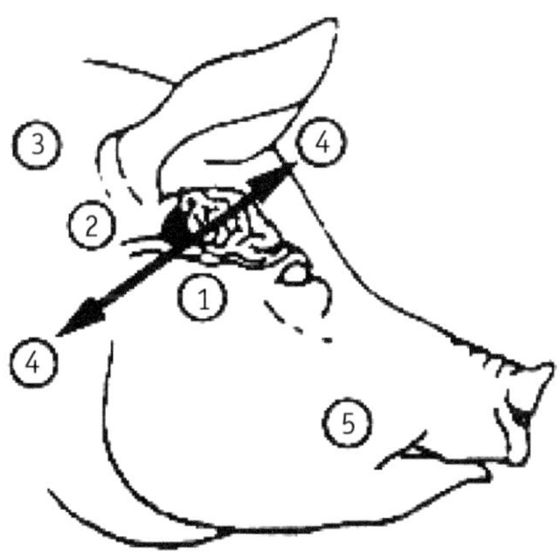


*Figure 2. Electrical stunning tong positions, example in pigs (Source: Anil and McKinstry, 1998) 1 = between the eye and ear on each side of the head, 2 = below the ear on each side of the head, 3 = on the neck, 4 = diagonal placement between the top and bottom of the head, 5 = on the snout or jaws. Tong positions 1, 2 and 4 span the brain and therefore acceptable. Positions 3 and 5 are unacceptable.*

From humane stunning and killing points of view, a 50 Hz sine wave alternating current (AC) should be used. The minimum current recommended for electrical stunning (method 1) or stun/killing (methods 2 and 3) of pigs is 1.3Amps, cattle 1.5 Amps, sheep and goat (including lamb and kid) is 1.0Amp.These currents should be applied for a minimum of 3 seconds in each current cycle, longer durations help to achieve better results. A minimum of 250 Volts (V) root mean square (RMS) would be necessary to quickly deliver the recommended amount of current to pigs, cattle, sheep and goat. The use of hand held high voltage electrical stunners require certain conditions to meet the heal and safety of operators, which includes operating on dry surface and wearing of personal protection equipment.

Implementation of constant current, rather than constant voltage, devices would improve animal welfare during stunning and killing. Electrodes should be appropriate to the species of animals and they should be cleaned regularly using wire brushes in order to improve electrical conductivity and minimise the risk of poor stunning. Maintaining good electrical contact between the electrodes and animals during current application is essential. Recovery of consciousness following head-only stunning will lead to poor welfare, e.g. due to the induction of potentially painful cardiac arrest, which should be prevented.

Electrical stun/killing by using methods 2 and 3 appears to be better than stunning because it eliminates the chances of recovery following stunning. It is known that successful induction of cardiac ventricular fibrilation (CVF) leads to cardiac arrest, which in turn reduces cardiac output to less than 30% of normal (Reed *et al.*, 2003). In humans, this level of cardiac output is inadequate to cope with myocardial and cerebral oxygen demands (Reed *et al.*, 2003). As coronary blood flow and myocardial perfusion falls, cellular high energy phosphates are depleted and lactate and hydrogen ions accumulate intra- and extra-cellularly. These changes in the physiological milieu of the heart would be sufficient to induce rapid onset of brain death.

Cardiac fibrillation threshold testing in experimental models suggests that cardiac tissue is most sensitive to sinusoidal stimulation between 30 and 60Hz current and increased stimulus duration increases efficacy (Weirich *et al.*, 1983). However, successful induction of CVF would depend upon the delivery of sufficient electrical current to the myocardium. The amount of current delivered will depend upon the voltage and total impedance in the pathway (i.e. between the electrodes). Scientific literature concerning termination of ventricular fibrillation (VF), in pig models, by electrical counter-shocks reveals that transthoracic impedance is affected by size of the electrodes (paddle size), applied pressure, the phase of respiration during which the shock is applied, use of coupling gel and its salt content and the distance between the electrodes, which is dependent upon the circumference of chest (Niemann *et al.*, 2003). Nevertheless, it has been reported, for example in pigs that transthoracic application of 90V of 50Hz AC for five seconds, using electrical stunning electrodes, resulted in CVF (Warriss and Wotton, 1981). Application of this amount of current will also be within the HSE guidelines for hand-held electrical apparatus.

There is a need to develop mobile electrical stunning and killing systems.

### 3.4 Gas mixtures

A potential animal welfare benefit of using gas mixtures, in comparison with the mechanical and electrical methods, is that they could be administered to small groups of animals contained in cages or crates without the need for restraint. Under this situation, killing with gas mixtures is better than stunning. Gas mixtures can be used to kill pigs, lamb and kids on farms during disease outbreaks.

Carbon dioxide in air is used widely to stun or kill pigs in slaughterhouses and anoxia induced with argon and/or nitrogen and mixtures of carbon dioxide and argon or nitrogen are being developed.

Gas mixtures can be applied to kill pigs in certain types of houses using a whole house gassing methods as described previously for poultry. The protocols developed for poultry species could be modified to suit species specific requirements.

Containerised gassing methods can also be employed on small to medium size pig farms and also for killing lamb and kids.Ideally, such a system should incorporate some good animal welfare principles, for example, as proposed by the FAWC for stunning or killing of pigs in slaughterhouses (www.fawc.org):
1. Animals should be maintained in a stable social group with the minimum of restraint.
2. Handling facilities should be designed to minimise stress.
3. The gas used to induce unconsciousness should be non-aversive.
4. All animals should be rendered rapidly unconscious in the gas.
5. An irreversible state of unconsciousness must be reached in all animals.
6. There should be adequate monitoring of the system, i.e. death should be confirmed before carcass disposal.

The exposure time required to killing pigs with a minimum of 70% by volume of carbon dioxide in air is reported to be about 5 minutes, which may be shortened by increasing the concentration of this gas (EFSA, 2004). The exposure times required to killing pigs with 90% by volume of argon in air or a mixture of 70% by volume of argon and 30% by volume of carbon dioxide are reported to be longer than 7 minutes and 5 minutes, respectively (Raj, 1999). However, it is suggested that system involving exposure of small groups of pigs (including piglets, boars and sows) to non aversive gas mixtures such as argon or nitrogen using a containerised gassing system for stunning followed by the induction of cardiac arrest using an electric current needs to be developed.

The exposure times required to killing lamb and kids with gas mixture are not known. However, it is suggested that these species of animals should be exposed to gas mixtures until they are dead, i.e. cessation of respiration and brain stem reflexes.

The concentration of gases should be monitored and controlled during killing. Animals exposed to gas mixtures should not be allowed to recover.

### 3.5 Lethal injection

Overdose of anaesthetic and sedative drug that cause death through the depression of the central nervous system (CNS) can be used to kill pigs and ruminants humanely. Barbiturates (and its derivates) are the drug of choice for this purpose.

The manufacturers' instruction should be carefully followed in determining the dose and route of administration. Intravenous injection is preferred, which may require prior sedation or severe restraint. Intraperitonial or intramuscular administration also require restraint and may be used if the drug is non-irritating. However, the onset of unconsciousness may be slow and may not be without the risk of compromising animal welfare, for example, animals may be prone to injury due to fall.

Some combinations of drugs and routes of administration can be painful and these should only be administered to unconscious animals.

## 4. Conclusions

### 4.1 What has been achieved?

For effective disease control many different approaches are developed. All of the applied methods have their specific impact on animal welfare. However, all described and legally accepted methods are applicable under different situations.

For poultry [ is until now the most preferable method because it requires minimal human-animal interaction. However, the impact on animal welfare depends strongly on how accurate the gassing procedure is executed. Moreover, controlling the process is crucial to quarantee animal welfare.

### 4.2 What has been neglected?

There is still no general agreement on the physiological signs of death or on how to assess death under practical on-farm situations. Furthermore, there are no protocols developed on how to asign death of animals during emergency killing.

### 4.2 What needs to be done?

Whole house gassing is not possible in all situations. For buildings that can't be sealed properly like free range situations the development of alternative whole flock treatments like the use of gas-filled foam should be further developed.

Pigs and ruminants are killed individually by using various methods. However, gas mixtures can be applied to kill pigs, lambs and kids in certain types of houses using a whole house gassing methods as described previously for poultry. The protocols developed for poultry species could be modified to suit species specific requirements.

## References

Anil, M.H. and McKinstry, J.L., 1998. Variations in electrical stunning tong placements and relative consequences in slaughter pigs. The Veterinary Journal 155, 85-90.

Anil, M.H., Preston, J., McKinstry, J.L. Rodway, R.G. and Brown, S.N., 1996. An assessment of stress caused in sheep by watching slaughter of other sheep. Animal Welfare 5, 435-441.

Anil, M.H., McKinstry, J.L., Field, M. and Rodway, R.G., 1997. Lack of evidence for stress being caused to pigs by witnessing the slaughter of conspecifics. Animal Welfare 6, 3-8.

Benson, E., Malone, G.W., Alphin, R.L., Dawsion, M.D., Pope, C.R., Van Wicken G.L., 2007. Foam-based mass emergency depopulation of floor-reared meat-type poultry operations. Poultry Science 86, 219-224.

Benson, E.R., Alpin, R.L., Dawson, M.D. and Malone, G.W., 2009. Use of water based foam to depopulate ducks and other species. Poultry Science 88, 904-910.

Brambell Committee, 1965. Report of the technical committee to enquire into the welfare of animals kept under intensive livestock husbandry systems. Command report 2836, Her Majesty's Stationary Office, London, UK.

EFSA, 2004. Welfare aspects of animal stunning and killing methods. AHAW/04-027.

Gerritzen, M.A., 2006. Acceptable methods for large scale on-farm killing of poultry for disease control. PhD Thesis, Utrecht University, Faculty of Veterinary Medicine, Utrecht, the Netherlands.

Gerritzen, M.A., 2007. Modelling $CO_2$ flow in poultry houses. Animal Sciences Group report 64. ASG, Lelystad, the Netherlands.

Gerritzen M.A. and Sparrey, J., 2008. A pilot study to assess whether high expansion $CO_2$-enriched foam is acceptable for on-farm emergency killing of poultry. Animal Welfare 17, 285-288.

Gerritzen, M.A., Lambooij, E., Stegeman, J.A. and Spruijt, B.M., 2006. Killing poultry during the 2003 avian influenza epidemic in the Netherlands. Veterinary Record 159, 39-42.

Gerritzen, M.A., Lambooij, B., Reimert, H., Stegeman, A. and Spruijt, B., 2007. A note on behaviour of poultry exposed to increasing carbon dioxide concentrations. Applied Animal Behaviour Science 108, 179-185.

Niemann, J.T., Garner, D. and Lewis, R.J., 2003. Transthoracic impedance does not decrease with rapidly repeated countershocks in a swine cardiac arrest model. Resuscitation 56, 91-95.

Raj, A.B.M., 1999. Behaviour of pigs exposed to mixtures of gases and the time required to stun and kill them: welfare implications. Veterinary Record 144, 165-168.

Raj, A.B.M., Sandilands, V. and Sparks N.H.C., 2006. Review of gaseous methods of killing poultry on farm for disease control purposes. Veterinary Record 159, 229-235.

Raj, A.B.M., Smith, C. and Hickman, G., 2008. Novel method of killing poultry in with dry foam created using nitrogen. Veterinary Record 162, 722-723.

Reed, M.J., Clegg, G.R. and Robertson, C.E., 2003. Analysing the ventricular fibrillation waveform. Resuscitation 57, 11-20.

Warriss, P.D. and Wotton, S.B., 1981. Effect of cardiac arrest on exsanguinations in pigs. Research in Veterinary Science 31, 82-86.

Weirich, J., Hohnloser, S. and Antoni, H., 1983. Factors determining the susceptibility of the isolated guinea pig heart to ventricular fibrillation induced by sinusoidal alternating current at frequencies from 1 to 1000Hz. Basic Research in Cardiology 78, 604-616.

Wotton, S.B. and Gregory, N.G., 1986. Pigs slaughtering procedures: time to loss of brain responsiveness after exsanguination or cardiac arrest. Research in Veterinary Science 40, 148-151.

# Animal welfare during hunting: the ethical perspective

**Rudolf Winkelmayer**
Veterinary Administration of Lower Austria, District Bruck/L., Bruck/L., Austria; rudolf.winkelmayer@noel.gv.at

## Summary

This contribution describes hunting strategies and situations typical for Central Europe, from an animal welfare point-of-view. Hunting is a very complex acitivity. Its contribution to the food supply, though essential in previous times, has in this day and age become of marginal importance. Still, there have always been other factors motivating the desire to hunt. To be justifiable as contributing to the provision of foods of animal origin, hunting activities have to be based on hunters' adherence to an ethical code of conduct resulting in minimizing adverse effects on game animals. Current practices to handle and kill wild game – be it in a free-range habitat or in fenced grounds – do not necessarily fulfil these ethical requirements and thus these may, in the long run, no longer be considered acceptable by society.

**Keywords**: hunting, 'pseudo-hunting', ethics, animal welfare, game

## 1. Introduction

In interpreting the contents of this chapter – hunting practices and in particular the associated ethical considerations – the reader should realise (1) that the author of this contribution is a veterinarian and dedicated hunter, (2) that the views expressed here are inspired by his personal ethical beliefs, and, finally, (3) that his framework of reference is predominantly based on experiences gained in and the literature available from Central European countries with a long established hunting tradition.

Obviously, worldwide there are several documents addressing the acceptability of hunting, the preferable mode of conduct during hunting and the ethical considerations that underpin these codices. In the framework of this contribution it would be impossible to review all these and discuss their similarities and differences. Instead, this contribution is meant to familiarise the less-informed reader with essential considerations on hunting, that pertain to the welfare of game animals and to highlight major points of concern.

## 2. General considerations on hunting: the pros and cons

The variety of today's hunting practices originates from ancient times, when hunting of wild game was an essential activity to secure valuable nutrients in the diet. In fact, in the Stone Age, a 'human being' was equivalent to a 'hunting being' (Ortega y Gasset, 1953). Since, the situation has changed and there are only very few regions or situations where hunting could be considered instrumental for the provision of food. Consequently, nowadays hunting is frequently associated with a 'luxury', a 'hobby', or – more controversially – with 'killing for pleasure', as a result of which objections are increasingly raised from the viewpoint of animal welfare.

In this context, two extreme positions can be observed in the discussion on the acceptability of hunting. Arguments in favour consider that hunting facilitates the stabilizing of ecosystems and hence represents a sort of 'applied nature conservation and wildlife protection'. Arguments against hunting are often based on the notion that the latter is largely driven by the hunters' need to secure a continuous supply with game animals through protection and multiplication of 'huntable' species, rather than by the desire to protect nature *per se*.

A fundamental question to be answered is if wild game has another moral status than farm animals and hence if – from a point-of-view of ethics – the killing of a wild animal is any different from killing farm animals during slaughter or sacrificing laboratory animals in the course of scientific experiments.

Animal rights activists take viewpoints that are largely centred on the notion of an 'intrinsic value' of animals. In general, in their reasoning they rely on three possible types of arguments i.e. the *physiocentric* (which gives reasons for respecting the 'good' of nature for its own sake), the *anthropocentric* (inspired by an anthropological viewpoint) and the *pathocentric* (primarily considering the suffering of animals). In the framework of the debate on hunting ethics, the latter two types are most relevant as they concentrate on a respectful handling of non-human beings and on minimizing pain when killing animals.

If it is to be considered if, from both viewpoints, killing of animals is principally justifiable, it is crucial to define which motivations and modes of killing can be accepted and which cannot.

A radical pathocentric position would be not to tolerate any killing of game animals, not even the killing of predators to protect huntable wild game or farm animals. On the other hand, up to now, no clearly formulated anthropocentric concept exists, which balances animal rights against humans' needs regarding food supply. Likewise, there is no ethical code, elaborated and generally accepted by hunters, which goes beyond the adherence to legal regulations on the hunting of game. To develop such a code, it is essential to define the nature of 'hunting'. Besides the mere 'killing of wild animals', the incentives for hunting are based on a number of other motivations, the societal and cultural implications of which have changed over the centuries. For instance, the killing of wild animals with a view to protect human food supply and human health could be classified as a sort of 'self defence', whilst – in the context of a Central European tradition – it can be considered a 'hobby', a 'sport', a 'social event', a means to satisfy some instincts, or simply resulting in pleasant feelings.

The mode and conditions of hunting have to be evaluated from an ethical viewpoint as well. Hunters have not always faced this challenge or have done so in a rather ambiguous fashion. There is of course, a – vaguely defined – set of rules, in German termed *'Weidgerechtigkeit'* (see below) and legislation on hunting indeed considers some animal welfare aspects. Still, excesses such as hunting exclusively for acquiring trophies, breeding and rearing of fenced game solely for the purpose of being released and killed as 'wild game', hunting in fenced areas, hunting with traps, etc. are not that infrequent. Although such practices, sometimes termed 'pseudo-hunting', may entirely be in conformity with legislation, they require critical evaluation from an ethical viewpoint.

Hunting, in the sense of Ortega y Gasset (1953), is more than the successful finalisation of a hunting exercise (or the fortunate escape of the hunted animal). It also includes tracking down, driving, flushing or stalking. These activities allow the hunter to synthesize motivational ('hobby-', or instinct-driven) aspects on the one hand, and the respect wild game deserve as living creatures on the other. By consequence, this would suggest that the excessive use of various technical items currently available for effectively tracking or killing wild game should be discouraged.

For hunting to remain a defendable activity still appropriate in the 21$^{st}$ century, it is essential to find a position between the extremes of a materialistic approach to hunting (i.e. efforts to maximize the number of trophies or kills) and a total ban of hunting. Such a balance can be achieved only, provided each individual hunter carries responsibility to implement ethics in her/his hunting practice (Kampits, 2008).

## 3. Definition of 'hunting'

Hunting, defined by Meyers (2008) as tracking, driving, flushing, stalking, killing or trapping of certain wild game species by authorized persons, has been practised for thousands of years. For a long period, killing or trapping of wild animals required enormous skills, strength, and cleverness. In the course of the centuries, a variety of technical devices has been developed to make hunting less demanding. It has always been necessary to impose rules on hunting practices so as not to excessively use the technological advantages against the wild animal. These rules constitute the so-called *'Weidgerechtigkeit'*, which is a sort of ethical codex (Bieger and Hoffer, 1941).

In the past few years, hunting and hunters in Europe are subject to increasing criticism. Mostly this is motivated by some hunters' attitudes (maximizing the hunting bag or 'trophy cult') or a result of ill-understood hunting traditions which are considered unacceptable in today's society. Consequently, Zeiler (1996) and Forstner *et al.* (2006) have presented codes of practice, which should ensure that hunting is not conflicting with today's paradigms and these authors provide a hierarchical model defining principles, criteria and indicators for a sustainable and socially acceptable hunting concept.

A clear definition of 'hunting', apart from having to be based on philosophical aspects and evolutionary biology, also needs to deal with tracking and killing techniques (which are commercially motivated) and with hunting farmed game. The latter activity should be clearly separated from hunting of free-range game, i.e. by a clear terminology ('pseudo-hunting' as opposed to 'hunting'). As for any other human activity, ethical criteria have to be observed (Winkelmayer and Hackländer, 2008).

### 3.1 'Weidgerechtigkeit' and hunting

Winkelmayer and Hackländer (2008) state that the book *'Mediations about hunting'* is probably the most frequently cited source when a definition of hunting is to be given. In this book the Spanish philosopher Ortega y Gasset argues that today, rather than being of existential significance, hunting has become a 'hobby', where the tracking and seizing hold, not the death of the animal is the core activity. Accordingly, the concept of 'hunting' is equivalent to one animal capturing another animal (alive or dead) with comparably inferior 'survival' functions, albeit that the hunter should not possess an absolute superiority over the hunted animal.

According to this definition, successful hunting does not necessarily have to result in game ending up being killed. Also, it should be considered that an overabundance of game would depreciate all tracking efforts. Tracking game is thus the first step in hunting. In indigenous hunting populations it is a tradition that rather than the hunter who killed, the successful tracker of the hunted animal receives the 'best part'. In several European countries, a single hunter's major aspiration is to pursue a hunt that includes the tracking, chasing and seizing/killing of an animal in high mountainous areas.

The US anthropologist Matt Cartmill discusses 'hunting' extensively (Cartmill, 1995), and identifies as its key elements that it must regard a special type of animal, to be killed by special techniques and for a special reason. Moreover, it is essential that the animal is wild, and on the free range, with the ability to escape or defend itself against human attacks.

Definitions in the standard German encyclopedias are relatively uniform, in that they describe different hunting strategies and legal restrictions, and do not deal with motivation and ethical implications (e.g. Meyers, 2008).

The term *'Weidgerechtigkeit'*, coined quite recently, was used only sporadically before 1900, and, for example not reported in Meyers Encyclopedia from 1909 (Meyers, 1909). After 1918, this term apparently gained more popularity and it first appears in Prussian legal documents soon followed by the rest of Germany in national German legislation of 1934. Still, a clear definition of this term was lacking. Although sounding very similar, the older term *'weidmännisch'* had a different meaning, as it did not consider pain of the animals, but above all the enhancement of the hunting experience while observing professional hunting practices (Winkelmayer, 1999).

The standard training book for the hunting license exam in Austria (Sternath, 2006), presents no definition of „hunting', but describes: (1) that for hunting to be sustainable (*'weidgerechte Jagdausübung'*) apart from fulfilling legal obligations also regional traditions have to be respected, (2) that a game animal has to be recognized as a living creature, (3) that all activities have to be ecologically acceptable, (4) that there is an accepted use for the hunted animal (this usually means food supply), and, finally (5) that animal welfare and nature conservation have to be observed.

Hunting societies have elaborated more or less detailed mission statements for 'Weidgerechtigkeit'. For example, the Carinthian hunting society (2004) defines the concept as ethics in hunting, which is characterised by: closeness to and responsilibity for nature, respect for all living creatures, and commitment to richness in wildlife species, protection of habitats and re-establishment of lost biotopes, commitment to sustainable use of animals and nature, and respecting traditions and the needs of persons which use other functions of the environment (recreation, forestry, agriculture). This implies continuing training and education, not only to optimise skills and the equipment (e.g. marksmanship, training of dogs, etc., processing of game meat), but also to obtain insight in ecology.

Rosenberger (2008) approaches 'Weidgerechtigkeit' from a theologian's viewpoint, and considers his work as a first attempt to do so. According to this author, hunting activities need to be governed by certain criteria the establishment of which is the task of ethicists. It is a peculiarity of hunting, that this activity is associated with strong, positive, pleasant feelings. This is acceptable, but some control and formation is required. Animals bear some intrinsic dignity, which lies in their ability to positively and negatively assess their environment and which entails some degree of self-determination and autonomy.

Schwenk (1997) points out that it is not only the relation of hunter to hunted animal, but also the conservation of nature, and sustainable use of renewable resources and the societal acceptance, which determine the future of hunting. In these terms, killing of farmed game animals which have been released solely for the purpose of being killed is not acceptable. Also, hunting in fenced areas can be considered problematic. Schwenk cites the IUCM (1990) statement on hunting, which considers both sustainable and wise use of natural resources and animal welfare.

The wildlife biologist Schmidt (2007), although not providing a definition of 'hunting', gives an insight into the motivations to hunt: it is the seizing of a trophy by using accepted methods. Wildlife will – at a gobal level and in the long run – not markedly be influenced irrespective of their being hunted or not. However, on a regional level, and in a short time frame, the consequences of hunting practices need to be assessed.

A recognized expert in the Austrian hunting society (Prossinagg, 1997) criticizes the tendency of wild game being provided supplementary feed, as is done for farm animals and that the magnitude of trophies is used to characterize the quality of hunting grounds. He also argues against the use of the term 'hobby'[1] in connection with hunting. Shooting of released game (e.g. pheasants) as 'living targets' and excessive trophy hunting abroad should not be accepted by hunting societies. Traditionally, such activities (pursued in previous centuries by noblemen) were not termed 'hunting', but *'Abschießung'* (shooting) or *'jagdliche Belustigung'* [hunting amusement] (Prossinagg and Haubenberger, 2007).

[1] In the German original, Prossinagg uses the word 'Sport', which, in the context of hunting, has a slightly different connotation than the British term 'sport'.

## 3.2 Legislation on hunting

Legislation on hunting differs between various countries. It seems that 'hunting' is defined rather indirectly *via* the prerequisites for hunting. The Austrian legislation may serve as an example for several Central European countries. According to the *Bundesverfassungsgesetz (B-VG 1920)* [Federal Constitutional law], hunting is governed by the nine Austrian provinces, which means that each province has to issue such legislation. The right to hunt rests – with no exceptions – with the ownership of land. However, this right can only be executed provided the land is of a suitable size (e.g. in Lower Austria a contiguous area of minimally 115 hectares). Smaller areas of different owners are combined to a shared hunting ground, '*Genossenschaftsjagdgebiete*' (co-operative hunting areas). These shared hunting grounds have to be leased, whereafter the right to hunt rests with the tenant. Hunting also includes the obligation to ensure that the wildlife population is rich in terms of species and that the animals are healthy *('Wildhege';* game preservation), and to achieve this under consideration of the needs of foresty and agriculture. All activities of tracking, stalking, driving and trapping/seizing wild game or parts/products thereof (e.g. antlers, eggs) are also considered hunting. Finally, these activities have to be performed in a generally recognized way *('Weidgerechtigkeit')* and according to principles of a structured hunting management (Lower Austria, Act on Hunting of 1974).

The Austrian [federal] Animal Welfare Act (BGBl. 118/2004) does not apply to hunting or fishing; therefore it has to include a definition, which activities categorise as 'hunting'. These include: keeping and training of animals which assist in hunting (dogs) and keeping of animals in fenced area for other than hunting purposes.

## 3.3 Hunting of large game (stand hunting, stalking, battue)

From both an animal welfare and food hygiene viewpoint, game, which is quiet and standing broadside, is ideally killed by a shot in the anterior chest *('Blattschuss'),* the neck or the head (see Figures 1 and 2). This will effect immediate death, and allow rapid evisceration and further processing of the carcass. This situation is quite typical for hunting from elevated hide-outs.

In contrast, during stalking the animal may have already recognised the hunter, will not stand broadside, and there will usually be no stable rest for the rifle. All these factors contribute to a higher risk of hits which are not instantly lethal. Battue hunting – where the animals are driven or chased by beating up the bushes where they reside and hunters are standing – is critical as it requires special skills to aim and correctly hit moving targets, and also, evisceration and further treatment of carcasses cannot start until after the hunt has ended. A German study (Krug, 1996) demonstrated, that the rate of correct hits decreased from 90% for hunting from a hidden position ('blind' hunting) to 25-35% for battue hunting. The latter mode of hunting, however, is very efficient in bushlands or in the forest and it is known that such single events represent less disturbance for game than repeated 'blind' hunting. It is a challenge for hunting management to ensure that time management and marksmanship are adequate to meet the requirements for animal welfare and food hygiene (Winkelmayer, 2005).

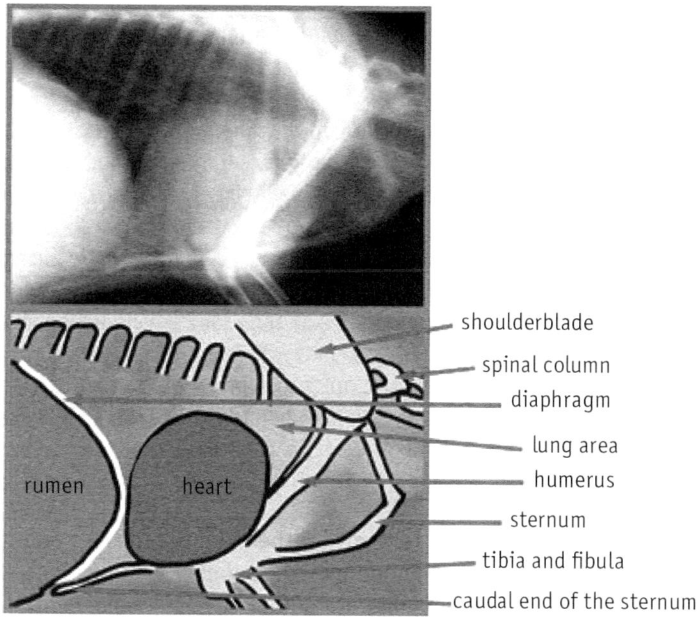


*Figure 1. The essential anatomical features in the anterior body cavity of a deer. [Note that in the grazing animal the diaphragm is generally positioned close to the heart]; (Winkelmayer* et al., *2005).*

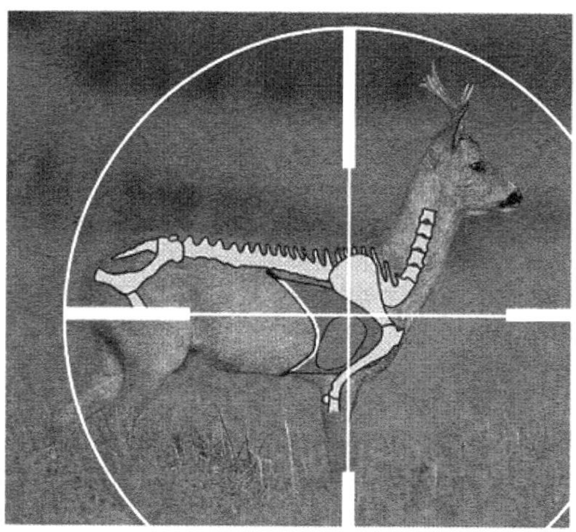

*Figure 2. The ideal target area when killing deer (Winkelmayer* et al., *2005).*

Many hunters still cherish the joining of forces in a shooting party, driving and killing the animal(s) and displaying the harvested game at the end of the day (*'Streckenlegung'* see Figure 3) to be one of the most significant hunting traditions. In this tradition all animals harvested by the end of the day are presented according to a ritual for which strict rules

[*'Wintergatter'* (winter enclosures) for red deer in some provinces in Austria], but can also function as breeding stations for trophy-carrying deer. When keeping game for the purpose of meat production, it is also practice to keep them in stables.

Fenced grounds destined solely for 'hunting' or 'pseudo-hunting' can be operated like unfenced hunting grouds (population density as in free-range and no supplementary feeding in areas which are restocked right before the killing event takes place).

With respect to the societal acceptance and ethical considerations, it must be noted that sustainable use of natural resources is mostly not possible in fenced grounds. To be justifiable, sustainable hunting in fenced grounds must be based on observing the following requirements:
1. Grounds should only be populated by autochtonous wild game species whose activity range complies with the size of the fenced area. This can be more than 800 hectares in the case of red deer.
2. Population density should not be higher than twice the density in the free range.
3. Feeding restricted to winter or periods when there is a substantial shortage in natural food supply.
4. Documentation of farmed game introduced for breeding (marked by ear tags and kept for at least one mating period) and of killed/culled/perished game.
5. Observance of hunting legislation (incl. close season).

Of particular concern is the transport of game. There are basically three scenarios when such a transport can be arguable:
1. farm-to-farm for breeding purposes (farmed game);
2. farm-to-slaughterhouse transport (farmed game);
3. farm-to-hunting ground for breeding.

However, there is evidence that large game is transported – within a country or even to foreign countries – between farms, fenced grounds, etc., and solely for the purpose of restocking hunting areas with 'living targets' or trophies. With respect to wild game, there are recommendations from the province of Bavaria, Germany *('Bayerisches Landesamt für Gesundheit und Lebensmittelsicherheit – Empfehlungen zum Transport von Rot-, Dam- und Sikawild';* recommendations for the transport of red, fallow and sika deer*) stating* that the transport of wild game to hunting grounds is justifiable only for the purpose of breeding and not for hunting.

There is no legal definition what 'justifiable' means, but there is agreement that in this context, it should reflect the viewpoint of educated people which are familiar with the concept of animal welfare. This viewpoint seems to be stricter than legal requirements are (Luy, 2008).

### 4.2 Releasing of wild birds

Farming and subsequent releasing of birds to the wild has a long tradition in Central Europe. Usually this is not done for restocking or introducing new species, but to provide 'living targets' for hunters who intend to kill as many game as possible.

Admittedly, in earlier times, wild game was 'imported' and released to increase the variety in species in Central Europe (notably pheasant, but also muntiak, sika, white-tail-deer), but only those which were interesting for hunters, 'survived', e.g. wild turkeys released in the Austrian-Hungarian monarchy disappeared in the 20$^{th}$ century. Rearing and releasing of wild game as 'living targets' is not exclusive for Central Europe, as evidenced by similar practices (e.g. grouse) in Scotland.

Of course, these types of hunting are not in agreement with the sustainable and prudent use of natural renewable resources – although this sustainability issue represents the only internationally accepted ethical argument for hunting.

## 5. Animal welfare for dogs which assist in hunting in training and work

Domestication of dogs dates back to the Stone Age. It is estimated that the diversification of dog races dates to 6,000-10,000 B.C. and that in these times, dogs assisted in hunting.

Zimen (1988) and Schöning (2001) report the first written evidence on training of dogs in ancient Greek and Roman culture. The work of Xenophon (426-355 B.C.) on dogs – *'Kynegeticus'* was a popular book up to the Middle ages. Some of the specialisations of dogs got lost in the course of the centuries (Krivy, 2004). In the 20$^{th}$ century, training received substantial inputs from behavioural scientists, as Lorenz, Scott and Fuller (Schöning, 2001).

Hunting dogs are subject to the Federal Act on Animal Welfare (BGBl I 2004/118), as far as training is concerned. This has implications for training techniques, devices used in training and for keeping of dogs (fennels, etc.). Their use as hunting facilitators is exempt from this legislation.

## 6. Conclusions

### 6.1 What has been achieved?

Legislation on hunting in the German speaking countries, but also in the entire European Union, defines primarily, that hunting is to be restricted to certain animal species, and to be done by approved or licensed persons under some spatial and temporal restrictions. Rules and regulations are less detailed about the conditions of tracking down, chasing, stalking and appropriation of wild game. From this definition, it can be concluded that killing of farm animals or crossbreeds of farm and wild animals, is not 'hunting', even when the mode of killing and the ante mortem phase (tracking down, etc.) would be similar to hunting procedures.

### 6.2 What has been neglected?

Current legal definitions allow that wild game reared in fenced areas (farmed game) are released in the wilderness and hunted as 'living targets' (this particularly applies to pheasants, partridges or ducks), and it is not to be expected that this will change in the near future. Austrian animal welfare legislation provides no definition of 'hunting', but defines which forms of animal husbandry are not related to hunting. This means, for example, that farmed game which is taken from the farm to the wilderness or to a fenced hunting ground, becomes 'wild game' and hence is no longer subject to animal welfare, but rather to hunting legislation. This practice is not uncommon, and it allows no distinction of 'wild game' reared in the wild from those bred in farms and then released for 'killing for trophy' purposes.

To term any killing of game as 'hunting' is not only a legal issue, but it also misleads the public, media, and animal welfare organisations. Arguably, this means that both killing of free range game for meat supply and 'killing for trophies' of released farmed game are not clearly differentiated in the view of these groups. An ethical statement would clarify the situation: killing of wild animals requires a justification. The appropriation of trophies as such can never be such a justification.

In absence of a complete and exhaustive definition of hunting, it is important to point out, what does not categorise as 'hunting', even when pursued by licensed hunters. So, any activity not compliant to the idea of sustainable hunting (Forstner *et al.*, 2006) should never be termed 'hunting'.

### 6.3 What needs to be done?

The future of hunting, as a part of human sociocultural history, will require a clear distinction of hunting from 'pseudo-hunting' activities, the latter providing no justification whatsoever for killing of game animals. This clear distinction is essential in all discussions about hunting. Ortega y Gasset and Cartmill give concise definitions for hunting, and these should be combined and amended to yield a definition which is acceptable for 'rational' thinking humans in the sense of Schmidt-Salomon (2006). Thus, activities related to hunting are not arbitrary or neutral, but are only acceptable when certain criteria are fulfilled (Rosenberger, 2008).

## References

Bieger, W., and W. Hoffer, 1941. Jagdgeschichte. In: Bieger, W. (Ed.) Handbuch der deutschen Jagd. Paul Parey, Berlin, Germany, pp. 525-529.

Binder, R., 2005. Zur Ausbildung von Jagdhunden im Kunstbau aus der Sicht des Tierschutzrechts. Cited by Fiala-Köck, B., 2009. Hundeausbildung und Tierschutz.

Binder, R., 2008. Tierquälerei und Tiermissbrauch – ein Thema für die Tierärzteschaft? Cited by Fiala-Köck, B., 2009. Hundeausbildung und Tierschutz.

Carinthian hunter's association – Kärntner Jägerschaft, 2004. Jagd ist Verantwortung – Jagd ist Freude! Das Leitbild der Kärntner Jäger. 2nd ed., Kärntner Jägerschaft, Klagenfurt.
Cartmill, M., 1995. Das Bambi-Syndrom. Jagdleidenschaft und Misanthropie in der Kulturgeschichte. Rowohlt Enzyklopädie Hamburg.
Fiala-Köck, B., 2009. Hundeausbildung und Tierschutz. Österreichische Jägertagung 2009, 15-20.
Forstner, M., Reimoser, F., Lexer, W., Heckl, F. and Hackl, J., 2006. Nachhaltigkeit der Jagd. Prinzipien, Kriterien und Indikatoren. avBUCH im Österreichischen Agrarverlag, Wien, Austria.
Kampits, P., 2008. Präambel zu: Der Begriff 'Jagd' – eine Differenzierung. Österreichs Weidwerk, 9, 10 und 11/2008.
Krivy, P., 2004. Die Hundeschule. Aufzucht, Haltung und Erziehung, Kosmos-Verlag, Stuttgart, Germany.
Krug, W., 1996. Jagd. Tierschutz. Wildbretqualität. Amtstierärztl. Dienst u. Lebensmittelkontrolle 3, II, 134-141.
Luy, J., 2008. Ethische Aspekte der Tiertötung. Fleischwirtschaft, 12/2008.
Meyers, 2008. Available at: http://lexikon.meyers.de/meyers/Jagd. Accessed: 17 January 2008.
Meyers Großes Konversations-Lexikon, 1909. Ein Nachschlagewerk des allgemeinen Wissens. 6., gänzlich neubearbeitete u. vermehrte Auflage. 20 Bde. Bibliographisches Institut, Leipzig u. Wien 1902-08. Ergänzungen u. Nachträge. Bd. 21.
Ortega y Gasset, J., 1953. Meditationen über die Jagd. Gustav Klipper Verlag, Stuttgart, Germany.
Paulsen, P., Nagy, J., Lazar, P. and Winkelmayer, R., 2005. Der Schrottschuss beim Feldhasen und mögliche Auswirkungen auf die Qualität und Haltbarkeit des Fleisches. Weidwerk, Heft 11/2005, p. 12-14
Prossinagg, H., 1997. Die Jagd in der pluralistischen Gesellschaft der Gegenwart. In: Hödl, G. and Pucker, H. (Eds.). Alles Jagd – eine Kulturgeschichte. Land Kärnten, Kulturabteilung Klagenfurt, Austria.
Prossinagg, H. and Haubenberger, G., 2007. Kaiserliche Jagdreviere in den Donau-Auen. Österreichischer Jagd- und Fischereiverlag, Wien, Austria.
Rosenberger, M., 2008. Grundzüge einer christlichen Ethik der Jagd. Österreichische Jägertagung, Gumpenstein, Austria.
Schmidt, K., 2007. Die Große Illusion. Die Presse, 9 November 2007.
Schmidt-Salomon, M., 2006. Von der illusorischen zur realen Freiheit. Forum Philosophicum Lech. Available at: http://www.philosophicum.com/2006/index.html. Accessed 22 Mai 2008.
Schöning, B., 2001. Hundeverhalten. Kosmos-Verlag, Stuttgart, Germany.
Schwenk, S., 1997. Jagdethik. In: Hödl, G. and Pucker H. (Eds.). Alles Jagd – eine Kulturgeschichte. Land Kärnten, Kulturabteilung Klagenfurt, Austria.
Sternath, M. (Ed.), 2006. Der Jagdprüfungsbehelf. 16. neubearbeitete Auflage. Österreichischer Jagd – und Fischerei-Verlag, Wien, Austria.
Welcker, H., 1991. Tierschutz und Jagd. Dt. Tierärztl. Wschr. 98, 28-30.
Winkelmayer, R., 1999. Tierschutz und Jagd. ÖGT – Österreichische Gesellschaft der Tierärzte – aktuell, 5/1999, pp. 12-15.
Winkelmayer, R., 2005. Ethische und ökologische Qualität von Wildfleisch; Internationale Tagung 'Fisch und Wild als Lebensmittel', Lebensmittel tierischer Herkunft Teil 2; Gesellschaft Österreichischer Chemiker AG Lebensmittel, Kosmetik und Tenside in Zusammenarbeit mit dem Institut für Analytische Chemie und Lebensmittelchemie der Universität Wien; Universität Wien, 21. und 22. Februar 2005; Tagungsband, ISBN 3-900-554-48-X, pp. 19-30.
Winkelmayer, R. and Hackländer, K., 2008. Der Begriff 'Jagd' – eine Differenzierung. Österreichs Weidwerk, 9, 10, 11/2008.
Winkelmayer, R., Malleczek, D., Paulsen, P. and Vodnansky, M., 2005. Radiological examination of the thoracal cavity of roe deer to identify the optimum aiming point with respect to animal welfare and meat hygiene. Wiener Tierärztliche Monatsschrift, Mai 2005.
Zeiler, H., 1996. Jagd und Nachhaltigkeit. Monographien Band 73. Umweltbundesamt, Wien und Jagd, Austria.

Zentralstelle Österreichischer Landesjagdverbände, 2008. Available at: http://www.ljv.at/start.htm. Accessed 27 January 2008.

Zimen, E., 1988. Der Hund – Abstammung, Verhalten, Mensch und Hund. Goldmann, Munchen, Germany.

# Part 2 – The assessment of animal welfare risks

# A risk assessment approach to animal welfare

**Bo Algers**
Department of Animal Environment and Health, Swedish University of Agricultural Sciences, P.O. Box 234, SE-532 23 Skara, Sweden; bo.algers@hmh.slu.se

## Summary

In response to the increased concern for animal welfare, risk managers are obliged to take action to ensure that the production of food of animal origin is in accordance with societal demands. As resources are limited, a risk manager must seek to invest these in those areas where they are most needed and could be put to use most efficiently. To this end, a risk based approach to identifying major risks for poor animal welfare would appear to be instrumental. The team responsible for conducting a risk assessment (RA) exercise must first agree upon a useful definition of animal welfare and subsequently list all possible hazards and associated adverse effects to animal welfare as it was defined. In this context it is important to discuss how the concept of 'health' is to be interpreted, as the latter is a significant component of well-being and hence must be properly considered. In veterinary medicine, health is generally regarded as the mere 'absence of disease'. However useful such a dichotomous definition may be (e.g. for conducting epidemiological studies of the occurrence of diseases in various animal populations), in discussions on its relevance as a co-determinant of animal welfare 'health' must be related to other welfare components such as an animal's ability to cope with its environment. As the first essential step in conducting an animal welfare risk assessment (AWRA), a team of experts with a relevant scientific background needs to be identified. These experts do not necessarily share the same opinion when it comes to their judgement of the seriousness of the adverse effects resulting from a particular hazard and this is largely explained by their various backgrounds. Consequently, the development of criteria by which AWRA experts should be selected is crucial and the arguments underpinning the validity of these criteria should – for reasons of transparency – be indicated in the RA report. Consensus is to be reached on which are the essential components of animal welfare, on their relative significance ('weighting'), and on whether each of these components should be considered separately or rather as co-contributing to 'overall welfare'. Only provided this is achieved, will the further development of a useful and practicable approach of AWRA be feasible.

**Keywords**: animal welfare, risk assessment, needs, experts

## 1. Introduction: risk assessment in animal welfare – why?

The keeping of animals, in particular those that we raise for food production, has undergone dramatic changes since the Second World War. With the advent of the industrialization of farming of animals, successful attempts have been made to radically increase the efficiency of producing milk, meat and eggs. These were based on developing distinct genetic lines, e.g. in cattle to improve either milk or meat production, in poultry for either egg or meat production, or in pork production for which fast growing pigs with increasingly improved feed conversion efficiency have been bred. Unfortunately, such genetic improvements have also caused unfavourable and undesirable effects on welfare (Rauw *et al.*, 1998). Hence, it is now generally accepted that, unless besides production criteria welfare considerations are also taken into account, the risk of poor welfare resulting from genetic selection will be increased. A major task of risk managers is therefore to ensure that production efficiency is not achieved at the expense of animal welfare. These concerns are further elaborated by Oltenacu (2009) elsewhere in this Volume.

In parallel to the genetic improvements, the housing of animals kept for food production has substantially changed. To shelter animals from extreme climatic conditions, to avoid their exposure to predators and to facilitate the introduction of more efficient feeding regimens allowing faster animal growth, indoor housing conditions have been greatly improved e.g. by applying more sophisticated ventilation and manure handling techniques and automated distribution of feed. This also enabled the reduction of labour costs, as a result of which the price of meat, milk and eggs could be lowered. Notwithstanding these desirable results, it is now recognised that during this process, many biological needs of the animals have been overlooked. Because most production animals are living in socially stable groups, which spend much of their time on foraging behaviour, the provision of their feed in the form of concentrates has seriously altered behavioural patterns. For instance, not only did concentrate feeding result in dramatically reduced eating times, but also in a reduction of exploratory behaviour. Furthermore, housing methods were developed, which did not take account of the needs for sows to perform nest building behaviour or for poultry to dust bathe. It has become clear that disregarding such innate biological needs may compromise the welfare of the animals. The effects of industrialized farming on animal welfare have been extensively described in the first section of this Volume.

With an increasing body of scientific evidence on how animal housing and management affects animal welfare becoming available, and the more the awareness and concern for these issues grows as a consequence, the more pressure is exerted on risk managers to take action to ensure that animal food production is in accordance with societal demands. Such risk managers are active on different political or implementing levels such as the EU Commission, national competent authorities, farmers or producers associations and individual farmers. For the EU, the Commission has taken action along the lines formulated in the Community Action plan on the Protection and Welfare of Animals 2006-2010 (EC, 2006), which not only stipulates the Community's objective to promote high animal welfare standards in the EU but also states that the coordination of existing resources should be enhanced while identifying future needs and ensuring a more consistent and coordinated approach to animal protection and welfare. This implies using resources where they are most needed and where they will give the most efficient output. In order to do so, a risk based strategy would seem to be the most appropriate approach.

Risk Assessment (RA) is a scientific activity that has been established in the areas of human and animal health (e.g. CAC, 1999; OIE, 2004a,b). Its objective is to assist those individuals responsible for Risk Management (RM) to prioritize their resources in their handling of a defined problem. The risk assessment includes four steps: Hazard Identification (HI), Hazard Characterization (HC), Exposure Assessment (EA) and Risk Characterization (RC). It has been defined as the 'evaluation of the likelihood of an adverse effect' (EC, 2000). Thus, a risk is a function of the severity and probability of the adverse effect to occur due to exposure to the hazard identified.

As, until recently, no suitable Risk Assessment approach for Animal Welfare (AW) was available, and a need was identified to develop methods that would allow a reliable, transparent, useful and practicable assessment of welfare risks, the European Food Safety Authority undertook a series of activities to promote such a development. The major achievements following these endeavours are described in the following three contributions (Smulders, 2009; Serratosa and Ribó, 2009; Ribó and Serratosa, 2009) and by Hultgren (2009) to be found in this Volume.

## 2. The needs of animals – a starting point for animal welfare risk assessment?

Obviously, to allow conducting an appropriate Animal Welfare Risk Assessment (AWRA) a useful definition of the concept 'welfare' must first be available. However, identifying such a definition is not as easy as one may think. For instance, inspired by Thorpe (1969) – who indicated that apart from their physiological needs, animals have 'behavioural needs' – it has been suggested that poor animal welfare is merely a violation of those behavioural needs. Interestingly, the discussion on behavioural needs has generated the viewpoint that both external and internal factors may be of crucial importance (Jensen and Toates, 1993). For example, if an animal is subjected to strong cold winds, it will seek shelter (e.g. Houseal and Olson, 1995). Hence an external factor like climate will affect the animal in such a way that it will carry out locomotion behaviour to seek shelter. Internal factors, like hormonal changes in relation to farrowing in sows, will affect the sow to conduct a series of motor behaviours to seek a nest site and build a farrowing nest (e.g. see Algers and Uvnäs-Moberg, 2007 and Figure 1 where this is further illustrated).

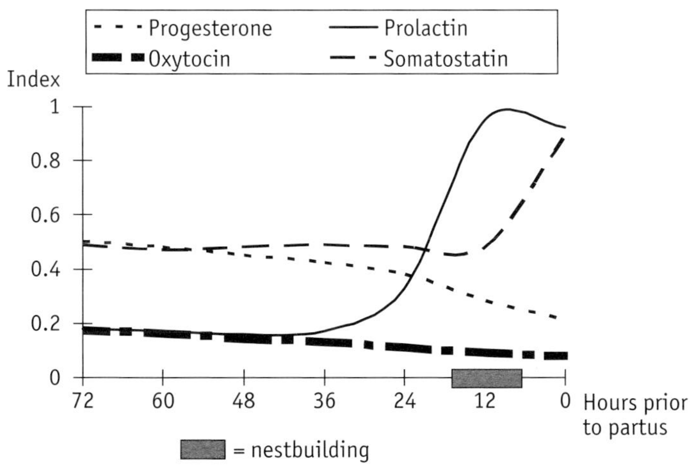


*Figure 1. The release of progesterone, prolactin, oxytocin and somatostatin during nest building by sows (From Castrén* et al.*, 1993).*

The question is, to what extent the needs that result from external factors are of more importance than those that are triggered by internal factors. Research on the consequences for the animals should these various needs not be fulfilled, may serve to help us understand how important they are for animal welfare. For instance, if an internally triggered behaviour is carried out as 'vacuum behaviour' (e.g. sham dust bathing in hens) and the behaviour in itself is rewarding, one might argue that the welfare of the animal is not compromised (provided it has space enough to perform the behaviour). If other signs of poor welfare are evident one might argue that welfare is indeed compromised. Examples of the latter situation include an increased cortisol release [as is the case when sows confined in crates build nests (Lawrence *et al.*, 1994)], or the preparedness of an animal to invest a substantial effort in accessing an area or a material where it can perform its behaviour [as is the case with the dust bathing hen (Gunnarsson *et al.*, 2000)].

Nordenfelt (2009) in this Volume discusses the use of 'needs' and the value of using such a concept in relation to animal welfare questions. He refers to Dawkins (1983) distinguishing between ultimate and proximate needs. Needs may be ultimate in that they are life saving and life supporting or proximate in that – if not fulfilled – animals will experience some sort of suffering. Although ultimate needs may sound as being more important (i.e. life supporting), the non-fulfilment of proximate needs may cause even greater impairment to welfare. Ultimately, such considerations may in some cases lead to the decision to euthanize very sick animals in order to reduce the compromising of their welfare.

To serve the purpose of risk assessment, it is important to define what a hazard really is. Is a hazard a circumstance that will cause disease? Will it cause pain? Will it contribute to poor growth? In the EFSA calf report (EFSA, 2006), the RA relied on the appreciation of animal needs that must be fulfilled and such needs are described as *'a requirement, which is a consequence of the biology of the animal, to obtain a particular resource or respond to a particular environmental or bodily stimulus'*. Some needs were regarded to require urgent fulfilment so as to avoid body functioning from becoming impaired in the medium or long term. Other animal needs relate to avoiding behavioural frustration e.g. excessive activities of an animal in an attempt to fulfil a need not properly considered in a particular husbandry system, the adverse welfare effects of which could be extreme and prolonged.

It is essential, therefore, that the group addressing animal welfare risks agrees on a common point of departure, i.e. what various components comprise animal welfare and, by consequence, which factors (potential hazards) affecting these components need to be addressed in the RA exercise?

If avoidance of pain, fear and anxiety is considered to represent important components of welfare, then any hazard that may cause those effects should be listed, whereas if, for instance, poor growth and/or disease are in focus of RA, factors causing such effects should be considered. In the animal welfare risk assessments conducted so far, ALL these components have been incorporated. It is crucial that it is not left to the reader to make his or her own personal judgement of how the concept of welfare could have been interpreted by the RA expert group, but rather that is becomes unmistakably clear what the RA group's starting point was. This transparency is absolutely necessary to avoid confusion and unnecessary debate on the conclusions and recommendations formulated in the final report.

## 3. Definitions of animal welfare for use in Animal Welfare Risk Assessment

To enable expressing animal welfare in more quantitative terms, based on the measuring of certain critical variables and indeed allowing ultimate verification if animal welfare has been properly considered, welfare scientists have sought for suitable definitions, including such variables as the prevalence, size and intensity of injuries, levels of stress hormones released, occurrence of aggression or proportion of animals showing clinical signs of disease. However, in parallel, philosophers have discussed what is morally relevant and, thus, how the various variables considered by scientists should be prioritised from an ethical point of view. Three principally different positions have emerged from such discussions (elaborated in more detail by Duncan and Fraser, 1997):

- The subjective experience approach. This approach claims that only animal feelings, such as suffering, pain or pleasure are of concern when the welfare status of the animal is being evaluated (e.g. Dawkins, 1988; Wolf, 1990, 1992; Duncan, 1993; Figure 2).

*Figure 2. Shoulder lesion on a sow is associated with pain but does this influence her biological functioning? (Photo: Rebecka Westin).*

- The biological functioning approach. This approach is based on the argument that the quality of life is good when the animal's biological systems are functioning in a normal or satisfactory manner (Figure 3). Broom (1991) suggested that good welfare depends on the animal's ability to cope successfully with its environment and to function normally from a biological perspective. Others have put more emphasis on high levels of growth and reproduction (McGlone, 1993).
- The natural living approach. Here, the animal's possibility to express its natural behaviour (Webster *et al.*, 1986) and living a 'natural' life according to its genetically encoded nature or 'telos' (Rollin, 1993) is in focus (Figure 4).

*Figure 3. A calf in good condition – physically [and mentally?], (Photo: Peter Stamp Enemark).*

*Figure 4. Dustbathing hens. This 'natural' behaviour is usually exhibited in the afternoons. (Photo: Anna Lundberg).*

It is rather likely that such morally different positions may colour the assessments made by various welfare experts and although these three positions briefly described above partly overlap, it is of importance to make a choice from these definitions and indicate in the RA report the arguments underpinning this choice.

As pointed out by Nordenfelt (2009, this Volume) definitions of AW may vary considerably. When the animal's well-being is assessed in the framework of an RA, any suffering of the animal is of ethical relevance. Hence, the feelings of the animal, such as pain, malaise, fear, anxiety and frustration should be concentrated on. However, it is difficult or sometimes impossible to record such feelings and, to date, it has proven to be very complicated to directly measure these. Instead, various signs that are regarded to be valid indications of such feelings are relied upon. Again, for reasons of transparency, those signs that are selected as 'indicators' and the arguments why they serve this purpose should clearly be included in the RA report. Commonly used indicators may include: limping, inactivity, behavioural stereotypies, increased levels of stress hormones, clinical signs of disease such as coughing or diarrhoea, etc.

In recent animal welfare risk assessments major welfare components have been identified, notably: pain, distress, fear, anxiety, malaise, frustration and behavioural disorders (see e.g. EFSA, 2007a). Although it is possible to consider the adverse effects of a particular hazard for each separate component, this approach is cumbersome and time-consuming. However, unless AW experts could agree to integrate the various components in a mathematical function, in which these are weighted, it is equally difficult to calculate a single 'overall' welfare outcome. Until now, no such efforts have been undertaken in risk assessment exercises, although attempts to follow such an approach are currently being made in the framework of the WelfareQuality® project. For future risk assessments, it should be seriously questioned whether seeking consensus on the weighting of the various welfare components (i.e. prior to conducting an RA) would not yield results that are of more use to the risk manager. Until then, it would be useful if risk managers would more clearly indicate which of the welfare components they consider of overriding importance for their purpose.

It is of great importance in AWRA exercises to agree on how the concept of health is interpreted, as health is a significant component of wellbeing (Broom, 1996). Lerner (2008) interviewed 8 different veterinary surgeons with various backgrounds, such as laboratory animal veterinarians, pathologists, cardiologists, zoo animal experts, horse practitioners etc. and found that they defined health in six different ways; (1) health as normal biological function, (2) health as freedom from disease, (3) health as performance, iv) health as being well or living in a sound way, (5) health as absence of diseases, injuries and unwanted mental states, and, finally (6) health as normal recovery processes and control of the situation. Also, their definition of animal welfare was varying.

Experts involved in an AWRA may have different appreciations of what health is. Usually, in veterinary medicine, health is regarded as the mere absence of disease. Such a dichotomous definition may be useful e.g. in epidemiological studies of the occurrence of disease in various populations. However, health as a co-determinant (component) of welfare should also be related to other welfare aspects such as the animal's ability to cope (Broom, 1996).

For instance, if the animal is shedding *Salmonella* bacteria, one may assume this bacterium may possibly affect the microbial ecology of the gut, which could ultimately even lead to disease. However, if there are no clinical signs, which occurs when the animal apparently copes with the organism and all other demands and becomes a symptomless 'carrier', its welfare is not compromised. Therefore, it would appear that a more clinically oriented definition of health is necessary. Gunnarsson (2006) examined 500 textbooks in veterinary medicine and found that only 8% had an explicit form of definition of health. The definitions used fell into five categories; (1) health as normality, (2) health as biological function, (3) health as homeostasis, (4) health as physical and psychological well-being, and (5) health as productivity including reproduction. It would seem useful, before starting any RA, to discuss within the expert group the concepts of health and its contribution to animal welfare, so as to facilitate the reliable estimation of adverse effects of a health hazard.

## 4. Animal-based, management-based and resource-based measures

As a result of an increased concern for animal welfare, legislators as well as the private sector have increased efforts to ensure a better animal welfare in animal husbandry. Legislators on national and international (EU) level as well as other international bodies (e.g. OIE) and the private industry have set up regulations or schemes that are to provide farming conditions that will promote animal welfare (see e.g. Vessier *et al.*, 2008). Almost all such regulations or schemes refer to conditions that should be provided in the animals' environment or through management procedures (Bock and Van Leeuwen, 2005). These may include space allowances, measures of air quality (such as relative humidity, dust, temperature, level of manure gasses), access to straw, appropriate feed, other resources such as nest boxes for laying hens etc. It could involve interventions such as tail docking, teeth clipping or debeaking. However, even if all these prevail, such does not in itself guarantee the welfare of the animal, it only facilitates to monitor the compliance with requirements.

Resource-based, as well as management-based measures have only indirect impact on the welfare of the animal, i.e. although they will eventually affect the state of welfare of the animal, many other factors such as the animal's ability to cope (e.g. immune status or age) should not be overlooked. In this respect, outcome-based parameters, such as lameness, flight response, injuries, agonistic interactions, diarrhoea, etc. are more suitable as these indicate the actual effect of the resource- and management-based measures and serve as indicators for the state of welfare of the animal as regards its attempts to cope (to use the definition of the OIE, 2009).

One might argue that one of the most important animal-based parameters to consider in the assessment of animal welfare would be the presence of pain. However, the measurement of pain is complicated and requires either taking samples for physiological analysis or rather sophisticated behavioural analyses (Weary *et al.*, 2006). Also, the scientists' attitudes to pain may differ. Phillips (1993) reported that a majority of animal researchers in New York City did not provide analgesics to laboratory animals recovering from surgery and they admitted no inclination to do so. Ninetysix percent of French veterinarians were reported to be concerned about recognition and alleviation of animal pain but, although 84% of them said they used analgesics following orthopaedic surgery, only 17% of the respondents used analgesics following castration (Hugonnard *et al.*, 2004). Thus, an outcome-based parameter such as pain may be difficult to recognize and the opinions on the importance of such a central parameter in relation to animal welfare may vary considerably.

## 5. Recruitment of a team for making a risk assessment

To be able to conduct a risk assessment for animal welfare, a team of experts with a relevant scientific background is needed. A risk assessor, i.e. a person with good knowledge and insights in risk assessment methodology, has to assure that the steps of the process are carefully followed and he or she should also provide the necessary mathematical and computer based tools for risk calculations. However, it is of importance that this risk assessor also has some insights into animal welfare science. As welfare covers a broad field such as presence of injuries, symptoms of disease, pathophysiological states and behavioural disorders [see Nordenfelt (2009, this Volume) and Broom (2009, this Volume)] and within these sciences scientific data ranges from well quantified to qualitative, an awareness of the methodological aspects of handling such data is crucial. Thus, the role of the risk assessor is also to coach the animal welfare experts in the process of conducting a risk assessment.

In addition, welfare scientists with particular expertise pertaining to the scope of the risk manager's question(s) are needed to identify relevant hazards, to define and assess the adverse effects on welfare of the animal and to estimate exposure of the target group in question.

### 5.1 The expert's basic education

Experts in animal welfare may have various forms of basic training which would appear to have primed their research focus. Experts may include, for instance, veterinarians, animal scientists or biologists having majored in ethology. Their research focus may vary from infectious diseases to social behaviour or from physiological stress markers to clinical pathology. Bracke *et al.* (2008) were able to demonstrate differences between veterinarians and ethologists in their assessment of calf welfare by analysing the working group's composition vs. the outcome of the 'Scientific opinion on the risks of poor welfare in intensive calf farming systems' (EFSA, 2006). In general, veterinarians scored the welfare of calves to be somewhat better for most housing systems, although it did not affect the ranking of the systems. Differences in scoring may relate to different appreciations of the importance of various elements of an adverse effect. However, in this particular scientific opinion the description of the severity of an adverse effect was not much elaborated. Severity was roughly described in 5 classes, ranging from 'slightly adverse', *via* 'adverse', 'moderately serious' and 'serious' to 'very serious'. Therefore, different experts may have had different appreciations of what the adverse effect on welfare really was.

If the adverse effect of the hazard is stated in the afore-mentioned general terms only, such leaves room for different interpretations. Whereas one expert may refer to the effect a hazard has on behaviour (e.g. frustrated sucking), another may refer to pain or impaired immuno-competence. Therefore, the more explicitly the adverse effects are described, the better the possibilities are for experts to appropriately address the impairment of welfare.

### 5.2 The expert's research area

In later reports such as the 'Scientific Report on animal health and welfare in fattening pigs in relation to housing and husbandry' (EFSA, 2007b), the adverse effects were more thoroughly described. (e.g. a 'severe' effect being described as 'Involving explicit pain, malaise, frustration, fear or anxiety – Strong stress reaction – dramatic change in motor behaviour – vocalization may occur'). It is likely that these more elaborate descriptions have prevented some of the variations in interpretations described earlier. Yet, in addition, it would seem advisable to ensure that those welfare scientists are recruited to participate in an RA team, who have sufficient research experience in the areas under scrutiny.

But even if this has been considered, experts may still vary in their appreciation of the adverse effect as a result of their various backgrounds. An expert having spent a career on studying behavioural frustrations may overtax such effects in comparison to a scientist who has in-depth knowledge about immunocompetence and *vice versa*. Hence, the development of a procedure based on well-defined criteria for selecting experts is crucial and such criteria should be presented for reasons of transparency.

### 5.3 Experts on exposure

Experts may be very familiar with the degree to which animals are exposed to certain hazards in their home countries, but are likely to have little if any information or experience from other countries or areas. Usually, reliable databases on exposure to most hazards are not available. In the framework of the EFSA calf welfare RA exercise (EFSA, 2006), a group of veterinarians – experts in clinical practice in calf production – was identified, in an attempt to provide more accurate estimates of the degree of exposure by basing these on the practitioners familiar with the various European calf production systems currently existing. Criteria for their invitation were: (1) predominantly engaged in clinical practice, (2) extensive clinical experience in calf medicine, and (3) covering various geographical areas where calf production is significant in the EU. Another important criterion was that the consultant should not be affiliated with the calf production industry, to avoid their assessment being tainted by other interests. In total 7 veterinarians accepted an invitation to assist in the exposure assessment. The experience of the individuals covered the various husbandry systems and important veal producing countries in Europe.

Without solid exposure data, the outcome of the risk assessment relies heavily on informed expert opinion. The uncertainty of the exposure estimate must then be made clear and the expert be made aware of the importance of not overtaxing the accuracy of their exposure estimates.

## 6. Getting people familiar with animal welfare risk assessment and the risk assessor familiar with animal welfare

As indicated earlier, properly discussing and agreeing upon a definition of welfare is the first thing to do when an RA group first meets. As hazards to welfare, resulting from identification of compromised animal needs are to be listed by the welfare scientists, questions are likely to be raised on which needs the animal species under scrutiny may have (on needs, see section 1 of this Volume). The chosen welfare definition largely determines these needs and both definition and needs must – for reasons of transparency – be clearly indicated in the RA report. Starting a risk assessment on anything less than consensus on these points is not productive. To reach this point of agreement is no small task, which should be undertaken by the chairperson of the expert group. Time and patience must be allocated to this procedure and an enduring leadership is a prerequisite for a successful completion of the RA exercise.

When a risk manager has asked for a risk assessment on animal welfare to be conducted, he should (by a clear formulation of his risk question) clarify the animal species under scrutiny, and more specifically indicate the target population or subpopulation (e.g. animal of a particular age group) to be addressed. This sounds easier than it is. There are several reasons for that. For instance, when risk managers are responsible for a large geographical area (e.g. the EU commission), substantial variation in climatic and cultural circumstances relevant to the animal's environment are to be expected. Examples of such variations include (1) animals in cold climates being kept indoors for a longer period, (2) marked differences in the scale of farming between various countries, (3) prevalence of specific breeds of an animal species used more in some countries than in others, (4) production methods differing among countries, as is the case for the production of veal calves, etc. Therefore, there is a risk that indiscriminately 'averaging' risks to serve a paneuropean purpose will likely yield meaningless risk estimates that are far from reality.

Such and other methodological problems are extensively discussed in the contribution by Smulders (2009, this Volume).

## 7. Developing a method that supports efficient risk management

As stated above, the understanding and appreciation of what animal welfare encompasses, and to what extent it should be of concern to farmers as well as society may vary. With growing attention and concern in society, efforts have been increased to develop a methodology for the scientific assessment of animal welfare through qualitative and semi-quantitative approaches. Contributions by Smulders (2009) and Ribó and Serratosa (2009) in this Volume describe the state-of-the-art in RA methodology.

Besides providing, ideally, a quantification of hazards and risks, probably the major benefit for a risk manager is the ranking of the various risks, which allows him to make a judgement on which of these should be addressed with some degree of priority. Given the budgetary and political constraints risk managers have to consider, the ranking of hazards and risks contained in RA reports serves as an important tool in decision making. Incidentally, this consideration has proven helpful to explain to less-informed welfare scientists about to be engaged in RA working groups what the ultimate goal of their efforts should be, i.e., in colloquial terms: *'if the available budget would be, say, 100 Euro, you are to indicate from your scientific perspective the justifiability of spending 70 on managing risk A, 20 on risk B, 10 on risk C and possibly 0 on risks D through F'*. Obviously, the ultimate decision on this issue is made by risk managers, who will also have to consider other 'legitimate' factors [e.g. see the contribution by Gavinelli and Ferrara (2009); this Volume].

In this context a word of caution is timely. It is important to consider that the European legislation (EC, 1998) stipulates that no practices should be allowed that would cause unnecessary pain, suffering or injury for the animal. Hence, although the target animal population at large may be rarely exposed (as indicated by a low score for 'exposure assessment') to a particular hazard that is considered major (as indicated by a high 'hazard magnitude') and, consequently, the risk estimate for this hazard is low, a risk manager cannot afford to disregard such a hazard, but rather might consider to take targeted action which should generally not be associated with a significant budgetary burden.

The World Trade Organisation (WTO) recognizes the Office International des Épizooties (OIE) as the international organisation to set standards relating to animal health. The OIE has recognised animal welfare as an important factor to be taken into account when animals are transported, slaughtered or killed. If animal welfare is to be discussed e.g. as part of WTO negotiations, firstly a generally accepted, repeatable, transparent and validated methodology for the assessment of risks for animal welfare must be in place. With a consensus view on what animals need, on what are relevant components of welfare and how to address these (i.e. separately or as co-contributing to 'overall' welfare), the further development of a useful and practicable method appears to be entirely feasible. Once this has been accomplished, the animal welfare risk assessment may become an important tool underpinning trade negotiations. At present there are no agreements made under the WTO that would allow a country to differentiate between similar products produced with different animal welfare standards. However, in negotiations that should ultimately lead to bilateral agreements, one should not underestimate the significance of properly conducted animal welfare risk assessments and the role they may play in the future.

## 8. Conclusions

### 8.1 What has been achieved?

The discussion on adequately defining the concept of welfare has intensified, and one has come to realise how the multi-factorial character of the concept of animal welfare complicates risk assessment methodologies for this area. A method for the risk assessment of animal welfare has been developed that is useful and practicable for identifying and ranking risks of poor welfare as well as for ranking the magnitude of various welfare hazards.

### 8.2 What has been neglected?

Notwithstanding these achievements, the welfare science community has neither reached consensus on a generally accepted definition of animal welfare, nor yet entered the discussion on whether or not the various components of welfare can possibly be 'weighted' to allow an 'overall' estimation of welfare. Also, the issue of which expertise or combination of expertises should be relied upon for conducting risk assessment exercises has not been given enough attention.

### 8.3 What needs to be done?

The process of conducting an animal welfare risk assessment should include a formalised identification of the point of departure in relation to which aspects (components) of welfare are of major concern. Only provided such is clearly communicated between risk managers and risk assessors prior to accepting an RA mandate, will RA exercises become efficient. Should consensus be reached on the feasibility of weighting welfare components, such would allow eliminating some of the methodological problems risk assessors are faced with. Finally, one must develop a formalised procedure to identify and select the individual experts to be recruited for an AWRA expert group.

### Acknowledgements

The author wishes to acknowledge the risk assessors and EFSA staff who have contributed substantially to discussions and thoughts on animal welfare risk assessment, notably; Dr. Frans Smulders, Dr. Telmo Nunes, Dr. Christine Müller-Graf, Dr. Jan Hultgren, Dr. Moez Saana and Dr. Oriol Ribó.

## References

Algers, B. and Uvnäs-Moberg, K., 2007. Maternal behaviour in pigs. Hormones and Behavior 52, p. 78-85.

Bock, B. and van Leeuwen, F., 2005. Review of socio-political and market development of animal welfare schemes. In: Roex, J. and Miele, M. (Eds.). Farm animal welfare concerns, consumers, retailers and producers. WelfareQuality® Report No. 1. Cardiff University, Cardiff, UK, pp. 115-167.

Bracke, M.B.M., Edwards, S.A., Engel, B., Buist, W.G. and Algers B., 2008. Expert opinion as 'validation' of risk assessment applied to calf welfare. Acta Veterinaria Scandinavica 50, 29.

Broom, D.M., 1991. Animal welfare: concepts and measurement. Journal of Animal Science 69, 4167-4175.

Broom, D., 1996. Animal welfare defined in terms of attempts to cope with the environment. Acta Agricultura Scandinavica (Section A) Animal Science, Suppl. 27, 22-28.

Broom, D.M., 2009. Animal welfare and legislation. In: Smulders, F.J.M. and Algers, B. (Eds.). Food safety assurance and veterinary public health, Vol.5, Welfare of production animals: assessment and management of risks. Wageningen Academic Publishers, Wageningen, the Netherlands, pp. 339-352.

CAC (Codex Alimentarius Commission), 1999. Principles and guidelines for the conduct of microbiological risk assessment. CAC/GL-30, 7 pp. Available at: http://www.codexalimentarius.net/download/standards/357/CXG_030e.pdf. Accessed 20 May 2009.

Castrén, H., Algers, B., de Passillé, A.-M., Rushen, J. and Uvnäs-Moberg, K., 1993. Preparturient variation in progresterone, prolactin, oxytocin and somatostatin in relation to nest building in sows. Applied Animal Behaviour Science 38, 91-102.

Dawkins, M.S., 1983. Battery hens name their price: consumer demand theory and the measurement of ethological 'needs'. Animal Behaviour 31, 1195-1205.

Dawkins, M.S., 1988. Behavioural deprivation: a central problem in animal welfare. Applied Animal Behaviour Science 20, 209-225.

Duncan, I.J.H., 1993. Welfare is to do with what animals feel. Journal of Agricultural and. Environmental Ethics 6, Suppl. 2, 8-14.

Duncan, I.J.H. and Fraser, D., 1997. Understanding animal welfare. In: Appleby, M.C. and Hughes, B.O. (Eds.). Animal welfare. CAB International, Wallingford, UK, pp. 19-31.

European Commission (EC), 1998. Council Directive 98/58/EC of 20 July 1998 concerning the protection of animals kept for farming purposes. Official Journal of the European Communities L 221, 23-27. Available at: http://eur-lex.europa.eu/LexUriServ/LexUriServ.do?uri= OJ:L:1998:221:0023:0027:EN:PDF

European Commission (EC), 2000. First report on the harmonisation of risk assessment procedures. Part 1: The report of the Scientific Steering Committee's Working Group on Harmonisation of Risk Assessment Procedures in the Scientific Committees advising the European Commission in the area of human and environmental health. European Commission, Brussels, Belgium.

European Commission (EC), 2006. Communication from the Commission to the European Parliament and the Council on a Community Action Plan on the Protection and Welfare of Animals 2006-2010, COM (2006) 13 final. Available at: http://ec.europa.eu/food/animal/ welfare/com_action_plan230106_en.pdf. Accessed 10 June 2009.

European Food Safety Authority (EFSA), 2006. Opinion of the Scientific Panel on Animal Health and Welfare (AHAW) on a request from the Commission related with the risks of poor welfare in intensive calf farming systems. Question No. EFSA-Q-2005-024, The EFSA Journal 366.

European Food Safety Authority (EFSA), 2007a. Scientific Opinion of the Panel on Animal Health and Welfare on a request from the Commission on the animal health and welfare aspects of different housing systems for adult breeding boars, pregnant, farrowing sows and unweaned piglets. The EFSA Journal 572, 1-13.

European Food Safety Authority (EFSA), 2007b. Scientific Opinion of the Panel on Animal Health and Welfare on a request from the Commission on animal health and welfare in fattening pigs. The EFSA Journal 564, 1-14.

Gavinelli, A. and Ferrara, M., 2009. Animal welfare – the risk manager's perspective. In: Smulders, F.J.M. and Algers, B. (Eds.). Food safety assurance and veterinary public health, Vol.5, Welfare of production animals: assessment and management of risks. Wageningen Academic Publishers, Wageningen, the Netherlands, pp. 353-369.

Gunnarsson, S., 2006. The conceptualization of health and disease in veterinary medicine. Acta Veterinaria Scandinavica 48, 20.

Gunnarsson, S., Matthews, L.R., Foster, T.M. and Temple, W., 2000. The demand for straw and feathers as litter substrates by laying hens. Applied Animal Behaviour Science 65, 321-330.

Houseal, G.A. and Olson, B.E, 1995, Cattle use of microclimates on a northern latitude winter range, Canadian Journal of Animal Science 75, 501-507.

Hultgren, J., 2009. Animal welfare risk assessment and management from a national perspective. In: Smulders, F.J.M. and Algers, B. (Eds.). Food safety assurance and veterinary public health, Vol.5, Welfare of production animals: assessment and management of risks. Wageningen Academic Publishers, Wageningen, the Netherlands, pp. 461-482.

Hugonnard, M., Leblond, A., Keroack, S., Cadore, J. and Troney, E., 2004. Attitudes and concerns of French veterinarians towards pain and analgesia in dogs and cats. Veterinary Anaesthesia and Analgesia 31, 154-163.

Jensen, P. and Toates, F.M., 1993. Who needs 'behavioural needs'? Motivational aspects of the needs of animals. Applied Animal Behaviour Science 37, 161-181.

Lawrence, A.B., Petherick, J.C., McLean, K.A., Deans, L.A., Chirnside, J., Vaughan, A., Clutton, E. and Terlouw, E.M.C., 1994. The effect of environment on behaviour, plasma cortisol and prolactin in periparturient sows. Applied Animal Behaviour Science 39, 313-330.

Lerner, H., 2008. The concepts of health, well-being and welfare as applied to animals. linköping studies in arts and science No. 438. Dissertations on Health and Society No. 13. Linköping University, Linkoping, Sweden. 212 pp.

McGlone, J.J., 1993. What is animal welfare? Journal of Agricultural and. Environmental Ethics 6, Suppl. 2, 26-36.

Nordenfelt, L., 2009. The concept of animal welfare; a philosopher's view. In: Smulders, F.J.M. and Algers, B. (Eds.). Food safety assurance and veterinary public health, Vol.5, Welfare of production animals: assessment and management of risks. Wageningen Academic Publishers, Wageningen, the Netherlands, pp. 29-44.

OIE (Office International des Épizooties), 2004a. Handbook on import risk analysis for animals and animal products. Volume 1. Introduction and qualitative risk analysis. OIE, Paris, France.

OIE (Office International des Épizooties), 2004b. Handbook on import risk analysis for animals and animal products. Volume 2. Quantitative Risk Asessment. OIE, Paris, France.

OIE (Office International des Épizooties), 2009. Terrestrial animal health code, chapter 7.1 article 7.1.1. Available at: www.oie.int/eng/normes/mcode/en_chapitre_1.7.1.htm. Accessed 10 June 2009.

Oltenacu, P.A., 2009. Genetic improvements for production and animal welfare. In: Smulders, F.J.M. and Algers, B. (Eds.). Food safety assurance and veterinary public health, Vol.5, Welfare of production animals: assessment and management of risks. Wageningen Academic Publishers, Wageningen, the Netherlands, pp. 417-441.

Phillips, M.T., 1993. Savages, drunks and lab animals: the researcher's perception of pain. Society and Animals 1, 61-81.

Rauw, W.M., Kanis, E., Noordhuizen-Stassen, E.N. and Grommers, F.J., 1998. Undesirable side effects of selection for high production efficiency in farm animals: a review. Livestock Production Science 56, 15-33.

Ribó, O. and Serratosa, J., 2009. History and procedural aspects of the animal welfare risk assessment at EFSA. In: Smulders, F.J.M. and Algers, B. (Eds.). Food safety assurance and veterinary public health, Vol.5, Welfare of production animals: assessment and management of risks. Wageningen Academic Publishers, Wageningen, the Netherlands, pp. 305-335.

Rollin, B.E., 1993. Animal welfare, science and value. Journal of Agricultural and. Environmental Ethics 6, Suppl. 2, 44-50.

Serratosa, J. and Ribó, O., 2009. International context and impact of EFSA activities in animal welfare in the European Union. In: Smulders, F.J.M. and Algers, B. (Eds.). Food safety assurance and veterinary public health, Vol.5, Welfare of production animals: assessment and management of risks. Wageningen Academic Publishers, Wageningen, the Netherlands, pp. 275-303.

Smulders, F., 2009. A practicable approach to assessing risks for animal welfare – methodological considerations. In: Smulders, F.J.M. and Algers, B. (Eds.). Food safety assurance and veterinary public health, Vol.5, Welfare of production animals: assessment and management of risks. Wageningen Academic Publishers, Wageningen, the Netherlands, pp. 239-274.

Thorpe, W.H., 1969. Welfare of domestic animals. Nature 244, 18-20.

Vessier, I., Butterworth, A., Bock, B. and Roe, E., 2008. European approaches to ensure good animal welfare. Applied Animal Behaviour Science 113, 279-297.

Weary, D.M., Niel, L., Flower, F.C. and Fraser, D., 2006. Identifying pain in animals. Applied Animal Behaviour Science 100, 64-76.

Webster, J., Saville, C. and Welchman, D., 1986. Improved husbandry systems for veal calves. Farm Animal Care Trust, London, UK.

Wolf, U., 1990. Das Tier in der Moral. Klostermann. Frankfurt am Main, Germany. 169 pp.

Wolf, J.-C., 1992. Tierethik: neue Perspektiven für Menschen und Tiere. Paulusverlag. Freiburg, Switzerland, 188 pp.

# A practicable approach to assessing risks for animal welfare – methodological considerations

**Frans J.M. Smulders**
Department of Production Animal Medicine and Veterinary Public Health, University of Veterinary Medicine, Vienna, Austria, e-mail: frans.smulders@vu-wien.ac.at.

## Summary

This contribution addresses the methodological approach to conducting a formal Risk Assessment (RA) for animal welfare, as developed in the past 5 years during various RA's as a result of mandates by the European Commission to the European Food Safety Authority (EFSA) and performed by a number of working groups instituted by its Scientific Panel on Animal Health and Welfare (AHAW). The model (variants) published to date in RA exercises on welfare risks of various animal species and/or their subpopulations (i.e. veal calves, sows/boars, fattening pigs, seals, salmon) produced under different pre-harvest production systems are critically reviewed, their limitations discussed and a 'generic' approach, considered suitable for Animal Welfare Risk Assessment (AWRA) in general, is presented. Finally, suggestions are made on how the HACCP concept could serve for identifying Risk Management (RM) options.

**Keywords**: risk assessment, animal welfare, methodologies, HACCP

## 1. Introduction

Risk analysis in the proper sense of the concept includes 3 components, i.e. Risk Assessment (RA), Risk Management (RM) and Risk Communication (RC). These three elements are not necessarily addressed by one and the same actor or faction. For instance, within the European Union, risk analysis of food production (including foods of animal origin) is the joint responsibility of the European Food Safety Authority (EFSA; addressing RA and RC) and the European Commission (addressing RM).

Risk assessment has been developed as a tool for decision makers to choose between different options in a given situation. It represents a set of rational and probability based methods, applicable to situations, where decisions need to be based on limited information. RA guidelines in the animal production area, aimed mainly at the control of contagious disease and food safety, have been formulated by the Codex Alimentarius Commission (CAC) and the Office International des Épizooties (OIE), whilst similar guidelines for plant production have been issued by the International Plant Protection Convention (IPPC). The joint responsibility for developing and continually harmonising the RA approach lies with these three organisations, as has been stipulated in the Sanitary and Phytosanitary (SPS) Agreement of the World Trade Organisation. Hence, RA is a widely used term in the areas of epidemiology and food safety (for which tried and commonly accepted methodologies have been established), introducing the concepts of *hazard* (an infectious agent or risk factor) and *risk* (a function of the probability of occurrence of a hazard and its consequences in case of occurrence).

Fundamental principles allowing substantiation of the outcome of risk assessment are that RA should be:

- *Science-based*: The process should be based on the best available evidence, i.e. on results that have been obtained by relying on recognized scientific methods.
- *Well-documented*: It should be assured that all available scientific information is considered and kept available for review.
- *Objective*: The problems to be addressed in RA should be clearly stated.
- *Repeatable:* A group of experts other than the one engaged in RA should reach the same conclusions on the basis of the information compiled.
- *Transparent*: The methodologies and data used for RA should be clearly documented and uncertainties should be clearly identified and taken into consideration in the final assessment.

Risks are preferably assessed quantitatively (provided enough data are available), semi-quantitatively (when data are scarce) or qualitatively (in absence of quantitative data). Though still useful for purposes of risk communication and -management, the outcome of a qualitative risk assessment is inevitably more subjective.

## 2. Choosing a suitable approach for assessing animal welfare risks

When it was first contemplated to conduct a formal RA exercise in the area of animal welfare it was carefully considered which of the various approaches (international guidelines; see above) would be most suitable. Although international standards and guidelines for risk analysis according to the SPS agreement may vary between CAC, OIE and IPPC, essentially the same structure is recognizable in all documents, i.e. the main questions to be addressed include (1) what can cause risk? (2) how does it cause risk? (3) what is the probability of a risk occurring? (4) what are the consequences? and (5) what are the prerequisites for risks to indeed occur?. In the framework of analyzing risks for animal welfare it is most appropriate to consider the definitions formulated by OIE and CAC.

Whereas OIE's main focus is on risks for animal and human health associated with animal imports and its definition of RA also includes the economic consequences of entry, CAC concentrates on domestic risks and exclusively addresses risks related to microbiological, chemical and physical agents of serious public health concern. To avoid animal welfare issues being confounded by economic considerations EFSA decided to follow the approach (and terminology/definitions) of CAC, i.e. along the lines suggested in its 22nd session (CAC, 2002). Incidentally, these guidelines have been characterized by the CAC as 'interim', because they are subject to modifications in the light of the developments in the science of RA and as a result of efforts to harmonize definitions across various disciplines.

For purposes of Animal Welfare Risk Assessement (AWRA) the definitions of CAC obviously needed to be adapted. The latter was essentially achieved by replacing the word 'public health' by 'animal welfare' wherever applicable. This has yielded a new set of definitions (see Table 1) that has since been further used in all AWRA exercises by EFSA.

*Table 1. Definitions used in Animal Welfare Risk Assessment (AWRA) partly adapted from Risk Assessment terminology formulated by the Codex Alimentarius Commission (CAC, 2002).*

| | |
|---|---|
| *Dose-response assessment*[1] | The determination of the relationship between the magnitude of exposure of animals to certain hazards and the severity and frequency of associated adverse effects on animal welfare |
| *Exposure assessment*[1] | The quantitative and qualitative evaluation of the likelihood of hazards to welfare occurring in a given animal population. |
| *Hazard*[1] | Any factor, occurring from birth to slaughter or the end of the animal's life, with the potential to cause an adverse effect on animal welfare. |
| *Hazard characterization*[1] | The qualitative and quantitative evaluation of the nature of the adverse effects associated with the hazard. Considering the scope of AWRA the concerns relate exclusively to animal welfare. |
| *Hazard identification*[1] | The identification of any factor, from birth to slaughter or the end of the animal's life, capable of causing adverse effects on animal welfare. |
| *Hazard magnitude* | The combination of intensity and duration of adverse effects. |
| *Risk*[1] | A function of the probability of an adverse effect and the severity of that effect, consequent to a hazard for animal welfare. |
| *Risk characterization*[1] | The process of determining the qualitative or quantitative estimation, including attendant uncertainties, of the probability of occurrence and severity of known or potential adverse effects on welfare in a given animal population based on hazard identification, hazard characterization, and exposure assessment. |
| *Quantitative risk assessment* | A risk assessment that provides numerical expressions of risk and an indication of the attendant uncertainties (stated in the 1995 expert consultation definition on risk analysis). |
| *Qualitative risk assessment* | A risk assessment based on data which, while forming an inadequate basis for numerical risk estimations, nevertheless, when conditioned by prior expert knowledge and identification of attendant uncertainties, permits risk ranking or separation into descriptive categories of risk. |
| *Risk analysis* | A process consisting of three components: risk assessment, risk management and risk communication. |
| *Risk assessment* | A scientifically based process consisting of the following steps: (1) hazard identification, (2) hazard characterization, (3) exposure assessment and iv) risk characterization. |
| *Risk communication* | The interactive exchange of information and opinions concerning the risk and risk management among risk assessors, risk managers, consumers and other interested parties. |
| *Risk estimate* | Output of risk characterisation. |
| *Risk management* | The process of weighing policy alternatives in the light of the results of risk assessment and, if required, selecting and implementing appropriate control options (i.e. prevention, elimination, or reduction of hazards and /or minimization of risks) options, including regulatory measures. |

*Table 1. Continued.*

| | |
|---|---|
| *Sensitivity analysis* | A method to examine the behaviour of a model by measuring the variation in its outputs resulting from changes to its inputs. |
| *Transparent* | Characteristics of a process where the rationale, the logic of development, constraints, assumptions, value judgements, decisions, limitations and uncertainties of the expressed determination are fully and systematically stated, documented, and accessible for review. |
| *Uncertainty analysis* | A method used to estimate the uncertainty associated with model inputs, assumptions and structure/form. |

[1] Definitions altered to suit AWRA purposes

## 3. A brief overview of the various methods used to date in EFSA's AWRA exercises and their limitations

In responding to various requests from the European Commission, EFSA has, over the past few years, issued a number of scientific opinions on the risks for animal welfare and for this purpose commissioned their Scientific Panel on Animal Health and Animal Welfare (AHAW) to form working groups that included – besides a large number of experts on animal welfare scientists with a background in RA. In this section four approaches to AWRA (3.1 through 3.4) documented in six scientific opinions that – when this contribution was conceived (May 2009) – had been published by EFSA are summarized and the limitations of each approach are indicated.

### 3.1 The risks of poor welfare in intensive calf farming systems (EFSA, 2006a)

Guided by a list of 'needs of animals' (Broom, 1996), all production factors to which calves (held under defined husbandry conditions in Europe) are subjected and which may represent a hazard to their welfare were identified and characterized in terms of severity [the latter assigned with a severity score (HC in the final calculation; see below)]. Subsequently, the working group estimated the exposure of the entire European calf population held under the defined production systems, cross-checked these estimations with a consultation group of field experts (see below) with the aim to reach a consensus opinion on exposure assessment (EA) scores. Then a risk estimate score (RE) was calculated (RE = HC x EA). The analysis, carried out separately for nine different production options, was semi-quantitative and based on the calculated RE's risks associated with each hazard were classified as 'negligible', 'minor' or 'major'. When published data were absent or considered insufficient to allow risk estimation, risks were classified as 'uncertain'. Executive summary charts provided graphical, more quantitative presentations for the scores of each production system.

#### *3.1.1 Hazard identification*

The relation between the hazards and impaired needs were listed in a table. Possible interactions between hazards were not considered.

### 3.1.2 Hazard characterization (HC)

The characterization of the impact of the hazards on animal welfare was based on expert opinion, using a 5-point qualitative scoring scale (see Table 2) substantiated where possible by published records.

### 3.1.3 Exposure assessment (EA)

The exposure (% of the population exposed to a given hazard) was scored on a 5-point scale based on a quintile division (i.e. increments of 20%; see Table 3). This was done by reaching consensus after consultation of an expert group of independent veterinarians with experience in the various husbandry systems practiced in different European countries.

### 3.1.4 Risk estimation and characterization

The risk for poor welfare was estimated by multiplication (HC×EA), which yielded a risk estimate score which was subsequently expressed by integrating both the hazard characterization and the exposure assessment using a classification matrix (see Table 4). Thus, qualitative risk expressions were generated. The ultimate risks were classified as 'negligible' (multiplied scores 1-8), 'minor' (multiplied scores 9-16) and 'major' (multiplied scores 20-25).

*Table 2. Hazard characterization score (calf welfare RA; from EFSA, 2006a).*

| Evaluation | Code | Score | Explanation |
|---|---|---|---|
| Hazard characterisation | SA | 1 | Slight adverse effect |
| | AE | 2 | Adverse effect |
| | MS | 3 | Moderately serious |
| | SE | 4 | Serious |
| | VS | 5 | Very serious |

*Table 3. Exposure assessment scores(calf welfare RA; from EFSA, 2006a).*

| Evaluation | Code | Score | Explanation |
|---|---|---|---|
| Exposure assessment | VR | 1 | Very rare (1 to 20%)[1] |
| | RA | 2 | Rare (21 to 40%) |
| | MF | 3 | Moderately frequent (41 to 60%) |
| | FR | 4 | Frequent (61 to 80%) |
| | VF | 5 | Very frequent (81 to 100) |

[1] Zero percentage is considered not applicable.

*Table 4. Risk classification matrix (calf welfare RA; from EFSA, 2006a).*

| Effect | Very rare (1-20%) | Rare (21-40) | Moderately frequent (41-60%) | Frequent (61-80%) | Very frequent (81-100%) |
|---|---|---|---|---|---|
| Slight adverse | Negligible | Negligible | Negligible | Negligible | Negligible |
| Adverse | Negligible | Negligible | Negligible | Negligible | Minor |
| Moderately serious | Negligible | Negligible | Minor | Minor | Minor |
| Serious | Negligible | Negligible | Minor | Minor | Major |
| Very serious | Negligible | Minor | Minor | Major | Major |

### *3.1.5 Uncertainty and variability*

When insufficient data were available for the exposure assessment, the risk was characterized as uncertain and when not enough data were available to support the view that a certain factor would constitute a hazard these were not further considered.

*Weaknesses/limitations of the approach – summarized*
- Does not allow for variation in severity or exposure (inherent to classification).
- Descriptors in classification tables not sufficiently transparent (open to interpretation).
- Quality (reliability) and availability of (published) data not considered.
- No Uncertainty Analysis in HC (based on quality of published or expert data). In EA mere indication of presence-or-absence of uncertainty.
- Duration of adverse effects was not separately scored although considered in the HC (severity score).
- As a consequence of HC and EA scores being discrete, the Risk Estimate (HC×EA) scale is discontinuous (remedied by designing a Risk outcome matrix).

## 3.2 Pig welfare risk assessment (EFSA 2007a,b,d)

Three similar RA's were conducted (sows/boars, fattening pigs, tail-biting). For each step of production, experts in the working groups were asked to individually fill out the tables for each of the three target populations, based on current scientific knowledge and published data. Their scores were compared and if there was disagreement, further data collection efforts were made (by revisiting the scientific literature and by compilation of exposure data from questionnaires sent to field experts in Europe) to allow coming to a consensus opinion.

The experts were divided in groups each conducting separate RA's for different target groups, (Sows and Boars, Fattening pigs), considering their life cycle (Outdoor vs. Indoor production system), thereby distinguishing between either the life stage of the animals (Sows and boars and fattening pigs report), production- (outdoor vs. indoor holding in fattening pigs report) or management-system (docked vs. undocked pigs in tail-biting report).

### 3.2.1 Hazard identification

Same procedure as described for calf welfare.

### 3.2.2 Hazard characterization

For each hazard a quantitative estimate of its adverse effect on the welfare of the individual animal (*'magnitude'*) was made, for which purpose both its *'severity'* (expressed by a score indicating the animal's physiological/behavioural response (see Table 5) and its *'duration'* (expressed in days) were separately considered.

In addition the *'likelihood'* [i.e. the probability of the individual animal suffering the adverse effect of a hazard, assuming exposure to a given scenario and expressed as %] was estimated and minimum, most likely and maximum values for duration of the effect indicated. Precise information on the latter was hardly available from literature, and so these ranges were generally provided by the experts as estimates that served as parameters of a Beta-Pert distribution expressing their uncertainty.

The numerical score characterizing the hazard was calculated as follows:

Magnitude = (severity score/4) × duration (in days)

### 3.2.3 Exposure assessment

For each of the factors the probability (in %) of the animal target population in Europe being exposed to the hazard at a defined intensity and duration was expressed as minimum, most likely and maximum values (Table 6).

*Table 5. Severity scores of the adverse effects (pig welfare RA; from EFSA, 2007a,b,d).*

| Severity of the adverse effect | Descriptive definition | Score |
|---|---|---|
| Critical | Fatal, death occurs either immediately or after some time | 4 |
| Severe | Involving explicit pain, malaise, frustration, fear or anxiety<br>Strong stress reaction, dramatic change in motor behaviour, vocalization may occur | 3 |
| Moderate | Some pain, malaise, frustration, fear or anxiety<br>Stress reaction, some change in motor behaviour, occasional vocalization may occur | 2 |
| Limited | Minor pain, malaise, frustration, fear or anxiety<br>Physiological effects may be recorded as well as moderate behavioural changes | 1 |
| Negligible | No pain, malaise, frustration, fear or anxiety | 0 |

*Table 6. Qualitative uncertainty scores for the likelihood and exposure (pig welfare RA; from EFSA, 2007a,b,d).*

| | |
|---|---|
| Low | Solid and complete data available |
| | Strong evidence provided in multiple refs |
| | Authors report similar conclusions |
| Medium | Some but no complete data available |
| | Evidence provided in small number of refs |
| | Authors' conclusions vary from one to other; |
| | Solid and complete data available from other species which can be extrapolated to the species considered |
| High | Scarce or no data available |
| | Rather evidence provided in unpublished reports, based on observations or personal communications |
| | Authors' conclusions vary considerably between them |

### *3.2.4 Risk characterization*

The numerical score estimating the risk of each hazard expresses the welfare burden for the animal target population in Europe. It was calculated as follows:

Risk estimate = magnitude × likelihood of effect × probability of exposure

### *3.2.5 Uncertainty and variability*

The uncertainty (i.e. both that related to the 'likelihood of effect prevailing in an individual animal' and to the 'probability of the target population being exposed') was expressed both qualitatively (via scoring the availability and reliability of literature or exposure data as shown in Table 7) and quantitatively (by indicating the ranges of the risk estimates).

*Weaknesses/limitations of the approach – summarized*
- Descriptors in classification tables not sufficiently transparent (open to interpretation).
- Formula for magnitude assumes linearity of the severity scores.

*Table 7. Uncertainty Classification matrix to assess the overall qualitative uncertainty (H, M or L) of a risk estimate (pig welfare RA; from EFSA, 2007a,b,d).*

| | | Exposure uncertainty | | |
|---|---|---|---|---|
| | | High (H) | Medium (M) | Low (L) |
| Likelihood uncertainty | High (H) | H | H | H |
| | Medium (M) | H | M | M |
| | Low (L) | H | M | L |

- In EA the intensity could only be expressed by the presence-or-absence of the factor. Only rarely could the hazard be defined quantitatively (e.g. concentration of ammonia in the range 25-49 ppm); partly remedied through introducing specific exposure scenario's describing defined combinations of EA intensities and durations.
- In HC both the uncertainty about severity and duration of the adverse effect not taken into account (not included in the tables).

## 3.3 Animal welfare aspects of the killing and skinning of seals (EFSA, 2007c)

Major practices for catching and killing seals (i.e. using nets, physical methods or firearms) were identified. The likely chain of events and the associated inherent hazards to welfare were described in various scenarios (in the report termed 'risk pathways'; see Figure 2 of Ribó and Serratosa, 2009) for each of which RA was conducted. A total of 11 scenarios' (also taking account of different weather and/or habitat conditions) were considered. As the number of publications on the issue was very limited the assessment primarily relied on data provided by a small number of experts and consequently the methodological approach was largely qualitative. For this reason expressing RA parameters in numerical or graphic terms was abandoned.

### *3.3.1 Hazard identification*

The hazards related to stunning, killing and skinning of seals were listed for each scenario. Interactions between hazards were not considered.

### *3.3.2 Hazard characterization*

The magnitude of the adverse effect was assessed relying on classifications of the severity (termed 'intensity' in the report) and duration, using 4-point scales for both. For severity (intensity) and duration, qualitative descriptions and quantitative time scales were used, respectively (Table 8).

In view of published data being hardly available, the magnitude of the effect was not numerically expressed but classified using the classification matrix in Table 9.

### *3.3.3 Exposure assessment*

Exposure was ranked in 4 classes: 'Very unlikely', 'Likely', 'Unlikely', 'Very likely'.

### *3.3.4 Risk characterization*

Rather than expressing risks in numerical or descriptive terms, they were characterized by merely presenting the magnitude and the likelihood of a given adverse effect. Since most of the data resulted from expert opinion, disagreements among experts were accounted for by listing the lower and upper limits of their estimates (see Figure 3 of Ribó and Serratosa, 2009). When factors like weather and habitat were considered to affect the likelihood of a given adverse effect, those scenarios were presented.

*Table 8. Severity ('intensity') and duration scales used in the HC of seal killing/skinning (from EFSA, 2007c).*

| Evaluation | | Code | Category |
|---|---|---|---|
| Hazard characterization – intensity (intensity of the adverse effect – recognition of pain and distress) | | Se | Severe |
| | | Mo | Moderate |
| | | Mi | Mild |
| | | Ne | Negligible |
| Hazard characterization – duration (duration of the adverse effect) | | | |
| | Physical methods and firearms | 4 | >1 min |
| | | 3 | 30-60 sec |
| | | 2 | 5-30 sec |
| | | 1 | <5 sec |
| | Netting | 4 | >30 min |
| | | 3 | 15-30 min |
| | | 2 | 5-15 min |
| | | 1 | <5 min |

*Table 9. Categories of magnitude of adverse welfare effect based on duration and severity (intensity) of adverse welfare effects used in seals risk assessment (from EFSA, 2007c).*

| Magnitude of adverse welfare effect | | Duration of adverse welfare effect | | | |
|---|---|---|---|---|---|
| | | 4 | 3 | 2 | 1 |
| Intensity of adverse welfare effect | Severe | Major | Major | Moderate | Moderate |
| | Moderate | Major | Moderate | Moderate | Minor |
| | Mild | Moderate | Moderate | Minor | Minor |
| | Negligible | Moderate | Minor | Minor | Negligible |

### *3.3.5 Uncertainty and variability*

Uncertainty was classified relying as indicated in Table 10. Two measures of uncertainty were identified (i.e. availability and the reliability of the data source) in a report annex including the tables with the scores of the various experts.

*Weaknesses/limitations of the approach – summarized*

- Fully applying the improved RA model was not considered [possibly for reasons of lack of data (see below)] allowing little if any quantification.
- Very restricted availability of published quality data and of experts, which generated high uncertainties.

*Table 10. Data source scores used in seals risk assessment (from EFSA, 2007c).*

| Data source score | Definition |
|---|---|
| 3 | Scarce or no data available |
| | Unpublished data or opinion |
| | Expert opinion |
| 2 | Technical opinion |
| | Peer reviewed publications with contradictory results |
| | Non-peer reviewed publication |
| | Some but incomplete data |
| | Comparative data from other species |
| 1 | Solid and complete data available |
| | International peer reviewed publication |

- The terms with which severity ('intensity') was described in HC are not transparent (open to interpretation); it was merely stated that they were based on pain and distress recognition. Note that the original term severity was changed into intensity.
- The descriptors for the EA classification were not defined.
- The criteria by which uncertainty should be classified were lacking.

## 3.4 Animal welfare aspects of husbandry systems for farmed Atlantic salmon (EFSA, 2008)

This represents the first of a series of reports on the welfare of fish and has recently been adopted by EFSA's AHAW panel. As was the case for seals, it was recognized that quantitative data related to production systems and effects are very limited so it was decided to follow a largely qualitative approach primarily relying on expert opinion. Five major groups of factors potentially impairing salmon welfare – i.e. environmental (biotic and abiotic), feed & feeding, husbandry, genetics, and, finally, the impact of disease and disease control measures – were ranked as to their priority in terms of representing risks and it was estimated which hazards are most important for each life stage to enable a comparison of the different production systems. Life stages were not compared due to their different length.

The aim of the exercise was to rank the risks to allow prioritization of research needs and legal measures to be contemplated. As opposed to other RA's where these had already been considered, the effects of genetics were not taken into account since those effects are by definition life long. The RA in salmon was carried out similarly as those discussed before (hazard identification; hazard characterisation; exposure assessment and risk characterisation). Experts were asked to score the different factors and scoring was then discussed within the panel.

The severity of the adverse effect was assessed as presented in Table 11. Scores were attributed according to scientific evidence of the level of physiological and behavioural responses.

The duration of the adverse effects were scored on a 0 to 100% scale considering the rest of the life of the fish and not just the particular life stage mentioned. The likelihood of adverse effects was scored as presented in Table 12. The uncertainty value (low, medium and high), estimating the degree of confidence in the information, was not used in the calculation but as extra-information to describe the context (Table 13).

The exposure assessment (frequency of exposure and the duration of the hazard) was scored according to Table 14.

The duration of the hazard as such (how long would the hazard prevail) as well as remaining after-effects were considered for the relevant life stage of the fish. For instance, whereas a predator attack might only last for a short time, a temperature change of a longer duration might ensue. The duration of the hazard during a life stage was indicated in % (0% to 100%). In cases where a certain hazard would lead to instant death and would therefore

*Table 11. Severity of adverse effect (from farmed Atlantic salmon RA; EFSA, 2008b).*

| Evaluation | Score | Explanation |
|---|---|---|
| Negligible | 0 | No pain, malaise, frustration, fear or anxiety as evidenced by measures of the normal range of behavioural observations, physiological measures and clinical signs for >95% of the species or strain/breed |
| Mild | 1 | Minor changes from normality and indicative of pain, malaise, fear or anxiety |
| Moderate | 2 | Moderate changes from normality and indicative of pain, malaise, fear or anxiety |
| Substantial | 3 | Substantial changes from normality and indicative of pain, malaise, fear or anxiety |
| Severe | 4 | Extreme changes from normality and indicative of pain, malaise, fear or anxiety, that if persist would be incompatible with life |

*Table 12. Likelihood of adverse effect occurring, i.e. proportion of population affected (from farmed Atlantic salmon RA; EFSA, 2008b).*

| Evaluation | Score | Explanation |
|---|---|---|
| Negligible | 0 | The event would almost certainly not occur |
| Extremely low | 1 | The event would be extremely unlikely to occur |
| Very low | 2 | The event would be very unlikely to occur |
| Low | 3 | The event would be unlikely to occur |
| Moderate | 4 | The event would occur with an even probability |
| High | 5 | The event would be very likely to occur |

*Table 13. Uncertainty (from farmed Atlantic salmon RA; EFSA, 2008b).*

| Evaluation | Score | Explanation |
|---|---|---|
| Low | 1 | Solid and complete data available: strong evidence in multiple references with most authors coming to the same conclusions (e.g. in a meta analysis). |
| Medium | 2 | Some or only incomplete data available: evidence provided in small number of references; authors' conclusions vary. |
| | | Solid and complete data available from other species which can be extrapolated to the species considered. |
| High | 3 | Scarce or no data available: evidence provided in unpublished reports, or based on observation or personal communications; authors' conclusions vary considerably |

*Table 14. Frequency of exposure (from farmed Atlantic salmon RA; EFSA, 2008b).*

| Evaluation | Score | Explanation |
|---|---|---|
| Negligible | 0 | The exposure would almost certainly not occur |
| Extremely low | 1 | The exposure would be extremely unlikely to occur |
| Very low | 2 | The exposure would be very unlikely to occur |
| Low | 3 | The exposure would be unlikely to occur |
| Moderate | 4 | The exposure would occur with an even probability |
| High | 5 | The exposure would be very likely to occur |

rule out subsequent adverse welfare effects this was indicated. The latter is relevant, firstly because some risk scores may conflict with the reader's intuition (e.g. large mortality is also considered a welfare problem) and secondly to ensure comparability with other risk assessments.

### *3.4.1 Risk characterization*

For each hazard a semi-quantitative risk score for each life stage in all of the production systems employed during this life stage was calculated as follows:

Risk score = (severity of adverse effect) × (duration of the adverse effects) × (likelihood of adverse effect) × (frequency of hazard) × (duration of hazard)

The scores of frequency of hazard, severity and likelihood of effect were standardized to give even weighting to the scores (frequency of hazard/5, severity/4, and likelihood of hazard/5). Duration of hazard and duration of effect were divided by 100. Finally, the risk score was multiplied by 100 to allow for their easier interpretation.

Uncertainty scores, not included in the risk estimate calculation, were indicated in the last column. This also indicates which areas need to be further researched. The two uncertainty scores for hazard characterization and exposure, respectively, were integrated in a single figure according to an uncertainty classification matrix (same as presented in Table 7; but in addition, H, M and L were associated with the scores 3, 2 and 1, respectively).

*Weaknesses/limitations of the approach- summarized*

- Largely qualitative exercise inspired by the stated purpose of RA, probably related to limited data availability (see below).
- Very limited amount of quantitative and good data from literature or experts available.
- Interaction between factors makes RA very difficult.
- Different life stages with very different conditions make a 'total' description of fish welfare difficult.
- Problem in scoring the 'duration of adverse effect' arises when the animal dies as a consequence of a particular hazard. This can be described in two different ways depending on how the concept of 'life time' is interpreted (see next bullet point).
- It is virtually impossible to consider death as a primary welfare problem. If the adverse effect is fatal then the duration before death (i.e. an animal would not be subjected to suffering) would be the key welfare issue, even though death itself might indicate a primary welfare problem. If life time is considered as the 'potential life time', rapid (or instant) death resulting from being exposed to a certain hazard has a very short (or practically no) duration. When defined in more absolute terms ('not being alive') the duration of the adverse effect (i.e. death) would be 100%. In the case of the fish welfare, it was decided to score the duration of the effect over the 'potential life time' of the animal, but to indicate if a hazard was so severe that it could lead to instant death.

## 4. Experiences and lessons learned from previous Risk Assessment excercises

The various approaches summarized above have largely been based on existing assessment methodologies published in Codex Alimentarius RA guidelines on food safety, after these were adopted for purposes of addressing animal welfare issues. The various risk assessors involved gradually improved the model taking into account as many elements of the afore-mentioned guidelines as – at the time – seemed feasible for the exercise.

Over the past few years a number of methodological problems have been identified, some of a general nature (and consequently equally relevant for e.g. the epidemiology/food safety area), some inherently related to AWRA. In the following these are discussed with a focus on those difficulties encountered in the latter.

## 4.1 The relevance of formulating proper RA questions and appreciating the limitations of RA

### *4.1.1 Statement of purpose of the RA exercise*

When authorities set out to develop effective strategies for the management of risks it is crucial that 'tailor-made' risk assessments become available. To allow for this it is paramount that risk managers carefully consider what exact purposes these RA's should serve so risk assessors can deliver a document specifically addressing these essentials ['the better the question, the better the risk assessment' (Maijala, 2006)]. These purposes (essentials) are to be stated clearly in any RA document.

### *4.1.2 A clear definition of animal welfare components on which RA needs to be concentrated*

In contrast with RM in the area of food safety for which, as a rule, risks resulting from a single clearly defined hazard are assessed (e.g. *Salmonella* in one particular product, produced by a defined processing method in a specified production unit; see section 9 of this contribution] risk questions in the animal welfare area tend to be formulated less precisely. The concept 'welfare' encompasses various 'welfare components' that need to be addressed to secure successful assurance of overall welfare. In animal welfare terms these components are often associated with animal needs [see e.g. 'The risks of poor welfare in intensive calf farming systems' (EFSA, 2006a)]. If these needs are not satisfied this may result in, more or less serious, adverse welfare effects. Major adverse effects identified include: pain, distress, fear, anxiety, malaise, frustration, behavioural disorders. Consequently, from a methodological point of view, it is virtually impossible to calculate a single 'overall welfare' risk outcome, unless animal welfare experts could agree to integrate the various components in a mathematical function in which these are weighted. The latter is an important consideration for deciding which modeling approach is feasible (see the comments under 'linearity'; below). Although a similar issue was successfully dealt with in the WelfareQuality® project by a large representation of EU welfare scientists (see Manteca *et al.*, 2009, this Volume) it is unlikely that general consensus on such an approach can be reached within the community of animal welfare experts, so it would be useful for risk managers to indicate which of the welfare components is/are considered of overriding importance for their purpose.

### *4.1.3 A clear definition of the animal target population*

Whilst the animal target population for which an RA is commissioned is often indicated in general terms (e.g. calves, dairy cows, pigs, fish) one must realize that the hazards these species are subjected to vary for different husbandry systems, animal age groups, geographical regions/climatic conditions, etc. Consequently, when an RA for the entirety of Europe is to be conducted, the number of production options and associated scenarios that need to be considered will be substantial. Particularly where a (semi)quantitative approach to RA is feasible in principle, this will inevitably lead to extremely time-consuming exercises, or, alternatively, to expert groups being discouraged to conduct these and instead rather being inclined to follow a more qualitative approach and/or restricting the number of production options (target populations) under scrutiny. Depending on the desired outcome risk managers are well-advised to take the latter into consideration.

### 4.2 The necessity of building on previous AWRA experience

In the various RA exercises by EFSA at least one scientist with a background in risk assessment was included. From the relatively primitive approach followed in the calf welfare report a more elegant RA model was gradually developed that appears to be suitable for conducting many welfare RA exercises. To assure that such a growing 'generic' RA model is based on past experience it is essential that the various risk assessors are aware of the methodological problems already addressed and/or partly solved and remain involved in those with which newly attracted assessors are faced. Although room for the application of alternative approaches (particularly in situations where data availability is limited) should remain, it appears counter-productive to allow risk assessments to be conducted without proper consideration of the already developed model or model variants.

Also, for reasons of transparency it would appear crucial that in further extending the methodology of welfare RA one relies on the existing terminology whenever possible and that before new terms are introduced these are carefully screened for correctness and/or compatibility with risk assessment approaches recommended by the responsible organizations (CAC, OIE, IPPC).

### 4.3 The availability and quality of published data and experts

The reliability of the outcome of every RA is dependent on the quality of the data on which it is based. For reasons of transparency the information on hazards and their adverse effects is preferably to be found in the published literature, its reliability being dependent of the following parameters (prerequisites or preferable conditions in parentheses):

- Methods of generating data (scientifically accepted).
- Availability of species-specific data (i.e. relating to the animal species addressed in RA).
- Completeness of data (all answers to relevant questions provided).
- Number of publications containing these data (multiple references).
- Solidity of data (in refereed 'quality' journals, recognized expert authors).
- Degree of consensus about data (similar conclusions in multiple references).

It should be noted that it would be helpful if agreement is found on what can be considered an adequate data base to allow a solid conclusion (how many publications can be considered 'multiple'?).

When there is a scarcity of data, when it is of lesser quality or is unpublished, one must rely on the opinion of experts. Inherently, for less well-publicized themes, only a limited number of scientists with knowledge of the area can be identified.

A qualitative expression of the uncertainty analysis is based on consulting an uncertainty classification table. The experience gained in the AWRA exercises discussed earlier has shown that working groups need to be continually reminded that uncertainties classified as 'medium' or 'low' (as defined in the uncertainty classification tables) must be substantiated as such by ensuring that references are included in the report ('open to review').

# 5. General factors complicating RA of animal welfare

## 5.1 Following an appropriate RA approach – use of models

Dransfield and Scheffer (1991) provided basic arguments for what approach might be most suitable in developing models. In the following two paragraphs the essentials are summarized.

Statistic (empirical, descriptive, input-output, black-box, stochastic) models essentially relate the 'output' or dependent variable (y) of the studied object to the 'input' or dependent variable (x) in an empirical way, as in the typical example of simple linear regression: y = ax + b, by which a straight line representing the best fit is plotted through the data. When no additional knowledge is available a straight line is the best choice. However, one should remain aware of the danger of drawing nonsensical conclusions that are sometimes suggested by a linear regression approach. When theoretical knowledge on contribution phenomena is available to substantiate this, one may choose to apply a multiple regression approach (i.e. a curvilinear relationship).

Deterministic (analytical, theoretical, process-orientated, interpretative) models essentially describe the mechanisms underlying a phenomenon in a mathematical function. Such functions are simplified abstractions of the real world, designed to study one special aspect only (complex hypotheses about how the system works). Such models need to be constantly subjected to careful scrutiny whether the theoretical assumptions on which the function is based still apply ('garbage in, garbage out'). Hence, such models can be applied more universally because the differentiated model structure allows for adaptation for all kinds of model situations. Designing computer-based programs ('expert systems') in which all available expert knowledge is compiled is a cumbersome process as it involves obtaining and classifying the relevant information from experts, many of which appear to be unable to explain how they arrived at their final conclusion.

In the RA of animal welfare, both stochastic and deterministic models can be considered. The multi-factorial character of animal welfare and the discussion within the animal welfare science community whether or not weighting factors for each of the various welfare components can be assigned in the first place, complicates the decision on what approach is most suitable. In the developing phase of welfare risk assessment the 'informed opinion' of welfare experts plays an important role as long as theoretical knowledge is incomplete. On the other hand, the RA exercises commissioned to expert working groups in the framework of mandates from the European Commission generate an important data base (to be continuously updated to assure its validity) that can be used for developing computer-based 'expert systems'. The results of 'semantic' models based on such systems [e.g. those developed by Bracke *et al.* (2008), addressing specific welfare problems for which certain factors (e.g. attributes of a housing system) are analyzed] are promising, and these will certainly contribute to improving the reliability of RA models particularly in terms of hazard characterization and to assessing the suitability of various production systems.

## 5.2 Particular problems associated with Uncertainty Analysis

An intrinsic part of risk assessment (RA) is the estimation of the degree of uncertainty of various model inputs, assumptions and of structure/form (CAC, 2002). In exercises dedicated to the assessment of animal welfare, published data supporting a quantitative (or semi-quantitative) analysis are not always available or are scarce. Hence, in general, the RA of animal welfare has a largely qualitative character, i.e. data (informed opinions) are solicited from scientists active in [or associated with ('consultants')] the RA working group who are particularly knowledgeable in the area and the formulation of a joint 'expert' opinion is sought.

Two principal uncertainty factors influence the outcome of RA, i.e. the likelihood of an individual animal's welfare being adversely affected assuming exposure to the particular hazard, and the proportion of the animal population subjected to the hazard in a geographical region under a certain specifically defined scenario of animal production. The former is preferably assessed by scrutinising scientific studies published in high quality, peer-reviewed publications, whereby the degree to which the results are corroborated in studies by fellow-scientists determines the degree of confidence in the correctness of the final statement. However, inherent to the general nature of scientific publications on animal welfare, few if any published data are available on the degree of exposure of animal populations to a particular hazard. In addition, one must realize (1) that considerable differences in exposure exist across Europe, dependent on the various animal production or processing systems, and (2) that welfare experts involved in a particular area are not necessarily aware of the situation outside their immediate activity arena.

Also, the scientific arguments underpinning a *consensus* opinion of experts have not necessarily been published in the literature, either because these are sometimes considered to be 'too obvious' to be specifically included in a scientific publication or because few if any other scientists, besides those already involved in RA, are active or have published in the specific field of research.

In many AWRA reports [e.g. those on the killing and skinning of seals (see above), or in the WRAPSTUN report on welfare aspects of the stunning and killing of animals (Algers *et al.*, 2009) the afore-mentioned particularly applies: few scientists world-wide are active in these fields. Consequently, relatively few ('corroborating') publications are available and reliable documents on the exposure of animals in Europe to the prevailing systems are scarce.

Regardless of the credibility of the experts *per se,* the risk assessor exclusively relying on the 'experience' of few has by consequence of the fundamental principles of RA (science-based; well documented, objective, repeatable, transparent) no other option than to attribute a high uncertainty score for many hazards and exposure rates. This will inevitably lead to a number of hazard entries ultimately generating risk estimates with a level of uncertainty which in the perception of some experts is higher than justified.

### 5.3 Other concerns

In the predominantly stochastic approach chosen by most welfare risk assessors, it was assumed: (1) that there is no interaction between hazards, (2) that there is linearity in the severity scores and (3) that hazards are mutually exclusive. However, it remains to be considered if such assumptions are justified.

Interaction between hazards is not unlikely to occur. For instance, exposure to dust and exposure to ammonia will together increase the risk of pneumonia and associated welfare problems. Consequently, an inherent danger of comparing risk estimates calculated for individually listed hazards is that their significance may be over- or underestimated, particularly when mechanisms of interaction are less known. Also, the adverse effects may interact (e.g. fear causing stress or acute stress reducing the perception of pain). Conversely, when there is consensus about major interactions and their underlying mechanism, these could be considered – e.g. by entering major scenarios prevailing in (industry) practice in RA tables – to enable reliable scoring of interactive hazards.

Often (i.e. particularly in the absence of evidence from literature suggesting otherwise) linearity of adverse effects is assumed so as not to unnecessarily complicate risk calculation. However, it is more than questionable if (see Table 11), for instance, 4 time units of severe pain (level 3) equals 6 time units of moderate pain (level 2) or 12 time units of minor pain (level 1). In this example it must also be considered that the pain experienced over time can be attenuated or aggravated dependent of the application or withholding of proper pain treatment. Finally, it should be realized that when multiple adverse effects (pain, frustration, behavioural disorders, etc.) result from a single hazard, the durations (duration-1; see below) of the different effects may be different, in some cases possibly even longer than the exposure to the hazard (duration-2; see below). Again, provided a clear understanding of the presence-or absence of linearity is given, these can be addressed by describing specific scenarios.

Clearly defining the terminology of RA is essential to ensure transparency of the methodological approach as well as to allow repeatability of the process when this is deemed useful at a later stage. To this end including a clearly formulated glossary of terms and tables with unmistakably defined descriptions in an RA is essential.

## 6. Composing and correctly interpreting RA tables

A typical RA table includes a Hazard Identification (HI), Hazard Characterization (HC), an Exposure Assessment (EA) and a Risk Characterization (RC) section. Table 15, taken from the Scientific Report on the effects of farming systems on dairy cow welfare and disease (EFSA, 2009b), illustrates which typical entries are included in such tables. In the sections 6.1 to 6.4 each of these entries (indicated in bold italics) is further explained.

It is essential to realize that AWRA starts by assessing adverse effects a hazard may have for the INDIVIDUAL animal. This is achieved by completing the hazard identification and characterization steps. Subsequently, in the exposure assessment step it is considered what

*Table 15. RA of the impact of housing in straw yards on metabolic and reproductive disorders in dairy cows (adapted from EFSA, 2009c).*

| Hazard identification | | Hazard characterization | | | |
|---|---|---|---|---|---|
| Hazard description | Hazard specification (scenario) | Adverse effect | Severity | Duration (-1) (in days) | Hazard magnitude |
| 1 Inadequate ventilation, inappropriate airflow, airspeed | too low ventilation (1/3 of winter period) | reduced feed intake, immunosuppression, less oestrus expression, reduced fertility, SARA[a], ketosis. | 2 | 20 | 10 |
| 2 Inadequate ventilation, inappropriate airflow, airspeed | too low ventilation (1/3 of indoor period + part of summer) | reduced feed intake, immunosuppression, less oestrus expression, reduced fertility, SARA, ketosis. | 3 | 80 | 60 |
| 3 Insufficient light level (day/night) | too dark (for both cows and stockperson) | reduced disease detection, reduced fertility, sub-optimal feed intake | 1 | 20 | 5 |
| 4 Light duration | too short (<14 h) | reduced fertility, sub-optimal feed intake | 1 | 25 | 6.25 |
| 5 Inadequate feeding installation | automatic feeder failure | reduced or excessive feed intake, acute ruminal acidosis | 4 | 7 | 7 |
| 6 Inadequate bedding | hygiene, composition and quantity (in zero-grazing) | SARA, ketosis, reduced fertility, pain | 1 | 60 | 15 |
| 7 Inadequate bedding | hygiene, composition and quantity (with access to pasture) | SARA, ketosis, reduced fertility, pain | 1 | 20 | 5 |
| 8 Inadequate floor (limited to passage ways, feeding and milking areas) | too slippery, too hard, injuring, too rough | ketosis, reduced fertility | 1 | 30 | 7.5 |
| 9 Lack of space, e.g. for exercising, social interactions and resting | -- | immunosuppression, less oestrus expression, reduced fertility, ketosis | 1 | 30 | 7.5 |
| 10 Lack of facilities for sick animals | -- | downer cow, (check calving problems in the rest of this table) | 2 | 7 | 3.5 |
| 11 Insufficient access to water | inappropriate system design | reduced dry matter intake, SARA, ketosis | 1 | 120 | 30 |

| | | | | Exposure assessment | | | | | | | Risk characterization | |
|---|---|---|---|---|---|---|---|---|---|---|---|---|
| Quantitative assessment of likelihood (%) | | | Qualitative assessment of the uncertainty | Duration (-2) (days per year) | Intensity (scenario) | Quantitative assessment of probability of exposure (%) | | | Qualitative assessment of the uncertainty | | Risk estimate | Overall qualitative uncertainty of the risk estimate |
| min | most likely | max | | | | min | most likely | max | | | | |
| 25 | 35 | 45 | M | 60 | <60 m3 air/500 kg LW[b]/h | 5 | 10 | 15 | M | | 0.35 | M |
| 25 | 35 | 45 | L | 120 | <60 m3 (300 in summer) air/500 kg LW/h | 5 | 10 | 15 | M | | 2.10 | M |
| 0 | 5 | 10 | H | 45 | <40 lux | 5 | 10 | 15 | M | | 0.03 | H |
| 0 | 2 | 4 | M | 60 | <14 h | 15 | 30 | 50 | M | | 0.04 | M |
| 30 | 50 | 70 | L | 1 | full exposure | 0.1 | 0.2 | 0.3 | M | | 0.01 | M |
| 20 | 30 | 40 | L | 365 | full exposure | 20 | 30 | 40 | M | | 1.35 | M |
| 20 | 30 | 40 | L | 180 | full exposure | 20 | 30 | 40 | M | | 0.45 | M |
| 1 | 5 | 10 | M | 365 | full exposure | 30 | 50 | 70 | M | | 0.19 | M |
| 5 | 10 | 15 | M | 365 | full exposure | 20 | 30 | 40 | H | | 0.23 | H |
| 0.5 | 1 | 1.5 | L | 365 | full exposure | 50 | 70 | 90 | H | | 0.02 | H |
| 30 | 50 | 70 | M | 365 | any cow can't access water for 5 h | 1 | 5 | 10 | H | | 0.75 | H |

*Table 15. Continued.*

| Hazard identification | | Hazard characterization | | | |
|---|---|---|---|---|---|
| Hazard description | Hazard specification (scenario) | Adverse effect | Severity | Duration (-1) (in days) | Hazard magnitude |
| 12 Insufficient access to water | broken system, management problem, etc… | dehydration, reduced dry matter intake, SARA | 2 | 1 | 0.5 |
| 13 Inadequate or lack of handling/restraining facilities | -- | reproductive failure, injuries | 3 | 10 | 7.5 |
| 14 Inappropriate temperature, humidity | see scenario under 'intensity' | reduced feed intake, ketosis, SARA, reproductive failure | 1 | 20 | 5 |
| 15 Inappropriate temperature, humidity | see scenario under 'intensity' | dehydration, reduced feed intake, ketosis, SARA, reproductive failure | 2 | 12 | 6 |
| 16 Inappropriate temperature, humidity | see scenario under 'intensity' | dehydration, reduced feed intake, ketosis, SARA, reproductive failure | 3 | 5 | 3.75 |
| 17 Poor calving conditions | absence of pen, pen design, facilities (inability to separate from other animals) | metritis, dystocia, reduced fertility, prolonged parturition, reduced feed intake, fatty liver, downer cows | 1 | 21 | 5.25 |
| 18 Too few feeding places in zero grazing systems | - | reduced feed intake, ketosis, SARA, reproductive failure | 2 | 60 | 30 |
| 19 Too few feeding places (cows with access to pasture) | - | reduced feed intake, ketosis, SARA, reproductive failure | 2 | 60 | 30 |

[a] SARA = subacute ruminal acidosis.
[b] LW = live weight.
[c] THI = temperature humidity index.

proportion of the entire animal population would indeed suffer the adverse effect if exposed to the specified hazard.

## 6.1 Hazard identification

For listing factors possibly representing hazards to animal welfare it is helpful to first identify various ***production/management/handling options*** that differ substantially in the type of

| | | | | Exposure assessment | | | | | | Risk characterization | |
|---|---|---|---|---|---|---|---|---|---|---|---|
| Quantitative assessment of likelihood (%) | | | Qualitative assessment of the uncertainty | Duration (-2) (days per year) | Intensity (scenario) | Quantitative assessment of probability of exposure (%) | | | Qualitative assessment of the uncertainty | Risk estimate | Overall qualitative uncertainty of the risk estimate |
| min | most likely | max | | | | min | most likely | max | | | |
| 70 | 80 | 90 | L | 1 | total lack of water | 0 | 0.1 | 0.2 | H | 0.00 | H |
| 0.01 | 0.1 | 0.2 | H | 10 | full exposure | 30 | 50 | 70 | H | 0.00 | H |
| 15 | 30 | 45 | M | 30 | 75<THIc<78 | 50 | 60 | 70 | M | 0.90 | M |
| 40 | 50 | 60 | M | 15 | 78<THI<83 | 40 | 50 | 60 | M | 1.50 | M |
| 60 | 80 | 100 | M | 5 | THI>83 | 10 | 15 | 20 | M | 0.45 | M |
| 10 | 15 | 20 | L | 7 | full exposure | 30 | 50 | 70 | H | 0.39 | H |
| 5 | 10 | 15 | L | 365 | >1 cow per place | 5 | 10 | 15 | H | 0.30 | H |
| 5 | 10 | 15 | L | 215 | >1 cow per place | 10 | 20 | 30 | H | 0.60 | H |

hazards to be considered [e.g. major husbandry-, transport- or lairage/stunning systems practiced for a particular species; in Table 15 for instance specified as the effects of housing for dairy cows held in straw yards]. Each of these is to be presented in a separate RA table, in which specific categories of interest (e.g. housing, nutrition, management, genetics etc.) are systematically analyzed for relevant hazards (e.g. hazards potentially leading to metabolic and reproductive disorders and related to housing conditions, as in Table 15). Obviously, it is to be ensured that all members of an AWRA working group are sufficiently familiar with the specifics of the various production options and related scenarios and how these are practiced *lege artis*. Only then can one afford to begin with the listing of potential hazards.

An approach which has proven useful for not overlooking relevant factors is through cross-checking ('challenging') production options with a list of 'animal needs' using 'guide words' [e.g. 'too high' or 'too low', or, preferably, more quantitative terms such as 'more than' or 'less than' a critical value, i.e. rather similar to the HAZOP technique applied in HACCP studies on food safety (Baird Parker, 1987)]. This procedure leads to a comprehensive list of ***hazard descriptions*** to be considered further.

In the process of discussing the various hazards, the working group may decide it is necessary to separately address more specific ***scenarios***. These are included either under the heading ***hazard specification*** [to more clearly distinguish the attributes that make a factor a potential hazard (e.g. 'poor feed quality' in two scenarios thus distinguishing 'poor nutritive value' and 'presence of pathogens', or – as in Table 15 – 'insufficient access to water' distinguished in scenarios 11 and 12] or under the heading Exposure Assessment/Intensity (see below) to distinguish between levels of exposure. Should one – for reasons of table lay-out – choose to include shorter and less clear terminology, it is important to add a reference to where in the report the specifications are more clearly discussed. The latter approach will inevitably result in RA tables becoming rather extended. By the same token, it serves to increase transparency and enables the confrontation of stakeholders, who might otherwise give their own, not necessarily correct interpretation of the results presented.

Specific scenarios should reflect conditions commonly prevailing in practice. For reasons of practicability their inclusion should be limited to the most important ones (e.g. addressing climatic conditions, breed-specifics or indeed situations where hazards interact) on the understanding that the effects of other scenarios may possibly be estimated through extrapolation or (in exceptional situations) should be considered in a targeted, more specific RA.

## 6.2 Hazard characterization

To specify the ***adverse effects*** a hazard may have on an individual animal, its ***magnitude***, [expressed by both its ***severity*** and its ***duration(-1)***][1] needs to be calculated using the following formula (assuming severity scores range from 0 to 4):

Magnitude = severity score/4 × duration-1 of the effect

Depending on the period considered in a RA (e.g. compare lifetime or lactation period with the very short period between unloading and stunning/killing) and the time units in which the duration of adverse effects are expressed (e.g. days vs. minutes) the scale of the hazard magnitude will vary significantly between various RA's. Consequently – as duration-1 is

[1] The indication '-1' in duration-1 has been attached not to confuse this variable with the duration of a scenario to which – over the considered period (e.g. life cycle, period of lactation) – the entire animal population is exposed (duration-2; see 'exposure assessment').

included in the calculation of a risk estimate – one should be very cautious in comparing the absolute values of risk estimates of different RA's.

The fact that overall welfare is a composite of many components complicates the scoring of severity. Hence, it is important to carefully list which animal needs are compromised, which of the related specific welfare consequences (pain, frustration, etc.) are involved and what are the severity scores of the adverse effects.

Although it is principally possible to quantitatively measure pain in terms of strength of escape behaviour, frustration in terms of 'operant effort' or 'recovery', etc., such data are usually not available for practical conditions. Alternatively, one must rely on data from experimental studies or turn to distinguishing severity scores by identifying animal-based (clinical) indicators by way of classification tables. In composing such tables it is important to assure transparency, so that unclear descriptions that are open to anthropomorphic interpretation are avoided. Different welfare components such as fear, pain, frustration, etc. should ideally be scored separately, i.e. more clearly delineating 'explicit' from 'some' or 'minor' pain or alternatively – where appropriate – for malaise, for stress or for behavioural disorders, scored in unmistakable terms. For reasons of practicability such an approach has not been adopted in the RA's conducted to date, probably because this procedure would inevitably lead to an almost limitless range of combinations. Instead, severity score tables refer to 'overall welfare' related to each of the hazards as for instance the one developed in the pig RA reports (EFSA, 2007a,b,d).

Severity score tables should be tailor-made. For instance, using the highest severity score '1' as defined in the pig welfare approach (see Table 5) would, in the case of assessing welfare risks during stunning and killing (see Algers *et al.*, 2009), be useless since death is the ultimate purpose of stunning/killing procedures.

A slightly more focused approach (followed by an RA group addressing dairy cow welfare) is to concentrate a RA on specific welfare categories, such as lameness ('leg and locomotion'), mastitis ('udder problems'), 'metabolic and reproductive disorders' (see Table 15) and 'behavioural disorders' and to indicate for each hazard in these categories which welfare components are of overriding importance (EFSA, 2009a,b,c,d). This procedure at least generates an outcome indicating which of these are most compromised.

Should a hazard compromise more than one welfare component one could consider to indicate the ones of overriding importance by underscoring these in RA tables.

The ***duration-1*** of an adverse effect is important to estimate the impact of a hazard on the individual animal's welfare. It is expressed in time units (e.g. days, or even minutes where appropriate). Again, one should realize that different welfare components may be experienced over different periods (e.g. a short period of pain but a long period of frustration). Where one considers that such has a significant bearing on the risk outcome, there is no alternative than to insert separate scores for different welfare components.

In RA's a measure for agreement on the solidity of the HC (a quantitative expression of the likelihood) needs to be included. The extent to which the working group members can agree generally depends on the quality of the data available and/or (in case of scarcity or absence of data) principally on the match of 'informed opinions' of experts. By requesting experts to express their degree of certainty of the proportion of animal population likely to be affected by the adverse effect [through having them indicate a range from ***minimum*** (say 40%), ***most likely*** (say 60%) and ***maximum*** proportion (say 80%)] the likelihood can be modelled using a Beta-Pert distribution (***quantitative assessment of likelihood***). Consequently, values will range from 0 to 100%. Unless there is documented proof to suggest otherwise, a symmetrical distribution is assumed.

The quantitative measure of uncertainty of HC estimates is NOT included in the mathematical equation that ultimately yields the risk estimate, but its numerical range is rather used to determine the ***qualitative assessment of uncertainty of the risk estimate*** [expressed as H (high uncertainty), M (medium uncertainty) or L (low uncertainty)] which co-determines the overall measure of uncertainty of the RA as indicated in a matrix (see below).

### 6.3 Exposure assessment

The proportion (in %) of the entire animal target population in a specified area (e.g. Europe or parts thereof) that is exposed to a certain scenario is estimated in the exposure assessment.

To more specifically define the scenario under scrutiny it is necessary to determine what part of the considered time period (e.g. the entire life-cycle, lactation period) of the scenario applies. This variable, called ***duration-2***, is expressed in appropriate time units (days, or even minutes). Also, each scenario must – wherever possible – include a definition of the ***intensity*** of the hazard in quantitative terms (e.g. as in the exposure to light expressed in Lux). When this is not meaningful (e.g. animals continuously being exposed to improper management) it suffices to insert 'full exposure'. Thus, hazards with the same hazard description can be distinguished as regards their intensity with which the animal population is confronted. Although important for delineating different scenarios of exposure, the variable 'duration-2' is not further considered in the calculation that renders the risk estimate (see below).

The level of uncertainty about the EA data (***quantitative assessment of probability of exposure***) – assessed as described for HC (i.e. ***minimum, most likely, maximum*** probabilities) – leads to a ***qualitative description of EA uncertainty*** (H, M or L), which co-determines the overall measure of uncertainty as indicated in a matrix (see below).

In general, it is difficult to access reliable data sources for purposes of EA. This applies to virtually all areas in which RA is conducted. Only in rare cases are exposure data systematically collected. The latter is for instance the case for EA data on zoonoses where a defined approach to Europe-wide (yearly) monitoring has been laid down in legislation and exposure data are documented in a standardized way (European Commission, 2003). For animal welfare such a monitoring system does not yet exist and it is even unclear if and which national agencies in Europe consider it their task to develop a monitoring system. The EU action plan for Animal Welfare clearly states that the EU Commission intends to develop indicators to allow on-farm assessment and certification. In this context it is important to realize that even merely screening the European animal welfare situation at regular periods of multiple years would render a data base from which more reliable risk estimates can be calculated. This would serve as a justifiable basis for introducing more targeted Europe-wide legislation. Obviously, the latter presumes that some of the afore-mentioned methodological issues have been solved [e.g. which welfare components to primarily consider, how to measure these, measured through which animal- (or associated production factor-) based welfare indicators, etc.]. The WelfareQuality® EU project (see Manteca *et al.*, 2009, this Volume) is currently developing the tools for this purpose.

For purposes of generating (more reliable) estimations on EA, it usually does not suffice to exclusively rely on the information of working group members, as these might not be fully aware of the situation in an area such as the EU. This particularly applies when the geographical area to be covered is large. Consequently – pending mandatory monitoring schemes laid down in European legislation – it is recommendable to rely on consulting field experts from the various sub-regions (countries) and to solicit more detailed information, e.g. through 'consultants meetings'. One must ensure that the individuals delegated to such meetings are independent, preferably have access to documents substantiating their statements (or, alternatively, can base their estimations on a thorough knowledge of the local situation and the production/processing option under scrutiny). Only then can unnecessary arguments with stakeholders be prevented.

### 6.4 Risk characterization

Using Monte Carlo simulations and Pert probability distributions a ***risk estimate*** for each hazard can be calculated and its animal welfare burden in the considered population expressed as follows (using the figures from the shaded columns in Table 15):

Risk = magnitude of the effect) × (likelihood of the effect given a scenario of exposure) × (probability of the considered scenario of exposure)

For instance, for hazard description 2 in Table 15 the risk is:

60 × 0.35 (most likely value) × 0.10 (most likely value) = 2.10

Finally, the ***overall qualitative uncertainty of the risk estimate*** – taking into account both the qualitative uncertainties of hazard characterization and of exposure assessment – is determined using an uncertainty classification matrix (see Table 7).

## 7. Usefulness of graphics for expressing RA outcomes

Graphically expressing the RA outcome (i.e. principally distinguishing hazard magnitude and risk estimate in one figure) serves an important purpose, in that it allows the experts involved to identify the correctness of unexpected ('odd') outcomes that would at first sight not appear to make sense. Past experience shows that some of these are indeed the result of failure to properly fill out the tables. With few exceptions, welfare experts are less familiar with the RA methodology and without proper guidance by a risk assessor inclined to misinterpret some procedures.

Figure 1, taken from the AWRA of dairy cows (EFSA, 2009c), represents an example of such a graphic.

Also, graphics serve to highlight the areas on which risk managers should consider concentrating their attention. Obviously, the more animals in a population that are affected the more immediacy is required in taking action. However, a word of caution is timely in this context.

Hazards with a high magnitude (i.e. very relevant as they produce a major adverse effect in an individual animal) but a low exposure (i.e. relatively few animals in the population being subjected to these hazards) yield risk estimates that – erroneously – would seem to suggest these are less relevant and consequently would not necessarily deserve receiving the risk managers' attention. However, as animal welfare legislation aims at avoiding unnecessary suffering and was certainly not exclusively formulated for the population level, it is essential that such scenarios are properly dealt with in welfare management. They would seem to be best addressed by assuring that producers are made aware that these 'extraordinary' circumstances DO occur (under circumstances to be specifically described) and can entirely be prevented by strict adherence to 'Good Practices'.

## 8. The feasibility of identifying 'welfare promoters' and options for benefit analysis

Animal welfare is a continuous variable on a scale from high to low. In more discrete terms welfare could be classified in certain ranges (e.g. 'optimal', 'suboptimal but acceptable' to 'unacceptably bad'). Under the limitations of current production systems animal welfare will rarely reach its maximum. These limitations are not necessarily only associated with economic considerations (or ignorance of the profitability of good welfare), but also with animal and public health concerns (see the contribution of Buncic *et al.*, 2009 in this Volume) as well as with environmental and energy conservation issues. In addition, in view of the multi-component character of animal welfare, optimally addressing one of those components may compromise another (compare free range animal production allowing a natural habitat and promoting natural behaviour with the increased risk of animals contracting parasitic or other diseases). In consequence, one must confront the fact that in animal production practice (and in nature for that matter) welfare may rarely reach its maximum.

The desirability to improve animal welfare beyond simply eliminating major welfare risks has inspired discussions on whether or not (parallel to avoiding adverse effects of identified hazards) 'promoters' of animal welfare should be identified as well, and in this framework the

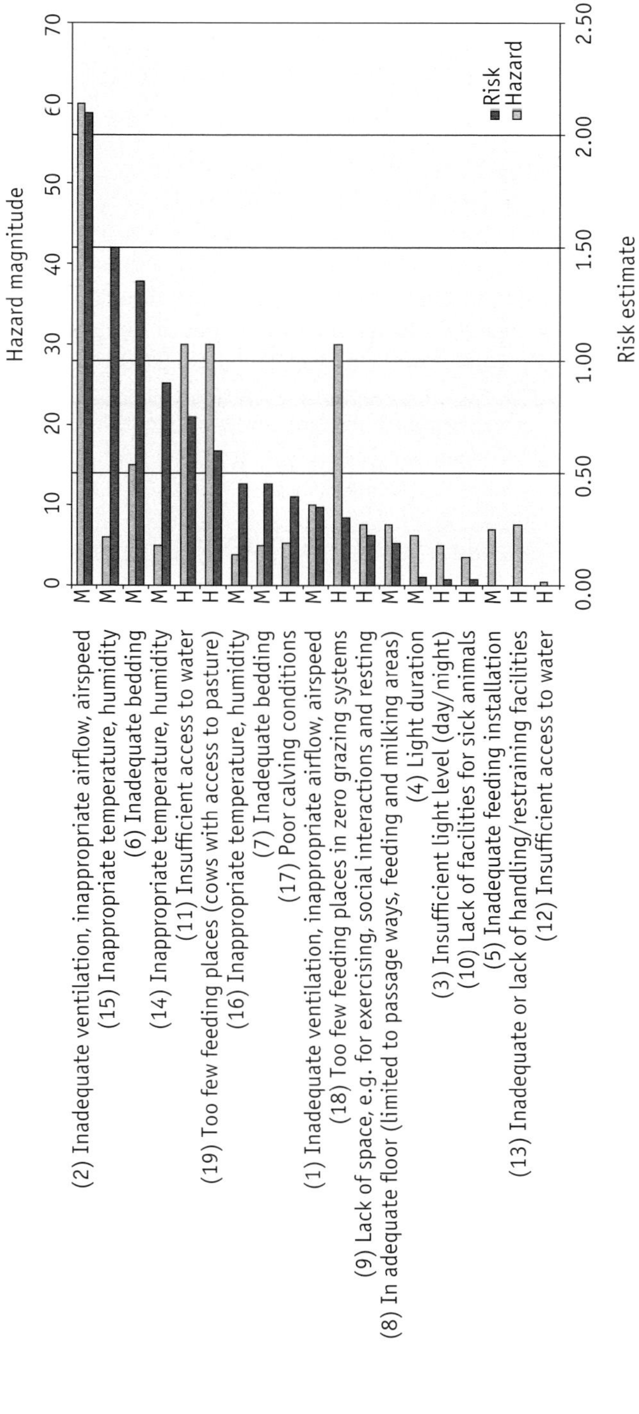


*Figure 1. Hazard magnitude and resulting risk estimates of the impact of housing in straw yards on metabolic and reproductive disorders in dairy cows; figures relate to scenarios indicated in Table 15, M, H denote medium or high uncertainties (adapted from EFSA, 2009c).*

term 'risk-benefit analysis' (RBA) has been used (e.g. EFSA, 2007a)[2]. This concept is known from food toxicology, where it is, for instance, applied for assessing the risks and benefits of including nutrient components at critical levels. A classic example is nitrite added to curing salt, which is known to effectively prevent the growth of *Clostridium* spp. in cured meats, but at levels in salt of higher than ca. 0.5% would represent a risk for the formation of the carcinogenic nitrosamine.

As the methodology of RBA is in its early development stage, guidelines for its proper application even in the food toxicology area are not available (EFSA, 2006b).

The most important question in this context is how – other than through determining (by relying on animal-based welfare indicators) that adverse welfare effects are absent – one can decide that a certain factor is a 'welfare promoter' without having to rely on anthropomorphic interpretation. Conversely, should benefits be interpreted in terms of 'absence of risks' the methodology described in this chapter serves the purpose of benefit analysis, in that neglecting those factors that decidedly promote animal welfare (and consequently are to be considered 'Good Practices') are defined as hazards.

## 9. From risk assessment to risk management using HACCP principles

RA is the scientific basis for implementing any decisions in terms of risk management (RM) and – as already indicated – these are not necessarily exclusively inspired by welfare considerations. Yet, the major recommendations included in AWRA reports of EFSA could be extended with suggestions for RM.

Analogous to the approach chosen in preventive quality assurance in the food safety area as outlined by the Codex Alimentarius, it should be considered how the principles of the widely used Hazard Analysis Critical Control Point (HACCP) system (CAC, 2001) could serve a similar purpose for the welfare management of animals.

### 9.1 HACCP – general principles

Various options for the management of risks exist, most notably the Hazard Analysis Critical Control Point (HACCP) system as explicitly indicated in European legislation for achieving food safety (e.g. European Commission, 2002, 2004a,b,c). In suggesting the adoption of HACCP as a useful approach, (be it targeted at food safety or animal welfare; see later), it is essential to realise that this only makes sense provided basic measures of quality and safety assurance are adhered to first. The latter are commonly called 'Good Practices' [e.g. 'Good Farming Practices' (GFP), 'Good Veterinary Practices' (GVP), 'Good Manufacturing Practices' (GMP, principally addressing the technical aspects of the whole production process) or 'Good Hygiene Practices' (GHP, concentrating on hygiene aspects of production)]. All these describe the 'best practices' that are universally applicable based on scientific, well-documented

[2] Comparing different production (husbandry) systems for assessing their potential to promote animal welfare (in which the presence-or-absence of various hazards occurring in one or the other system is carefully considered) is sometimes also referred to as 'benefit analysis'. However, the use of this term in this context is confusing.

proof, and which represent absolute prerequisites. They serve as the 'foundations' of risk management, in that they principally prevent external hazards being introduced in a production environment. Additionally relying on a HACCP system essentially serves to significantly reduce or even eliminate those hazards that, despite strict adherence to best practices, may still prevail.

This concept, repeatedly stressed in the literature on quality and safety assurance (e.g. ICMSF, 1988; Mayes and Baird-Parker, 1992; NACMCF, 1998; Buncic, 2006; Smulders, 2007), has been discussed by Untermann (1998) who compared the HACCP risk management system with a house, where good practices and adequate structural facilities represent its foundations and walls without which the 'HACCP roof' would collapse. In addition, Fellner and Riedl (2009) have extended this concept by stressing that truly effective 'longitudinal' risk management according to HACCP principles can only be realised by integrating the various risk management efforts all along the production chain. Consequently, such an approach is compared with a row of houses (the various subsequent links of the production chain) with a common HACCP roof. In terms of food animal production this means that for each link in the chain a HACCP system is adopted concentrating on those risks that cannot be eliminated further down the line and assuring that the joint efforts of the various links generate the best possible result.

HACCP was principally developed for purposes of food safety assurance and is based on 7 principles (CAC, 2001; FAO/WHO 2003), viz.:
1. Conduct a *Hazard Analysis* (Hazard Identification and Hazard Characterisation),
2. Determine *Critical Control Points* [CCP's, i.e. carefully determine at which point in production hazards can be significantly reduced (CCP-2) or even eliminated (CCP-1)],
3. Determine *Critical Limits*, i.e. define criteria above or beneath which the situation is acceptable/unacceptable,
4. Adopt a system for *Monitoring* whether or not the CCP is in check,
5. Establish the *Corrective Action* necessary when CCP's are not under control,
6. Take measures for *Verification* if the HACCP system functions,
7. Establish a reliable procedure for *Documentation* that shows that all the above principles and associated protocols have been taken into consideration.

It is also important to realise that a properly executed HACCP is characterised by the following:
1. It is *systematic*, i.e. following a structured approach and considering all relevant factors, thereby assuring that nothing is overlooked,
2. It is *product-specific*, i.e. only targeted at one specific product,
3. It is *hazard-specific*, i.e. only addresses the hazards that are associated with the specific product (see 2) and a particular situation (see 4 and 5),
4. It is *process-specific*, i.e. considers only the hazards associated with a well-defined way of processing,
5. It is *enterprise-specific*, i.e. considers only the circumstances of a particular processing unit.

It is important to appreciate the difference between risk assessment and HACCP (a risk management system). Only provided the former is available can an HACCP team take a decision on whether or not a hazard may constitute a risk which consequently should be properly managed. Whereas in some cases risk assessments are available to the HACCP team from the literature, in other cases these need to be conducted by the team itself. In summary: RA is an integral part of the HACCP system, HACCP is not an element of RA.

## 9.2 HACCP applied in primary animal production including for purposes of animal welfare management

Whereas the HACCP system is well-established for risk management in food processing, one has only begun to introduce it in primary animal production, largely because its full application still meets with methodological and other difficulties [e.g. how can verification and auditing be achieved?, how can the system be applied in an economic way? how to assure a longitudinal approach including all relevant links of the production chain? (Sofos, 2002)]. In the realisation that HACCP in the ante mortem production phase is still 'work in progress' the European legislature has not (yet) made HACCP – in the strict sense of the concept – mandatory for primary production. However, in European legislation (e.g. European Commission, 2002, 2004) member states are prompted to adopt 'HACCP-like' plans to meet the issues of food safety, public health and animal health and welfare. Consequently, many such plans suggested in the literature are still largely based on adhering to Good Practices (e.g. Noordhuizen *et al.*, 2008).

For purposes of using the HACCP concept in the area of welfare management, the food-related terminology needs to be adapted. This could essentially be achieved by defining hazards as factors adversely affecting animal welfare and exchanging the words 'product' and 'process' for (a particular) 'animal species' and (a particular) 'animal handling procedure' (e.g. a husbandry-, transport-, lairage- or stunning/killing system).

*Table 16. Good stunning/killing practices (GSKP) and critical control points (CCP) in slaughter cattle; during stunning for disease control; lethal injection (adapted from Algers* et al.*, 2009).*

| Hazard | Factor | Do's | Don'ts | GSKP/CCP? |
|---|---|---|---|---|
| Operator | Untrained, unlicensed, careless | | Do not allow unqualified operators | GSKP |
| Drug/agent | Wrong choice, dose | Choose appropriate drug/dose | Do not allow wrong/ insufficient dose | GSKP |
| Injection method | Wrong injection site, missed target, extravascular injection | Identify operator qualified to carry out injection | Do not allow unqualified operators | GSKP |
| | Delay before death | Check drug/dose; Use back up killing | Do not allow wrong/ insufficient dose | GSKP |
| Induction | Successful application | Produce immediate collapse | Do not continue if recovery signs present | CCP |
| Signs of good stun | Successful application | Affirm: NO Corneal reflex, NO Rhythmic breathing | Do not continue if recovery signs present | CCP |

Table 16 presents an example taken from a recent report submitted to EFSA (Algers *et al.*, 2009; on welfare aspects of stunning and killing of various animal species to which the reader is referred for further information). By distinguishing between things to do ('Do's') and things to prevent ('Don'ts') those process steps that are monitorable, correctable, verifiable and documentable (and could therefore represent CCP's), risk management options can more clearly be identified.

## 10. Conclusions

### 10.1 What has been achieved?

Over the past 5 years a generic semi-quantitative AWRA model gradually came into being, which promises to be a good basis for future risk assessment exercises. This has been the result of the dedicated efforts of animal welfare scientists and risk assessors alike. At the date of release of this Volume (September 2009) 32 scientific opinions based on RA have been made available *via* EFSA, 23 of them with a formal semi-quantitative character (see Table 1 in Ribó and Serratosa, 2009).

### 10.2 What has been neglected?

Although the various risk assessors engaged in AWRA exercises have been in contact, a more regular feed-back and -forward of information on methodological problems encountered and how to best solve these would seem advisable. Also, one must confront the fact that, unless risk questions are well-formulated they should not be accepted for methodological reasons. Finally, the issue of AWRA model validation (sensitivity analysis) has thus far only partly been addressed at least on an international Europe-wide level.

### 10.3 What needs to be done?

There is significant room for improvement particularly as regards the making availability of reliable data to serve exposure assessment, collected through methodical monitoring of (preferably animal-based) welfare indicators. Current Europe-wide research efforts sponsored by the European Union hope to inspire the legislature on how such would be best achieved, so monitoring programmes enforced through legislation could be established in due time.

## Acknowledgements

The author acknowledges the efforts of his fellow risk assessors and EFSA Staff, who have made significant contributions to the development of the approach described in this chapter, notably: Dr. Denise Candiani, Dr. Jan Hultgren, Dr. Christine Müller-Graf, Dr. Telmo Nunes, Dr. Oriol Ribó and Dr. Moez Saana. Finally, I am grateful to the EFSA Directorate for their kind permission to reproduce sizeable sections of previous EFSA publications and to Prof. Dr. Marten Scheffer (Wageningen University and Research Centre, the Netherlands) for proofreading, and for his valuable comments and suggestions for improvement, of the draft version of this contribution.

## References

Algers B., Anil, H., Blokhuis, H., Fuchs, K., Hultgren, J., Lambooij, B., Nunes, T., Paulsen, P. and Smulders, F.J.M., 2009. WRAPSTUN, Final report on the project to develop animal welfare risk assessment guidelines on stunning and killing, CFP/EFSA/AHAW/ 2007/01.

Baird-Parker, A.C., 1987. The application of preventive quality assurance. In: Smulders, F.J.M. (Ed.). Elimination of pathogenic organisms from meat and poultry. Publ. Elsevier, Amsterdam, the Netherlands, pp. 149-161.

Bracke, M.B.M., Edwards, S.A., Engel, B., Buist, W.G. and Algers B., 2008. Expert opinion as 'validation' of risk assessment applied to calf welfare. Acta Veterinaria Scandinavia 50, 29.

Broom, D.M., 1996. Animal welfare in terms of attempt to cope with the environment. Acta Agriculturae Scandinavia (Section A, Animal Science) 27, 22-29.

Buncic, S., 2006. GHP and HACCP principles. In: Integrated food safety and veterinary public health, Chapter 8.1., Cabi, Wallingford, UK, pp. 262-269.

Buncic, S., Collins, J.D., Smulders, F.J.M. and Colin, P., 2009. Biological food safety in relation to animal welfare. In: Smulders, F.J.M. and Algers, B. (Eds.). Food safety assurance and veterinary public health, Vol.5, Welfare of production animals: assessment and management of risks. Wageningen Academic Publishers, Wageningen, the Netherlands, pp. 485-532.

Codex Alimentarius Commission (CAC), 2001. Food hygiene – basic texts. FAO/WHO, Rome, Italy.

Codex Alimentarius Commission (CAC), 2002. Principles and guidelines for the conduct of microbiological risk assessment. Document CAC/GL 30. FAO/WHO, Rome, Italy.

Dransfield, E. and Scheffer, M., 1991. Mathematical modelling in the meat industry. In: Smulders, F.J.M. (Ed.). The European meat industry in the 1990's. Publ. Audet/ECCEAMST, Utrecht, the Netherlands, pp. 91-115.

European Commission (EC), 2002. Regulation (EC) No. 178/2002, laying down the general principles and requirements of food law, establishing the European Food Safety Authority and laying down procedures in matters of food safety. Official Journal of the European Union L 31, 1.

European Commission (EC), 2003. Council Directive 2003/99/EC of the European Parliament and the Council of the 17th November 2003 on the monitoring of zoonoses and zoonotic agents amending Council Decision 90/424/EEC and repealing Council Directive 92/117/EC. Official Journal of the European Union L 325, 1.

European Commission (EC), 2004a. Regulation (EC) No. 852/2004 of the European Parliament and the Council of 29th of April 2004 on the hygiene of foodstuffs. Official Journal of the European Union L 138, 1.

European Commission (EC), 2004b. Regulation (EC) No. 853/2004 of the European Parliament and the Council of 29th of April 2004 containing specific rules for the hygiene of foodstuffs. Official Journal of the European Union L 139, 55.

European Commission (EC), 2004c. Regulation (EC) No. 854/2004 of the European Parliament and the Council of 29th April 2004, containing specific rules concerning official controls on products of animal origin. Official Journal of the European Union L 155, 206.

European Food Safety Authority (EFSA), 2006a Opinion of the Scientific Panel on Animal Health and Welfare (AHAW) on a request from the Commission related with the risks of poor welfare in intensive calf farming systems. Question No. EFSA-Q-2005-024. The EFSA Journal 366.

European Food Safety Authority (EFSA), 2006b. Risk-Benefit Analysis of foods – Methods and approaches. EFSA Scientific Colloquium Summary Report, EFSA, Parma, Italy, 153 pp.

European Food Safety Authority (EFSA), 2007a. Scientific Opinion of the Panel on Animal Health and Welfare on a request from the Commission on animal health and welfare in fattening pigs. The EFSA Journal 564, 1-14.

European Food Safety Authority (EFSA), 2007b. Scientific Opinion of the Panel on Animal Health and Welfare on a request from the Commission on the animal health and welfare aspects of different housing systems for adult breeding boars, pregnant, farrowing sows and unweaned piglets. The EFSA Journal 572, 1-13.

European Food Safety Authority (EFSA), 2007c. Scientific Opinion of the Panel on Animal Health and Welfare on a request from the Commission on the Animal Welfare aspects of the killing and skinning of seals (Question No EFSA-Q-2007-118). The EFSA Journal 610, 1-123.

European Food Safety Authority (EFSA), 2007d. Scientific Opinion of the Panel on Animal Health and Welfare on a request from the Commission on the risks associated with tail-biting in pigs and possible means to reduce the need for tail-biting considering the different housing and husbandry systems. The EFSA Journal 611, 1-13.

European Food Safety Authority (EFSA), 2008. Scientific opinion of the Panel on Animal Health and Welfare on a request b EFSA (European Food Safety Authority), y the Commission on animal welfare aspects of husbandry systems for farmed Atlantic salmon, The EFSA Journal 736, 1-31.

European Food Safety Authority (EFSA), 2009a. Scientific Opinion of the Panel on Animal Health and Welfare on a request of the Commission on the impact of housing, nutrition and feeding, management and genetic selection on leg and locomotion problems in dairy cows, EFSA-Q-2008-337. The EFSA Journal 1142, 1-57.

European Food Safety Authority (EFSA), 2009b. Scientific Opinion of the Panel on Animal Health and Welfare on a request of the Commission on the impact of housing, nutrition and feeding, management and genetic selection on udder problems in dairy cows, EFSA-Q-2008-338. The EFSA Journal 1141, 1-60.

European Food Safety Authority (EFSA), 2009c. Scientific Opinion of the Panel on Animal Health and Welfare on a request of the Commission on the impact of housing, nutrition and feeding, management and genetic selection on metabolic and reproductive disorders in dairy cows, EFSA-Q-2008-339. The EFSA Journal 1140, 1-75.

European Food Safety Authority (EFSA), 2009d. Scientific Opinion of the Panel on Animal Health and Welfare on a request of the Commission on the impact of housing, nutrition and feeding, management and genetic selection on behaviour, fear and pain in dairy cows, EFSA-Q-2008-340. The EFSA Journal 1130, 1-68.

FAO/WHO, 2003. Guidance to governments on the application of HACCP in small and less-developed food business. FAO food & nutrition paper N° 86. FAO, Rome, Italy.

Fellner, C. and Riedl, R., 2009. HACCP nach dem FAO/WHO-Codex Alimentarius, Theoretische Grundlagen und praxisbezogene Hilfestellungen zur korrekten Umsetzung des HACCP-Konzeptes. 2. überarbeitete und ergänzte Auflage. Verlag Österreich, Wien, Austria, p. 162.

International Commission on the Microbiological Specification for Foods (ICMSF), 1988. Micro-organisms in Foods Vol. 4. Application of the Hazard Analysis Critical Control Point (HACCP) system to ensure microbiological safety and quality. Blackwell Science Publishers, Oxford, UK.

Maijala, R., 2006. Risk assessment as a tool for evaluating risk management options for food safety. In: F.J.M. Smulders (Ed.). Food Safety Assurance and Veterinary Public Health, Vol.4, Towards a risk-based chain control, Wageningen Academic Publishers, Wageningen, the Netherlands, pp. 19-32.

Manteca, X., Velarde, A. and Jones, B., 2009. Animal welfare components. In: Smulders, F.J.M. and Algers, B. (Eds.). Food safety assurance and veterinary public health, Vol.5, Welfare of production animals: assessment and management of risks. Wageningen Academic Publishers, Wageningen, the Netherlands, pp. 61-77.

Mayes, T. and Baird-Parker, A.C., 1992. Hazard analysis studies. In: Smulders, F.J.M., Toldrá, F., Flores, J. and Prieto, M. (Eds.). New technologies for meat products – Fermentation and starter cultures, muscle enzymology and meat ageing, quality control systems. Publ. ECCEAMST/Audet, Nijmegen, the Netherlands, pp. 263-272.

National Advisory Committee on Microbiological Criteria for Foods (NACMCF), 1998. Hazard analysis and critical control point principles and application guidelines. Journal of Food Protection 61, 762-775.

agreements. The significance of such documents lies with their inclusion of enforcing measures and – in the case of incompliance with set standards – of sanctions to be imposed by controlling agencies, by either governmental or private.

Whereas animal health is currently predominantly expressed in terms of (the absence of) animal diseases, animal welfare has purposely been defined by the Council of Europe (CoE) and the OIE as incorporating more complex issues with important scientific, ethical, economic and political dimensions (see sections 1.1 and 1.3 of this contribution). The impact of such approach may have consequences for the international trade in the light of WTO agreements. The different definitions of animal disease and animal welfare allow drafting distinctly separate, rather than integrated pieces of legislation, which potentially has serious disadvantages. From the risk assessment perspective there is an obvious correlation between risks associated with animal disease or physiological/behavioural reactions, or those associated with the well-being of the animals and in some cases may have food safety- (see section 4 of this contribution and the contribution of Buncic *et al.*, 2009, this Volume) or environmental ramifications. Animal welfare has been defined by several institutions/ experts in different ways (see also contributions of Nordenfelt, 2009 and Broom, 2009 in this Volume). Some of the official definitions used by the various organizations are described in this contribution. Interestingly, in addressing public health, WHO defines health as *'a state of complete physical, mental and social well-being and not merely the absence of disease or infirmity'* (WHO, 2009).

### 1.1 Council of Europe

The Council of Europe (CoE) has a long tradition in animal welfare. Ever since 1968 it has launched many relevant documents, including conventions and recommendations on animal welfare issues (CoE, 2008). Most of these have been adopted in the EU legislation subject to later modifications to accommodate new scientific knowledge having become available (CoE, 2009). However, the Council of Europe, consisting of governmental experts, has sought close collaboration with non-governmental observers participating in its meetings, who represent the different fields concerned, such as veterinarians, farmers, animal protection associations, animal behaviour specialists and zootechnicians.

In its 45$^{th}$ Meeting in Strasbourg, 25-27 November 2003, the standing Committee of the European convention for the protection of animals kept for farming purposes (T-AP) agreed on the definition of the expression 'essential needs'. It was stated that for the purpose of the European Convention for the protection of animals kept for farming purposes (ETS, 87), these represent *'those needs which have to be fulfilled to enable an individual animal to maintain its health and welfare by safeguarding its physiological and mental functions, including performing satisfactorily species-specific behaviour. They should be defined taking into account established experience and scientific knowledge concerning biological characteristics and genetic type, sex, age and the physiological stage of the animal'.*

Since 1968 the Council of Europe has been adopting Conventions on animal protection and animal welfare, which have been ratified by the EU and are in constant evolution and revision. Main conditions of the adopted conventions are explained below:

1. The European Convention for the protection of animals during international transport in 1968 lays down the general conditions for the international transport of animals from their preparation to loading to their unloading, such as design of transport means, fitness to transport of the animals, veterinary controls, handling of animals, certificates, etc. and sets out special conditions for the transport of animals by road, air, sea and rail. Detailed recommendations on the transport of respectively horses, pigs, cattle, sheep and goats, and poultry were adopted by the Committee of Ministers of the Council of Europe. To each of these Recommendations is appended a very detailed Code of Conduct. In 1992, a Protocol of Amendment to the Convention (ETS 145) was opened for signature. This Protocol provides for the extension of the scope of the Convention to the breeding of animals produced as a result of genetic modifications or novel genetic combinations. In 1988, the EU ratified the Council of Europe convention and the Member States were expected to give effect to the recommendations. Further modifications have been updating the initial document (2003). The EU signed the CoE Convention on international animal transport in June 2004 (EC, 1997, 2005b, 2006).
2. The European Convention for the protection of animals kept for farming purposes (1976; ETS 87) is a 'framework convention' which gives principles for the keeping, care and housing of animals, in particular in intensive breeding systems (EC, 1978, 1998, 2007d).
3. The European Convention for the protection of vertebrate animals used for experimental and other scientific purposes (1986, ETS 123) concerns the use of animals in procedures (experiments). Its provisions cover areas such as care and accommodation, conduct of experiments, humane killing, authorisation procedures, control of breeding or supplying and user establishments, education and training, and statistical information. The Convention includes two technical Appendices presenting guidelines for accommodation and care of animals (Appendix A) and tables for the presentation of statistical data on the use of animals for scientific purposes, respectively (EC, 1986).
4. The European Convention on the protection of pet animals (1987, ETS125) contains provisions to protect pet animals, and seeks to establish a basic common standard of attitude and practice towards pet ownership. Provisions are included on breeding, boarding, keeping, etc. The Convention also aims at regulating trading in and breeding of pet animals, at prohibiting the modification of their natural appearance and at reducing the number of stray animals.
5. The European Convention for the protection of animals for slaughter (1979; ETS 102) applies to the movement, lairaging, restraint, stunning and slaughter of domestic solipeds, ruminants, pigs, rabbits and poultry. The provisions have been adapted and elaborated in a Recommendation and a Code of conduct, which the Committee of Ministers adopted in 1991 (EC, 1988, 1993, 2008a).

Additionally, a number of recommendations (CoE, 2009) developed for each one of the working groups responsible for the conventions, have been agreed and published in the framework of the CoE (e.g. recommendations adopted by the Standing Committee of the European Convention for the protection of animals kept for farming purposes on different species). Most of these recommendations have been adapted and adopted in the EU legislation, such as the protection of domestic ducks and domestic geese and of fur animals, pigs (EC, 1991), farmed fish, turkeys, fur animals, ratites, domestic fowl (EC 1999b, 2007b,c), calves, goats, sheep, cats and dogs (EC, 2007b) wild animals (EC, 1999a) and cattle (EC, 2008b).

Although all above documents have been drafted using the best available expertise in the field, the principles of hazard identification, characterisation and exposure assessment, allowing risk estimation (now understood as risk assessment methodologies) were not applied. An in-depth scientific revision could take place, but the main purpose of those documents was not only to identify the biggest risks but also to draft applicable recommendations in the practice, meaning risk management.

### 1.2 World Trade Organization

The World Trade Organisation (WTO) was created by the Final Act of the Uruguay Round of Multilateral Trade Negotiations signed in Marrakech on 15 April 1994. The WTO replaced the General Agreement on Tariffs and Trade, typically abbreviated GATT (GATT, 1994), but this agreement still exists as the legal frame for trade in goods. Article 20(b) of GATT allows governments to apply exceptions to free circulation of goods in order to protect human, animal or plant life or health, provided they do not use this prerogative as disguised protectionism. The 'Agreement on the Application of Sanitary and Phytosanitary Measures' (WTO, 1994) also known as the 'SPS Agreement' was created and entered into force simultaneously with the establishment of the World Trade Organization (WTO) on 1 January 1995.

The Agreement on the Application of SPS Measures has the main objective to improve the human health, animal health and phytosanitary situation for all WTO Members, and to ensure that their SPS measures do not arbitrarily or unjustifiably discriminate between WTO Members where identical or similar conditions prevail, including between their own territory and that of other WTO Members. SPS measures shall not be applied in a manner which would constitute a disguised restriction on international trade. Regarding harmonisation, Article 3 states that SPS measures has to conform to international standards, guidelines or recommendations and shall be deemed to be necessary to protect human, animal or plant life or health.

Also, WTO Members may introduce or maintain SPS measures which result in a higher level of SPS protection than would be achieved by measures based on the relevant international standards, guidelines or recommendations, if there is a scientific justification. Article 5 of the 'SPS Agreement' (WTO, 1994) describes that the assessment of Risk and Determination of the Appropriate Level should be based on an assessment, as appropriate to the circumstances, of the risks to human, animal or plant life or health, taking into account risk assessment techniques developed by the relevant international organizations. In the assessment of risks, WTO Members shall take into account available scientific evidence; relevant processes and production methods; relevant inspection, sampling and testing methods; prevalence of specific diseases or pests; existence of pest- or disease-free areas; relevant ecological and environmental conditions; and quarantine or other treatment. In assessing the risk to animal or plant life or health and determining the measure to be applied for achieving the appropriate level of SPS protection from such risk, Members shall take into account as relevant economic factors: the potential damage in terms of loss of production or sales in the event of the entry, establishment or spread of a pest or disease; the costs of control or eradication in the territory of the importing WTO Member; and the relative cost-effectiveness of alternative approaches to limiting risks. However Members should take into account the objective of minimizing negative trade effects.

In cases where relevant scientific evidence is insufficient, a Member may provisionally adopt SPS measures on the basis of available pertinent information, including that from the relevant international organizations as well as from SPS measures applied by other Members. In such circumstances, WTO Members shall seek to obtain the additional information necessary for a more objective assessment of risk and review the SPS measure accordingly within a reasonable period of time (Article 5 of SPS measures; WTO, 1994).

Principles and definitions are defined and developed in the SPS agreement like Transparency, Control, Inspection and Approval Procedures, Implementation and Technical Assistance. It is also stated that where substantial investments are required in order for an exporting developing country Member to fulfil the SPS requirements of an importing Member, the latter shall consider providing such technical assistance as will permit the developing country Member to maintain and expand its market access opportunities for the product involved.

In the chapter of definitions, Risk Assessment is defined as 'the evaluation of the likelihood of entry, establishment or spread of a pest or disease within the territory of an importing Member according to the SPS measures which might be applied, and of the associated potential biological and economic consequences; or the evaluation of the potential for adverse effects on human or animal health arising from the presence of additives, contaminants, toxins or disease-causing organisms in food, beverages or feedstuffs'. Also, the appropriate level of SPS protection deemed appropriate by the Member establishing a SPS measure to protect human, animal or plant life or health within its territory is defined. Many Members may refer to this concept as the 'acceptable level of risk'.

In conclusion, in spite of the fact that the SPS agreement contains plenty of principles to be used when assessing risks and drafting legislation for the food safety, animal health and plant health SPS, also protects the health of fish and wild fauna, as well as of forests and wild flora. However, the measures for environmental protection, to protect consumer interests, or for the welfare of animals are not covered by the SPS Agreement. SPS acknowledges that for animal health and zoonoses, the standards, guidelines and recommendations developed under the auspices of the OIE (Office International des Épizoooties or recently named World Organisation of Animal Health) apply.

The Council approved, on behalf of the European Community, the Final Act of the Uruguay Round of Multilateral Trade Negotiations. This act includes the 'Agreement on the Application of SPS' also known as the SPS Agreement (WTO, 1994). Therefore, since January 1995, this legislation applies to the EU.

The European Commission has undertaken to raise the issue of animal welfare in the new WTO negotiating round. However, during the negotiations in Seattle in 1999 and Doha in November 2001, there was no support for the approach proposed by the Community. The Commission continued nonetheless to work towards international acceptance of animal welfare. It was stated that trade does not undermine the efforts to improve the protection of the welfare of the animals. It was not the intention to create new trade barriers; however, it was stressed that the increased liberalisation of the trade should not lead to competitive reductions in animal welfare standards (EC, 1994).

In November 2002 the Commission adopted a Communication to the Council and European Parliament (EC, 2002b), comparing animal welfare standards in the European Union with those in third country trading partners. The Communication shows that there is no international consensus on the role of animal welfare and the measures in place in the EU cannot be readily compared with the standards in third countries. A key issue for the Communication was whether competitive disadvantages arise from disparities in animal welfare measures. The available evidence suggests that competitive distortions are most likely to arise in the more intensive forms of agricultural production, notably in the pig and poultry sectors.

Starting from the assumption that competitive distortions (whether to the advantage or disadvantage of EU producers) arising from differences in standards have the clear potential to undermine higher animal welfare, the Communication investigates a number of channels to prevent such a development like (1) market mechanisms, reason being that consumers are increasingly willing to pay more for 'ethical' products; (2) dialogue at the international level aiming at greater recognition of animal welfare, in particular in the framework of the OIE (World Organisation for Animal Health) and the Council of Europe and WTO; (3) promotion of animal welfare standards in trade arrangements; (4) improvement of labelling regimes to respond to consumer demands for higher standards; (5) Strengthening the position of animal welfare in EU agricultural policy as part of the reform of the Common Agricultural Policy focusing on 'quality rather than quantity'.

The pros and cons of each path are analysed in the Communication with a particular focus on the implications of animal welfare as regards animal health, food safety and consumer concerns. The Report concludes that efforts must concentrate on all fronts in order to cope with the complexity of animal welfare and its ethical and cultural dimension.

The rules of international trade may have an immense impact on animal welfare, and mostly on the welfare of farmed animals. The development of trade agreements can either undermine important legislative achievements, or offer relevant opportunities to promote animal welfare legislation and standards at the international level. Similarly, legislation approved by single countries and policies adopted by either farmers or retailers can have a major impact on the way farm animals are treated. The European Commission has organized two relevant meetings (International Forum on global aspects of Farm Animal Welfare on 22-23 April 2008 and a Conference on Global Trade and farm Animal welfare on 20-21 January 2009) with the main stakeholders on these issues which should be considered to balance all the relevant factors on animal welfare and trade (EC, 2008c). The main objective of the Global Trade Conference was to bring together different constituents from all continents, who presented positive experiences of inclusion of animal welfare in the trade environment. Main conclusions from the conference were that animal welfare (AW) is a clear demand from the consumers, which have a clear perception of what AW means. It is necessary to set up AW standards for animal transport. The AW model should be promoted at international level and the compliance of the AW standards should have to be ensured at national level. The improvement of animal welfare is financially affordable, however assistance is needed. In relation to that, the purchase of products only produced under AW standards should be encouraged.

### 1.3 World Organization for Animal Health (OIE)

The 'Office International des Épizooties' or 'World Organization for Animal Health' has acknowledged that for animal health and zoonoses, the standards, guidelines and recommendations developed represent the worldwide reference point for substantiating the risk analysis and the legislation (WTO, 1994).

Although the welfare of animals is not covered by the SPS Agreement, animal welfare was first identified as a priority in the OIE Strategic Plan 2001-2005 (OIE, 2001). OIE Member Countries and Territories mandated the organisation to take the lead internationally on animal welfare and, as the international reference organisation for animal health, to elaborate recommendations and guidelines covering animal welfare practices, reaffirming that animal health is a key component of animal welfare (OIE, 2008).

The OIE Animal Welfare Working Group was inaugurated at the 70$^{th}$ General Session of the OIE in May 2002 and the first recommendations of the Working Group were adopted one year later. The OIE Guiding Principles on Animal Welfare were included in the OIE Terrestrial Animal Health Code [Terrestrial Code] in 2004 (OIE, 2004a,b).

In particular, it examined the best options for incorporating the views of non-governmental organisations (NGO's), including industry, in its work on animal welfare. The International Committee of the OIE adopted in May 2003 the resolution No XXVI that supports the work program of the OIE in this field and the organisation of the first global conference on animal welfare and calls for support to the initiative.

The OIE convened a First Global Conference on Animal Welfare in February 2004. As well as targeting the Veterinary Services in OIE Member Countries and Territories, the Conference addressed livestock producers and actors in the meat sector, veterinary practitioners and international non-governmental organisations (NGOs) working in animal welfare. The main objective of the Conference was to raise awareness of, and to explain, the OIE's animal welfare initiative.

At its 73$^{rd}$ General Session, in May 2005, the OIE International Committee (comprising the National Delegates of the 173 Member Countries and Territories) adopted the following five animal welfare standards for inclusion in the OIE 'Terrestrial Code – Introduction to the recommendations for animal welfare': (1) transport of animals by land; (2) transport of animals by sea; (3) transport of animals by air; (4) slaughter of animals for human consumption; and (5) killing of animals for disease control purposes.

The International Committee of OIE has recognised that animal welfare is a complex, multi-faceted public policy issue which includes scientific, ethical, economic and political dimensions. In May 2008 the OIE in the annual session with 172 Members defined animal welfare as *'means how an animal is coping with the conditions in which it lives. An animal is in a good state of welfare if (as indicated by scientific evidence) it is healthy, comfortable, well nourished, safe, able to express innate behaviour, and if it is not suffering from unpleasant states such as pain, fear, and distress. Good animal welfare requires disease prevention and veterinary treatment, appropriate shelter, management, nutrition, humane handling and humane slaughter/ killing. Animal welfare refers to the state of the animal; the treatment that an animal receives is covered by other terms such as animal care, animal husbandry and humane treatment'* (OIE, 2008).

Under the current Strategic Plan, for the period 2006-2010 (OIE, 2005), the OIE will continue to develop international standards and guidelines in accordance with the direction established in 2004. The OIE has also established four ad hoc groups to develop animal welfare guidelines concerning the land and sea transport of animals, the slaughter of animals for human consumption and the humane killing of animals for disease control. The guidelines developed by these *ad hoc* groups were adopted at the May 2005 OIE General Session, as well as a Resolution concerning future OIE activities on animal welfare (OIE, 2005).

## 1.4 CODEX and FAO

The Codex Alimentarius is a collection of international food safety standards that have been adopted by the Codex Alimentarius Commission (the 'Codex'). The Codex is based in Rome and funded jointly by the FAO and the WHO (WHO, 1999). Importantly, the SPS Agreement cites Codex's food safety standards, guidelines and recommendations for facilitating international trade and protecting public health. The FAO/WHO CODEX Alimentarius has no competences at all in animal welfare but on food safety and nutrition. However, the Codex has played a very important role describing the methodologies of risk assessment which have been taken as a departure point/model for some risk assessment applied in animal welfare (EFSA, 2006e; Ribó and Serratosa, 2009). The Codex Alimentarius is a science-based organization with independent experts and specialists in a wide range of disciplines who have contributed to its work to ensure its standards. The work of the Codex Alimentarius Commission, together with that of FAO and WHO in their supportive roles, has provided a focal point for food-related scientific research and investigation, and has become an important international reference organisation (WTO, 1994) for the exchange of scientific information about the safety of food. For this reason EFSA maintains a strong relationship with the Codex Alimentarius Commission on all food safety issues and for the interest of animal welfare the CODEX risk assessment methodologies have to be considered (CAC, 2009).

According to the statement concerning the role of science in the Codex decision-making process and the extent to which other factors are taken into account, the food standards, guidelines and other recommendations of Codex Alimentarius shall be based on the principle of sound scientific analysis and evidence, involving a thorough review of all relevant information, in order that the standards assure the quality and safety of the food supply. The decisions should be based on risk assessment, and taking into account, where appropriate, other legitimate factors relevant for the health protection of consumers and for the promotion of fair practices in food trade, in accordance with the Criteria for the Consideration of the Other Factors Referred to in the Second Statement of Principles. In this regard it is noted that food labelling plays an important role for other legitimate factors relevant for health protection. Fair trade practices may be identified in the risk management process, and risk managers should indicate how these factors affect the selection of risk management options and the development of standards, guidelines and related texts; the integration of other legitimate factors in risk management should not create unjustified barriers to trade;

particular attention should be given to the impact on developing countries of the inclusion of such other factors (CAC, 2008).

In summary, CODEX develops standards, guidelines or recommendations as a result of a risk analysis, on the basis of scientific evidence (risk assessment), and in regard of risk management options that where appropriate consider other legitimate factors relevant to the health of consumers and the promotion of fair practices in the food trade.

The Food and Agriculture Organization of the United Nations leads international efforts to defeat hunger. Serving both developed and developing countries, FAO acts as a neutral forum where all nations meet as equals to negotiate agreements and debate policy. FAO is also a source of knowledge and information. FAO helps developing countries and countries in transition to modernize and improve agriculture, forestry and fisheries practices and ensure good nutrition for all. Since the founding of FAO in 1945, it has focused special attention on developing rural areas, home to 70 percent of the world's poor and hungry people (FAO, 2009a).

FAO, stated in his first meeting on AW capacity building (FAO, 2009b) that the welfare of humans and the welfare of animals are closely linked. *FAO 'recognises the importance of animal welfare practices that lead to benefits for both people and their animals and supports their implementation'*. In many regions, a secure supply of food for people depends on the health and productivity of animals, and these in turn depend on the care and nutrition that animals receive. The massive increase in animal production of the last decades has raised a wide range of ethical issues, including concern for animal welfare, which has to be considered alongside environmental sustainability and secure access to food. Animal welfare was recognized by the experts as a core component of a responsible livestock sector.

The main conclusions are that good animal practices include prevention and treatment of disease and injury; prevention and mitigation of pain and distress and other negative states; and providing diets and living conditions that are suited to the need and nature of animals. Good animal welfare practices will contribute to the sustainability, increase productivity, food safety and human health. FAO promotes that animal welfare assessment should be done with the full participation of the people involved in the field. It is stressed that the capacity building for implementing good animal welfare practices involves (1) education to create awareness, (2) engagement to foster active involvement, (3) training in specific procedures and (4) communication among the different involved partners of the food production and government. Furthermore, scientific research on animal welfare provides the scientific evidence behind many animal welfare practices and standards, and FAO should consider working with centres of expertise in animal welfare science to facilitate access by members countries to the findings of animal welfare research and to encourage research on issues to countries with developing economies. FAO is strongly encouraged to cooperate with OIE through its regional Commissions, the WHO and other institutions engaged at international level. It should also work together with academic, producers organizations, NGO's, financial institutions and private sector (FAO, 2009a,b).

It is accepted to be integral to programmes that improve animal health, increase livestock productivity, respond to natural disasters where animals are involved, and to be instrumental in defining the fit between the genetic makeup of animals and the environments in which they are kept. FAO realizes, on the other hand, that animal welfare practices, despite their evident positive impacts, are insufficiently applied throughout the sector, both in traditional and modern systems. FAO has decided to enable giving more explicit and strategic attention to animal welfare and its main streaming into practice, within FAO's capacity building activities, to set an animal welfare web[1] (portal) as a powerful instrument to disseminates useful information and to build synergies worldwide.

## 2. Role of EFSA within the EU strategy on animal health and welfare

The Treaty of Amsterdam through its protocol on the protection and welfare of animals ensures improved protection and respect for the welfare of animals as sentient beings, providing precise indications: '*In formulating and implementing the Community's agriculture, transport, internal market and research policies, the Community and the Member States shall pay full regard to the welfare requirements of animals, while respecting the legislative or administrative provisions and customs of the Member States relating in particular to religious rites, cultural traditions and regional heritage*' (EC, 1997a).

Regulation (EC) No 178/2002 (EC, 2002a) states that in order to ensure the safety of food, it is necessary to consider all aspects of the food production chain as a continuum from and including primary production, because each element may have a direct or indirect potential impact on food safety. It is also laid down that the Food Law shall pursue one or more of the general objectives of a high level of protection of human life and health and the protection of consumers' interests, including fair practices in food trade, taking account of, where appropriate, the protection of animal health and welfare, plant health and the environment. When referring to the feed, any adverse effect on animal health should have also to be considered. Therefore, the Regulation establishes that the European Food Safety Authority (EFSA) should provide comprehensive independent scientific view of the safety and other aspects of the whole food and feed supply chains, including issues which have a direct or indirect impact on the safety of the food and feed supply chains, animal health, animal welfare and plant health.

The implementation of the EU legislation on animal welfare is the responsibility of the member states, and any event suspected to be a breach of the Community legislation should be communicated at first to the competent authority of the member state concerned. If the competent authority of a member State fails to apply Community rules, the Commission may consider opening an infringement procedure under article 226 of the Treaty against the member state concerned. The Commission may initiate an infringement procedure based on consistent, sufficient and reliable data. Information concerning the failure of a Member State to apply Community rules can originate from the services of the Commission as an official mission report from the Food and Veterinary Office (FVO) or from complains by organizations or individuals. FVO produce specific reports on their missions to assess the compliances and identified risks, for instance, in 2007, eight inspections were carried out in pigs and laying

[1] http://www.fao.org/ag/againfo/programmes/animal-welfare/en/

hens, 3 missions on slaughter and killing, 4 missions on transport of equidae and slaughter. It is stated that further missions will be undertaken during 2008, to monitor progress inside the EU Member States (FVO, 2009).

The mission of EFSA is to provide scientific and technical advice for the Community's legislation and policies in all fields which have a direct or indirect impact on food and feed safety, including animal health and welfare (EC, 2002a). The Animal Health and Animal Welfare (AHAW) Panel of EFSA delivers scientific opinions on all aspects of animal diseases and animal welfare, primarily relating to food producing animals including fish. The mission includes providing independent information on all matters within these fields and communicating on risks.

EFSA provides independent assessment on all matters within its remit by using a Risk Assessment (RA) approach, whenever possible. RA is the process of evaluating the likelihood and severity of an adverse event occurring to humans, animals or to the environment following exposure under defined conditions to a specific hazard. Guidelines for the conduct of RA have been developed to assess the risk of animal disease import (OIE, 2004a,b) and the risk of microbiological hazards in food (CAC, 2002). However, for animal welfare it has been concluded, that there are no available standard methodologies on Risk assessment (EFSA, 2006a). Therefore, efforts have been devoted to develop such methodologies.

The first European legislation on farm animal welfare was adopted in 1974 and concerned the stunning of animals before slaughter, and the European Commission has been taking the leadership to promote it internationally, supporting initiatives of the Council of Europe and more recently of the OIE. The principles inspiring the legislation were related to respect the basic five freedoms (freedom from discomfort, from hunger and thirst, from fear and distress, from pain, injury and disease and freedom to express natural behaviour) as stated in 1979 by the Council of Europe (CoE, 2009).

The Community Action Plan on the Protection and Welfare of Animals 2006-2010 (EC, 2005a), adopted by the European Commission on 23 January 2006, embodies the Commission's commitment to EU citizens, stakeholders, the European Parliament and the Council for a clear map of the Commission's planned animal welfare initiatives for the coming years. The action plan has identified five main areas: (1) upgrading existing minimum standards for animal protection and welfare; (2) giving priority to promoting policy oriented future research on animal protection and welfare; (3) introducing standardised animal welfare indicators; (4) ensuring that animal keepers/handlers as well as the general public are more involved and informed on current standards; and (5) continuing to support and initiate further international initiatives to raise awareness and create consensus on animal welfare.

In line with the European Commission 'Future Strategy on Animal Health for 2007-2013' (EC, 2007a) and Community Action Plan on the Protection and Welfare of Animals (2006-2010) (EC, 2005a) EFSA has accepted a statement fixing the Role of EFSA in contributing to the improvement of animal health in Europe (EFSA, 2008f) in order to support the Animal Health Strategy (EC, 2007a).

It is stated that the concept of animal health covers not only the diseases of animals, but also the critical relationship between animal welfare and animal disease, as well as its relevance to public health. While some animal diseases are important for animal welfare or their impact on trade (e.g. Bluetongue) many diseases also include a zoonotic aspect, i.e. they are relevant to public health. High standards of animal health are thus an essential pillar for safeguarding public health, ensuring food safety, and securing animal welfare. On the other hand, diseases often decrease the welfare of animals. For example, stress may increase susceptibility to infections, which in turn may lead to illness in animals or increased shedding of pathogens and contaminants into the food chain. On the other hand, whereas outdoor farming systems may improve animal welfare, they may also increase the exposure of animals to certain biological hazards (Blokhuis *et al.*, 2008; Ribó *et al.*, 2008a, 2009).

High standards of animal health are thus an essential pillar for safeguarding public health, ensuring food safety, and securing animal welfare. The EU's Animal Health Strategy (AHS) for 2007-2013 (EC, 2007a) encompasses a 6-year programme of work, divided into four different pillars: (1) prioritisation of the EU intervention; (2) a modern and appropriate animal health framework; (3) better prevention, surveillance and crisis preparedness; and (4) science, innovation and research. EFSA is the keystone of the EU risk assessment regarding food and feed safety and the related fields of zoonoses, diseases in wild and domesticated animals, animal welfare and plant health. In close collaboration with scientists as well as with national authorities and in open consultation with its stakeholders, EFSA provides independent scientific advice and communication on existing and emerging risks. Since EFSA's scientific advice serves to inform risk managers, a large part of EFSA's work is undertaken in response to requests from the EC, the European Parliament (EP) and EU Member States (MS). EFSA also undertakes scientific work on its own initiative, so-called self-tasking.

Concerning the activities of the first pillar of the Prioritisation of EU intervention '*The new Animal Health strategy must be seen as an integrated risk assessment and management strategy focusing on biological and chemical risks of EU relevance*' EFSA plays a key role in this area by providing scientific advice and risk assessments. The usefulness of this work depends on the approach taken to requesting scientific advice, the urgency of the question, the availability of pertinent data and their quality, and the risk assessment approach used to address the question in hand. Since decisions on prioritisation of interventions must be based on sound science and appropriate risk assessment (Pillar 4), close collaboration between EFSA, EC and MS is essential in order to obtain useful scientific advice and risk assessment for decision making. Risk assessment must be based on sound data; this cooperation and data exchange between EFSA, EC, Member States and Community Reference Laboratories, in association with EMEA, ECDC and other relevant EU organizations as well as with international organisations such as OIE and WHO, enable EFSA to provide high-quality scientific advice in a timely manner.

The second pillar refers to a modern and appropriate animal health framework which means going '*Towards a single regulatory framework, with a greater focus on incentives rather than penalties, consistent with other EU policies and converging to international standards*'. A modern animal health legislative framework relies on scientific information and modern systems to assess this information. Using risk assessment methods that are recognised internationally, EFSA will continue to develop recognition for its work within the international risk assessment community. On the other hand, developing a single regulatory framework may have an effect on the animal nutrition policies as well as policies linked to animal diseases, animal welfare and public health and thereby would evidently influence on the practices in EU. These changes need to be taken into account also in future risk assessments.

The third pillar of the animal health strategy '*Identifying problems before they take hold and being ready to manage outbreaks and crises*', allows that the risk assessments and scientific advice provided by EFSA can serve as a useful tool to evaluate the effect of various risk management options on risk reduction in line with the principle 'prevention is better than cure'. This advice together with the other legitimate factors (to be considered by the risk managers) can be used for planning the prevention or control of current and emerging animal diseases in domesticated or wild animals, for animal welfare, animal nutrition and/or for public health. Furthermore, EFSA can contribute to surveillance for animal diseases in the EU. A key factor in enabling effective data evaluation is the harmonization of various aspects of these programmes. This can take place through (1) recommendations in EFSA opinions on topics to be covered in the surveillance programmes; and (2) development of indicators for animal welfare that could be used for surveillance purposes.

It should be noted that it is essential that crisis preparedness be developed in 'peace time' so that, when the need to use the data and conduct data analyses arises, adequate and efficient data exchange is already ensured. Support to risk managers in setting up surveillance systems may provide a good basis for smooth collaboration during crises. EFSA can also provide urgent scientific advice by using fast-track responses or statements (as defined in the Management Board decision in September 11$^{th}$ 2007) during crises and by providing technical support for data collection and analysis.

The fourth pillar bases the actions on Science, Innovation and Research in order to '*To stimulate and coordinate risk analysis, science, innovation and research, hence contributing to a high level of public health and to the competitiveness of EU animal health businesses*'. While the scientific activities of EFSA are of an applied nature, EFSA aims to use the best available science to carry out its tasks. Therefore, it is essential for EFSA to mobilise and coordinate scientific resources throughout the EU to provide high-quality and independent scientific advice and risk assessments. In practice, this takes place by means of its scientific panels, working groups, task forces, grants and the contracting-out of scientific work, as well as in other ways of networking with scientists. EFSA is also actively collaborating with DG Research on the research needs identified during risk assessment work and with the European Medicines Agency (EMEA), the European Centre for Disease Prevention and Control (ECDC) and other relevant EU agencies.

It is very relevant to stress the effort carried out by DG Research RTD supporting the community Action Plan on the protection of Animals 2006-2010 in funding projects Reports of studies like the 'Welfare Quality: improving animal welfare in the food quality chain'. The improvement on animal welfare may positively affect product quality, pathology and disease resistance, aspects with a direct impact on food quality and safety. The main objective of the WelfareQuality® Project is the integration of animal welfare in the food quality chain, through the development of reliable on-farm monitoring systems for welfare assessment, product information systems, and practical species-specific strategies to improve the welfare of cattle (beef and dairy), pigs, and poultry (broiler chickens and laying hens) (EC, 2004a).

## 3. Overview of EFSA/AHAW activities on risk assessment and scientific advice in animal welfare (2003-2009)

The EFSA's Panel on animal health and welfare is composed by 21 independent scientific experts from which one third are experts on animal welfare issues, one third on animal diseases and one third in animal health and welfare horizontal issues related to risk assessment methodologies, epidemiology and modelling. The panel is responsible for all adopted scientific opinion and receives the full support of AHAW Unit.

The main role of EFSA on animal welfare is to provide assessments and address animal welfare issues from a scientific point of view. The risk assessment approach should be followed whenever possible (EC, 2002a; EFSA, 2007e; Müller-Graf *et al.*, 2008). In the case of experimental animals, EFSA adopted a pro-active animal welfare approach in the Management Board document produced during its meeting of June 2004, saying that '*... while recognizing that animal testing cannot be eliminated at present, EFSA should make every effort to stimulate and participate in the development of new food and feed assessment approaches that would minimize the use of experimental animals and would reduce, to the extent possible, the level of suffering of animals...*' (EFSA, 2004a).

The EFSA Animal Health and Welfare Panel has delivered during the period 2003-June 2009, 68 scientific opinions on a variety of animal diseases and welfare issues, while other panels and EFSA units have also been delivering opinions related to animal health and welfare (Biological hazards, Feed additives, Contaminants and Zoonosis Unit). The AHAW panel has been increasing the number of activities and scientific opinions since 2003. A total of 32 scientific opinions have been adopted on animal welfare mainly of food producing animals (Figure 1).

The scientific opinions (SO) have been focused on pig welfare (5 SO), fish welfare (5 SO), stunning and killing of animals (3 SO), animal transport (2 SO), calves' welfare, import of wild birds, rabbits, laying hens, laboratory animals and the 'Framework for EFSA AHAW Risk Assessments' (EFSA, 2007e). Also, animal welfare aspects were requested to be considered in the scientific opinions on animal diseases, such as, avian influenza (EFSA, 2005b), import of captive birds (EFSA, 2006b) and avian influenza vaccines (EFSA, 2007d). It is important to point out that the European Commission when committing an animal welfare mandate, requests to take into account, where relevant, animal disease and food safety aspects (EFSA, 2004,a,b, 2005a,b, 2006a,b,c,d, 2007a,b,c,d, 2008 a,b,c,d,e,f).

The AHAW Panel has recently adopted (June 09) five scientific opinions on the welfare aspects of Dairy Cows. Four of these scientific opinions are based on four Risk Assessments of the impact of housing, nutrition and feeding, management and genetic selection on (1) leg and locomotion problems; (2) metabolic and reproductive problems; (3) udder problems and (4) behavioural, fear and pain problems in dairy cows. The fifth opinion is an overall welfare navigation systems should also help to improve the level of compliance with journey time requirements.

### 5.3 Other criteria for inspections in slaughterhouses

Regulation (EC) No 854/2004 clarifies that animal welfare is a task for the OV who is responsible for the slaughterhouse and adds to the controls laid down in Directive 93/119/EC by requiring the results of inspections to be recorded and evaluated. In relation to long distance transport where the destination is a slaughterhouse, Regulation (EC) No 1/2005 additionally lays down that, as well as targeted checks, the CA shall also carry out checks on a random basis to verify that travel times comply with the limits.

## 6. Defining the scope of inspection

Article 4 of the Official Control Regulation requires the competent authorities to ensure the effectiveness and appropriateness of official controls. The nature or scope of inspections is therefore an important aspect which needs to be considered. As monitoring of animal welfare is often carried out at the same time as checks for other purposes, defining the scope of inspections does help to improve their overall effectiveness, and provision of simple written procedures is useful for inspectors who are dealing with a considerable body of legislation, not just that concerned with animal welfare.

### 6.1 Scope of farm welfare inspections

Farm inspections do not necessarily have to be an exhaustive check of every legal requirement and Decision 2006/778/EC allows the CA to choose four or five issues which should be checked. The scope of the inspection can then be specified in the checklist provided where the key points which must be covered are highlighted. Checklists should also be designed to facilitate the subsequent recording of the results of the inspection so that an accurate overview of the monitoring activities is obtained.

Insufficient monitoring of conditions on farms has occurred where CAs have defined the scope of inspection on the basis of a single piece of legislation and omitted to integrate other relevant requirements. This has been the case in several Member States where all the technical details from Directive 91/630/EEC concerning the welfare of pigs were listed in checklists, but requirements from Directive 98/58/EC were overlooked such as the important welfare aspect of how farmers dealt with sick or injured pigs.

For controls on laying hen farms, assessments have been made using checklists which only contained the requirements from Directive 1999/74/EC. As a result inspectors have measured cage dimensions and inspected a number of cages for overstocking but did not assess the overall extent of overstocking. This requires the maximum number of birds which can be kept in the building to be calculated and compared with the data available from flock records on the number of birds which were introduced. The CAs who verify the data in the register required by Directive 2002/4/EC do include this point. Such inspections result in more effective enforcement action.

## 6.2 Scope of transport inspections

Checks of transport may be targeted on clearly identified issues, for example, issues which have been dealt with when vehicles are approved do not have to be subjected to the same degree of scrutiny during checks of the day to day operations. The site where inspections are organised also determines how effectively certain issues can be assessed.

### *6.2.1 Scope of checks at places of departure*

When assessing animal transport operations, inspectors should be able to rely on the data which has already been evaluated at the time of vehicle approval. So issues which have been dealt with when vehicles are approved, such as the capacity of the ventilation, the floor area of vehicles and the installation of temperature recording equipment do not have to be subjected to the same degree of scrutiny during checks of the day to day operations of these vehicles. On the spot checks of long distance transport should then focus on whether each fan is still operational, whether the equipment for temperature monitoring is functional and data is being recorded, that the condition of the animals is satisfactory and that sufficient space has been provided.

### *6.2.2 Scope of checks during transport*

The scope of inspection also varies depending on the volume of transport at a particular location. For example, at one port in France a sample of vehicles is checked, as many trucks arrive at the same time to unload animals, which are then loaded as part of a much bigger consignment onto a fitted livestock vessel. On the other hand at ports with RoRo ferry traffic, detailed checks on a much higher percentage of trucks is possible as not as many vehicles converge on the port at the same time [DG(SANCO)/7580/2006].

Livestock markets are also locations where many vehicles converge and certain CAs have adopted the approach of monitoring the overall standard of vehicles and then targeting certain vehicles for a more detailed inspection [DG(SANCO)/7690/2008].

The siting of inspections also determines how effectively certain issues can be assessed. When vehicles are stopped at roadside checks, and the vehicle is fully loaded with animals, issues such as the fitness of animals for transport are difficult to assess. Roadside checks, as already discussed, usually involve the police with technical advice provided by the CA. Arrangements should be made to unload animals at a location in the vicinity where the check takes place and it may also be necessary to weigh vehicles both before and after unloading to obtain evidence of overstocking. The local CA in districts with significant sea ports have an important role to play in controls of long distance transport as vehicles waiting to board ferries provide an opportunity for the CA to carry out checks and delays at ports can pose problems for animal welfare. A particular problem at ports is that there are often no appropriate facilities to unload the animals in the vicinity of the port [DG(SANCO)8042/2006, DG(SANCO)/7690/2008 and DG(SANCO)/7768/2008] and although the transporter is primarily responsible for contingency arrangements, the CA too is obliged to have plans to deal with such situations when they arise.

Animal welfare concerns may not be the principle objective at locations determined by police checks. In certain regions the police have carried out major operations targeting all commercial transport carried out on a given day so that transporters cannot easily evade checks. Again provided the police can carry out rudimentary welfare inspections and the CA provide advice as required, significant findings have been detected.

*6.2.3 Scope of transport checks at slaughterhouses*

For checks of animals arriving at slaughterhouses, timing is obviously important as transport often takes place at specific times of the day, or night, and there may not be a veterinary presence in the lairages of the slaughterhouse at these times, as a veterinarian is often only present to carry out *ante mortem* inspections. The central level of the CA nearly always places major emphasis on checks at slaughterhouses as a fundamental part of their programme of transport inspections, as they have a daily presence in slaughterhouses, so a failure to provide adequate instructions to its staff to carry out checks at such locations seriously undermines their system of control. The handling of animals during unloading and the condition of the vehicle may not be a priority for an OV unless clear instructions, with minimum targets, have been issued. Some basic planning by the local level of the CA should ensure that all transporters who regularly deliver animals at slaughterhouses are inspected at least once in every round of inspections.

### 6.3 Scope of inspections within slaughterhouses

A recurring problem for welfare at slaughterhouses is that certain CAs do not indicate which issues OVs should monitor in relation to animal welfare. Although *ante mortem* inspections are made to determine the fitness of the animal for human consumption this should also serve the purpose of detecting many serious welfare problems. Although a small number of Member States continue to inadequately utilise this opportunity to ensure animal welfare requirements, most CAs now provide a simple aide-memoire for daily inspections, which ensures that animal welfare is considered at each location in the slaughter process. On a less frequent basis, e.g. every six months but again the frequency may be determined by risk assessment, the CA then carry out an inspection with a wider scope including checks of the maintenance of equipment and the training of slaughterhouse personnel.

## 7. Reporting the results of inspections

There is a general requirement in the Official Control Regulation regarding drawing up reports on the controls carried out, and the notification of the operator is a basic function of an inspection report. When non-compliances are detected these should be notified in writing to operators, but it is usually the case that oral warnings are given for more minor deficiencies and written notification is reserved for more serious issues or repeated offences.

It is important that not only is the operator notified of the action required but that reports of controls are available to the CA so that the results can be collated and evaluated, as the available data can be used to prioritise future inspections and improve methodologies. There are specific requirements in animal welfare Directives and Regulations on reporting.

### 7.1 Farm inspection reports

Decision 2006/778/EC provides some fifteen categories for classifying deficiencies detected at farm inspections. It is therefore important that the format of checklists facilitate collating the results of inspections, and that inspectors report deficiencies and follow-up actions in a standard way. If the CA does not give clear guidance how to categorise the commonly occurring deficiencies, deficiencies such as inappropriate pens for livestock might be recorded in one local office as an infringement of 'freedom of movement' and as an infringement of requirements for 'buildings and accommodation' by another. This results in summary reports with incoherent data.

### 7.2 Transport inspection reports

In order to facilitate reporting the results of inspections of long distance transport to the CA responsible for the place of origin, inspectors who detect infringements can notify the CA responsible for checks at the place of origin via the TRACES system, the information exchange system referred to in Article 20 of Directive 90/425/EEC. This is beneficial for speeding up the exchange of information and ensures that it gets to the relevant people at the place of departure. The simplified and codified format for reporting deficiencies also helps overcome linguistic difficulties. The designation of national contact points for dealing with such issues also helps to facilitate reporting, but all levels of the CA need to be made aware of the contact point and its function so that appropriate cases get communicated. Co-ordination within a CA needs to be good if contact points are to function effectively. So that CAs adopt a more strategic approach to transport inspections, Regulation (EC) No 1/2005 requires them not only to provide a report on the outcome of checks but to provide an action plan to address the most common occurring deficiencies.

### 7.3 Slaughterhouse inspection reports

The results of animal welfare checks in slaughterhouses must be recorded and evaluated by the OV. These reports, however, rarely specify incidents relating to animal welfare, and the other levels of the CA may not be adequately informed of the issues which have arisen. Failure to provide an adequate mechanism to record and report incidents, even though these may be isolated, prevents the CA from getting an adequate picture of the level of compliance, such as the incidence of injured animals arriving at slaughterhouses. As a result the next levels of the CA may not be adequately aware of major animal welfare problems and do not have an overview which in turn could be used to develop guidance, training and enforcement.

Reports of inspections need to be sufficiently detailed to provide a basis for the next level of the CA to supervise that tasks have been carried out.

# 8. Ensuring the quality of checks

Article 8 of the Official Control Regulation states that the CA must have procedures in place to verify the effectiveness of official controls, to ensure effectiveness of corrective action and to update documentation where needed.

By analysing the results of inspections, the CA can see whether targets have been met and assess differences in the outcome of the checks in different regions. Deficiencies detected may be randomly distributed or there may be clustering in specific regions or at specific times of the year. A low detection rate in one region may merit just as much further investigation as a higher rate in another region. The detection rate of infringements may depend on the quality of the checks just as much as on the actual level of compliance. The subsequent investigation by the CA should seek to establish the cause of these differences: competence of inspectors, resources to enforce requirements, effectiveness and efficiency of sanctioning procedures, and management of these by the local CA. Supervision therefore provides a basis for updating instructions, identifying training needs and addressing any other weaknesses in controls.

Audits, which are required by Article 4 of the Official Control Regulation, are also being used by CAs to ensure that their controls are achieving their objective. Audits should generally not be a substitute for the more regular hierarchical supervision. Auditors should not audit areas or activities for which they have direct responsibility, as indicated in Commission Decision 2006/677/EC. In the field of animal welfare it is sometimes difficult to reconcile the requirement for independence of the auditor with the requirement to provide the necessary expertise to evaluate the impact of the controls on animal welfare. This is generally satisfied by having lead auditors who are from dedicated audit units, and the necessary expertise on animal welfare is obtained by involving an expert from a region other than that under audit.

## 8.1 Verification of farm checks

On the spot supervision of inspectors is part of the system seen in relation to cross-compliance controls [DG(SANCO)/7980/2008] and the best practice seen in relation to supervision usually involves accompanying inspectors on their farm visits. During FVO missions the approach has always been to ask the local inspector to carry out their controls and make their conclusions before any discussion takes place on the merits of the inspection procedure. The outcome of such visits needs to be put into a wider context by evaluating the reports of other inspections on other farms and the subsequent follow-up actions. The results of inspections reported from each local office of the CA, to contribute to the report required by Decision 2006/778/EC, can also be compared to verify whether the various offices are adopting a similar approach. The audit approach required by Article 4 of the Official Control Regulation has not been widely used to date in relation to animal welfare checks, but where it has been used it is showing promising results in validating the effectiveness of inspections and whether the procedures in place are suitable.

## 8.2 Verification of transport checks

The arrangement of checks in relation to the export of cattle where subsidies are involved [Regulation (EC) No 639/2003] provides for verification of the effectiveness of controls. Checks of transport conditions at exit points of the EU and in third countries of destination provide information not only about the compliance of the transporter at this point in the journey but also provide an indication regarding the quality of the check which took place at departure.

Prior to the entry into force of Regulation (EC) No 1/2005 there was already exchange of information taking place between the different CAs in the Member States regarding incidents arising during the long distance transport of animals. This has been improved by the additional mechanisms of reporting which are now in place via the contact point required by Regulation (EC) No 1/2005 to be designated by each CA and via TRACES.

Prior to FVO missions, the inspection team will have evaluated data from the TRACES database and use this to verify the effectiveness of official controls on journey times. During the mission the inspection team request the local CA to retrieve the corresponding journey logs for a number of consignments. These are then assessed to see how well the CA carried out controls before they approved the journey log. The inspection team will also have evaluated any animal welfare infringements which have been reported via TRACES and will ask the local CA responsible for checks at the place of departure what actions have been taken in response to this information. These simple supervisory exercises could also have been used on a more frequent basis by the local CA's hierarchy to evaluate the quality of the controls carried out by the local level.

## 8.3 Verification of checks at slaughterhouses

On the spot visits to verify the effectiveness of controls in slaughterhouses are usually carried out by the CA with hierarchal responsibility for the work of the OV. As has been mentioned in relation to OVs carrying out daily checks, the priority during these visits may not be animal welfare and so this may just get a cursory examination. On the spot visits to verify the effectiveness of daily checks do not however have to be carried out directly by those with hierarchal responsibility. In two regions of Germany [DG(SANCO)/7980/2008], two examples were seen where effective arrangements to verify checks in slaughterhouses had been made. In one district there was a parallel system of inspections where the unit of the local CA normally responsible for animal welfare checks outside slaughterhouses visited each slaughterhouse every six months and verified the effectiveness of daily controls by their colleagues in the meat hygiene unit. In another region, there was an equally effective system to verify that stunning of animals was appropriate. Here a technician from the CA carried out an annual visit, and aspects which had been overlooked by the OV, or for which they were not fully capable of assessing, were addressed. By using such arrangements for supervision, the hierarchy of the CA can effectively assess whether staff are giving sufficient attention to animal welfare issues during their daily checks.

## 9. Follow-up to inspections

Operators on farms, of transport and of slaughterhouses need to understand that there are consequences when non-compliances are detected during inspections. In order to ensure that corrective actions are taken monitoring activities by the CA may be sufficient as many operators are sufficiently amenable to an advisory approach. However the CA needs to have recourse to measures to deal with more recalcitrant operators who do not co-operate sufficiently or who are persistent offenders. Article 54 of the Official Control Regulation requires a CA which identifies non-compliance to take appropriate action to ensure that the operator/business remedies the situation. Article 55 states that Member States shall lay down the rules on sanctions and take all measures necessary to ensure that they are implemented. The sanctions provided for must be effective, proportionate and dissuasive.

A lack of adequate enforcement can give rise to a general malaise in official controls, as staff who don't have recourse to proper enforcement measures feel frustrated or tolerate so-called 'minor deficiencies'. As a result the quality of inspections suffers. Lack of priority by prosecution officials and courts has given rise to low levels of enforcement. This has occurred where certain requirements might be deemed as technical non-compliances and the CAs are reluctant to initiate a prosecution unless they think that unnecessary suffering can be proven. Prosecution officials need to be closely involved if they are to better understand the impact of the breaches of legislation on animal welfare.

Working groups have been created in several Member States to allow inspectors to agree the type of action to take for the commonly occurring deficiencies. Administrative actions following inspections are also important and best practice includes the use of model letters to notify the other services involved in enforcement. Procedures, sometimes in the form of decision trees, also assist the local CA in deciding the action to take and help to ensure that sanctions are more uniformly applied.

Enforcement policies need to be established so that the local CAs are clear on that they are supported in following certain actions, and this has a direct influence in the quality of inspections and the way evidence is collected. In Denmark only one written warning can be given for the same infringement, then the official veterinarian must initiate enforcement action [DG(SANCO)8039/2006]. Similarly in Germany a warning letter indicates a deadline for corrective measures and the sanction which will be imposed if the operator does not meet this deadline [DG(SANCO)/ 7980/2008].

## 10. Conclusions

### 10.1 What has been achieved?

All the CAs of the EU Member states have systems to monitor animal welfare. Animal welfare controls have received a higher priority than was previously the case; most recently the implementation of the Official Control Regulation and the system of cross-compliance has helped to bring about greater levels of co-ordination and co-operation between the various services involved.

Overall CA procedures for carrying out checks are effective in identifying deficiencies. The multi-disciplinary approach adopted in the majority of Member States has also led to improvements in relation to monitoring conditions of animal welfare during transport, which has been a problematic area.

By analysing the results of previous inspections certain CAs have already identified important risk criteria and subsequently used these to better target and time inspections.

### 10.2 What has been neglected?

There are significant differences between the CAs of the EU Member states in the effectiveness of the systems to monitor and enforce animal welfare requirements. Although most CAs effectively identify deficiencies during inspections, insufficient arrangements are in place to prioritise inspections and frequently there are insufficient procedures to ensure that corrective actions are made when problems are detected.

Although the level of training on animal welfare issues has increased in recent years, there are certain specialist areas which have often been neglected. In particular in relation to the assessment of equipment for vehicles involved in long distance transport, including sea vessels, and in relation to stunning of poultry.

In relation to checks of farms, the scope of inspections has not always been sufficient such as to conclude on the extent of overstocking on laying hen farms. In relation to transport checks, the requirement to increase the number of inspections where it is found that transporters are not respecting the rules, has not been systematically implemented, even though this is a simple measure which does bring about better levels of compliance.

Monitoring of journey times of long distance transport is often not effective as data from journey logs and drivers' records have not been sufficiently scrutinised by the CA. Arrangements for carrying out checks of animal welfare during long distance transport have also frequently been inadequate as locations, such as sea ports where such controls can be easily carried out lack appropriate facilities to unload the animals. It is also a requirement for the CA to have such arrangements at hand so that if a transporter fails to deal with a significant animal welfare problem, at least the CA can step in to resolve the situation.

Effective monitoring of animal welfare at slaughter is mostly compromised where there is a singular focus by all levels of the CA on food safety, the principle reason for the high level of veterinary presence in slaughterhouses, and as a consequence animal welfare issues are marginalised. This has caused serious welfare problems, such as the transport of injured animals to slaughterhouses, not to be addressed. The requirements for approval of slaughterhouses similarly do not adequately address animal welfare issues and systems are installed for restraining and stunning animals without any prior approval.

Reporting of the outcome of inspections is also neglected as it is too often seen as a bureaucratic exercise and the CAs do not have adequate data to analyse. This can be used to monitor the implementation of inspection programmes and also to improve the quality of future inspections.

### 10.3 What needs to be done?

Risk assessment needs to be further developed so that CAs can target their resources more effectively. Not only does the CA need to define the severity of animal welfare problems in each livestock sector and at each stage of production, transport and slaughter, but the CA also needs to better identify those particular operators, or establishments, which pose the greatest risk of serious non-compliance. Data already available from operators 'own checks' should be included to a greater extent in risk analysis so that the animal welfare issues can be better defined and a higher degree of surveillance carried out by the CA on these processes. The likelihood of non-compliance and the consequence of this on the welfare of the animals should be used to determine the frequency and intensity of inspections. Models of risk assessment which help identify areas for more frequent or intense controls would be invaluable to authorities who need to decide how best to deploy their limited inspection resources.

As farms often receive a lower frequency of inspections than transport operations or slaughterhouses the CA should not miss the opportunity to properly assess basic requirements such as care of sick animals and stocking rates. Clearer instructions and better supervision would ensure more comprehensive inspections.

Requiring data to be transmitted from a vehicle to a database accessible to all CAs, as proposed in an amendment to EU legislation, should help to simplify the monitoring of journey times for long distance transport. Future EU legislation may also require installations used in slaughterhouses for handling, stunning and bleeding animals to be assessed and approved before they are brought into use.

Improving reporting and verification procedures is already being addressed through the framework of the Official Control Regulation. This provides a sound basis to establish risk criteria, revise procedures, identify training needs and facilitate better quality inspections.

## Acknowledgements

This report is based on the work of the competent authorities of the EU Member States through whose assistance and co-operation the individual FVO missions were carried out. I am grateful also to all the colleagues, both past and present, in the FVO who helped in carrying out the inspections on which this report is based. I would also like to thank those who were involved in drafting or providing compliance and legal checks on the individual mission reports and to the assistants who dealt with the various administrative tasks involved.

## References


Codex Alimentarius, 2003. Recommended international code of practice, general principles of food hygiene – Annex on HACCP System and guidelines for its application. Available at: http://www.fao.org/docrep/w6419e/w6419e03.htm.

Decision 2006/677/EC of 29 September 2006 setting out the guidelines laying down criteria for the conduct of audits under Regulation (EC) No 882/2004 of the European Parliament and of the Council on official controls to verify compliance with feed and food law, animal health and animal welfare rules OJ L 278, 10.10.2006, p. 15.

Decision 2006/778/EC of 14 November 2006 concerning minimum requirements for the collection of information during the inspections of production sites on which certain animals are kept for farming purposes OJ L 314, 15.11.2006, p. 39-47.

Directive 90/425/EEC of 26 June 1990 concerning veterinary and zootechnical checks applicable in intra-Community trade in certain live animals and products with a view to the completion of the internal market, OJ L 224, 18.8.1990, p. 29-41.

Directive 91/629/EEC of 19 November 1991 laying down minimum standards for the protection of calves, OJ L 340 of 11.12.1991, p. 28.

Directive 91/630/EEC of 19 November 1991 laying down minimum standards for the protection of pigs, OJ L 340 of 11.12.1991, p.33.

Directive 93/119/EEC of 22 December 1993 on the protection of animals at the time of slaughter or killing, OJ L 340, 31.12.1993, p.21.

Directive 98/58/EC of 20 July 1998 concerning the protection of animals kept for farming purposes, OJ L 221 of 8.8.1998, p. 23.

Directive 1999/74/EC of 19 July 1999, laying down minimum standards for the protection of laying hens, OJ L 203, 3.8.1999, p.53.

Directive 2002/4/EC of 30 January 2002 on the registration of establishments keeping laying hens, covered by Council Directive 1999/74/EC, OJ L 30, 31.1.2002, p.44.

DG(SANCO)/7512/2005, report of a mission carried out in the Netherlands from 31 January to 4 February 2005 concerning animal welfare on farms. Available at: http://ec.europa.eu/food/fvo/index_en.htm.

DG(SANCO)8039/2006, report of a mission carried out in Denmark from 06/02/2006 to 10/02/2006 in order to assess animal welfare at slaughter. Available at: http://ec.europa.eu/food/fvo/index_en.htm.

DG(SANCO)8042/2006, report of a mission carried out in Greece from 21/02/2006 to 01/03/2006 concerning animal welfare during transport and at the time of slaughter. Available at: http://ec.europa.eu/food/fvo/index_en.htm.

DG(SANCO)/7580/2007, report of a mission carried out in France from 17 to 21 September 2007 in order to assess animal welfare during transport as linked to Regulation (EC) No 639/2003 on the granting of refunds for the export of live bovine animals. Available at: http://ec.europa.eu/food/fvo/index_en.htm.

DG(SANCO)/7690/2008, report of a mission report of a mission carried out in the UK from 18/02/2006 to 22/02/2006 concerning animal welfare during transport. Available at: http://ec.europa.eu/food/fvo/index_en.htm.

DG(SANCO)/7768/2008, report of a specific audit carried out in Ireland from 8 to 13 September 2008 to evaluate the implementation of controls for animal welfare on farms, during transport and at the time of slaughter. Available at: http://ec.europa.eu/food/fvo/index_en.htm.

DG(SANCO)/7980/2008, report of a specific audit carried out in Germany from 6 to 10 October 2008 to evaluate the implementation of controls for animal welfare on farms, during transport and at the time of slaughter. Available at: http://ec.europa.eu/food/fvo/index_en.htm.

EFSA, 2007. Opinion of the Scientific Panel on animal health and welfare on a self mandate on the Framework for EFSA AHAW Risk Assessments, The EFSA Journal 550, 1-46.

Regulation (EC) No 3821/85 of 20 December 1985 on recording equipment in road transport, OJ L 370, 31.12.1985, p. 8. Regulation as last amended by Commission Regulation (EC) No 432/2004 OJ L 71, 10.3.2004, p. 3.

Regulation (EC) 639/2003 of 9 April 2003 [hereafter: Regulation (EC) No 639/2003] laying down detailed rules pursuant to Council Regulation (EC) No 1254/1999 as regards requirements for the granting of export refunds related to the welfare of live bovine animals during transport OJ L 93, 10.04.2003, p. 10.

Regulation (EC) No 796/2004/EC of 21 April 2004 laying down detailed rules for the implementation of cross-compliance, modulation and the integrated administration and control system provided for Council Regulation (EC) No 1782/2003 establishing common rules for direct support schemes under the common agricultural policy and establishing certain support schemes for farmers OJ L 141, 30.4.2004 corrected in OJ L 37, 10.2.2005, p. 22-23.

Regulation (EC) No 854/2004 of the European Parliament and of the Council of 29 April 2004 laying down specific rules for the organisation of official controls on products of animal origin intended for human consumption, OJ L 139, 30.04.2004 corrected and republished in OJ L 226, 25.06.2004 p.83.

Regulation (EC) No. 882/2004 of the European Parliament and of the Council of 29 April 2004 on official controls performed to ensure the verification of compliance with feed and food law, animal health and animal welfare rules, OJ L 165, 30.04.2004, corrected and republished in OJ L 191, 28.05.2004, p.1.

Regulation (EC) No 1/2005 of 22 December 2004 on the protection of animals during transport and related operations and amending Directives 64/432/EEC and 93/119/EC and Regulation (EC) No 1255/97, OJ L 3, 5.1.2005, p. 1.

# Animal welfare risk assessment and management from a national perspective

**Jan Hultgren**
Department of Animal Environment and Health, Swedish University of Agricultural Sciences, P.O. Box 234, SE-53223 Skara, Sweden; jan.hultgren@hmh.slu.se

## Summary

This chapter describes the Swedish view on animal welfare, existing disease monitoring and quality assurance schemes, labelling programmes and the organisation and implementation of risk-based official control. Through a number of monitoring schemes and investigations, valuable data on e.g. hazard exposure, disease occurrence and other welfare indicators are regularly fed into extensive Swedish registers and databases. Swedish animal welfare legislation in many respects stretches far beyond EU directives. Beside regular inspections of animal operations, pre-testing of new technologies and prior approval of livestock buildings are key components of official control. In response to research and increased public awareness and concern for animal welfare issues, the Government has initiated work to modernise national legislation. As from 2006, EU Member States shall ensure that national official animal welfare control accounts for identified risks at different animal operations. Sweden, like other countries, is developing control policies and routines to meet this requirement under national conditions. The Swedish Board of Agriculture is the national expert authority in the field of agricultural and food policy, centrally responsible for the welfare of animals. A national animal welfare control database has been established, with information about all known animal owners and control sites in Sweden, including site-specific data such as type of animal husbandry, numbers of animals of different categories and established non-compliance or offense against legislation. As from the beginning of 2009, the responsibility for local control was transferred from 290 municipalities to 21 county administrative boards. This contribution also gives a brief critical appraisal of some fundamental principles of animal welfare risk assessment and management. Neglected areas are identified and further work suggested.

**Keywords**: monitoring, official control, quality assurance, risk classification, Sweden

## 1. Introduction: a Swedish perspective on animal welfare

Animal production matters to us all because it influences the environment, our daily food, and our relationship to animal fellow-creatures – all vital to our own welfare. Producers and their organisations need to communicate specific quality aspects of the animal production chain to compete on the international food market. Because health and well-being of food-producing animals are also relevant to consumers, good animal welfare contributes to the perceived quality of food products of animal origin (Blokhuis *et al.*, 2003; European Commission, 2007). National animal welfare legislation sometimes goes beyond that of the EU, reflecting different societal demands on national authorities in regulating animal husbandry. There is thus an incentive for national and local stakeholders to develop methods to assess and monitor animal welfare status and animal welfare risks for production animals.

EC Regulation No. 882/2004 and related legislation lay down the EU policy regarding official controls to verify compliance with animal health and welfare regulations in the food and feed chain. Member States shall prepare a control plan, and ensure that official controls are carried out regularly, on a risk basis and with appropriate frequency, taking account of identified risks that can influence animal health or welfare. The frequency of official controls shall be proportionate to the estimated risk. Risk-based official control shall thus try to identify animal operations (control sites) where the animals run a high risk for impaired welfare and/or non-compliance with relevant legislation, and to use this information when allocating control resources and planning inspection visits, thus securing the economic efficiency of the control scheme. The principal objective of official animal welfare control is to check compliance with relevant national and international legislation, aiming to secure animal welfare. Official control must therefore involve the use of both animal-based and resource/management-based welfare measures. Specifically, as dictated by Regulation No. 882/2004, risk-based control is required to use site-specific information to assess the risk for non-compliance. Regular Swedish systems for assessing welfare in food-producing animals on the farm level have been incorporated e.g. in quality assurance schemes for broilers (Berg and Algers, 2004) and laying hens, in pre-testing of new technologies, and in prior approval of livestock buildings, as discussed later in this contribution. However, so far there is no common system of operational animal welfare standards. Nor is there any system for animal welfare risk standards.

In Sweden, the Swedish Board of Agriculture is the Government's expert authority in the field of agricultural and food policy, centrally responsible for the agricultural and horticultural sectors, including the welfare of all animal species. Between 2004 and 2007, animal welfare fell under the responsibility of a separate authority, the Swedish Animal Welfare Agency (Berg and Hammarström, 2006). Sweden has for a long time considered animal welfare an important aspect of animal husbandry and food production. Swedish legislation in this area is extensive and there are detailed regulations on animal housing, management and feeding of animals, as well as official reports, government bills and communications from the Riksdag explaining the scientific background to those regulations. The first ban on animal abuse was issued in 1857 and in 1944, a complete animal welfare legislation came into force (Ekesbo, 2003). The current animal welfare act was issued in 1988. The Swedish legislation in many ways stretches far beyond the EU directives. The use of antibiotics in feed to promote animal growth has been prohibited since 1986. There are detailed requirements for climate and air quality in buildings for farm animals. Conventional battery cages for laying hens were banned in 1988 with a 10-year transition period, followed by a 4-year period when exemptions were made for some individual producers. At the beginning of the 21$^{st}$ century, most egg producers had changed to either aviary or barn systems or modified cages with perches, dust baths and nests. Local animal markets are rare and most farm animals in Sweden are transported no more than once or twice in their entire lives. Few animals are sent abroad for slaughter, with a few important exceptions. Nevertheless, transport duration within Sweden can sometimes be more than 8 hours. Slaughter is another debated subject. At the beginning of 2009, there were 27 Swedish abattoirs for cattle, sheep or pigs, denoted as large-scale establishments. Slaughter of un-stunned animals is prohibited in all circumstances except extreme emergencies. Mobile slaughter is practised to some extent for poultry and is discussed for other species. Increasing numbers of wildlife are kept in captivity, including native birds and mammals kept in large fenced-in enclosures. Many exotic species can now be purchased on the open market and are kept as pets. These include a large number of fish, reptiles and birds, of which the owners sometimes have very limited knowledge.

As a consequence of natural geographical isolation and systematic disease control, Sweden was for long considered free from many contagious animal diseases and zoonoses. National animal disease control programmes have a long tradition. As one of the first countries worldwide, Sweden was declared free from bovine tuberculosis in 1958. Through strict disease monitoring, a restricted use of antibiotics to animals, and other systematic control measures, *Salmonella* in farm animals decreased drastically from the 1960s and Swedish animal production is basically *Salmonella* free (Boqvist and Vågsholm, 2005; National Veterinary Institute, 2009). More recently, Sweden started a control programme against BVD (bovine virus diarrhoea) in 1993 and is presently almost free from the disease, together with other Nordic countries. However, increased international trade, travelling, and climate changes constitute potential animal disease hazards. From the end of the 1990s, several new contagious diseases have been introduced or re-introduced. There is some concern that increasing herd sizes and high turnover rates of farm animals will cause or aggravate health problems.

The Swedish Government has initiated work to modernise the national animal welfare legislation while maintaining the level of animal protection. In 2008, the Ministry of Agriculture commissioned the Swedish Board of Agriculture to investigate (1) how animal welfare assurance schemes can be applied, (2) to develop and implement a system for standardised and risk-based official animal welfare control, and (3) to revise the existing scheme for prior approval of livestock buildings. In relation to standardised and risk-based animal welfare control, the Board identified four tasks, namely to investigate (1) the possibility to introduce digitalised inspection checklists, (2) how existing sector-specific quality assurance schemes could be developed and used, (3) how risk-based control could be achieved, and (4) consequences of the preceding points, and their compatibility with EG legislation (Barchiesi *et al.*, 2008). It was felt that the control scheme should be characterised by clarity, transparency, fairness, robustness and flexibility, and – to achieve efficiency – acceptance among Swedish stakeholders and consumers, and a potential for gradual development. An important step was the transfer of the local control responsibility from the 290 municipalities to 21 county administrative boards from the beginning of 2009, as further described below. During 2008 and 2009, the Swedish Board of Agriculture conducted extensive work to develop common policies, routines, forms, checklists and databases, collaborating with county representatives, industry players, researchers and other key persons. The authorities also felt a special need to secure standardised assessments through education of affected staff and through detailed guidelines and instructions for the completion of inspection checklists.

As of 2009, the Swedish Board of Agriculture administers a national database for official animal welfare control, regulated by law. The database contains information about all (disclosed) physical and juridical animal owners and control sites in Sweden, holding site-specific information relevant for the control such as type of animal husbandry, number of animals of different categories and established non-compliance or offense against legislation. Data may be processed for planning, coordination, administration, follow-up and reporting of the operative control on the regional and national levels.

## 2. Monitoring and assurance of animal welfare

Quality assurance has been defined as 'the systematic monitoring and evaluation of the various aspects of a project, service, or facility to ensure that standards of quality are being met' (Merriam-Webster Online Dictionary). Fraser (2006) describes how different quality assurance programmes in food production have been developed since the end of the 20$^{th}$ century. Most quality assurance programmes were set up for food safety reasons, but some have later developed to include animal welfare (Edge and Barnett, 2008). Governmental regulations are sometimes used to provide assurance about animal welfare, but European assurance schemes covering animal welfare often go beyond common legislation (Veissier *et al.*, 2008).

According to Fraser (2006), the effectiveness of a given welfare assurance scheme will depend on several factors including support received by industry players, ease of maintenance and enforcement, coverage, and public confidence. Fraser groups specific requirements into four main types, dealing with: (1) the fulfilment of basic needs, e.g. space allowance and species-specific feeding, (2) the avoidance of unpleasant states of pain, distress and hunger, (3) the expression of natural behaviour, e.g. perching and nesting in poultry and exploratory behaviour in pigs, and (4) the possibilities to lead a 'natural life', e.g. access to outdoor areas. According to Fraser, the first three requirement categories have so far had the strongest scientific support, while the last three types have most easily gained public trust. One of the earliest responses to public animal welfare concern was the creation of non-mandatory codes and guidelines for animal production, generally taking the form of recommendations. Such codes were usually formulated by the industry and based on available scientific research combined with practical experience. Fraser points out that non-mandatory welfare codes and guidelines provide little assurance to the public, unless some form of monitoring is practised. Auditing or inspection programmes have been designed to follow up recommended guidelines. Compliance with a quality assurance scheme is sometimes controlled by an external expert body.

Since long, Swedish farmer associations, food industry organisations, breeding companies, insurance companies, animal clinics, abattoirs and authorities have maintained databases containing extensive information on e.g. animal descent, production, reproduction, disease recordings, insurance cases and veterinary care on both the animal and the herd/flock level. Examples are the central database of bovine animals, the pig register, the slaughter register, and the national database of veterinary disease reports (Olsson *et al.*, 2001; Swedish Board of Agriculture, 2008). Although most of these databases have been created for reasons other than animal welfare monitoring, they contain information that could prove valuable when establishing risk-based animal welfare control.

## 2.1 Animal welfare codes and quality assurance schemes

The Federation of Swedish Farmers offers a web-based tool (*Miljöhusesyn*) for farmers to carry out self-control of compliance with current legislation on animal welfare and on other areas. The tool contains relevant legislation and a checklist adapted to the type of animal production concerned. Approximately 27,000 Swedish farmers (37%) applied the tool to their operations in 2008 – numbers being higher for animal farmers and large farmers – in this way checking compliance with current legislation (Karlsson, A., Federation of Swedish Farmers, 2009, pers. comm.). Some food processing companies and quality assurance schemes require of their farmers to use the tool. There are no external controls or audits. The mission of the Federation of Swedish Farmers is to contribute to the development of companies and industries based on farming and forestry, thereby enabling individual members to realise their goals in terms of profitability, growth and quality of life.

Different animal welfare codes, disease control programmes and schemes for quality assurance in Swedish animal husbandry have been launched and are managed by the industry or other stakeholders. In 2008, the Swedish Board of Agriculture judged only two of these schemes, for poultry meat and egg producers, to fulfil minimum requirements allowing for flexible legislative requirements based on their classification systems (Barchiesi *et al.*, 2008).

### *2.1.1 Poultry meat production*

The broiler meat industry is by far the most standardised sector of Swedish animal production, with the highest standards of contagious disease control. The Swedish Poultry Meat Association represents 98% of all production in Sweden, covering the total production chain from import of day-old grand-parent chickens to abattoirs, including hatcheries, growers, and feed companies (Swedish Poultry Meat Association, 2008). The association manages a quality assurance scheme for its broiler chicken and turkey producers, aiming to differentiate population density while securing animal welfare (Berg and Algers, 2004). The more capable a poultry grower is, judged by e.g. the prevalence of foot pad dermatitis, the higher the stocking density he/she is allowed to apply. The basic population density limit by legislation is 20 $kg/m^2$ of floor area but growers who satisfy the requirements of the scheme's classification system are allowed up to a maximum of 36 $kg/m^2$, depending on the number of assessment points assigned to the operation. Thirty-one control points are checked, covering housing, technical equipment and management. Yearly classification is done by a National Standards Officer, employed by the Agria Animal Insurance company and reporting to the official *Salmonella* control veterinarian appointed by the Swedish Board of Agriculture. Changes of the classification system must be approved by the Board. The system contains a foot health scoring scheme, in which feet from a number of animals in each flock are scored for lesions in connection to slaughter. Although the scheme is non-mandatory, in practice all growers affiliated to the association are enrolled (Swedish Poultry Meat Association, 2008). The association also runs control programmes for *Campylobacter*, coccidiosis, and clostridiosis.

### *2.1.2 Egg production*

The Swedish Egg Producers Association organises hatcheries, pullet rearers, egg producers, packeries, and feed companies (Swedish Egg Producers Association, 2009). The association has a quality assurance scheme covering animal welfare, hygienic standards, contagious disease control, food safety and technical standards. In a classification system, 25 control points are checked, and minimum scores at 12 of these points are required for approval. The basic population density limit by legislation is 7.5 to 20 free-range hens/m$^2$ of floor area, depending on the type of system and the size of the birds, but producers who satisfy the requirements of the classification system are allowed up to a maximum of 20 hens/m$^2$ of floor area. As in the case of poultry meat production, classification of operations is done by a National Standards Officer, sometimes in collaboration with a local animal welfare inspector, and changes of the classification system must be approved by the Swedish Board of Agriculture. The scheme is non-mandatory but the majority of producers are enrolled to be able to deliver eggs and take out an insurance (Swedish Egg Producers Association, 2009). In 2009, the association was also aiming to develop a separate scheme for pullet rearers. The association also runs a control programme for intestinal nematodes.

### *2.1.3 Other animal production*

The Swedish Animal Health Service provides services to all Swedish breeders of pigs, beef and sheep, aiming for sound production of healthy animals on an economically competitive basis. Farmers and veterinarians have demanded guarantees that animal trade does not spread infectious disease and organised animal health control started already in the 1940's. The organisation guarantees that animal trade deals only with healthy animals. A main part of the work is health control of cattle and weaner pigs. Presently, the organisation runs programmes aiming at national control or eradication of *Salmonella*, Johne's disease in cattle, PRRS (Porcine Reproductive and Respiratory Syndrome), AD (Aujeszky's disease) and atrophic rhinitis in pigs, Maedi-Visna in sheep, CAE (caprine arthritis-encephalitis) in goats, and bovine tuberculosis in farmed deer (Swedish Animal Health Service, 2009). Normally, disease control programmes are voluntary initially, but might enter a mandatory phase to achieve eradication. In 2009, the Swedish Association of Beef Producers was trying to establish a quality assurance scheme for range beef, awaiting results from current research.

A quality assurance scheme for milk production is run by the Swedish Dairy Association, aimed to secure good hygiene and food safety. Inspections are focused on cow welfare and housing environment. Since 2004, the association has also conducted research to develop a system for quality-assured milk production, in order to provide good animal welfare and strengthen consumer trust. Hallén Sandgren *et al.* (2009) found data extracted from the national dairy database to classify up to 77% of herds correctly with a sensitivity of 0.77, scoring herds as either having or not having poor welfare by using nine animal-based measures developed by the EU-funded WelfareQuality® project. The association has run control programmes to eradicate EBL (enzootic bovine leucosis) and BVD, and coordinates efforts to control bluetongue in ruminants and hoof diseases in dairy cattle.

Swedish *Salmonella* control is globally unique. It covers import, feed production, farm-level detection and eradication, slaughter, meat cutting, food processing and food deliverance. Findings of *Salmonella* have been recorded since 1968 in Swedish broiler flocks. Regulations have been issued by the Swedish Board of Agriculture and the National Food Administration. To detect cases in food-producing animals, faecal sampling is mandatory in poultry flocks and hatcheries over a stipulated minimum size (practically all commercial flocks). If *Salmonella* is found in a poultry flock, all animals are killed and the building is sanitised. If found in a cattle or pig herd, mandatory restrictions are put on the operation until it has been cleaned up. Non-mandatory control is organised by the industry. *Salmonella* is found only sporadically in Swedish production animals (Swedish Board of Agriculture, 2008). Growers affiliated to the Swedish Poultry Meat Association are required to enrol in the non-mandatory *Salmonella* control programme, and are only allowed to buy chickens from hatcheries which are enrolled (Swedish Poultry Meat Association, 2008).

## 2.2 Auditing and labelling programmes

Animal welfare assurance may also involve more pronounced labelling programmes, where animal products produced according to defined methods or standards are labelled and sold at a premium price (Fraser, 2006). The programmes can be established by producers, retailers or independent bodies such as animal welfare organisations or organic certification agencies (Fraser, 2006; Veissier *et al.*, 2008). Programmes set up by the industry are most often intended to enable companies to attract a specific group of consumers and enter a specific market (Veissier *et al.*, 2008). For example, Scan, a large actor in the Swedish slaughter and meat processing industry, has implemented a set of welfare standards for animals at slaughter in addition to their general quality assurance scheme. The company performs internal audits and, following the company's animal welfare training, staff and transporters delivering animals to the slaughter plants are certified (Scan, 2008). This type of scheme has been described as general quality assurance that includes some welfare requirements but it mainly focuses on other aspects of food quality, such as microbiological safety, traceability, and organoleptic qualities (Veissier *et al.*, 2008).

The Swedish Seal of Quality (*Svenskt Sigill*) is another food quality label, assuring production on farms which follow requirements of safety, animal welfare, responsibility for the environment and a living countryside. The fundamentals of Swedish Seal are a set of rules, differing between food production sectors and developed in co-operation with experts, researchers, the sector-specific industry and producer organisations. An independent certification company reviews compliance with the requirements. In 1996, the Swedish Seal was developed by Sweden's major cereal cooperative, the Swedish Farmers Supply and Crop Marketing Association, aiming to assure the quality of cereals. In 2002, the Federation of Swedish Farmers bought the logo as a means of creating a profile for all types of Swedish raw produce and to promote the logo to Swedish consumers. Seal Quality Systems Ltd. was formed to administer and co-ordinate marketing, criteria development and inspections for the new label, carried out through self-control and audits by a certification company. Currently, dairy, pig, beef and broiler farming can be certified for animal welfare. The requirements are based on Swedish legislation for environmental protection, food safety, animal welfare and work environment, as described in the *Miljöhusesyn* tool described earlier, and on EU cross compliance legislation, with complementary regulations in some areas. In 2009, for example, the scheme required daily management routines in writing, plans for emergency killing in writing, and claw trimming of dairy cows performed at least once yearly (Swedish Seal of Quality, 2007).

The KRAV association was founded in 1985 to develop organic standards and to promote a label for Swedish organic production (KRAV, 2009). At present, KRAV has 28 members representing farming, processing, trade, and also consumer, environmental and animal welfare interests. The mandatory requirements meet the IFOAM Basic Standards and EC Regulations Nos. 834/2007 and 967/2008 on organic production and labelling, but have also come to involve far-reaching standards for animal welfare both on farm and at slaughter. KRAV owns a subsidiary, Aranea Certification AB, which is one of four certification bodies that carry out regular inspections according to KRAV standards. In some aspects these standards go beyond the legal requirements of Swedish animal welfare legislation. The standards generally prioritize good animal care, organic feeding in accordance with the animals' needs, limited concentrate feeding to ruminants, restrictive preventive medication, and access to outdoor areas. The KRAV logo and the EU organic logo are used together.

## 3. Official animal welfare control

### 3.1 Organisation

Until December 2008, the Swedish official animal welfare control was carried out locally by 290 municipalities in a regular, directed or project-based way. County administrative boards (21 counties) were responsible for the control on the regional level. The county boards also decided on injunctions, prohibitions against having animals in care, charge taking of animals, and prior approval of livestock buildings. From January 2009, Swedish official control was transferred to the state and the county administrative boards thus became operative control authorities also on the local level.

Swedish official control covers many different types of animal operations, establishments and activities holding or handling numerous animal species, e.g. farms, reindeer operations, transports, airports, ports, fairs, zoological gardens, dealers, competitions, exhibitions, circuses, fish farms, laboratories, and abattoirs. Regulations issued by the Swedish Animal Welfare Agency in 2007 stipulate how official control of the compliance with current legislation shall be executed. Minimum inspection intervals for different types of operations are recommended (Table 1). These regulations also recommend minimum frequencies of animal transport inspections, e.g. ten inspections at loading or unloading of slaughter transports per municipality and year.

Veterinary Officers at abattoirs, employed by the National Food Administration, are responsible for official animal welfare control in connection to animal transports to abattoirs and during stunning and killing at abattoirs to the extent prescribed in EC Regulation No. 854/2004. Remaining animal welfare control at abattoirs is conducted by the county administrative boards.

*Table 1. Minimum inspection intervals in official animal welfare control recommended by the Swedish Animal Welfare Agency in 2007.*

| | |
|---|---|
| Abattoirs, animal wholesale dealers, border inspection posts, live animal intermediaries | 6 months |
| Zoological gardens and other public animal exhibitions, quarantines, stationary reindeer operations, security companies, winter quarters of circuses, zoological shops | 12 months |
| Persons prohibited against having animals in care | 12 months |
| Circuses on tour, reindeer operations during grazing | Once per season |
| Riding-schools, wildlife farmers, wildlife rehabilitators | 24 months |
| Farm operations with cattle, pigs, sheep, goats, poultry, horses, fish, or fur animals, commercial establishments for rearing/keeping/selling pet animals, breeding companies, animal operations for educational purposes | 36 months |
| Animal fairs | at every occasion |
| Animal laboratories and other permanent establishments for using/rearing/keeping/selling experimental animals | 6 months |

## 3.2 Pre-testing of new technologies

Since many years, Sweden has a system for pre-testing of new farming technologies, which means that any equipment, machine or appliance meant for housing or handling of horses or animals kept for production of food, wool, skins or fur has to be evaluated and/or tested scientifically before it can be sold to farmers or other animal caretakers (Ekesbo and Van den Weghe, 1998; Algers, 1999). Applications from manufacturers or importers to the Swedish Board of Agriculture are divided into three categories: (1) decisions based on the application, (2) decisions taken after a minor literature review or expert consultation, and (3) decisions taken after scientific testing. Since 2004, technologies sufficiently approved in other countries within the EU or the European Economic Area (EEA) or in Turkey do not require pre-testing. Examples of pre-tested technologies are aviary systems for laying hens (Gunnarsson, 2000), mechanical broiler harvesters, automatic milking systems for cows, rubber slatted flooring for tied cows (Hultgren and Bergsten, 2001) and early weaning of piglets (Ekesbo and Van den Weghe, 1998).

## 3.3 Prior approval of livestock buildings

Official prior approval of animal buildings was introduced in Sweden 1973 as a preventive means to improve animal health and welfare standards in Swedish farm operations for horses or animals kept for production of food, wool, skins or fur, and for animals kept for educational purposes. Before a building or enclosure for a certain minimum number of animals is erected, modified or expanded, the county administrative board shall check compliance with relevant animal welfare legislation and approve the plans. Normally, a building adviser and/or the Head Veterinary Officer at the county administrative board examines plans and drawings with respect to e.g. dimensions, fittings, walking areas, lying surfaces, technical equipment and solutions for feeding, watering, manure removal, ventilation, heating, alarms, and fire evacuation. Before the new building or enclosure comes into use, it should be inspected and found to accord with the approved plans. In approximately 20 to 30% of the cases, a second inspection is needed (Barchiesi *et al.*, 2008). Comments, requirements and injunctions may be issued during the process.

Due to perceived differences in judgement between counties, a need to differentiate approval with respect to scope and complexity, and a wish to incorporate e.g. aspects of contagious disease control, the scheme of prior building approval is under revision since 2008.

### 3.4 Results from official control

In 2008, a total of 86,689 Swedish animal operations were known to responsible control authorities (Barchiesi *et al.*, 2008). Yearly summary data from Swedish official animal welfare control are produced by the responsible national authority, previously the Swedish Animal Welfare Agency and presently the Swedish Board of Agriculture. Before January 2009, local and regional authorities reported on relevant events and decisions. These reports comprised the amount of resources used, the number of animal operations (control objects), the number of operations visited, the number of visits made, and the number of visits resulting in corrective actions such as comments, injunctions, charge taking of animals, prohibitions against having animals in care, handing over to prosecutor, etc. They also contained the number of animal transport vehicles of different types inspected, the number of animals of different species transported, and the number of inspected vehicles not complying with legislation for different reasons. On account of EU reporting requirements, additional information about operations with calves, pigs and laying hens was supplied. Table 2 contains summary statistics of official control 2001-2006 (Barchiesi *et al.*, 2007). Due to uneven distribution of animal populations and local political priorities, the resources available and conditions for official control varied greatly between municipalities. Efforts were often disproportionate to the number of animal operations located in the area and some municipalities even lacked a systematic control. An average municipality had 319 known animal operations and carried out 66-86 inspections from 2002 to 2007, i.e. operations were visited at intervals of on average 4 to 5 years. During the same period, remarks were given at 50-60% of inspections, and the proportion of inspections resulting in charge taking of animals increasing to 4%. During 2007, 1,843 building plans were examined by regional authorities and 1,238 inspections were made (Barchiesi *et al.*, 2008).

In 2007, the National Food Administration conducted 694 animal welfare inspections at 93 abattoirs, of which 292 at 25 large-scale abattoirs and 380 at 22 poultry abattoirs. The administration reported unsatisfactory conditions to the municipalities in 128 cases of farm operations or transports, and in 17 cases of abattoirs (Barchiesi *et al.*, 2008). According to the Swedish Board of Agriculture, the municipalities did not prioritise animal transports and abattoirs sufficiently, thus neglecting these parts of the official control (Barchiesi *et al.*, 2008).

*Table 2. Swedish official animal welfare control 2001-2006 (Barchiesi* et al.*, 2007).*

| | 2001[3] | 2002[3] | 2003[3] | 2004[3] | 2005 | 2006 |
|---|---|---|---|---|---|---|
| Number of known operations[1] | n.i. | 78,083 | 75,334 | 83,756 | 88,739 | 86,689 |
| Number of inspected operations[1] | 14,947 | 17,016 | 17,844 | 19,096 | 18,756 | 17,631 |
| (% of known operations) | n.i. | (22) | (24) | (23) | (21) | (20) |
| Number of inspections[1] | 20,043 | 20,700 | 18,909 | 25,467 | 24,581 | 23,120 |
| Number of inspections without comments[1] | 10,332 | 7,960 | 6,108 | 7,655 | 8,257 | 8,798 |
| (% of inspections) | (69) | (47) | (34) | (40) | (44) | (50) |
| Number of comments or advice[1] | n.i. | 5,935 | 6,662 | 7,931 | 8,349 | 7,961 |
| Number of injunctions, bans or charge-takings, locally[1] | 870 | 859 | 1,048 | 1,003 | 1,039 | 1,212 |
| Number of charge takings, regionally[2] | 211 | 176 | 209 | 248 | 316 | 336 |
| Number of prohibitions against having animals in care[1] | 77 | 62 | 74 | 80 | 124 | 145 |
| Number of cases handed over to prosecutor | | | | | | |
| locally[1] | 162 | 157 | 189 | 170 | 178 | 164 |
| regionally[2] | n.i. | 7 | 12 | 4 | 13 | 6 |
| Number of examined building plans[2] | n.i. | n.i. | n.i. | n.i. | 1,424 | 1,642 |
| Number of inspections after prior approval of livestock buildings[2] | n.i. | n.i. | n.i. | n.i. | 1,080 | 1,205 |

[1] Reported by municipalities.
[2] Reported by county administrative boards.
[3] n.i. = no information available.

# 4. A critical view on animal welfare risk assessment and management

## 4.1 Fundamental considerations

Because of the lack of an unambiguous definition of animal welfare, value-based judgements will inevitably influence the way different aspects of welfare are weighted and combined. Implications of this were discussed by Mason and Mendl (1993), Fraser *et al.* (1997) and Lassen *et al.* (2006). Botreau and co-workers, reviewing existing methods to aggregate welfare measures (Botreau *et al.*, 2007a), suggested a practical approach to sequential aggregation as agreed in the WelfareQuality® project (Botreau *et al.*, 2007b) and developed a methodology for on-farm assessment of cattle welfare based on multiple criteria (Botreau, 2008). The lack of a unified concept also poses a problem if clarity and transparency of the risk assessment process are reduced, as also discussed by Smulders (2009). From a risk management perspective, it is desirable to express the total impact on animal welfare of e.g. a husbandry system in units that are understandable intuitively and possible to compare with costs for actions to improve animal husbandry. Fair comparisons between different operations and over time require a standardised methodology. A poor common understanding of the welfare concept prevents estimations of the monetary value of welfare improvements and, eventually, rational decisions based on risk assessment. Hence an acceptable definition should always be chosen explicitly and methods for combining different welfare components should be declared.

Alternatively, different components of welfare might be considered separately. This option should be considered if the belief is strong that there is no unifying concept of animal welfare underlying the different aspects. An observed lack of correspondence between the different components, i.e. differing hazards and no strong statistical correlation, might be conducive to such a notion. One should realise that considering welfare components separately by no means justifies omitting an explicit definition of each component. The choice of approach may of course vary according to the specific goal of the assessment exercise. For instance, a need to communicate the rationale of the risk model to stakeholders and the public calls for a conceptually rather simple approach, possibly at the cost of scientific fidelity.

## 4.2 Validity of the risk model

An overly simplistic animal welfare risk model might be problematic from a validity perspective. In other applications of the risk assessment methodology, there is usually a well-defined negative event, such as the occurrence of a toxic substance in food or a road accident (Vose, 2008). The negative consequences of that event can then be identified and their cost estimated. In animal welfare risk assessment, the hazards, the central negative event and the consequences are not very easily distinguishable. Invariably, a series of parallel events jointly contribute to reduced animal welfare, but it is not obvious which of these events are hazards, which can be regarded as negative events, and which are the consequences. For example, a certain floor design might be identified as a welfare hazard to farm animals because it is known to increase the occurrence of slipping, injuries and premature death. These three latter events can thus be denoted negative welfare effects and the joint probability of their occurrence in the population, dependent on the presence of the floor design, is estimated as a random distribution. In assessments conducted hitherto, the intensity and duration of instances of slipping, injury and death have also been estimated jointly. Such an approach is probably seriously over-simplistic. In reality, the inappropriate floor design might cause abnormal gait which increases the probability of slipping, which in turn increases the probability of injuries, which in turn increases the probabilities of pain, of distress and of premature culling. Each one of these probabilities can rightfully be described as a random variable and all the events from abnormal gait to premature death can be called negative welfare effects. Relationships between events might be linear or of any other type. Moreover, these random probabilities are probably to some degree correlated with each other, possibly heavily so, and with the random probability of hazard occurrence in the population, which also might require consideration.

Animal welfare risk estimates published to date possibly do not account sufficiently for the occurrence of poor welfare in the absence of hazards. When assessing risks associated with things like food contamination or road accidents, negative consequences in the absence of the contamination or accident might be negligible and disregarded. Animal welfare, on the other hand, at all instances ranges from good to bad. Certain events or conditions will decrease welfare to some extent – or some aspects of it – while others will increase it. The important question is to what extent welfare is reduced through exposure to identified hazards, which can only be derived from knowledge about both the welfare status with hazard exposure and the status without such exposure. The way risk models have been applied in EFSA assessments (2006; 2007a; 2007b; 2007c; 2008), it is not clear whether the conditional probability (likelihood) of the negative welfare effect was specified consistently as the excess effect due to hazard exposure. Failure to do so will logically overestimate the risk.

## 5. Towards a standardised and risk-based official control

### 5.1 What is risk-based control?

EC Regulation No. 882/2004 stipulates that official national control shall take account of identified risks that can influence animal health or welfare. In some respects, Swedish official animal welfare control has been risk-based already since long, albeit not in a thorough and systematic way. Decisions regarding inspections have been taken on the basis of knowledge gained from earlier inspections and have been inspired by priorities of the responsible authorities or inspectors, implying a risk perspective. Recommendations from the Swedish Animal Welfare Agency on minimum inspection intervals (Table 1) are based on the idea that activities with a high flow of animals potentially affect more animals per time unit and thus also reflect perceived welfare risks (Barchiesi *et al.*, 2008).

In contrast to animal welfare risk assessment as applied by e.g. EFSA, as described by Candiani *et al.* (2009) and Smulders (2009), risk classification intended for national risk management does not estimate the overall welfare status or risk in a single target population of animals or animal operations. Instead, it aims at estimating the risk at each one of the individual animal operations in the control scheme at regular intervals. Each operation can be regarded as sampled from an imaginary target population of operations with an identical set of hazards.

Different operations are compared with respect to the risk for impaired welfare, or to the risk for non-compliance with current legislation. These two risk aspects are closely related but not equivalent. Impaired welfare can be defined ad hoc according to current requirements and preferences, while non-compliance is defined strictly by legislative requirements. Normally, some of the requirements correspond to a minimum acceptable level of animal welfare (usually applying animal-based criteria) while others express necessary preventive actions to secure future welfare, i.e. to reduce welfare risks (also using resource- and management-based measures). Although the core purpose of official control is merely to check compliance with legislation, control authorities and inspectors can be expected to emphasise and prioritise – deliberately or not – legal articles, paragraphs and measures which they feel are more important to animal welfare than others. Interpretation of regulations is often subject to value-based judgements. Furthermore, legislation and control develops, gradually incorporating new perspectives and methods to assess welfare status and prevent impaired welfare, which requires knowledge about welfare risks in a deeper sense than what is merely defined by current legislation. Hence, both risk perspectives are relevant to the development of official animal welfare control. To simplify the discussion, however, both these risk aspects are in the following referred to by the term non-compliance.

The actual risk level of an operation can be expressed in different ways. One option would be to assign a value or score on a numerical (continuous or discrete ordinal) scale, denoting a risk estimate. This would imply an absolute scale of welfare risk (one operation has a risk estimate *a* units higher than another operation, or a *b* times higher risk estimate). Another option is to rank operations from the highest to the lowest welfare risk, which will generate a relative scale. The choice of scale influences the usefulness of risk estimates, but also the possibilities to obtain good accuracy and precision in the assessment. Perfect assessment on an absolute scale, e.g. from 0 to 100 or from 'very poor' to 'very good', would allow us to compare risk estimates to any other operation or time point – within reasonable limits. Obviously, it is more difficult to assign values on an absolute scale than to assign ranks. And if the task is only to identify, say, the 5% poorest performing operations, it becomes even easier to achieve a correct assessment. Most people might agree on which animal operations have the poorest animal welfare, possibly irrespective of the basis for their judgement. It might even be argued that neither the selection of welfare measures and their possible combination nor the choice of scale is crucial, and that different methods will generate approximately the same result. However, so far, the scientific support for such a notion is scarce or non-existing. Nonetheless, given the complex nature of the animal welfare concept, changes in perception over time, financial constraints, and numerous technical difficulties, assessment on a globally agreed absolute scale does not appear realistic, as also discussed briefly by Smulders (2009). However, it is important to recognise that even the simplest relative assessment requires agreement on the basis for judgements of different animal operations. If one aims at achieving high validity of the assessment scheme, it appears appropriate to define precisely what leads to a judgement of risks as higher at one operation than at another. An obvious way around this dilemma is to find a reasonably practical definition of the scale and to obtain sufficient agreement or consensus among stakeholders regarding the chosen definition. In other words, expressing animal welfare risks in a meaningful way, it is probably imperative to find a good enough definition for most people to agree on, irrespective of their motives.

## 5.2 Risk variability

When any given animal operation is risk classified, knowledge about that operation increases and, inevitably, variability decreases. Generally, knowledge increases step by step by adding new pieces of relevant information. The more we know about the occurrence of important welfare hazards at different operations, the more certain we will be about the risks. In theory, the uncertainty is low immediately after an inspection visit to an operation, when knowledge is at a maximum. Uncertainty increases with time post inspection because findings become obsolete or irrelevant. Knowledge from animal-based observations of welfare status is likely to become obsolete quite rapidly, while information about housing facilities and management routines logically remain valid and relevant for longer periods. Due to variability (composed of variation and uncertainty), the risk estimate of an operation can be expressed as a random variable, i.e. a probability distribution around the most likely risk value. The proportion of the distribution that lies above a defined cut-off value corresponds with the confidence that the operation exhibits a risk higher than the cut-off. A decision to apply a certain control measure to a particular operation, e.g. to carry out inspection, can therefore be based on the proportion of the estimated risk distribution that lies above a certain risk level. As illustrated in Figure 1, the confidence of a certain welfare risk level at an animal operation will depend on the shape of the risk distribution at the site. A higher cut-off and/or confidence will result in fewer operations classified as 'risky'. If control resources are limited, the choice of cut-off and minimum confidence level will probably be influenced by the risk estimates of all sites in the control scheme. Optimisation of a control scheme implies a trade-off between the benefit/cost of obtaining sufficient operation-specific information and of classifying operations correctly.

It is undoubtedly a challenging task to communicate risk variability – due to conditions of e.g. specific production sectors, husbandry systems or animal breeds – to animal caretakers and other stakeholders and thus maintain confidence in the risk classification system. If the rationale for risk classification is not communicated well, the control might be perceived as unfair by some industry players. For instance, caged layers have a more standardised animal environment and thus probably require less management skills than loose-housed hens, which might motivate a less intensive control and less frequent inspections, even if the most-likely risk estimates of the two systems are virtually equal, as for operations A and B in Figure 1.

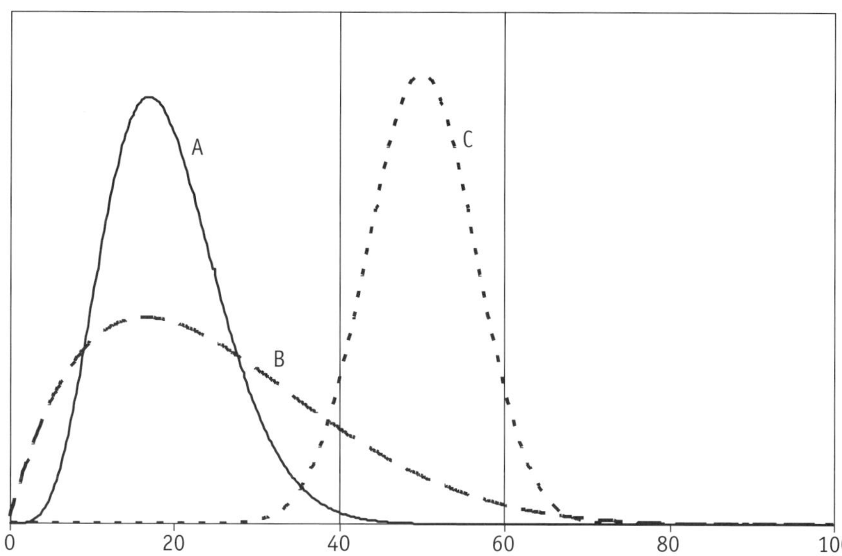


*Figure 1. Hypothetical example of welfare risk estimates of three animal operations (A, B, C). Risk level assumes values from 0 to 100 (unit not specified). With a maximum acceptable risk of 40, the operations are thought to exhibit unacceptable welfare with a confidence of 0.5, 16 and 94%, respectively. With a cut-off risk of 60, expected confidences will change to 0, 1.8 and 6.0%. If the minimum confidence required for control inspection is set to 5%, operations B and C will be inspected in the first case, but only operation C in the second case. The most likely risk is 17, 17 and 50, respectively, for the three operation.*

## 5.3 Animal welfare risk classification

Preferably, risk classification should be done at the highest hierarchical level of the control scheme, because it allows each operation to be compared with all the other operations. All players in the scheme will share a common set of regulations and common policies to allocate resources. Typically, this would be the national level. Such central risk classification will also provide a basis for decisions to allocate resources to different regions within the scheme, e.g. to different counties, in relation to the estimated number of animal operations with a high welfare risk estimate. Regional authorities will then be able to make priorities within their separate budgets, most likely at the start of each financial year. It is probably not possible to risk classify temporary animal activities, e.g. transports, exhibitions, fairs, and circuses, together with permanent establishments such as farm operations. Logically, such activities must be risk classified separately.

In the control scheme, information from different sources should be accounted for and used to arrive at useful risk estimates. Efficiency demands that the least possible effort be made to obtain estimates which are sufficiently accurate. The way different pieces of information are used and combined with other information constitutes the risk model and will influence the risk classification of individual operations. Potentially valuable and useful information can be obtained from many different databases or registers administered by the responsible control authorities, by other authorities or by the industry. These include national animal databases, veterinary-reported disease diagnoses, milk recordings, reproductive events recordings, disease control programmes, quality assurance schemes, audit systems, results from self control, and results from earlier official control including corrective actions taken. Furthermore, research and evaluations carried out by independent bodies such as EFSA or the EU Food and Veterinary Office should be used to improve risk estimation. Generally, the value and usefulness of information to the control scheme can be expected to depend on the following aspects:

- relevance to the scheme;
- validity and reliability, i.e. if the information can be trusted;
- spatial and temporal dissolution and coverage, i.e. how detailed and specific the information is (individual or grouped data; continuous recordings or compilations at regular or irregular intervals);
- uniqueness, i.e. if the information cannot be substituted by other information;
- accessibility, i.e. if the information is readily available to the control authority.

After having consulted stakeholders and researchers, the Swedish Board of Agriculture outlined a scheme for official animal welfare control (Barchiesi *et al.*, 2008), comparable to the food safety control scheme applied by the National Food Administration (Bäcklund Stålenheim *et al.*, 2008) in primary production. The Board pointed out the importance of a flexible model, which can be complemented and expanded successively, and proposed principles for risk classification of animal operations, presented as a series of classification matrices supported by guidelines and definitions. Three risk modules were suggested, classifying animal operations based on, respectively:

- the presence of scientifically established or experience-based hazards related to readily available site characteristics;

- enrolment in reliable quality assurance schemes organised by the industry;
- detailed operation conditions, obtained from earlier control, quality assurance or self control schemes, and complaints or other indications of non-compliance.

Objectively assessed results from earlier control were proposed to have a strong influence on risk classification. Barchiesi *et al.* (2008) also proposed the risk classification procedure to follow consecutive steps, including the risk classification of each sector (based on known hazards, the probability of a given scenario, the probability of a welfare effect in a scenario, and the impact of the welfare effect, resulting in a risk score with three classes), scoring of quality assurance schemes run by the industry (four classes), combination of the adjusted sector score and the scheme score (resulting in 12 classes), and adjustment for operation-specific information. The final risk estimate was thought to range from 1 to 5 (five classes), expressing each operation's welfare risk in relation to the overall average of the respective production sector. The limited number of discrete risk classes was motivated by the risk of misclassification if more classes were applied.

Producer representatives argued that the content and scale of existing non-mandatory husbandry guidelines and quality assurance schemes should be considered when allocating resources for official animal welfare control to different sectors. The Swedish Board of Agriculture (Barchiesi *et al.*, 2008) proposed the following characteristics of a quality assurance scheme to be important for classification:

- coverage with respect to legislation;
- coverage with respect to the animal population and to the production chain;
- relevance for animal welfare;
- reliability and validity of used measures;
- transparency and availability to control authorities;
- organisation;
- personnel competence requirements;
- strategic support to secure scheme quality, e.g. trend analyses, training, instructions, and upgrading;
- priority given to certain action plans of operations or follow-up of scheme results;
- type and quality of documentation;
- degree of independence, e.g. self-assessment, external assessment, or external auditing;
- congruence with official control;
- corrective actions when deficiencies are found, e.g. reporting to control authorities, internal follow-up, or sanctions applied.

An obvious major obstacle in the development of a reliable and trustworthy risk classification system is the limited scientific basis regarding the way the magnitude or impact of a welfare effect shall be incorporated into risk estimates. For instance, should a 10,000-chicken flock be considered a higher risk than a stable with five horses, all other aspects being equal? How can a high welfare risk during the short time period of a transport or at the abattoir be compared to a low long-term risk on-farm? What happens to animal welfare considerations if they are incorporated in a broader concept of Food Safety – will the number of human consumers possibly affected by animal products have a major influence on the risk estimates assigned to the producers? These and other questions will affect how resources for public control are allocated, as well as the overall efficiency of the official control. Although research can clarify the difference between different options, the final answers should not be sought in science, but in political spheres. A broad acceptance of a scheme is likely to require agreement between all parties concerned.

### 5.4 Allocation of control resources

EC Regulation No. 882/2004 mentions monitoring, surveillance, verification, audit, inspection, sampling, and analysis as relevant control methods and techniques. Traditionally, direct inspection of animal operations has played a major role in Swedish official animal welfare control, complemented by checks of written material and records of e.g. carcass classifications at slaughter, sampling, or investigations. Normally, inspections are conducted through physical visits to the operation but, theoretically and in special cases, real-time or recorded video transmissions are also possible. Obviously, some results from inspections are of a short-time value, i.e. the information acquired at the day of inspection in certain respects does not say much about the operation some time earlier or later. On the other hand, the inspection might be a high-resolution snapshot of the conditions on that very day. The value of inspection results logically depends on the type of operation, the skills and carefulness of the inspector, the personal contact between the inspector and the animal owner or caretaker, and the extent to which the inspection has been foreseen. Announced or otherwise predictable inspections are of a limited value. Finally, inspection intervals will inevitably influence the picture obtained by control authorities, also with respect to animal welfare risks. Barchiesi *et al.* (2008) advocated total time spent on control as more relevant to risk classification than the number of inspections. It might be argued that verification of information acquired from self-control or quality assurance schemes, or from audits, are likely to give a more composed picture, also accounting for variability over time.

Basically, official animal welfare inspections can be (1) risk-based, based on regular risk classification, (2) random, (3) acute, based on complaints or other indications of non-compliance occurring after the most recent risk classification, or (4) cross compliance inspections. There are at least two good reasons for some random selection of control objects. Firstly, comparisons between random and risk-based control offers a means to validate the control scheme. Secondly, animal owners or caretakers will not be able to anticipate random control and thereby have a chance to prepare. In cross compliance inspections, 20-25% of the operations shall be selected at random, and the remainder according to criteria relevant for cross compliance, i.e. related to the protection of the environment, of public, animal and plant health, and of animal welfare. Table 3 illustrates how different control methods might be chosen depending on the estimated welfare risk.

## 6. Conclusions

### 6.1 What has been achieved?

During recent decades, a number of research initiatives and exercises have provided insights valuable to the assessment and management of animal welfare risks. For a number

*Table 3. Different official animal welfare control methods tentatively applicable at different animal welfare risk levels (modified from Barchiesi* et al., *2008).*

| Risk level | Control method | | | |
|---|---|---|---|---|
| | Audit | Regular animal welfare inspection | Acute inspection upon complaints or other indications | Checking of documents and records[1] |
| High | x | x | x | x |
| Moderate-high | x | x | x | x |
| Moderate | x | x | x | x |
| Low | x (partial) | x (partial or random) | x | x |
| Insignificant | | x (random) | x | x |

[1] Including information from cross compliance inspections.

of production sectors and specific animal health problems, Swedish monitoring schemes and investigations regularly feed welfare-related data into registers and databases, in most cases owned and administered by authorities or the industry. Resource- and management-based animal welfare indicators are since long applied in Swedish legislation and used by control authorities. Complementary animal-based indicators suitable for on-farm assessment of animal welfare status are currently being proposed and tested. Swedish authorities collaborate with industry players and academia in this area to reach agreement on key issues.

### 6.2 What has been neglected?

It should be more generally recognised that critical decisions on strategies for animal welfare risk assessment and management need to be value-based, although research can help to clarify different options and their consequences. Animal welfare does not fit readily into the risk assessment framework developed for food safety and contagious disease control. Models published to date have underestimated the complexity of animal welfare. Issues of risk model validity have been largely neglected. Risk quantification might require the combination and weighting of different welfare components, based on considerations related to risk management and communication.

### 6.3 What needs to be done?

Risk management needs adaptation to specific regional or national conditions. In Sweden and elsewhere, sector-specific quality assurance schemes should be further developed and utilised in animal welfare risk assessment and management, requiring collaboration between producer organisations and authorities aiming for practical solutions in data accessibility and transfer. A national animal welfare risk classification system must be designed and gradually developed for application in official control. Despite great collaborative efforts by risk assessors and experts on animal husbandry, health and welfare in exercises conducted by

EFSA, important methodological issues remain to be solved. Data on animal welfare status and hazard exposure on animal and group levels – simulated or acquired e.g. from official control records and existing databases – can be used to compare alternative sampling or inspection strategies and check assumptions regarding welfare definitions, model structure, variable categorisations and model parameters. Sensitivity and scenario analyses should prove useful in this context. A common view on animal welfare risk assessment and management among different stakeholders is highly desirable, to promote optimal allocation of control resources and contribute to improved welfare of production animals.

## Acknowledgements

The author is grateful for contributions and support from colleagues and co-workers Dr. Alexandre Barchiesi at the Swedish Board of Agriculture and Prof. Bo Algers, Prof. Linda Keeling, Prof. Harry Blokhuis and Dr. Stefan Gunnarsson at the Swedish University of Agricultural Sciences.

## References

Algers, B., 1999. Pre-testing of new farm animal management technologies in Sweden – scientific aspects. Proc. International seminar on pre-testing of new farm animal management technologies, Jönköping, Sweden, 20-21 October 1999, p. 1-5.

Barchiesi, A., Andersson, K. and Lundmark, F., 2007. Lokal och regional offentlig djurskyddskontroll under år 2006 (Local and regional official animal welfare control during 2006). Swedish Animal Welfare Agency, Skara, Sweden. Dnr 2006-2376, Report.

Barchiesi, A., Lindkvist-Ottow, E., Loberg, J. and Palmqvist, G., 2008. Ett djurskydd i förändring – genom tillämpning av djuromsorgsprogram, likvärdiga och riskbaserade controller samt en utvecklad förprövning (Animal protection in change). Swedish Board of Agriculture, Jönköping, Sweden. Report 2008:24, pp. 173.

Berg, L. and Algers, B., 2004. Using welfare outcomes to control intensification: the Swedish model. In: Weeks, C.A., Butterworth, A., (Eds.) Measuring and auditing broiler welfare. CABI Publishing, Wallingford, UK and Cambridge, MA, USA, pp. 223-229.

Berg, L. and Hammarström, M., 2006. The process of building a new governmental authority based on public demands for improved animal welfare. Livestock Science 103, 297-302.

Blokhuis, H.J., Jones, R.B., Geers, R., Miele, M. and Veissier, I., 2003. Measuring and monitoring animal welfare: transparency in the food product quality chain. Animal Welfare 12, p. 445-455.

Boqvist, S. and Vågsholm, I., 2005. Risk factors for hazard of release from *Salmonella*-control restriction on Swedish cattle farms from 1993 to 2002. Preventive Veterinary Medicine 71, 35-44.

Botreau, R., 2008. Multicriteria evaluation of animal welfare at farm level: an application of MCDA methodologies (Évaluation multicritère du bien-être animal: exemple des vaches laitières en ferme). Agro Paris Tech, Paris. Doctoral dissertation.

Botreau, R., Bonde, M., Butterworth, A., Perny, P., Bracke, M.B.M., Capdeville, J. and Veissier, I., 2007a. Aggregation of measures to produce an overall assessment of animal welfare. Part 1: a review of existing methods. Animal 1, 1179-1187.

when there is a big danger of slipping. Dairy cows prefer floors with higher skid resistance and slippery floors have negative effects on locomotion, on the manifestation of oestrus behaviour, self-grooming behaviour [caudal licking] and the general activity, and slipping during self-grooming or mounting is higher (Sommer and Troxler, 1986; Phillips and Morris, 2001, 2002; Benz, 2002; Bendel, 2005; Haufe *et al.*, 2009). Coverage with slurry can increase slipperiness and cows prefer a dry floor over a floor covered with slurry (Phillips and Morris, 2000; Rushen *et al.*, 2007) which stresses the importance of frequent cleaning.

The hardness of the walking alleys determines the load on legs, joints and claws. Hard floors are often too abrasive or too slippery, which can result in claw damage or leg fractures (Haufe *et al.*, 2009). Dairy cows prefer softer floor for walking and standing (Tucker *et al.*, 2006). Soft floors, i.e. rubber covered, have positive effects on locomotion: step length and speed increases and cows walk more with their head upright which can be interpreted as a sign of secure footing; step length was similar to pasture (Benz, 2002; Telezhenko and Bergsten, 2005; Tucker *et al.*, 2006; Flower *et al.*, 2007; Haufe *et al.*, 2009). On mastic asphalt, which is hard and has a higher skid resistance than a concrete (but lower than rubber covered) floor, cows showed a stride length intermediate between concrete and rubber (Haufe *et al.*, 2009).

In cattle the possibility to walk or stand at least part of the time on soft surfaces seems to be beneficial for claw health. Prolonged standing on concrete is an important predisposing factor for lameness (Bergsten and Frank, 1996). In cubicle loose housing occurrence of lameness is generally higher than with deep litter (Somers *et al.*, 2001). The effects of soft rubber floors on claw health are controversial: some studies found improved claw health while others report no relevant positive effect (Benz, 2002; Boyle *et al.*, 2007; Kremer *et al.*, 2006). These conflicting findings point at the complex interactions of the whole system including management (e.g. cleanliness, claw trimming). Especially the bedding and dimensions of cubicles may play a role and interact with floor properties (Mülleder and Waiblinger, 2004). When cows are allowed to stand with all four legs in cubicles with soft bedding, i.e. by positioning the neck rail position farther from the rear kerb, lameness and claw disorders decrease (Bernardi *et al.*, 2009).

Slatted floors pose a higher risk for traumatic claw lesions than do solid floors, especially when floors are deficient [slippery, rickety] and they are associated with a higher prevalence of lameness in dairy cows (Mülleder and Waiblinger, 2004). A large contact area between claw and floor, as seen in the solid surface floor, poses less mechanical stress to the claw and thus is preferable (Hinterhofer *et al.*, 2006). Lameness, injuries to the foot and leg weakness problems have also been reported to be higher in pigs housed on slatted in comparison to solid floor or deep litter (Scott *et al.*, 2007; O'Conell, 2009; Marchant-Forde, 2009). However, if fully solid floors with no bedding material in place get wet and dirty, the incidence of claw disorders may increase due to slipping (Jørgensen, 2003). In such cases sectioning the pen in a solid and a slatted area (of around 25%) may be advantageous in terms of preventing fouling of the solid area (O'Conell, 2009). The slot width in relation to foot size is an important factor determining potential floor-related foot injuries (Geyer, 1979a,b). For the newborn piglets' feet, the slatted floor in farrowing crates constitutes a high mechanical stress: the prevalence of sole haemorrhages was found range from 55 to 96% on 39 farms with different type of slatted floors (Putz, 2002). Further important factors include the absence of sharp or

jagged edges, and the provision of non-slippery, non-abrasive surfaces. In accordance with this, studies on weaner pigs have shown that foot injuries were reduced substantially by the use of plastic coating compared with uncoated expanded metal and woven wire flooring (O'Conell, 2009).

Deep-bedding generally offers a non-slippery soft floor that can be beneficial for foot health. However, there is a risk for excessive hoof growth, leading to lameness (Geyer, 1979b). Problems with infectious claw diseases may increase in badly maintained, wet and dirty, deep-bedding. Thus, again, good management is essential both in terms of providing sufficient use of bedding to keep it dry and clean, and in terms of hoof care (sufficient control and trimming).

### *3.5.2 Floors for lying*

Besides dimensions of the lying area, its surface plays an extremely important role in terms of welfare. Surfaces should be soft, skid-resistant, dry, provide thermal comfort and avoid injuries. However, many animals are housed on a hard floor or synthetic material with increased risk of injuries.

Cattle prefer soft lying surfaces (e.g. Fulwider and Palmer, 2004). A well maintained, dry deep litter or straw flow system offers optimal conditions. In cubicles, a straw-mattress (a compact mattress of straw and dry dung, composting after a while) provides both lying comfort and best avoids injuries at the hock and carpal joints. In temperate conditions, soft lying mats are equivalent to straw-mattresses as to their effects on cow behaviour [lying down, raising, lying time], but cows develop hock lesions on synthetic mats and mattresses (Wechsler *et al.*, 2000) and prefer straw bedding in (cold) winter (Manninen *et al.*, 2002). Hock lesions were comparable on soft synthetic or rubber mats and hard rubber mats, and only somewhat lower and less severe on loose straw; a straw mattress was the only floor type where cows develop nearly no hock lesions (Buchwalder *et al.*, 2000; Wechsler *et al.*, 2000). Sand is often recommended in cubicles to avoid mastitis, but cows prefer straw and soft lying mats (Manninen *et al.*, 2002) and the incidence of hock lesions is higher. The number of cows without any lesion can be summarised as follows: straw mattress 89%; compost 59%; solid matters with separation from liquid manure 59%; sand 52%; small quantity of loose straw 32%, comfort mats 15%; rubber mats 10% (Buchwalder *et al.*, 2000; Wechsler *et al.*, 2000; Zähner *et al.*, 2008). In tie stalls a higher amount of straw or shavings increased lying time (Tucker *et al.*, 2009).

Also in heifers and fattening bulls a hard floor impairs welfare. Animals on fully slatted concrete avoid lying down and rising due to pain so that lying bout length is increased (Gygax *et al.*, 2007a) and they develop more leg and tail tip lesions (Schrader *et al.*, 2001, Schulze Westerath *et al.*, 2007, Mülleder *et al.*, 2008). Rubber mats covering the concrete improve the situation and also decrease slipperiness, but still the welfare is compromised as compared with straw based systems [provided good management, i.e. sufficient litter is in place] (Gygax *et al.*, 2007a; Absmanner *et al.*, 2009).

Also in pigs, straw bedding is beneficial for lying comfort and for avoiding injuries. In straw bedded pens pigs take less time for lying down, spend more time lying laterally and sows show increased total lying time. There are also beneficial effects on injuries. With increased amount of straw and decreased proportion of slatted area bursitis incidence decreased (deep straw<little straw<partly-slatted<fully-slatted, with an Odds ratio of 18 for the fully-slatted) (KilBride *et al.*, 2008). The same study found that the risk of capped hock was lower in deep straw than with the other three floorings. Considering the findings on foot disorders, solid floors seem to be advantageous for pig welfare. O'Conell (2009) reports there may also be higher production when pigs are housed on solid floors or partly solid instead of fully slatted floors.

In crated sows with different types of slatted floors teat lesions (including teat end loss or complete teat losses) were found to be highly prevalent in sows in crates [on average 68% of sows had at least one teat lesion] (Putz, 2002). The negative impact of the floor may be augmented by the confined situation and related problems of leg disorders and difficulties in rising.

Newborn piglets often develop abrasions on their front legs from contact with the floor during suckling activity (Furniss *et al.*, 1986). The problem seems to be smaller on plastic-coated slatted- compared to concrete solid floors with little straw, which is possibly due to the less abrasive surface and the better hold on slatted floors (Baumgartner *et al.*, 2008).

### 3.6 Enrichment – substrates for manipulation

Animals in intensive housing systems often are subjected to a very barren environment where key stimuli for fulfilling the behavioural needs are lacking. This often causes particular behavioural problems, which are related to feeding motivation. In nature or in a semi-natural environment farm animal species spend a large time of the day with feeding behaviour in a complex, stimuli-rich environment, which constitutes a strong contrast to intensive production systems. In pig and poultry production, feed is generally provided in a concentrated, high-energetic form so feeding takes little time. Nevertheless, the motivation to show food searching behaviour remains.

Enrichment in terms of providing material that can be manipulated alleviates the situation. Straw offers excellent possibilities for diverse manipulation: to root or scratch in, to chew and eat it. Pigs with straw spend a large proportion of their time manipulating it, while pigs without straw are less active and spend more time manipulating the pen hardware (Scott *et al.*, 2007). Aggression of sows housed in systems with electronic sow feeder show less aggression when housed on solid floors and straw-bedded lying area than in partly slatted system without bedding (Jensen *et al.*,2000). Similarly, when sows had the possibility to root in mushroom compost suspended on racks this enrichment reduced aggressive interactions and injuries (Durell *et al.*, 1997, cited in Marchant-Forde, 2009).

When pigs have access to straw, tail and ear biting is less likely to occur (Cagienard *et al.*, 2005), although it cannot be absolutely prevented by bedding due to its multi-factorial nature. The conditions early in life seem to be important for later tail biting and the barren

environment in the farrowing pen may be one contributing factor (Smulders *et al.*, 2008). The development of feather-pecking in hens is also multi-factorial, but as mentioned before adequate rearing and later housing conditions (i.e. providing appropriate substrate such as straw) seems essential (Appleby *et al.*, 2004).

Provision of fresh straw often elicits play behaviour in [adult] cattle and pigs supporting the benefits of such enrichment in terms of positive emotions (Boissy *et al.*, 2007b).

### 3.7 Access to outdoor areas

Current husbandry systems vary greatly with respect to access to outdoor areas, ranging from zero- to permanent access. The outdoor area can be categorized in outside yards and pasture. Outside yards can have different floor qualities (slatted or solid, concrete or mastic asphalt, dirt ground, wood shavings) and different degrees of roofing.

Access to an outside yard offers several environmental characteristics that benefit welfare, such as additional space and structuring, climatic (fresh air, sun) and visual stimuli, all of which contribute to a more enriched environment. Cattle use the possibility to 'sun-bathe' [standing at right angles to the sun to get the most radiation] during the colder seasons; on sunny days in autumn the use of an outside run was especially high (Arnold and Dudzinski, 1978; Krötzl and Hauser, 1997). Access to sunlight supports vitamin D supply in cattle, which is important for regulating calcium homoeostasis and enhancing immune function (for review see EFSA, 2009, p. 111). But cattle also choose to go out in the rain (Krötzl and Hauser, 1997). In loose housed dairy cows, access to an outside yard reduces the frequency of agonistic interactions (Menke *et al.*, 2000), positively affects locomotion and was associated with reduced prevalence of lameness (Mülleder and Waiblinger, 2004). In calves, the occurrence of cross-sucking was lower in animals housed in outdoor group hutches compared to those kept indoors in deep litter pens and the risk of inter-sucking in heifers was more than 3 times higher when animals had no open-space contact [i.e. having access to outdoor yard, pasture or being housed in open-sided pens], most probably due to the more enriched environment (Keil *et al.*, 2000, 2002).

Access to pasture or other free-range areas (forests) offers an even more complex and enriched environment with many features resembling the natural habitat and with fewer restrictions in behaviour. In cattle even adult animals that are released to pasture for the first time after winter show play behaviour – bucking, galloping, head play – which can be interpreted as signs of positive emotions (Boissy *et al.*, 2007b). Pigs compensate limited access to pasture by more intensive use. This suggests they are motivated to use an outdoor area (Van der Mheen and Spoolder, 2004). Behaviour of many functional systems is unrestricted on pasture or much less restricted compared to usual housing systems, in other words: natural behaviour is possible in most functional behavioural systems. Accordingly, several studies found positive effects of access to pasture on health of dairy cows, especially a lower incidence of lameness, mastitis and mortality (e.g. Bendixen *et al.*, 1988b; Washburn *et al.*, 2002; Haskell *et al.*, 2006; Thomson *et al.*, 2006; Hernandez-Mendo *et al.*, 2007).

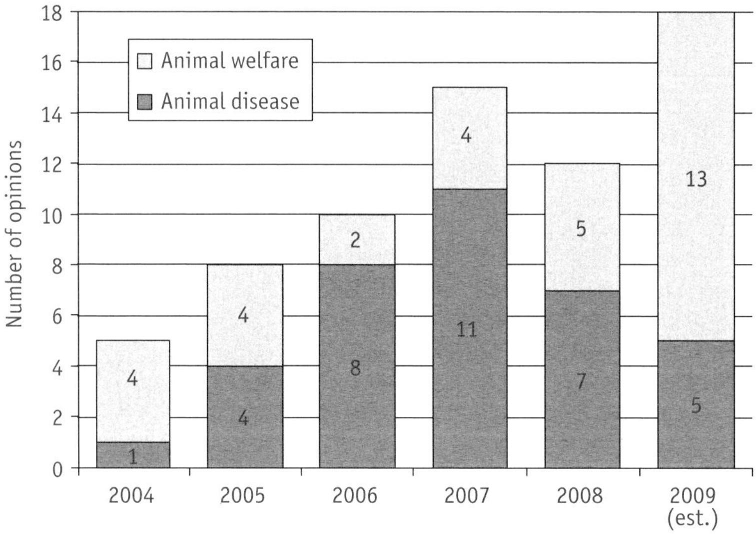


*Figure 1. Number of scientific opinions adopted by the AHAW panel on animal diseases and animal welfare (updated June 2009).*

assessment of the effects of farming systems on dairy cows' welfare and disease integrating the conclusions from the 4 risk assessments and the conclusion from the data compiled in a scientific report. Additional scientific opinions on welfare aspects of the stunning and killing of farmed fish, including 8 fish species (salmon, trout, carp, eel, tuna, sea bass, sea bream and turbot) have been adopted during in the first semester 2009 (EFSA, 2009a,b,c,d,e,f,g,h,i,j,k,l).

The evolution of the methodologies of the scientific opinions has been addressed in two directions, in the first period (2003-2006) the questions were broad, requesting to deal with many concepts and for various species, which resulted in a large scientific opinions primarily relying on a qualitative risk assessment approach (Candiani *et al.*, 2007, 2009; Bracke *et al.*, 2008). However, the second period (2006-2009) has been with a more targeted objectives [questions and species] allowing a more in deep analysis and a refined assessment methodology (Müller-Graf *et al.*, 2008; Ribó *et al.*, 2008b).

Each scientific opinion has been addressed by a collective of on average 7-8 *ad hoc* experts, chaired by one member of the AHAW Panel supported administratively and scientifically by AHAW staff. The AHAW panel has been collaborating in the scientific opinion with the EMEA (e.g. fish welfare) and the ECDC [e.g. avian influenza]. Furthermore EFSA and the Panel had permanently accepted an observer from the World organisation for animal health (OIE), and the World Organisation for Food and Agriculture (FAO) in order to promote scientific communication and to facilitate an adequate exchange of information and to promote networking between EFSA and these organizations in order to avoid duplication of effort and to improve efficiency and validity of the data for risk assessment purposes.

The networking in the European member states has been strongly encouraged through the EFSA's Advisory Forum, a special group on animal health and welfare. Two meetings were held in 2008 and 2009, and one document (EFSA, 2009m) collect all the contact points in the EU related to the experts involved in animal welfare risk assessment. In order to promote the European networking and take a proactive attitude in a preparatory work for scientific opinions, scientific and technical assistance and collection of data, EFSA has launched several public calls in the framework of article 36 of the Regulation 178/2002, specifically addressed to competent scientific/research organisations designated by the Member States. These calls have the objective to facilitate a scientific cooperation framework, the exchange of information and the development of joint projects for which EFSA may be eligible for financial support. Efforts have been made to identify data relevant to carry risk assessment of exotic diseases and in the welfare the efforts have been concentrated in developing guidelines for animal welfare risk assessment stunning and killing, animal transport and housing and management (Table 1).

*Table 1. Overview of projects launched by EFSA in animal health and welfare unit.*

| | Starting date | Completion |
|---|---|---|
| Animal disease | | |
| Classical swine fever, African swine fever, African horse sickness and arthropod disease vectors (consortium of 8 partners) | December 2007 | November 2008 |
| Epidemiology of different agents causing disease in aquatic animals (3 partners) | December 2008 | June 2009 |
| Tuberculosis in wildlife in the EU (5 partners) | December 2008 | July 2009 |
| Crimean-Congo Hemorrhagic Fever and Epizootic Hemorrhagic Disease (1 contractor) | December 2008 | September 2009 |
| Emerging animal diseases in the EU (AMU in collaboration with AHAW) | To be published in 2009 | |
| Fish: collecting data on emerging and exotic pathogens (interaction with wild) | To be published in 2009 | |
| Animal welfare | | |
| Development of Animal Welfare risk assessment guidelines on Stunning and Killing | September 2007 | December 2008 |
| Development of Animal Welfare risk assessment guidelines on Transport | September 2008 | November 2009 |
| Development of Animal Welfare risk assessment guidelines on Husbandry and Management | Call proposals June 09 | 2010 |
| Data collection and assessment on the influence of genetics on the health and welfare of commercial broilers | Call proposals June 09 | 2010 |

## 4. Interactions between animal welfare, animal disease and food safety in EFSA's scientific opinions on animal welfare

Whenever relevant and needed, the possible interactions and implications for food safety and animal disease have been considered in most of the animal welfare scientific opinions, involving other areas of expertise in EFSA, like Biohazards, Contaminants, and Plant Health. Some examples of this interaction are illustrated in Table 2.

In the scientific opinion on welfare aspects of various systems of keeping laying hens (EFSA, 2004b) it was concluded that keeping laying hens in non-cages systems will improve their welfare. However, it was also concluded that the risk of contamination with *Salmonella* spp. might be higher when eggs are produced in some non-cage systems because of the greater exposure of layers and their eggs to environmental contamination.

In the scientific assessment related with the migratory birds and their possible role in the spread of highly pathogenic avian influenza (EFSA, 2006a), it was concluded that allowing chickens to open air access, and thus improving their welfare, may increase the risk of exposure to contagious diseases such as Avian Influenza.

Several examples of the interactions between animal welfare and risk of disease can be also found on the scientific opinions on the welfare of pigs (EFSA, 2005a, 2007a,b,c). For instance, in the scientific opinion about the effect of different space allowances and floor types in pigs it was concluded that floor type may have an influence on enzootic multi-factorial diseases, including enteric and respiratory infections and that types of floor leading to bad urine drainage and to insufficient faeces removal increase the risk of infections, especially those leading to enteric disorders in pigs (EFSA, 2005a). The bedding material has been considered in each pig scientific opinion, concluding that using straw as bedding material might act as vehicle of diseases such as FMD or Aujeszky's disease. As mentioned for poultry, keeping the pigs outdoors with possible contact with the wild fauna, mainly wild boars, is believed to be a source of [ in domestic pigs (EFSA, 2007a,b,c).

In relation to the interaction between animal welfare and food safety aspects, the use of 'penetrating captive bolt' stunning system has been recognised as one of the most appropriate systems in terms on animal welfare, when properly applied. However, the potential dissemination of central nervous system emboli into edible tissues with consequent exposure of humans to BSE can not be excluded (EFSA, 2004e, 2006c).

Considering the available literature on the importance of interactions between animal welfare, animal disease and food safety (EC, 2000a, 2007a; Blokhuis *et al.*, 2008; Ribó *et al.*, 2009), and considering all above-mentioned aspects and the role of EFSA on defining animal welfare in relation to animal health, it was concluded that, for assessing the overall impact of animal welfare, it is important to also study factors which possibly affect the incidence of animal diseases and food safety. Finally, the evaluation of this overall impact and interaction between animal welfare, animal disease and food safety should help the development of control and monitoring plans at farm level.

*Table 2. Example of interactions between animal welfare, animal health and food safety from EFSA Scientific opinions (updated from Blokhuis et al., 2008; Ribó* et al.*, 2009).*

| Species (sc. opinion) | Animal welfare issue | Impact on animal disease | Impact on food safety |
|---|---|---|---|
| Cattle (2004c) | Related stress in transport | Shipping fever *(Pasteurella* spp., BRSV, IBR, *E. Coli* and *Salmonella* spp.) | Meat contamination with *Salmonella* (BIOHAZ SO inside EFSA, 2004c) |
| All species (2004c) | Mixing animals at staging points | Spread of infectious diseases (e.g. *E. Coli, Salmonella*) | Meat contamination with *Salmonella* (BIOHAZ SO inside EFSA, 2004c) |
| Ruminants (2004e; 2006c) | Stunning method: Penetrative Captive bolt for stunning | Not applicable | Dissemination of contaminated nervous tissue into edible parts<br>Risk of BSE/CJD (EFSA, 2004c) |
| Laying hens (2004b) | Non-cage systems | Risk of egg contamination with *Salmonella (S. enteritidis)* | Egg contamination with *Salmonella* (BIOHAZ and CONTAM SO inside EFSA, 2004b) |
| Poultry (2005b, 2006a) | Outdoor systems | Highly Pathogenic AI (HPAI) | HPAI in humans (EFSA, 2006e) |
| Calves (2006c) | Stress, large groups, space allowance, mixing from different sources, restocking | Increased susceptibility, cross-contamination, pathogens spread | Risk of foodborne pathogen contamination (i.e. *Salmonella*) (BIOHAZ SO inside EFSA, 2006c) |
| Pig (2005a) | Bedding materials (e.g. straw). Floor types (e.g. floors causing abrasion to feet or snout) | Vehicle of FMD or Aujeszky disease.<br>Spread of FMD, Vesicular Stomatitis. | (BIOHAZ SO inside EFSA, 2005a) |
| Pig (2005a, 2007a,b,c) | Outdoor systems | Wildlife pathogens (i.e. *Toxoplasma, Salmonella, Brucella*) | Foodborne pathogens<br>Risk of Toxoplasmosis, Salmonellosis and Brucellosis (BIOHAZ SO inside EFSA, 2005a; EFSA, 2007e) |
| Dairy cows (2009h,i,j,k,l) | Pasture system (reduced lameness, fertility and metabolic problems, etc.) | Risk of inclement weather, flies and access to toxic plants.<br>High Disease Impact (i.e. anthrax, bluetongue, leptospirosis, echinococcosis, FMD, Q-fever, anaplasmosis, babesiosis and liver fluke) | BIOHAZ scientific opinion to be adopted. |

## 5. Impact of the EFSA scientific assessment on the EU policy on animal welfare

The final aim of the scientific assessments on animal welfare by EFSA is to support animal welfare EU legislation based on scientific evidence. In this sense, there are many examples which illustrate how EFSA's scientific opinions are taken into consideration when legislative measures are proposed by the European Commission (EC, 2002a; Blokhuis *et al.*, 2008; Ribó *et al.*, 2008c).

In Council Regulation (EC) No 1/2005 (EC, 2005a) on the protection of animals during transport, amending previous EU legislation on Animal Welfare during transport, some of the measures are based on the conclusions and recommendations contained in previous EFSA Scientific Opinions on Animal Transport (EFSA, 2004c,d) which are based on scientific evidence. Table 3 presents some examples of the impact these EFSA documents have had.

The EFSA opinion on the animal health and welfare risks associated with the import of wild birds other than poultry into the European Union (EFSA, 2006b) was adopted in October 2006. Following some of the recommendations of the Scientific Opinion, the European Commission adopted in November 2006 a Commission Decision (EC, 2006a) which, in summary, permanently banned the import of 'pet birds' and wild caught birds, allowing only the import of captive-bred birds only from authorised countries.

The AHAW Panel adopted in 2004 and 2006 two scientific opinions (EFSA, 2004e, 2006c) on the welfare aspects of the main systems of stunning and killing [main commercial species: cattle, sheep, pigs, poultry, horses and farmed fish and deer, goats, rabbits, ostriches, ducks, geese and quail, respectively] with consideration of Directive 93/119/EC (EC, 1993). The opinions focused on the welfare at the point of application of the main stunning and stun/killing methods. Killing of animals without stunning and stunning and killing methods for disease control were also considered. The two EFSA opinions on slaughter and killing, taken together, address stunning and killing of most species kept for farming purposes within the EU. The Commission proposal for a Council Regulation on the protection of animals at the time of killing made on the 18$^{th}$ of September 2008 (EC, 2008a), clearly states that EU legislation in this area should be updated to take into account those EFSA scientific opinions. It is also written in the proposal that fish species are not included, as further scientific opinions and economic evaluation on farmed fish are also needed. Furthermore, EFSA received a mandate on the welfare aspects of the stunning and killing of farmed fish on June 2008 and the respective 7 scientific opinions were adopted in 2009 (EFSA, 2009a,b,c,d,e,f,g).

EFSA scientific opinions on animal welfare have an impact on the legislative tasks of the European Commission. Science based conclusions and recommendations are used to justify management measures. Table 4 includes an overview of different EFSA scientific opinion, the previous legislation and the new proposals for amendments after considering the new scientific evidences.

*Table 3. Adoption of (elements of) EFSA scientific conclusions and recommendations in Council Regulation (EC) No 1/2005 (EC, 2005a) on the protection of animals during transport (Ribó* et al.*, 2008c).*

| Conclusions and recommendations | Council Regulation (1/2005) |
|---|---|
| The education of personnel involved in handling animals, moving animals, or driving vehicles can greatly improve animal welfare. | Article 6. Transporters. 4. Transporters shall entrust the handling of the animals to personnel who have received training on the relevant provisions on Annexes I and II. Article 10. Requirements for transporters authorisation. Article 16. Training of staff and equipment of the competent authority. Article 17. Training courses and certificate of competence |
| Animals unfit to travel because they are injured or diseased or because their physiological state (might produce an egg during the transport period or mammals in the last five days of pregnancy) should not be transported (a list of situations is provided) | Annex I. Chapter I. Fitness for transport. 2. Animals that are injured or that present physiological weakness or pathological processes shall not be considered for transport and in particular..... |
| Behaviour modifying drugs should not be given to animals in place of good practice for the transport of any animal. Using some drugs when animals are transported by air may be harmful. | Annex I. Chapter I. Fitness for transport. 5. Sedatives shall not be used on animals to be transported unless strictly necessary to ensure the welfare of the animals and shall only be used under veterinary supervision. |
| Many animals are forced to move up or down ramps which are too steep during loading and unloading. Systems which eliminate the need for ramps or purpose built ramps no steeper than 10° are desirable. No animal should be required to negotiate a ramp steeper than 20°. | Annex I. Chapter III. Transport Practices. 1. Loading, unloading handling. 1.4 (a) Ramps shall not be steeper that an angle of 20°, ...Where the slope is steeper than 10°, ..., ramps shall be fitted with a system, ...which ensure that the animals climb or go down without risks or difficulties. |
| Animals which may fight if mixed (i.e. have previously lived in different pens) should not be mixed. | Annex I. Chapter III. Transport Practices. Separation 1.12 Animals shall be handled and transported separately in the following cases: a) animals of different species; b) animals of significantly different sizes and cages.... |
| Food and water should be provided at different time periods after a journey commences depending on the specie or, for instance, at the driver's break when the temperature is 20° C or above. | Annex I. Chapter V. Watering and feeding interval, journey times and resting periods. 1. Domestic equidae, domestic animals of bovine, ovine, caprine and porcine species. 2. Other species. |
| Animals have to be kept inside the vehicle at a stocking density which meets their movement and resting needs, including the maintenance of their balance when driving (space allowance specifications per specie are given). | Annex I. Chapter VII. Space Allowances. A. Domestic equidae, B. Bovine animals. C. Sheep/Goats. D. Pigs. E. Poultry. |

*Table 4. Overview of Legislative impact of some EFSA scientific opinions on animal welfare.*

| Scientific Opinion (year) | Previous legislation | Amendment or new legislation |
|---|---|---|
| Welfare of animals during transport (2004); Standards for the microclimate inside animal road transport vehicles (2004) (SCAHAW, 2002) | Regulation 1/2005 on the protection of the animals during transport | Specific provisions for poultry, cats and dogs will be set up in appropriate proposals when the relevant opinions of EFSA will be available |
| Stunning/killing of main commercial species (2004); Stunning/killing of minor species (2006) (SVC, 1996a) | Council Directive 74/577/EC on stunning animals before slaughter, replaced by Council Directive 93/119/EEC | Proposal for a Council Regulation on the protection of animals at the time of killing (18 September 2008)<br>'Community legislation in this area should be updated to take into account those scientific EFSA opinions'. |
| Welfare aspects of the main systems of stunning and killing farmed fish (7 Scientific Opinions) (2009) | | 'Recommendations on farm fish were not included in the Council Regulation proposal because there was a need for further scientific opinion and economic evaluation in this field'. |
| Welfare aspects of various systems of keeping laying hens (2005)<br>Welfare of laying hens (SCV, 1996b) | Directive 1999/74 laying down minimum standards for the protection of laying hens | Communication of the Commission (January 2008) on the various systems of rearing laying hens: 'Based on the scientific studies it is not recommended to amend the current provisions of Directive 1999/74/EC'. |
| Castration of piglets (2004); Pig space allowance/floor types (2005); Welfare of sows and boars (2007); Welfare of fattening pigs (2007); Pig tail biting (2007) (SVC, 1997) | Council Directive 2001/88/EC<br>Directive 2001/93/EC | Further initiatives on the welfare of pigs are foreseen in the framework of the action plan (2009) based on the outcomes of the EFSA opinions. |
| Animal health and welfare aspects of avian influenza (2005) | Directive 95/2004 | Decision 2006/7: suspending the import of no-treated feathers from certain third countries |
| Animal health and welfare risks of the import of captive birds from third countries (2006) | Commission Decision 2005/760. | Commission Decision 2007/21.<br>Permanent ban for 'pet birds' and wild caught birds<br>Import of captive-bred birds only from authorised countries |
| Risks of poor welfare in intensive calf farming systems (2006); Welfare of calves (SVC, 1995) | Directive 91/629 minimum standards on the protection of calves amended by 97/2/EC and 2008/119/EC. | The Commission shall submit to the Council a report, drawn up on the basis of a scientific opinion, on the intensive farming systems which comply with the requirements of the well-being of calves |

## 6. EFSA's future challenges in animal health and welfare

The main goals of future EFSA activities in the area of animal health and disease prevention including addressing the critical relationship between animal welfare and animal disease, as well as its relevance to public health are:

1. To deliver the best scientific advice at the right time and in the most appropriate manner thus supporting the decision making of risk managers; applying, when appropriate, the integrated approach to risk assessment associated with animal nutrition[2], animal diseases, animal welfare, and public health (2008-2013); further improving the methods of risk assessment in animal welfare in combination with animal diseases and food safety (2008-2012); defining and assessing indicators for animal welfare for food producing animals that could be used in a harmonised European monitoring system (2008-2010); establishing a formal internal and external quality improvement system for risk assessment at EFSA (2008-2009) and improving the provision of scientific input for risk-benefit analyses (2009-2010).
2. To decrease the time needed for risk assessment and scientific advice on animal health and welfare by enabling the rapid access and use of already collected data by data exchange between EFSA and the EC including FVO, MS, CRL and NRL, in association with EMEA, ECDC and other relevant EU institutions. Further development of the collaboration on data exchange for the purposes of data analyses and risk assessment in the animal health area by EFSA (2008-2010) including both the urgent issues/emerging risks as well as other scientific advice.
3. To provide scientific support for EU surveillance programmes for animal diseases, zoonoses and animal welfare as well as analysing the results of surveillance; Provide scientific input for the design of data collection schemes including diagnostic techniques in order to enhance the suitability of the data collected for risk assessment purposes in the area of animal health (2008-2013).
4. To provide scientific support for EU crisis preparedness by implementing the EFSA Management Board decision of 11$^{th}$ of September 2007 for accelerated procedures of scientific advice for urgent matters by using fast-track response or statements when relevant and without compromising the quality of the scientific opinions delivered (2008-2013) and providing technical support for data collection in the area of animal health and data sharing during crisis situations from a scientific point of view for EC and MS (2008-2013).
5. To avoid unnecessary divergence in opinions between EFSA and competent authorities in the various Member States (MS) or other relevant EU institutions (e.g. EMEA and ECDC). Identifying the reasoning behind the different opinions, if any, in a transparent manner and in accordance with Article 30 of Regulation 178/2002. Introducing an effective methodology to compare different risk assessments (2008-2009). Enhancing collaboration between animal health and welfare risk assessment bodies during all the steps of the risk assessment process (2008-2010), i.e. – *at the initiation of risk assessments* – by notifying MS and other EU institutions and providing information on the selection of experts and on adopting consistent approaches in line with best practice, – *during risk assessments*

[2] In respect to the animal nutrition, it is important to note that the Regulation (EC) No 1831/2003 on additives for use in animal nutrition, in it is Art. 5(3), establishes the functions of the feed additives, one of them being directly related to animal welfare, i.e.: *'The feed additive shall favorably affect animal production performance or welfare'*.

– when necessary by requesting feedback and -forward of information to be used in the risk assessment exercise; and of information on the progress of risk assessments in EFSA and MS; and – *upon completion of risk assessment* – by sharing of scientific opinions under embargo with the objective of detecting and resolving unnecessary divergence at an early stage and by agreeing an appropriate communication plan when divergence of opinion is inevitable the scientific rationale for the differing views need to be explained.

6. To mobilise and coordinate scientific expertise throughout the EU on issues within the remit of EFSA, networking with national, EU and international organizations *via* the scientific panels and their working groups, the Advisory Forum and its Focal Points, and other networks in the area of animal health; Creating a database of experts; contracting scientific work *via* Article 36 of EFSA's founding regulation and grants; organising scientific colloquia and other scientific events to review knowledge and stimulate debate on future scientific developments in the area of animal health and welfare; Follow up closely the science development and new research and collaborating with the EC for the identification of research needs as well as enhancing the use of research results in risk assessment in the area of animal health and welfare

## 7. Conclusions

### 7.1 What has been achieved?

The RA process has several benefits. The major advantage is transparency as scientific evidence is provided through the data used, through the definition of the risk pathways and assumptions and the description of the RA approach. RA can support the prioritisation of areas for intervention [risk management] and give information on further data needs (recommendations for future research).

### 7.2 What has been neglected?

Comprehensive animal welfare databases, which could give important support for EFSA risk assessment and provide information on the relevant factors possibly influencing animal health status, are currently not available at EU level. There is also a lack of harmonization of indicators for animal welfare despite the fact that there are already some good examples of relevant and reliable indicators available in some member states and industry.

### 7.3 What needs to be done?

To perform good science based risk assessment in the area of animal welfare, involvement is needed of welfare scientists (research, data from the field) and cooperation with the governmental bodies and stakeholders, including profit and non profit organizations, is crucial.

EFSA, in line with mandates received from the EC, will start to assess the indicators for the welfare of food-producing animals with a view to developing a harmonised European monitoring system assessing the quality of farming systems and their impact on the diseases and welfare of animals according to the outcome-based approach described in the Community

Action Plan on the Protection and Welfare of Animals (2006-2010). Such indicators could be integrated into future legislative proposals to provide the basis for the development of an EU label for animal welfare proposed by the EC. In this context, EFSA should provide to the future European Centre for the Protection and Welfare of Animals the scientific advice (based in risk assessment), needed to implement the activities described in the Action Plan and to ensure the integration with animal health policies.

## Acknowledgements

EFSA acknowledges the efforts of the 32 working groups (190 experts) who have been actively dealing with every issue related to scientific opinions on animal welfare, and of the AHAW Panel Members chairing the WG and/or reviewing these opinions (notably, Bo Algers, Harry J. Blokhuis, Donald M. Broom, Ilaria Capua, Stefano Cinotti, Patrizia Costa, Mariano Domingo, Mathias Greiner, Daniel Guemene, Michael Gunn, Joerg Hartung, Trevor Hastings, Per Have, Frank Koenen, Francisco Javier Manteca Vilanova, David B. Morton, Christine Müller-Graf, Albert Osterhaus, Michel Pepin, Dirk U. Pfeiffer, Ronald J. Roberts, Jose Manuel Moez Sanaa, Sanchez Vizcaino, Mo Salman, Alejandro Schudel, J. Michael Sharp, Georgios Theodoropoulos, Philippe Vannier, Marina Verga, Martin Wierup, Marion Wooldridge) and finally EFSA's AHAW staff supporting these individuals in their work.

The authors are also particularly grateful to Dr. Philippe Vannier – chairman of the AHAW panel – who, in the period 2003-2009, coordinated all the activities of the Panel including welfare, and to Prof. Dr Donald Broom for his devotion to animal welfare and his support in ensuring that minority opinions were expressed, thus granting the scientific independence of the experts. Finally, the Commission Officials, particularly Andrea Gavinelli and his team at the Unit D5 – Animal Welfare, Health and Consumer Directorate General are acknowledged for submission of their mandates and for their enthusiasm.

## References

Blokhuis, H.J., Keeling, L.J., Gavinelli, A. and Serratosa J., 2008. Animal welfare's impact in the food chain. Trends in Food Science & Technology 19, 75-83.

Bracke, M.B.M., Edwards, S.A., Engel, B., Buist, W.G. and Algers, B., 2008. Expert opinion as 'validation' of risk assessment applied to calf welfare. Acta Veterinaria Scandinavia 50, 29.

Broom, D.M., 2009. Animal welfare and legislation. In: Smulders, F.J.M. and Algers, B. (Eds.). Food safety assurance and veterinary public health, Vol.5, Welfare of production animals: assessment and management of risks. Wageningen Academic Publishers, Wageningen, the Netherlands, pp. 339-352.

Buncic, S., Collins, J.D., Smulders, F.J.M. and Colin, P., 2009. Biological food safety in relation to animal welfare. In: Smulders, F.J.M. and Algers, B. (Eds.). Food safety assurance and veterinary public health, Vol.5, Welfare of production animals: assessment and management of risks. Wageningen Academic Publishers, Wageningen, the Netherlands, pp. 485-53.

Codex Alimentarius Commission (CAC), 2002. Principles and guidelines for the conduct of microbiological risk assessment. Document CAC/GL 30.

Codex Alimentarius Commission (CAC), 2008. Codex Alimentarius Commission, procedural manual. Available at: ftp.fao.org/codex /Publications/ProcManuals/Manual_18e.pdf.

Codex Alimentarius Commission (CAC), 2009. Codex Alimentarius Commission. Available at: http://www.codexalimentarius.net.

Candiani, D., Ribó, O., Afonso, A., Aiassa, E., Correia, S., De Massis, F., Pujols, J. and Serratosa J., 2007. Risk assessment challenges in the field of animal welfare. In: Proceedings of the XIII international congress in animal hygiene, ISAH, June 17-21, Tartu, Estonia, pp. 587-581.

Candiani, D., Ribó, O., Barbieri, S., Afonso, A., Grudnik, T., Berthe, F. and Serratosa J., 2009. Development of a risk assessment methodology for animal welfare in EFSA's scientific opinions. In: Aland, A. and Madec, F. (Eds.). Sustainable animal production. The challenges and potential developments for professional farming. Wageningen Academic Publishers, Wageningen, the Netherlands, pp. 421-434.

Council of Europe (CoE), 2008. Available at: http://www.coe.int/t/e/legal_affairs/legal_co-operation/biological_safety%2C_use_of_animals/farming/A_texts_documents.asp.

Council of Europe (CoE), 2009. Website. Available at: http://www.coe.int/defaultEN.asp.

European Commission (EC), 1978. 78/923/EEC: Council Decision of 19 June 1978 concerning the conclusion of the European Convention for the protection of animals kept for farming purposes. Official Journal L 323, 12-13.

European Commission (EC), 1986. Council Directive 86/609/EEC of 24 November 1986 on the approximation of laws, regulations and administrative provisions of the Member States regarding the protection of animals used for experimental and other scientific purposes. Official Journal L 358, 1-28.

European Commission (EC), 1988. European Convention for the protection of animals for slaughter 88/306/EEC: Council Decision of 16 May 1988 on the conclusion of the European Convention for the Protection of Animals for Slaughter. Official Journal L 137, 27-38.

European Commission (EC), 1991. Council Directive 91/630/EEC of 19 November 1991 laying down minimum standards for the protection of pigs. Official Journal L 340, 33-38.

European Commission (EC), 1993. Council Directive 93/119/EC of 22 December 1993 on the protection of animals at the time of slaughter or killing. Official Journal L 340, 21-34.

European Commission (EC), 1994; Council Decision 94/800 (of 22 December 1994) concerning the conclusion on behalf of the European Community, as regards matters within its competence, of the agreements reached in the Uruguay Round multilateral negotiations (1986-1994). Official Journal L 336, 1-2.

European Commission (EC), 1997a. European Commission. Treaty of Amsterdam – Protocol on protection and welfare of animals. Official Journal C 340, 110.

European Commission (EC), 1997b. Council Regulation (EC) No 1255/97 of 25 June 1997 concerning Community criteria for staging points and amending the route plan referred to in the Annex to Directive 91/628/EEC. Official Journal L 174, 1.

European Commission (EC), 1998. Council Directive 98/58/EC of 20 July 1998 concerning the protection of animals kept for farming purposes. Official Journal L 221, 23-27.

European Commission (EC), 1999a. Council Directive 1999/22/EC of 29 March 1999 relating to the keeping of wild animals in zoos. Official Journal L 094, 24-26.

European Commission (EC), 1999b. Council Directive 1999/74/EC of 19 July 1999 laying down minimum standards for the protection of laying hens. Official Journal L 203, 53-57.

European Commission (EC), 2002a. Regulation (EC) No. 178/2002 of 28 January 2002, laying down the general principles and requirements of food law, establishing the European Food Safety Authority and laying down procedures in matters of food safety. Official Journal L 31, 1-24.

European Commission (EC), 2002b. Communication from the Commission to the Council and the European Parliament on Animal Welfare Legislation on farmed animals in Third Countries and the Implications for the EU Brussels, 18.11.2002, COM(2002) 626 final.

European Commission (EC), 2003. European Union website. Available at: http://ec.europa.eu/food/animal/welfare/index_en.htm.
European Commission (EC), 2004. Welfare quality project. Available at: http://www.welfarequality.net/.
European Commission (EC), 2005a. Community Action Plan on the Protection and Welfare of Animals (2006-2010). Available at: http://ec.europa.eu/food/animal/welfare/actionplan/actionplan_ en.htm.
European Commission (EC), 2005b. Council Regulation (EC) No 1/2005 of 22 December 2004, on the protection of animals during transport and related operations and amending Directives 64/432/EEC and 93/119/EC and Regulation (EC) No 1255/97. Official Journal L 3, 1-44.
European Commission (EC), 2006. Council Decision (2004/544/EC) of 21 June 2004 on the signing of the European Convention for the protection of animals during international transport. Official Journal L 241, 21.
European Commission (EC), 2007a. Future Strategy on Animal Health for 2007-2013. Available at: http://ec.europa.eu/food/animal/diseases/strategy/final_report_en.htm.
European Commission (EC), 2007b. Regulation (EC) No 1523/2007 of the European Parliament and of the Council of 11 December 2007 banning the placing on the market and the import to, or export from, the Community of cat and dog fur, and products containing such fur Text with EEA relevance. Official Journal L 343, 1-4.
European Commission (EC), 2007c. Council Directive 2007/43 of 28 June 2007 laying down minimum rules for the protection of chickens kept for meat production. Official Journal L 182, 19-28.
European Commission (EC), 2007d. Commission Decision (2006/778/EC) 14 November 2006 amending Decision 2000/50/EC concerning minimum requirements for the collection of information during the inspections of production sites on which certain animals are kept for farming purposes (Text with EEA relevance). Official Journal L 314, 39-47.
European Commission (EC), 2008a. Commission proposal COM 2008/553 of 18 September, 2008 for a Council Regulation the protection of animals at the time of killing.
European Commission (EC), 2008b. Council Directive 2008/119/EC of 18 December 2008, laying down minimum standards for the protection of calves (codified version). Official Journal L 010, 7-13.
European Commission (EC), 2008c. Conference and Forum on Global Trade and Farm Animal Welfare. Available at: http://www.animalwelfareandtrade.com/.
European Food Safety Authority (EFSA), 2003. Website. Available at: http://www.efsa.europa.eu.
European Food Safety Authority (EFSA), 2004a. Opinion of the Scientific Panel on Animal Health and Welfare related to the welfare of animals during transport. The EFSA Journal 44, 1-36.
European Food Safety Authority (EFSA), 2004b. Scientific Opinion of the AHAW Panel on the welfare aspects of the main systems of stunning and killing the main commercial species of animals. The EFSA Journal 45, 1-29.
European Food Safety Authority (EFSA), 2004c. Scientific Opinion of the Biohazard Panel on BSE risk from dissemination of brain particles in blood and carcass following stunning. The EFSA Journal 123, 1-4.
European Food Safety Authority (EFSA), 2005a. Opinion of the scientific panel on animal health and welfare (AHAW) on a request from the commission related to the Impact of the current housing and husbandry systems on the health and welfare of farmed domestic rabbits. The EFSA Journal 267, 1-31.
European Food Safety Authority (EFSA), 2005b. Opinion of the scientific panel on animal health and welfare (AHAW) on a request from the commission related to aspects of the biology and welfare of animals used for experimental and other scientific purposes. The EFSA Journal 292, 1-46.
European Food Safety Authority (EFSA), 2006a. Scientific Colloquium 'Principles of Risk Assessment of Food Producing Animals: Current and future approaches', http://www.efsa.europa.eu/en/science/colloquium_series/no4_animal_diseases.html.
European Food Safety Authority (EFSA), 2006b. Scientific Opinion of the AHAW Panel on the welfare aspects of the main systems of stunning and killing applied to commercially farmed deer, goats, rabbits, ostriches, ducks, geese and quail. The EFSA Journal 326, 1-18.

European Food Safety Authority (EFSA), 2006c. Opinion of the Scientific Panel on Animal Health and Welfare (AHAW) on a request from the Commission related with the risks of poor welfare in intensive calf farming systems. The EFSA Journal 366, 1-36.

European Food Safety Authority (EFSA), 2006d. Animal health and welfare risks associated with the import of wild birds other than poultry into the European Union

European Food Safety Authority (EFSA), 2006e. Scientific report of the Scientific Panel on Biological Hazards on 'Food as a possible source of infection with highly pathogenic avian influenza viruses for humans and other mammals'. The EFSA Journal 74, 1-29.

European Food Safety Authority (EFSA), 2007a. Scientific Opinion of the Panel on Animal Health and Welfare on a request from the Commission on animal health and welfare in fattening pigs in relation to housing and husbandry. The EFSA Journal 564, 1-14.

European Food Safety Authority (EFSA), 2007b. Scientific Opinion of the Panel on Animal Health and Welfare on a request from the Commission on the animal health and welfare aspects of different housing and husbandry systems for adult breeding boars, pregnant, farrowing sows and unweaned piglets. The EFSA Journal 572, 1-13.

European Food Safety Authority (EFSA), 2007c. Scientific Opinion of the Panel on Animal Health and Welfare on a request from the Commission on the risks associated with tail-biting in pigs and possible means to reduce the need for tail-biting considering the different housing and husbandry systems. The EFSA Journal 611, 1-13.

European Food Safety Authority (EFSA), 2007d. Scientific Opinion of the AHAW Panel on the welfare aspects of the killing and skinning of seals. The EFSA Journal 610, 1-122.

European Food Safety Authority (EFSA), 2007e. Scientific Opinion of the Panel on Biological Hazards on Food safety aspects of different pig housing and husbandry systems relating to Opinions of the Scientific Panel on Animal Health and Welfare: (1) The risks associated with tail biting in pigs and possible means to reduce the need for tail docking considering the different housing and husbandry systems; (2) Animal health and welfare aspects on different housing and husbandry systems for adult breeding boars, pregnant, farrowing sows and unweaned piglets; (3) Animal health and welfare in fattening pigs in relation to housing and husbandry. The EFSA Journal 613, 1-20.

European Food Safety Authority (EFSA), 2008a. Scientific opinion of the Panel on Animal Health and Welfare on animal welfare aspects of husbandry systems for farmed Atlantic salmon. The EFSA Journal 736, 1-31.

European Food Safety Authority (EFSA), 2008b. Scientific opinion of the Panel on Animal Health and Welfare on animal welfare aspects of husbandry systems for farmed trout. The EFSA Journal 796, 1-22.

European Food Safety Authority (EFSA), 2008c. Scientific opinion of the Panel on Animal Health and Welfare on animal welfare aspects of husbandry systems for farmed European eel. The EFSA Journal 809, 1-17.

European Food Safety Authority (EFSA), 2008d. Scientific opinion of the Panel on Animal Health and Welfare on animal welfare aspects of husbandry systems for farmed European sea bass and sea bream. The EFSA Journal 844, 1-21.

European Food Safety Authority (EFSA), 2008e. Scientific opinion of the Panel on Animal Health and Welfare on animal welfare aspects of husbandry systems for farmed common carp. The EFSA Journal 843, 1-28.

European Food Safety Authority (EFSA), 2008f. Role of EFSA in contributing to the improvement of animal health in Europe. 5 November 2008, www.efsa.europa.eu.

European Food Safety Authority (EFSA), 2009a. Scientific Opinion of the Panel on Animal Health and Welfare on species specific welfare aspects of the main systems of stunning and killing of farmed Atlantic salmon. The EFSA Journal 2012, 1-77.

European Food Safety Authority (EFSA), 2009b. Scientific Opinion of the Panel on Animal Health and Welfare on species specific welfare aspects of the main systems of stunning and killing of farmed rainbow trout. The EFSA Journal 1013, 1-55.

# History and procedural aspects of the animal welfare risk assessment at EFSA

**Oriol Ribó and Jordi Serratosa**
Animal Health and Welfare (AHAW) Unit, European Food Safety Authority (EFSA), Largo N. Palli, 5A, 43100 Parma, Italy; oriol.ribo@efsa.europa.eu

## Summary

EFSA was created in 2002 with the mission to provide scientific and technical support for the Community legislation in all fields with direct or indirect impact on food and feed safety, plant health, environment and animal health and welfare. When providing independent science-based advice, the risk assessment (RA) approach should be followed, whenever possible. This contribution reviews the work done by the AHAW Panel of EFSA in the provision of scientific advice on the welfare of food producing animals and describes the progress of the methodology of animal welfare risk assessment (AWRA), starting from the scientific reports of the former Scientific Veterinary Committee (SVC) and the Scientific Committee on Animal Health and Welfare (SCAHAW). Since 2004, the AHAW Panel of EFSA has adopted 32 scientific opinions on animal welfare. Whereas the first scientific reports adopted by the SVC and SCAHAW did not include AWRA, in the first opinions adopted by EFSA (2003-2005) the potential hazards as well as the risk pathways were identified. A first formal (semi-quantitative) RA including hazard identification and characterisation, exposure assessment and risk estimation was conducted for the welfare of calves. Subsequently, the concepts of intensity and severity were introduced in the RA model used both in pig and fish welfare assessments. Finally, the opportunity to challenge the utility of the current AWRA model by feeding it with solid and more abundantly available scientific data was taken in the dairy cow welfare assessments. Despite the lack of scientific evidence in many welfare areas – which represents one of the main limitations of the methodology – the approach followed in these last RA's promises to represent a valid method for ranking risks and identifying areas of concern and future research. The results of the implementation of welfare monitoring plans at farm level should provide feedback on the reliability of this RA approach, leading to further improvements of the methodology.

**Keywords**: scientific opinion, food producing animals, hazard, adverse welfare effect, risk

## 1. Introduction: international state of play of the risk assessment in animal welfare

As stipulated in Regulation (EC) No 178/2002 (EC, 2002), the main mission of the European Food Safety Authority (EFSA) is to provide comprehensive scientific and technical support for the Community legislation on the safety and other aspects of the whole food and feed supply chain, including animal health and animal welfare. Accordingly, EFSA provides independent information regarding risks associated with food and feed, plant health, environment, animal

health and animal welfare by following, whenever possible, a Risk Assessment (RA) approach. Another task of the Authority is to promote and coordinate the development of harmonised risk assessment methodologies in the above mentioned fields.

The Treaty of Amsterdam (EU, 1997), through its Protocol on the Protection and Welfare of Animals, obliges the European Institutions to pay full regard to the welfare requirements when formulating and implementing European Union (EU) legislation. EFSA issues scientific opinions relying, whenever possible, on an RA approach, in response to questions posed by various European Commission services as well as by the European Parliament, and Members States or as a result of EFSA self-mandates.

The key area of action of the Community Action Plan on the Protection and Welfare of Animals [2006-2010] (EC, 2005) is to promote policies for high animal welfare standards in the EU and at international level. The introduction of standardised animal welfare indicators was identified as one of the main areas of action, together with the upgrading of the existing minimum standards for animal protection and welfare in line with new scientific evidence and socio-economic assessments. In this context, the future Strategy on Animal Health for 2007-2013 (EC, 2006) will address the promotion of farm practices that comply with animal welfare standards and are risk analysis-based.

The Animal Health and Welfare (AHAW) Panel of EFSA, during its XV Plenary Meeting (September 2005), recognised the lack of standardised guidelines and therefore the need to harmonise the risk assessment on animal welfare, and suggested EFSA to take the appropriate steps to incorporate this approach in its work program for 2006. In December 2005, an EFSA Scientific Colloquium 'Principles of Risk Assessment of Food Producing Animals: Current and future approaches' was held in Parma to discuss the state of art regarding RA of food producing animals (EFSA, 2006a).

The main objectives of the Scientific Colloquium were to have an open scientific debate on how to conduct RA (i.e. which components of RA of diseases in food animals and of animal welfare are essential) and to explore options for making guidelines in these areas available. During the colloquium, the RA methodology, the need for reliable data, the need for setting up data collection systems, ways to identify the necessary expertise and other resource requirements and the best options to communicate results of an RA were discussed. Methodological aspects of animal welfare risk assessment (AWRA) were equally debated, notably: (1) whether such RA exercises should exclusively relate to animal welfare or might also include elements related to animal disease, (2) how in this context the concept 'hazard' should be interpreted, and, (3) the need for criteria to identify parameters ('indicators') for measuring animal welfare specific for animal species, breed, age, physiological state and the currently practised production systems. In relation to the data needs, it was agreed that for animal welfare to be addressed appropriately, it should be measured directly on the animals, for which (currently unavailable) quantitative data would be necessary. To acquire such data, it was recommended to rely on the use of physiological parameters, rather than on observations related to health and behaviour. In addition – as an increasing proportion of the food of animal origin is produced outside the EU – the global food chain, including

domesticated and wild animals in various regions of the world outside the EU, might need to be considered when performing a risk assessment (EFSA, 2006a).

One of the main conclusions of the Colloquium highlighted that although different approaches exist for the RA related to food microbiological (WHO, 1999) and animal health issues (OIE, 2004a; OIE, 2004b), 'no specific standardized methodology exists in the field of the Animal Welfare Risk Assessment'. The colloquium recommended that EFSA should consider developing guidelines in this area and that it would be worthwhile to set up a working group to further investigate methodologies for risk assessment of animal welfare (EFSA, 2006a).

Following this recommendation, an EFSA workshop on 'RA methodology in Animal Welfare' was held in Vienna in June 2007, during which past and current experiences in the development of risk assessments in animal welfare were reviewed, with the final objective to identify flaws in the current AWRA methodology and options for its further improvement and development. Main problems discussed during the workshop were: (1) the use and validation of the expert opinion mainly relating to exposure assessment, (2) the interaction between hazards and their cumulative adverse effect on the animal, (3) the interaction between animal welfare, animal disease and food safety, (4) the possible consideration of the positive effects of factors ('welfare promoters') on animal welfare, and (5) how to deal with 'death' as an animal welfare issue. The workshop concluded that the use of the risk assessment approach for purposes of formulating scientific opinions on animal welfare should be considered wherever appropriate, and that a systematic AWRA methodology would increase the validity and reliability of the animal welfare assessments.

Since its creation in 2003, EFSA's AHAW Panel has adopted a total of 32 Scientific Opinions on animal welfare, dealing – among others – with laboratory animals, stunning and killing methods, animal transport, the welfare of calves, the welfare risks of the import of captive birds, the welfare of pigs, fish welfare, welfare aspects of fish stunning and killing and dairy cow welfare (Ribó *et al.*, 2009a). During the development of these Scientific Opinions, various AWRA approaches have been followed. Prior to the creation of EFSA, the Scientific Veterinary Committee (SVC) and the Scientific Committee on Animal Health and Animal Welfare (SCAHAW) of the General Directorate of Health and Consumer Protection (DG SANCO) were in charge of the provision of Scientific Opinions on Animal Health and Welfare issues.

This contribution reviews the work done by the Animal Health and Welfare (AHAW) Panel of EFSA in the provision of scientific advice on the welfare of [mainly] food producing animals and how the methodological approach has gradually improved in the process of formulating the various scientific opinions starting from those issued by SVC and SCAHAW.

## 2. Existing risk assessment methodologies and their adaptation to animal welfare

Animal welfare is nowadays a real concern and there is common agreement that standards for animal welfare and a reliable monitoring system for its assessment at all levels of the food chain need to be established. The main constraint in the development of an overall welfare

assessment is that some welfare aspects are not easily described in an objective way, either qualitatively or quantitatively (Müller-Graf *et al.*, 2008; Blokhuis *et al.*, 2008).

RA has been commonly used to describe and quantify the risks of introduction of infections, toxi-infections or residues of veterinary medicines or risks resulting from the import of live animals and their products in the animal food chain (Müller-Graf *et al.*, 2008), for which purpose specific international guidelines – produced by the World Health Organization (WHO, 1999) and by the World Organization for Animal Health (OIE, 2004a,b) – have become available. Hence, the RA terminology is well established, although there are minor differences in terms of definitions (Vose *et al.*, 2001). For AWRA purposes, OIE and Codex Alimentarius Commission (CAC, 2001; CAC, 2002) definitions have been considered in the EFSA scientific opinions on animal welfare (www.efsa.europa.eu). Ultimately, the CAC terminology was considered most appropriate and through replacing the term 'public health' by 'animal welfare' the latest CAC guidelines have been adapted to suit purposes of AWRA (see Smulders, 2009; Table 1).

The AWRA approach followed by EFSA has evolved consistently. In the first EFSA scientific opinions (EFSA, 2004a; EFSA, 2004b) RA was limited to the listing of hazards which may compromise animal welfare and the definition of risk pathways. Further developments in AWRA methodology included the estimation of the magnitude of the adverse effects (depending on their severity and duration) associated with the hazards and of the probability of occurrence of the latter in the animal population, which has allowed a more reliable calculation of risk estimates. Table 1 shows the animal welfare scientific opinions adopted since 2003 and the evolution of the RA methodology.

There is general agreement that the current (qualitative/semi-quantitative) AWRA approach, through the ranking of risk estimates and hazard magnitudes, allows risk managers to identify areas where intervention is timely and for determining which issues need to be further clarified through future research (Müller-Graf, *et al.*, 2008, Ribó *et al.*, 2008; Candiani *et al.*, 2009).

## 3. Scientific reports on animal welfare without a risk assessment approach (SCAHAW and SVC)

The former Scientific Veterinary Committee (SVC) and Scientific Committee on Animal Health and Welfare (SCAHAW) of the Directorate General of Agriculture of the European Commission adopted several reports on animal welfare between 1996 and 2002. In these reports the current literature was reviewed with special reference to housing systems and management procedures, which were directly related to housing, feeding and the maintenance of good welfare. Usually, preventive measures to control infectious disease were either not considered in these reports or only where this was pertinent to the comparison of the different housing and management systems (SVC, 1995; SVC, 1997). For instance, in the report on slaughter and killing of animals it was stated that no stunning method can fulfil all welfare requirements, so decisions will have to be made on the basis of the available scientific evidence. Consequently, a summary of animal welfare criteria and concerns associated with the available stunning methods was presented, based on the analysis of the available information (SVC, 1996a).

*Table 1. Evolution of the RA approach amongst the AHAW Opinions (2003-2009) [modified from Müller-Graf* et al.*, 2008]; HI =hazard identification; Qual RA: qualitative RA; Semi-Qt RA: semi-quantitative RA.*

| AHAW opinion on animal welfare | Year | RA | HI | Qual. RA | Semi-Qt RA |
|---|---|---|---|---|---|
| Welfare of animals during transport | 2004 | - | - | - | - |
| Welfare aspects of various systems of keeping laying hens | 2004 | - | x | x | - |
| Impact of the current housing and husbandry systems on the health and welfare of farmed domestic rabbits | 2005 | - | x | x | - |
| Welfare of weaners and rearing pigs: effects of different space allowances and floor types | 2005 | - | x | x | - |
| Biology and welfare of animals used for experimental and other scientific purposes | 2005 | - | x | x | - |
| Welfare aspects of the main systems of stunning and killing applied to commercially farmed deer, goats, rabbits, ostriches, ducks, geese and quail | 2006 | - | x | x | - |
| The risks of poor welfare in intensive calf farming systems | 2006 | x | x | - | x |
| Animal health and welfare risks associated with the import of wild birds other than poultry into the European Union | 2006 | x | x | - | x |
| Welfare of pigs (Sows and boars, Fattening pigs and Tail biting) (3 scientific opinions) | 2007 | x | x | - | x |
| Stunning and killing methods for seals | 2007 | x | x | x | - |
| Welfare of fish (Salmon, Trout, Eel, Sea Bass-Sea Bream, Carp) (5 scientific opinions) | 2008 | x | x | - | x |
| Stunning and killing methods of fish (Salmon, Trout, Eel, Sea Bass-Sea Bream, Carp, Turbot and Tuna) (7 scientific opinions) | 2009 | x | x | - | x |
| Welfare assessment of Dairy Cows Welfare (Leg and locomotion, udder, metabolic and reproductive and behaviour) (5 SOs) | 2009 | x | x | - | x |

In each of these reports, conclusions and recommendations for AW improvement are predominantly based on the current state of knowledge and practical experience in the area under scrutiny and all aim to identify the main AW concerns within the EU (SVC, 1996b).

No systematic approach was followed by the SVC to produce these conclusions and associated recommendations, which were mainly based on the interpretation of the available scientific information and not on a systematic identification of hazards. As a result the repeatability and reliability of the analysis could be challenged. For risk managers this represents a major problem, when trying to identify the major areas of concern for formulating legislation and corrective measures.

# 4. Scientific opinions on animal welfare with some risk assessment elements (EFSA 2003-2006)

## 4.1 Scientific opinions with identification of hazards

### *4.1.1 Welfare aspects of the main systems of stunning and killing animals (EFSA, 2004b, 2006b)*

The AHAW Panel adopted two scientific opinions on the welfare aspects of the main systems of stunning and killing animals, with consideration of Directive 93/119/EC. The first opinion, adopted on the 15th of June 2004 (EFSA, 2004b), considered those welfare aspects pertaining to the major commercial species (cattle, sheep, pigs, poultry, horses and farmed fish), whilst the second, adopted on 13th February 2006 (EFSA, 2006b), dealt with other species such as deer, goats, rabbits, ostriches, ducks, geese and quail. For the latter species, it was clearly stated that published scientific information was very scarce, identifying an urgent need for further research.

The opinions focused on the welfare at the point of application of the main stunning and stun/killing methods under commercial slaughterhouse and farm conditions in Europe (Table 2).

Killing of animals without stunning, and stun/killing methods for disease control were also considered. The main objective of these scientific opinions was to recommend procedures appropriate for the species and related minimum requirements such that unconsciousness and insensibility was induced without causing avoidable pain, suffering and distress. Consequently, methods were described together with the principles of functioning of each stunning and stun/killing option. This detailed description allowed (1) the identification

*Table 2. Stunning and stun/killing methods used under commercial conditions (EFSA, 2004b).*

| Species of animals | Stunning methods | Stun/kill methods |
|---|---|---|
| Cattle and calves | Mechanical<br>Electrical | Electrical |
| Sheep | Mechanical<br>Electrical | Electrical |
| Pigs | Mechanical<br>Electrical<br>Gas mixtures | Electrical<br>Gas mixtures |
| Poultry | Electrical | Mechanical<br>Electrical<br>Gas mixtures |
| Horses | | [ |
| Fish | Electrical | Percussive<br>Mechanically applied spiking<br>Electrical<br>Shooting/electric harpoon (large tuna) |

of possible hazards affecting the proper application (often similar for many species) of the stunning method, (2) the listing of advantages and disadvantages of each method, and (3) the identification of some points at which animal welfare could be monitored.

Conclusions of both scientific opinions were drawn considering the different steps to stun or stun/kill an animal and applying the principle that a stun/killing method should induce unconsciousness and death either simultaneously or sequentially. Recommendations about the most appropriate procedures and their related minimum requirements to induce prompt unconsciousness and insensibility and minimise poor welfare were outlined. For example, – as the duration of unconsciousness and insensibility varies among methods, species and animals and unless sticking is carried out promptly, the stunned animal has the potential to regain consciousness – it was recommended that the stun-stick interval should be sufficiently short to induce death through blood loss before the animal recovers.

Anil and Lambooij (2009) and Gerritzen and Raj (2009) provide more detail on animal welfare aspects of stunning and killing.

*4.1.2. Welfare of animals during transport*

Although the EFSA Scientific Opinion adopted on the 30$^{th}$ of March 2004 (EFSA, 2004a), considered the same aspects as a previous SCAHAW Opinion (SCAHAW, 2002), it related to the transport of other animal species. It was emphasized that the welfare of animals can be substantially affected during and after transport, the effect being dependent of the species and transport characteristics.

The steps during transport include pre-transport preparation of animals and vehicles, planning for emergencies, inspection, loading, providing good vehicle conditions, monitoring of animals and conditions, and unloading with subsequent handling. Welfare consequences of animal handling and transport include, for instance, increased mortality, morbidity or other disturbance in the transported animals or increased disease incidence in other animals. In addition, when evaluating the impact of the transport on the animals, a variety of stressors strongly contribute to poor welfare in transported animals, adding to the increase in susceptibility to infection. To identify all hazards which may contribute to stress, four different transport stages or conditions were identified: loading/unloading, journey duration, trained operators and mixing of animals during and after transport as a source for disease transmission (Table 3).

These main hazards were analysed to allow conclusions to be drawn and recommendations to be made. In addition to those included in Table 3, the scientific opinion gives special recommendations for the transport of fish, which normally should be loaded and unloaded without getting into contact with air, should be provided with high oxygenated water and maintained at an adequate stocking density. The scientific opinion finally recommended that, in general, animals unfit to travel because of injury, disease or distress, or which might produce an egg during the transport period or in the case of mammals in the last five days of pregnancy, should not be transported.

*Table 3. Main hazards affecting animal welfare during transport (EFSA, 2004a).*

| | |
|---|---|
| Loading and unloading | Loading of unaccustomed animals |
| | Loaded/unloaded not carefully |
| | Forced to move up or down ramps which are too steep (>20°) |
| | Slippery ramps |
| | Loading a high density of animals |
| Journey duration | Too long journeys |
| | Inadequate roof height to meet needs |
| | Kept at a higher stocking density |
| | Inadequately fed, watered and rested (lack of normal movements) |
| | Bad driving (avoid keep balance, lateral acceleration, sudden braking) |
| | Mix of animals that may fight (without pens) |
| | Inadequate temperature and humidity of the air, air quality, ventilation and insulation of the surrounding walls, floor and roof. |
| Trained operators | Lack of proper training |
| | Lack of route plan |
| | Lack of information about whom to contact in emergency |
| | Absence of animal inspection before the journey |
| | Absence of animal inspection at adequate intervals |
| | Use of behaviour modifying drugs |
| Disease transmission and mixing of animals | Lack of clinical inspection before transport |
| | Inadequate biosecurity measures |
| | Lack/inadequate cleaning and disinfection of vehicles and equipment |
| | Facilitate contact between transported animals and other farm animals |
| | Avoid quarantine or isolation periods on farms after transport |
| | Use of markets for slaughter animals |
| | Transporting animals with different health status together |

Hartung and Springorum (2009) deal in more detail with animal welfare effects of transport.

## 4.2 Scientific opinions with identification of hazards and risk profiles

### *4.2.1 Health and welfare of farmed domestic rabbits*

The Scientific Opinion on 'The Impact of the current housing and husbandry systems on the health and welfare of farmed domestic rabbits' was adopted by the AHAW Panel on the 14$^{th}$ of September 2005 (EFSA, 2005a). During its development an attempt was made to carry out a risk assessment by relying on defining so-called 'risk profiles'. The identified risk profiles for disease and husbandry practices allowed drawing associated conclusions and recommendations included in the scientific opinion.

The existing general information on the recognition and assessment of good and poor welfare in rabbits was considered together with the knowledge about the biology and behaviour

of rabbits that may enable or disable them to respond and to adapt to their environmental circumstances. Subsequently, systems of rabbit production were described. This approach formed the basis for assessing the risk for poor rabbit welfare under specific farming practices in the following areas: housing and husbandry, nutritional aspects, reproductive practices, weaning practices and opportunities for the use of genetic selection that may make strains more susceptible or might improve welfare. The risk questions posed for each area are shown in Table 4.

For each risk question, poor welfare circumstances were recognised and their detection and incidence assessed. Whenever possible the factors affecting the exposure incidence of good welfare was also assessed and the factors affecting exposure in each situation were described. The consequences of the identified risks for animal welfare and the good agricultural practices preventing these risks were taken as a basis for formulating conclusions and recommendations. A major conclusion was that, even though the description of risk profiles was a relatively new way to assess animal welfare issues, this approach could be helpful to understand some of the critical hazards that may have an impact on rabbits' health and welfare.

### *4.2.2 Welfare of animals used for experimental purposes*

The Scientific Opinion on 'Aspects of the biology and welfare of animals used for experimental and other scientific purposes' was adopted by the AHAW Panel on the 14th of November 2005 (EFSA, 2005b). The Commission's mandate included several questions to be answered, i.e.: (1) the sentience and capacity to experience pain, suffering and distress of all invertebrate

*Table 4. Risk questions for each housing and management issue in rabbits (EFSA, 2005a).*

| Housing and management issue | Risk questions |
|---|---|
| Housing | The impact of flooring on welfare |
| | The impact of restricted space and social isolation on welfare |
| | The height of growers' cages |
| | The available surface area of growers' cages |
| Nutrition | Impact of poor nutrition on welfare and health, |
| | Impact of dietary toxins on health and welfare |
| | Impact of feed additives on health and welfare |
| Reproduction strategies | Is welfare jeopardised by intensive breeding systems? |
| | Is welfare affected by hormone treatments? |
| | Is poor welfare associated with artificial insemination? |
| | Is poor welfare caused by biostimulation? |
| Weaning practices | What is the impact of poor nursing and inadequate milk supply on the welfare? |
| | What is the impact of weaning on the welfare of the kits? |
| | What contribution does abnormal maternal behaviour make to poor welfare of kits? |
| Genetic selection | What are the welfare consequences in the selection of future breeders? |

*Table 7. Hazards which may have an impact on animal welfare or on the scientific quality of research when using non-purpose bred animals for experimental purposes (EFSA, 2005b).*

| |
|---|
| Impact on animal welfare |
| Poor environment |
| Exposure to inappropriate temperatures |
| Exposure to inappropriate Relative Humidity |
| Inadequate ventilation |
| Exposure to inappropriate noise |
| Failure to provide appropriate nesting/bedding materials |
| Failure to provide appropriate lighting |
| Failure to provide suitable social and environmental enrichment |
| Inadequate breeding management |
| Failure to implement suitable breeding strategies |
| Inadequate health management |
| Failure to maintain an appropriate health management programme |
| Unsuitable diet |
| Failure to provide an appropriate diet for the animals |
| Insufficient and inadequately trained staff |
| Genetically altered animals (GAA) |
| Lack of standardisation on animals used (e.g. health, genetic fidelity, microbial status, nutrition, environment) leading to an increase in the number of animals (see below). |
| Prolonged journey times |
| Capture in the wild |
| Overproduction of animals |
| Over exploitation of breeding animals and confinement for long periods |
| Impact on scientific quality |
| Lack of standardisation on animals used (e.g. health, genetic fidelity, microbial status, nutrition and environment, wild animals). |
| Insufficient animals of suitable quality |
| Extrapolation of data |

## 5. First attempts to conduct formal (semi-) quantitative AWRA

### 5.1 The calves' welfare approach

The European commission requested EFSA to issue a scientific opinion on the animal health and welfare aspects of intensive calf farming systems, including also an update of the Report on the welfare of calves of the 9th of November 1995 (SVC, 1995) of the Scientific Veterinary Committee (Animal Welfare Section), in light of more recent data on this issue. The AHAW Panel of EFSA adopted on the 24th of May 2006 the Scientific Opinion on 'The risks of poor welfare in intensive calf farming systems' (EFSA, 2006c).

This was EFSA's first scientific opinion on animal welfare for which a systematic RA methodology was followed. To this end, the definitions of the Codex Alimentarius Commission (CAC, 2002) – developed to assess microbiological risks for food safety – were adapted to the animal welfare field. To avoid overlooking any factor that would compromise the welfare of calves, the needs of calves were firstly identified and described, whereafter it was determined if factors (potential hazards) related to nutrition, housing and management as prevalent in various – separately considered – husbandry systems (see Table 8) would indeed compromise these needs. A total of 36 such hazards were identified and their probability of occurrence in the entire population assessed. The methodological specifics are described by Smulders (2009).

To ensure that the exposure estimates of the working group members indeed reflected the on-farm situations encountered in the various European calf production systems, a group of 'consultants' were asked to challenge these estimates. Consultants were veterinarians, and all experts in clinical practice in calf production (See also Algers, 2009). Where the exposure estimates of working group members and the consultants did not correspond, the opinion of the latter was interpreted to represent the factual. The practical experience of consultants proved to be very useful to more reliably determine exposure rates, which are rarely reported in literature (see Table 9).

Risks were semi-quantitatively estimated, i.e. both hazard characterisation (in terms of severity) and exposure assessment were expressed in numerical terms to allow calculating a risk estimate by multiplication of these scores. Ultimately, on the basis of these numerical scores, risks were classified as 'major', 'minor' and 'negligible' enabling their ('qualitative') ranking. In some cases, i.e. where the exposure could not be estimated due to lack of data the risks were labeled 'uncertain' (see Table 9).

A summary of the methodological aspects (and an analysis of its weaknesses and limitations) can be found on pages 242-244 of this Volume.

*Table 8. Calves farming systems considered in the Risk Assessment (EFSA, 2006c).*

| Animal type | Calves' farming systems and rearing categories |
|---|---|
| Veal | A.a. White veal in small groups, bucket fed (i.e. not suckling) |
| | A.b. White veal in larger groups, automatic feeding system (not suckling) |
| | A.c. Pink veal in small groups, bucket fed + solid foods, not suckling |
| | A.d. White veal in small groups, suckling |
| Replacement dairy calves | B.a. Small groups, bucket fed (not suckling)+ solid foods, weaned at 2-3 months |
| | B.b. Groups with an automatic feeding system (not suckling) + solid foods, weaned at 2-3 months |
| | B.c. Feed lots (high density groups within outside pens) |
| | B.d. Hutch outside, bucket fed (not suckling)+solid foods, weaned at 2-3 months |
| Beef calves | C.a. Suckler calves in small groups kept inside, led twice a day to the dam for suckling up to 6-9 months |

*Table 9. Major (M), minor (m) and negligible risks (empty cell) for the welfare of calves, held in various husbandry systems [indicated in Table 8] (EFSA, 2006c); L = lack of data.*

| Hazards | Husbandry systems in calves | | | | | | | | |
|---|---|---|---|---|---|---|---|---|---|
| | Aa | Ab | Ac | Ad | Ba | Bb | Bc | Bd | Ca |
| Inadequate colostrum intake – quantity | M | m | M | m | m | m | m | m | m |
| Inadequate colostrum intake – quality | M | m | M | m | m | m | m | m | m |
| Inadequate colostrum intake – duration | M | M | M | | M | M | M | M | |
| Iron deficiency resulting in haemoglobin levels below 4.5 mmol/l | L | L | | | | | | | |
| Insufficient access to water | M | m | | m | | | | | |
| Allergenic proteins | L | L | L | L | L | L | L | L | L |
| Insufficient balanced solid food | M | m | | m | | | | | m |
| Overfeeding (too rich diet) | L | L | L | L | L | L | L | L | L |
| High humidity | M | m | M | m | | | | | |
| Indoor draughts | M | m | M | m | | | | | |
| Inadequate ventilation, inappropriate airflow, airspeed, temperature | M | M | M | M | M | M | | | m |
| Poor air quality (ammonia, bio-aerosols and dust)) | | m | | | | m | | | |
| Poor floor conditions | | | | | | | | | |
| Gap too large; | M | m | M | m | m | m | | | m |
| Too slippery; | M | m | M | m | | | | | |
| Wet floor for lying; | M | m | M | m | | | | | |
| No bedding | M | m | M | m | | | | | |
| Insufficient floor space allowance | L | L | L | L | | | | | |
| Insufficient light for response to visual stimuli | M | m | M | m | | | | | |
| Exposure to pathogens causing respiratory and gastrointestinal disorders | M | M | M | m | m | m | M | | m |
| Inadequate clinical health monitoring | L | L | L | L | L | L | L | L | L |
| Inadequate haemoglobin monitoring | L | L | L | L | | | | | |
| Continuous restocking no 'all in - all out' policy | | | | | M | M | | | |
| Poor response of farmer to health problems, especially necessary dietary changes | M | m | M | m | | | | | |
| Lack of maternal care | M | m | M | | m | m | m | m | |
| Mixing calves from different sources | M | M | M | | | | M | | |
| Separation from the dam | M | m | M | | m | m | m | m | |

The biggest improvement in comparison to former RA exercises is that this approach allowed for a ranking of the various risks and hence should provide the basis for risk managers to decide on the elimination or reduction of which risks one should first concentrate. It should be noted, however, that the approach did not allow for discriminating whether a particular hazard prevails for a short or rather a long period or for how long the entire population is

subjected to the hazard. For this reason the concepts of 'duration' of the adverse effect (part of hazard characterisation) and the 'intensity' of exposure to the hazard (part of exposure assessment) were introduced for consideration in subsequent RA exercises (see below).

## 5.2 Integration of the duration and severity/intensity concepts

### *5.2.1 Pig welfare*

The European Commission was requested to submit a report to the Council concerning various aspects of housing and husbandry systems for farmed pigs in agreement with Council Directives 2001/88/EC[1] and 91/630/EEC[2] laying down minimum standards for the protection of pigs. EFSA was requested to submit a scientific opinion concerning animal health and welfare aspects of different housing and husbandry systems for adult breeding boars, farrowing and pregnant sows and farmed fattening pigs, which should also address the risks associated with pig tail biting and possible means to reduce the need for tail docking considering the various housing and husbandry systems. Three scientific opinions on adult breeding boars, farrowing and pregnant sows (EFSA, 2007b), farmed fattening pigs (EFSA, 2007a) and pig tail biting (EFSA, 2007c) were adopted by the AHAW Panel of EFSA.

The first step was to identify the target populations to be considered, depending on the animal categories (and life stages) and the different husbandry systems (Table 10).

*Table 10. Target populations considered for the risk assessment on pig welfare (EFSA, 2007a,b,c).*

| Animal category | Life stages and husbandry systems |
|---|---|
| Sows and Boars | Dry sows – from weaning to 4 weeks of service |
| | Pregnant sows (from 4 weeks after service) |
| | Farrowing sows |
| | Boars |
| | Piglets (up to weaning) |
| Fattening | Weaning (up to 10 weeks) indoor |
| | Weaning (up to 10 weeks) outdoor |
| | Growing (from 10 weeks onwards) indoor |
| | Growing (from 10 weeks onwards) outdoor |
| | Fattening pigs (over 110 kg) |
| Tail biting | Docked pigs |
| | Undocked pigs |

[1] E.C.O.J. n L316 of 1/12/2001. p. 1.

[2] E.C.O.J. n L340 of 11/12/1991. p. 33.

Similar as in the previous RA, those factors (hazards) which may compromise pigs' needs were identified separately for each population (Candiani *et al.*, 2007). For each hazard the consequent impact on animal welfare was reviewed and the adverse effect on the animal defined (Table 11). Hazards resulting in different adverse effects were considered as different hazards and scored separately. The methodology followed and its weaknesses and limitations are described on pages 244-247 of this Volume.

Besides graphically expressing the risk estimate (denoting the probability that the hazard prevails and is indeed relevant for the entire population), the magnitude of the adverse effect of each particular hazard (i.e. relevant for the individual animal) was included in the same graph. Figure 1 shows the risk estimates (black bar) and the magnitude of the adverse effect (white bar) for each considered hazard. It should be pointed out that this methodology does not allow assessing adverse effects of hazards acting in concert for which the synergistic effects cannot be estimated. The value of including such graphics in AWRA reports is further discussed by Smulders (2009).

Hazards have been ranked by decreasing risk estimate scores, which allows distinguishing those hazards with a high probability of occurrence in the animal population but with a medium or minor hazard magnitude from those that are less likely to occur but which have a high hazard magnitude. At this point it is up to the risk managers to make the decision on which of these to focus their attention.

*5.2.2 Fish Welfare*

EFSA was requested by the European Commission to assess the welfare aspects of husbandry systems for the main farmed fish species in the EU. Due to the great diversity of species it was proposed that separate scientific opinions on species or sets of similar species would be more adequate and effective. Therefore, the AHAW Panel adopted in 2008 five species-specific opinions on farmed Atlantic salmon, Trout, European eel, European Seabass and Gilthead Seabream, and Common carp (EFSA, 2008 a,b,c,d,e). For each species, different life stage and production systems (Table 12), and potential risks for fish welfare were identified including environmental conditions, feeding, husbandry practices, genetics and disease. A

*Table 11. Examples of hazards related to animal needs with related adverse effects (Candiani* et al., *2009).*

| Need | Hazards | Adverse effect |
|---|---|---|
| Nutrition: | Difficult access to water | Thirst |
| to drink, to thermoregulate | Insufficient feed | Hunger |
| | Too low milk | Stress, anxiety |
| Housing: | Sliding floors | Lameness |
| to rest, to exercise | Inappropriate ventilation | Pain, malaise |
| Management: | Staff without experience | Stereotypes |
| to avoid fear, to have proper social interactions | Mixing of unfamiliar animals | Fear |
| | | Stress |

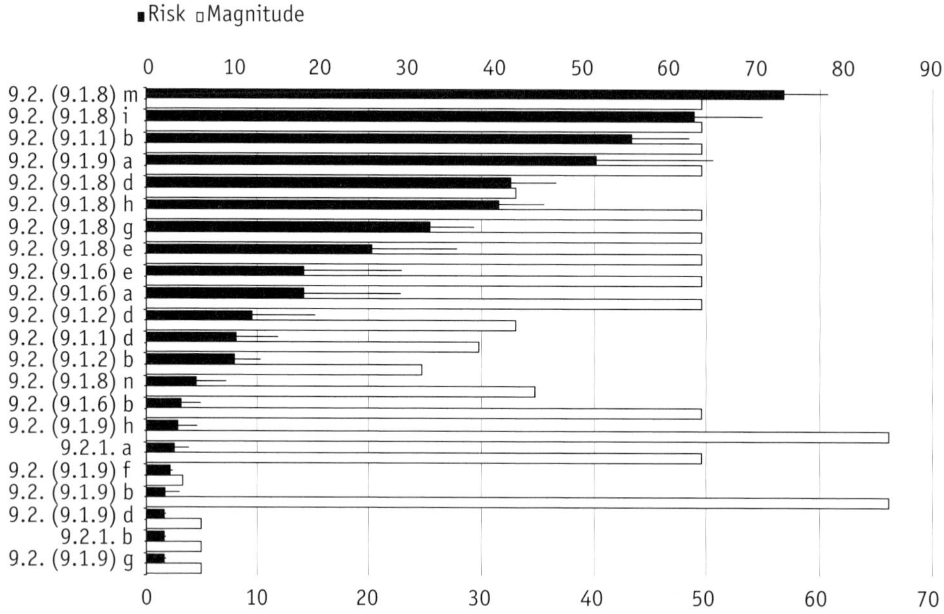


*Figure 1. Risk estimate, and magnitude of the adverse effect of hazards affecting pregnant sows (adapted from EFSA, 2007b).*

*Table 12. Production systems by life stages in Salmon (EFSA, 2008a).*

| Production systems | Life stages | | | | | | |
|---|---|---|---|---|---|---|---|
| | Egg | Alevin | Fry | Parr | Smolt | Ongrowing | Broodstock |
| Trays | X | X | | | | | |
| Cylinders | X | | | | | | |
| Tanks (recirculated) | | | X | X | X | | |
| Tanks (flowthrough) | | | X | X | X | | X |
| Freshwater cages | | | | X | X | | |
| Sea-water cages | | | | | | X | X |

description of the methodological approach including its weaknesses and limitations can be found on pages 249-252 of this Volume.

*5.2.3 Welfare assessment in case of impaired welfare: stunning and killing of animals*

*Animal Welfare aspects of the killing and skinning of seals.* After the recommendation from the Council of Europe on seal hunting, inviting to ban all cruel seal hunting methods, the

European Parliament adopted a written declaration requesting the Commission to draft a regulation to ban the import, export and sale of seal products. EFSA was then requested to issue a scientific opinion on this matter, as well so as to assess the most appropriate killing methods which reduce unnecessary pain, distress and suffering. The scientific opinion (EFSA, 2007d), adopted by the AHAW Panel on the 6th December of 2007, concluded that many seals can be, and indeed are, killed rapidly and effectively without causing avoidable pain, distress, fear and other forms of suffering. However, there is strong evidence that, in practice, effective killing is not always achieved although the degree to which this occurs has been difficult to assess.

The purpose of the qualitative risk assessment was to identify the hazards which may prevail when ineffective stunning and killing occurs and focused on the adverse animal effects that can take place during stunning, killing, monitoring for life and consciousness, bleeding-out, and skinning. Different pathways were identified, one for hunting using netting and one for stunning using physical methods and firearms. Figure 2 shows the steps in the risk pathway during killing and skinning using physical methods and firearms.

It should be pointed out that re-stunning can occur at different positions in the procedure if seals recover consciousness. Once the risk pathways were defined, depending on the

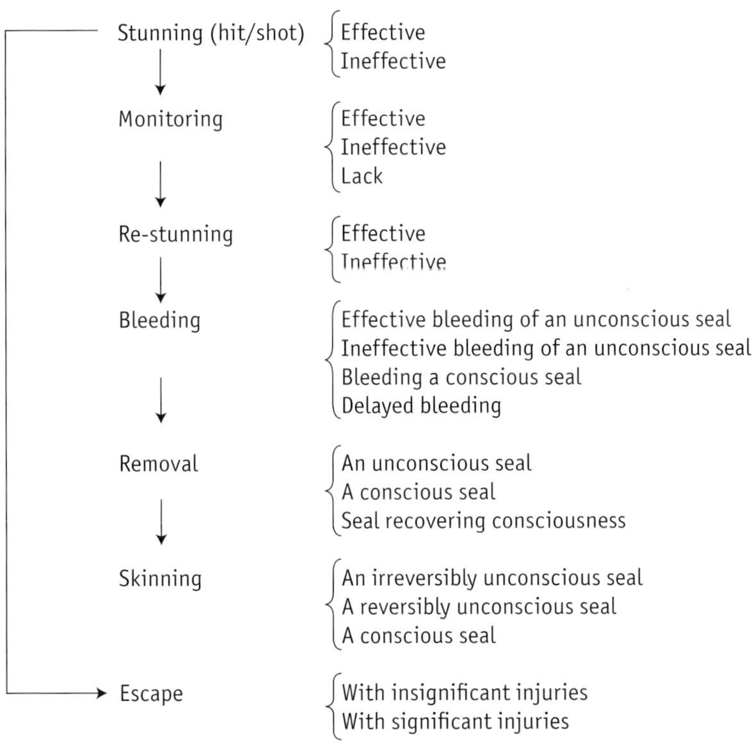


*Figure 2. Steps in the risk pathway for Risk Assessment for poor animal welfare during killing and skinning using physical methods and firearms (EFSA, 2007d).*

stunning and killing method used, 25 hazard outcomes in the stunning, killing and skinning of seals were identified (Table 13) and divided into categories corresponding to the different steps in the killing and skinning of seals. For example, after an ineffective hit or shot, seals have to be hit or shot again resulting in avoidable pain, distress, fear and other forms of suffering. Different weather and habitat conditions might result in different frequencies of the hazard outcomes to occur. Therefore, the likelihood of experiencing an adverse effect when being exposed to the hazards was estimated for two weather conditions (good and bad) in combination with two habitat conditions [good and bad]. It was assumed that the duration and intensity of the adverse welfare effect was the same for all these conditions.

Four categories were used for hazard characterisation (duration and intensity), exposure assessment and data source. As the duration of adverse effects might vary considerably depending on the hunting methods, two scoring scales for duration were defined (one for

*Table 13. List of potential hazards and possible outcomes considered in the Risk Assessment for adverse welfare effects of seals during killing and skinning (EFSA, 2007d).*

| Hunting method | Hazard | Outcome | |
|---|---|---|---|
| | | No. | Description |
| Netting: underwater seal net | Trapping | 1 | Trapping leading to suffocation without injuries |
| | | 2 | Trapping leading to suffocation with major injuries |
| | Escaping | 3 | Escapes with minor injuries |
| | | 4 | Escapes with major injuries affecting the function |
| | | 5 | Escapes with major injuries or lead to physical injury affecting the function in a long term |
| Physical methods: Hakapik, slagkrok and club | Stunning | 6 | Effective hit or series of rapid hits without delay leading to irreversible unconsciousness or death |
| | | 7 | Ineffective hit/s causing injury in a conscious seal or seal that regains consciousness |
| | Re-stunning | 8 | Effective hit or series of rapid hits without delay leading to irreversible unconsciousness or death |
| | | 9 | Ineffective hit/s causing injury in a conscious seal or seal that regains consciousness |
| Firearms: shotgun (slug) and rifle followed by physical methods | Stunning | 10 | Effective shot, e.g. brain or upper neck and possible use of a physical method to ensure death |
| | | 11 | Ineffective shot(s) without loss of consciousness or regains consciousness and without being bled |
| | Re-stunning | 12 | Effective shot, e.g. brain or upper neck and possible use of physical method to ensure death |
| | | 13 | Ineffective shot(s) without loss of consciousness or regains consciousness and without being bled |

*Table 13. Continued.*

| Hunting method | Hazard | Outcome | |
|---|---|---|---|
| | | No. | Description |
| Common hazards for physical methods and firearms | Monitoring | 14 | Effective monitoring (sealer/inspector checking for death or for irreversible unconsciousness) |
| | | 15 | Ineffective monitoring that does not ensure death or irreversible unconsciousness |
| | Bleeding | 16 | Effective bleeding of an unconscious seal |
| | | 17 | Ineffective/no attempt to bleed-out a seal to prevent recovery of consciousness and to ensure death |
| | | 18 | Bleeding-out of a conscious seal or one that regains consciousness during bleeding-out causing death |
| | Removal (hooking, dragging, etc) | 19 | Removal of an irreversibly unconscious or dead seal |
| | | 20 | Removal of a conscious seal or one that regains consciousness (bled or unbled) |
| | Skinning | 21 | Skinning an irreversibly unconscious or dead seal |
| | | 22 | Skinning a conscious seal or a reversibly unconscious seal that recovers consciousness |
| | Escaping | 23 | Struck, killed and lost |
| | | 24 | Escape after hit, without significant injury that could affect function or cause pain |
| | | 25 | Escape after hit, with significant injury that affects function and may cause significant pain |

netting and one for physical methods and firearms). The magnitude of an adverse welfare effect was defined by combining the duration and intensity of the adverse welfare effect. An example of the results of the hazard characterisation and exposure assessment performed by the working group is illustrated in Figure 3 (Hakapik).

Scores were agreed upon by the working group experts. However, in cases where agreement was not reached, the number of experts giving each qualitative score was indicated. Furthermore, for a number of hazard outcomes the experts could only agree upon a range of scores instead of one specific value for intensity, duration (the combination is the magnitude) and/or the likelihood. Therefore, two different evaluations appear in the Figures and for each hazard outcome, the lower bar indicates the lowest likelihood combined with the smallest magnitude and the upper bar indicates the largest likelihood combined with the largest magnitude (see Figure 3).

The first conclusion of the scientific opinion states very clearly that only a very limited number of studies published in peer-reviewed journals are available, which could be used to evaluate – with a high degree of certainty and quantitatively – the efficacy of the various

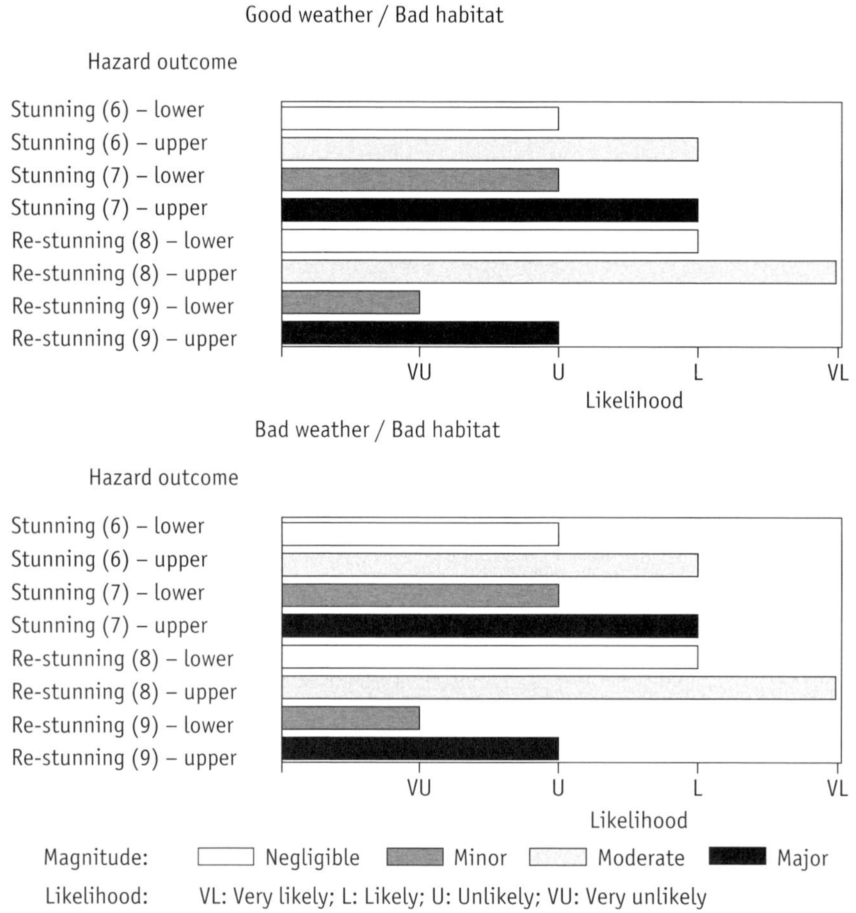


*Figure 3. Magnitude of adverse welfare effects and likelihood that the seal is exposed to the hakapik stunning hazards (EFSA, 2007d).*

killing methods used in seal hunts around the world (EFSA, 2007d). Consequently, this RA, largely based on expert opinion, must be classified as qualitative.

*Welfare aspects of the main systems of stunning and killing of farmed fish.* Recently, the AHAW Panel of EFSA adopted 7 species-specific opinions on the welfare aspects of stunning and killing methods for farmed fish. The individual scientific opinions concerned bluefin tuna, common carp, European eel, Atlantic salmon, rainbow trout, European turbot, European seabass and gilthead seabream (EFSA 2009 a,b,c,d,e,f,g). A semi-quantitative risk assessment approach mainly based on expert opinion was used to assess the risk to welfare of farmed fish at the time of killing and to rank the risks of poor welfare associated with the different commercially applied stunning and killing methods for farmed fish.

To assess the importance of a hazard three parameters were scored: the intensity and duration of the adverse effect and the probability of exposure to the hazard. The population in question was the fish killed in the EU by the selected method of stunning and slaughter. The probability of exposure to the hazard corresponds to the percentage of all fish exposed to the hazard (i.e. if 4% of the fish killed by a method are exposed to a hazard, the probability that any randomly selected fish within that population is exposed is 0.04). The consequence of the exposure can be assessed by scoring the intensity and the duration of the adverse effect in the individual. In addition, the risk assessment was based on the assumptions that all fish exposed to the hazard experience the adverse effect and that, if there is no evidence to the contrary, all fish exposed to the hazard experienced the same intensity and duration of the adverse effect.

Based on a review of the literature and field observations, the hazard identification identified potential welfare hazards associated with each stunning and killing method. The relationship between the time from applying a stun method, unconsciousness and the point at which the killing method was applied are illustrated graphically (Figure 4) and various scenarios (A to E) in which hazards may arise were identified as follows: 'A' where a fish is killed in some potentially painful way (asphyxia, bleeding out) while it is conscious; 'B' represents a fish that has been stunned and is killed or it dies after it is unconscious; 'C' where a fish has been stunned but it regains consciousness and is killed in some potentially painful way (asphyxia, bleeding out). 'D' represents a fish that, like B, is killed in some potentially painful way (asphyxia, bleeding out) while it is conscious, but has also suffered from the aversive nature of the stunning method; and 'E' represents a fish that has been stunned and is killed or it dies after it is unconscious but has also suffered from the aversive nature of the stunning method. For each method of stunning and killing, a risk pathway was described and the hazards which may be present in each of the defined scenarios in Figure 4 were listed.

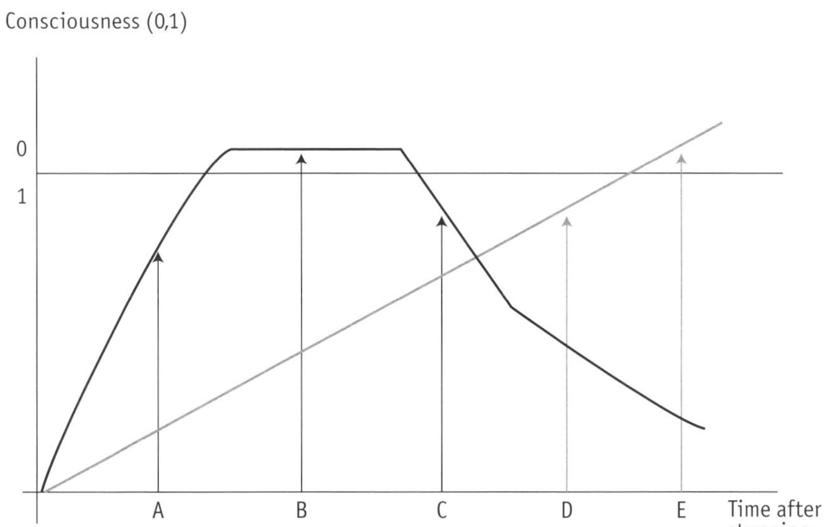


*Figure 4. Time to unconsciousness (insensibility) following stunning/killing (EFSA, 2009a,b,c,d,e,f,g,).*

For each hazard, severity and duration of the adverse effect as well as the probability of exposure were scored. Generic guidelines for defining intensity categories for pre-slaughter hazards and slaughter hazards were given. Mild and severe categories were defined and characterised while the moderate is defined as being neither mild nor severe. The time during which an animal will, on average, experience the adverse effect was estimated in minutes. The duration of an adverse effect can be longer than the duration of the hazard, for example a miss-stun takes a fraction of a second but the adverse effect lasts until the animal is unconscious or dies. Thus the duration of the hazard is included in the duration of the adverse effect. The exposure assessment was performed by assessing the proportion of the population of interest (i.e. all fish in the EU being killed by the method in question) that is likely to experience the hazard, which may vary depending on the farm of origin and slaughterhouse. Estimates of the most likely, maximum and minimum values for this proportion are required.

All three parameters (probability of exposure to the hazard; intensity of adverse effect and duration of adverse effect), were included in calculating the final risk estimate of a hazard. The magnitude provides an indication of the impact of the hazard on the fish, which are exposed to the hazard and experience the adverse effect. Thus a hazard that causes a prolonged and severe adverse effect but which affects only a small proportion of the population will have a low risk estimate but a high magnitude of severity score (Table 14 and Figure 5). The converse is also true.

*Table 14. Risk and magnitude scores for welfare hazards associated with electrical stunning of carp in Europe (adapted from EFSA, 2009c).*

| Hazard ID | Slaughter hazards | Description of adverse effects | Risk estimate | Magnitude |
|---|---|---|---|---|
| 10 | drop of fish into tank with water | stress, | 16.67 | 17 |
| 11 | new adverse environment (when transferred from lairage tank to stunner) | stress, aversion | 16.67 | 17 |
| 12 | insufficient current/voltage leading to miss-stun (normal time) | pain, distress, broken bones, muscle bleeding | 5.00 | 100 |
| 13 | prolonged exposure to insufficient current/voltage (delayed stun) | pain, distress, broken bones, muscle bleeding, exhaustion | 47.50 | 50 |
| 14 | further handling and processing after miss-stun; still conscious | severe trauma, pain, stress | 2.00 | 100 |
| 15 | further handling and processing after stun; regained consciousness | severe trauma, pain, stress | 47.50 | 100 |

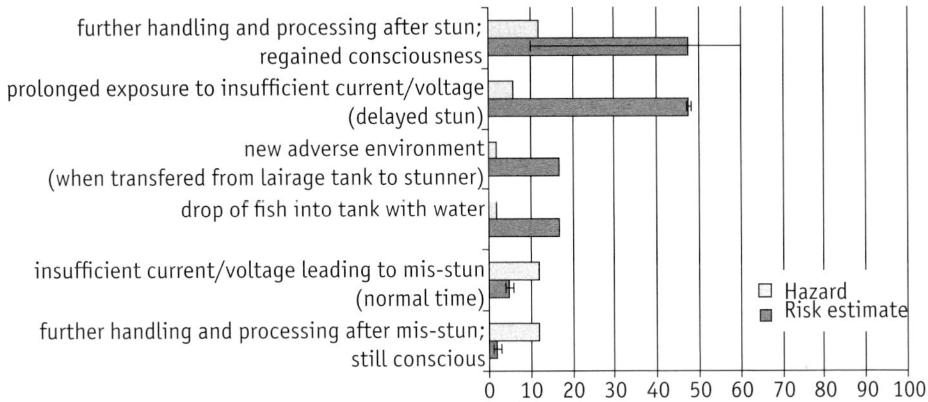


*Figure 5. Risk estimate and magnitude of adverse welfare effect for individual hazards associated with electrical stunning of carp (adapted from EFSA, 2009c). Hazards are ranked by risk estimate. Black bars show the estimated minimum and maximum values for the risk, reflecting the uncertainty about the probability of exposure to the hazard.*

Similar to the procedure followed in the pig welfare RA the major hazards affecting fish welfare during stunning and killing could be ranked and prioritised, which will allow risk managers to implement appropriate measures. The followed approach and its outcomes, allow risk managers to decide to prioritise measures aimed at reducing risks in most of the population, independently of the magnitude of the adverse effect, or concentrate their efforts on eliminating big suffering albeit this regards a small portion of the population. In addition, the outcomes helped the identification of areas with evident lack of data, i.e. where future research is encouraged.

## 6. First opportunity to challenge the utility of the current RA model by feeding it with solid data: AWRA of current farming and husbandry systems for dairy cows

Following a request from the European Commission, the AHAW Panel of EFSA adopted on the 5th of June 2009 five scientific opinions on the welfare of dairy cows (EFSA, 2009h,i,j,k,l). The scientific assessments considered whether current farming and husbandry systems comply with the requirements and welfare of dairy cows from the pathological, zootechnical, physiological and behavioural points of view. An exhaustive scientific report was compiled which included all available scientific evidence and peer reviewed data in relation to the welfare of dairy cows (summarised by Ribó *et al.*, 2009b).

Due to the great diversity of topics and the huge amount of scientific data, it was proposed that separating scientific opinions on different welfare subjects would be more adequate and effective. It was therefore agreed to subdivide the risk assessment process into four subjects: (1) metabolic and reproductive disorders, (2) udder disorders, (3) leg and locomotion problems and (4) behaviour, fear and pain. Since physiological and some other aspects of poor welfare

in dairy cows were not evaluated in the risk assessments, a fifth scientific opinion was also produced as a global ('overall') assessment on dairy cows welfare including these aspects.

The RA methodology adopted for the deployment of these four opinions was analogous to the one carried out for the three opinions on pig welfare (EFSA, 2007a,b,c, see 5.2.1). Once the hazards which have important consequences for dairy cow welfare were identified, RA tables were created where all hazards were listed. The working group members then gave scores to the parameters requested for the RA, such as duration, severity and likelihood of the adverse effects as well as duration, intensity and likelihood of the exposure to the hazard by the animal. Hazards were categorized into housing (e.g. space allowance, pen design, floor conditions, indoor temperature), feeding [e.g. liquid feed, water quality, transition feeding], management (e.g. grouping, preventive medicine, health and welfare plans, improper management of downer cows) and genetic selection [e.g. genetic selection for high production ignoring other traits of animal fitness]. Moreover, a different RA was developed for each of the most common husbandry systems in Europe – cubicle houses, tie stalls, straw yards and pasture – in order to enable risk managers to make a system comparison. Once all scores were given, the risk estimate and the magnitude of the adverse effect were calculated for each hazard, allowing the creation of graphs where hazards are presented as histograms. The methodology followed and its weaknesses and limitations are extensively described in section 6 of the contribution of Smulders (2009, this Volume).

A separate graph has been created for each hazard category within each production system and hazards in the graphs have been ordered by decreasing risk estimate scores. Figure 6 shows an example of an RA outcome, i.e. the five hazards related to the housing conditions recognized as the most important for the development of leg and locomotion disorders in dairy cows, are presented for cubicle houses and tie stalls.

The 'overall' assessment of dairy cow welfare (see above) aimed at combining conclusions and recommendations from the scientific report with the outcomes of the four separate risk assessments. Table 15 shows an illustrative schematisation of the integration of conclusions among the different opinions.

It should be noted that, particularly in the various dairy cow RA's, the availability of reliable scientific data and experts in this area, in many cases allowed to confirm the conclusions reached from the scientific literature thus enhancing the robustness of the conclusion and, *vice versa*, indicating the usefulness of the used RA approach.

For instance, the RA outcome on leg and locomotion disorders showed that in cubicles the greatest magnitudes of the adverse effects and risk estimates are associated with inadequate floor in the walking area, poor cubicle design and inadequate bedding. From the scientific report it was concluded that standing and walking for prolonged periods on concrete floors, or floors that are wet or covered in slurry cause severe foot disorders. It was also concluded that concrete flooring has a higher risk of claw disorders than pasture and straw-yards. Therefore the recommendation was made that dairy cattle should not be caused to stand or walk for prolonged periods on concrete floors or floors that are wet or covered in slurry.

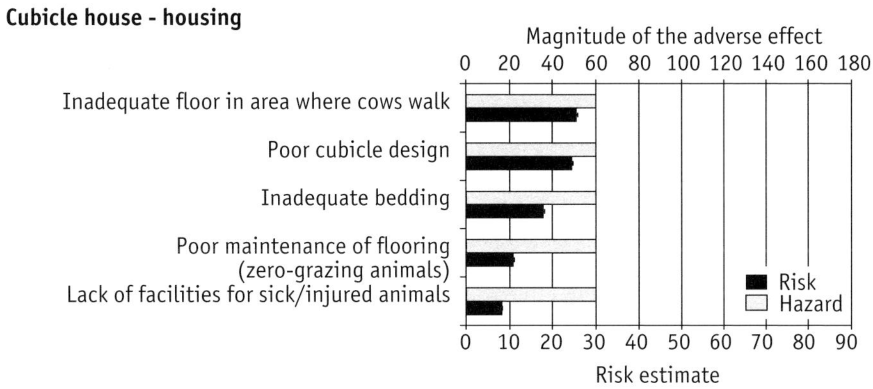


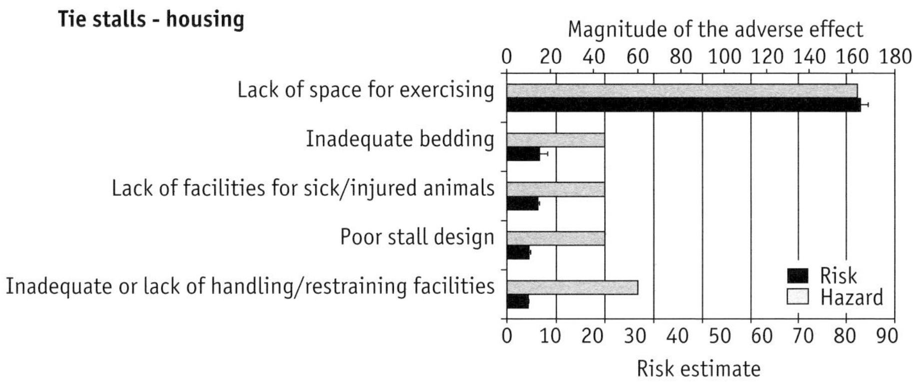


*Figure 6. Top five hazards for dairy cows in relation to leg and locomotion disorders (EFSA, 2009h).*

*Table 15. Integration of the conclusions and recommendations from the scientific report with the outcomes of the different specific risk assessments on dairy cow welfare.*

| Scientific report/opinions | Conclusions and recommendations (C&R) |
|---|---|
| 1. Scientific report on the effects of farming and husbandry systems on dairy cow welfare and disease | C1&R1 from scientific literature |
| 2. Scientific opinion on leg and locomotion problems in dairy cows based on a RA | C2&R2= C&R from RA + C1&R1 on leg and locomotion |
| 3. Scientific opinion on udder problems in dairy cows based on a RA | C3&R3= C&R from RA + C1&R1 on udder problems |
| 4. Scientific opinion on metabolic and reproductive problems in dairy cows based on a RA | C4&R4= C&R from RA + C1&R1 on met-rep problems |
| 5. Scientific opinion on behavioural problems, fear and pain in dairy cows based on a RA | C5&R5= C&R from RA + C1&R1 on behaviour |
| 6. Scientific opinion on overall assessment of dairy cows welfare | C1&R1 not related to any subject + most relevant C&R from 2, 3, 4, 5 |

## 7. Conclusions

### 7.1 What has been achieved?

Animal Welfare is a well established scientific discipline although its scientific study is relatively young. The start of animal welfare as a science can be traced back to the 1960's and has been developing and expanding since then. Animal welfare research activities have increased significantly and haven been targeted at finding ways to measure the welfare of food producing animals at farm level, during transport and at slaughter (Blokhuis *et al.*, 2007).

For the development of AHAW scientific opinions different RA methodologies with various weaknesses and limitations have been used, which indicates the need of standardised guidelines and working procedures to conduct AWRA. Whilst specific guidelines for assessing risks for animal diseases (OIE, 2004a; OIE, 2004b) or for biological, chemical and physical hazards occurring in foods of animal origin (WHO, 1999) are available, no specific standardised methodology exists in the field of the Animal Welfare Risk Assessment (EFSA, 2006a). By launching a self-mandate EFSA has taken the initiative to develop such guidelines where three main animal welfare issues have been identified: (1) stunning and killing; (2) transport; and (3) housing and management. The development of the guidelines has already started and should ultimately include a classification of the main animal welfare issues and animal categories to be considered and, for each AW issue, the risk factors [hazards] and the consequences (adverse effects) potentially affecting the animals. A harmonised definition of animal welfare, including the relationship with animal disease, is also going to be addressed in the framework of this self mandate. The current AWRA model has already proven its value in that it significantly supports the identification of scientific priorities in welfare science and risk management to serve further implementation of animal welfare standards. The first version of the AWRA guidelines is foreseen to be released by 2010.

### 7.2 What has been neglected?

The mandates requesting scientific assessments on animal welfare received by EFSA included very broad questions (i.e. welfare of pigs, welfare of fish, welfare of dairy cows). The quantitative assessment of welfare is therefore a problem, as necessary data are only available for certain animal categories and hazards, but not for others. Consequently, it is important that EFSA is provided with future mandates in which the risk questions are more precisely formulated both in terms of the welfare components considered to be most important by risk managers and in terms of the animal subpopulations, their life stages, husbandry systems under which they are held and preferably the geographical areas to be specifically addressed. This would allow conducting more precise and concise AWRA'S, as argued by Algers (2009) and Smulders (2009).

The European Commission has explicitly requested EFSA to include measurable welfare indicators, whenever possible, in the conclusions and recommendations of the future EFSA scientific opinions on animal welfare. Animal welfare scientific indicators will be useful for the monitoring of the farm practices, detection of poor welfare situation and application of corrective measures. Unfortunately, to date, such indicators have not yet conclusively

been identified, although a large European (the WelfareQuality®) project is likely to provide suggestions for these purposes (see Manteca *et al.*, 2009). The outcomes of the risk assessment methodology together with the identification of scientific indicators for welfare will be useful for the establishment and implementation of welfare control and monitoring plans at farm level. The control and monitoring plans at farm level should provide a valuable feedback on the reliability of this risk assessment approach, and should lead to further improvement of the proposed methodology.

### 7.3 What needs to be done?

The RA approach applied to animal welfare is work in progress. As all RA's, AWRA needs to be science-based, well-documented, objective, repeatable and transparent. Due to the limited amount of quantitative published data on the adverse effects of animal welfare hazards and exposure rates, AWRA's are often qualitative (largely based on expert opinion) or, at best, semi-quantitative. Therefore, the majority of the currently available RA reports fail to give precise numerical estimates of the risk attributed to certain hazards. Nevertheless, this approach needs to be followed to ensure that the major areas of welfare concern, research priorities and intervention options for animal (sub)populations are identified and further improvements of the AWRA methodology promoted.

## Acknowledgements

EFSA acknowledges the efforts and significant contributions of the working group experts, risk assessors and Members of the AHAW Panel on the development of the different scientific opinions using the risk assessment in animal welfare approach under development. Dr. Telmo Nunes, Dr. Moez Saana, Dr. Christine Müller-Graf and Dr. Frans Smulders are specially acknowledged as the initiators of the first attempts to conduct semi-quantitative RA's never done before in animal welfare. Dr. Denise Candiani and Dr. Jordi Tarrés from EFSA's AHAW Unit are acknowledged for the revision of the paper.

## References

Algers, B., 2009. A risk assessment approach to animal welfare. In: Smulders, F.J.M. and Algers, B. (Eds.). Food safety assurance and veterinary public health, Vol.5, Welfare of production animals: assessment and management of risks. Wageningen Academic Publishers, Wageningen, the Netherlands, pp. 223-237.

Blokhuis, H.J., Keeling, L.J., Gavinelli, A. and Serratosa, J., 2008. Animal welfare's impact in the food chain. Trends in Food Science & Technology 19 (1), 75-83.

CAC (Codex Alimentarius Commission), 2001. Food Hygiene, Basic texts, FAO/WHO, Rome, Italy.

CAC (Codex Alimentarius Commission), 2002. Principles and Guidelines for the Conduct of Microbiological Risk Assessment. Document CAC/GL 30.

Candiani, D., Ribó, O., Afonso, A., Aiassa, E., Correia, S., De Massis, F., Pujols, J. and Serratosa J., 2007. Risk assessment challenges in the field of animal welfare. In: Proceedings of the XIII international congress in animal hygiene, ISAH, June 17-21, Tartu, Estonia, 587-581.

Candiani, D., Ribó, O., Barbieri, S., Afonso, A., Grudnik, T., Berthe, F. and Serratosa, J., 2009. Development of a risk assessment methodology for animal welfare in EFSA's scientific opinions. In: Aland, A. and Madec, F. (Eds.). Sustainable Animal Production. The challenges and potential developments for professional farming. Wageningen Academic Publishers, Wageningen, the Netherlands, pp. 421-434.

EC (European Commission), 2002. Regulation (EC) No. 178/2002 of 28 January 2002, laying down the general principles and requirements of food law, establishing the European Food Safety Authority and laying down procedures in matters of food safety. Official Journal L31, 1/2/2002, p. 1-24.

EC (European Commission), 2005. Community Action Plan on the Protection and Welfare of Animals (2006-2010). http://ec.europa.eu/food/animal/welfare/actionplan/actionplan_ en.htm.

EC (European Commission), 2006. Future Strategy on Animal Health for 2007-2013 http://ec.europa.eu/food/animal/diseases/strategy/final_report_en.htm.

EFSA (European Food safety Authority), 2004a. Opinion of the Scientific Panel on Animal Health and Welfare related to the welfare of animals during transport. The EFSA Journal 44, 1-36.

EFSA (European Food Safety Authority), 2004b. Scientific Opinion of the AHAW Panel on the welfare aspects of the main systems of stunning and killing the main commercial species of animals. The EFSA Journal 45, 1-29.

EFSA (European Food safety Authority), 2005a. Opinion of the scientific panel on animal health and welfare (AHAW) on a request from the commission related to the Impact of the current housing and husbandry systems on the health and welfare of farmed domestic rabbits. The EFSA Journal 267, 1-31.

EFSA (European Food safety Authority), 2005b. Opinion of the scientific panel on animal health and welfare (AHAW) on a request from the commission related to aspects of the biology and welfare of animals used for experimental and other scientific purposes. The EFSA Journal 292, 1-46.

EFSA (European Food Safety Authority), 2006a. Scientific Colloquium 'Principles of Risk Assessment of Food Producing Animals: Current and future approaches', http://www.efsa.europa.eu/en/science/colloquium_series/no4_animal_diseases.html.

EFSA (European Food Safety Authority), 2006b. Scientific Opinion of the AHAW Panel on the welfare aspects of the main systems of stunning and killing applied to commercially farmed deer, goats, rabbits, ostriches, ducks, geese and quail. The EFSA Journal 326, 1-18.

EFSA (European Food Safety Authority), 2006c. Opinion of the Scientific Panel on Animal Health and Welfare (AHAW) on a request from the Commission related with the risks of poor welfare in intensive calf farming systems. The EFSA Journal 366, 1-36.

EFSA (European Food Safety Authority), 2007a. Scientific Opinion of the Panel on Animal Health and Welfare on a request from the Commission on animal health and welfare in fattening pigs in relation to housing and husbandry. The EFSA Journal 564, 1-14.

EFSA (European Food Safety Authority), 2007b. Scientific Opinion of the Panel on Animal Health and Welfare on a request from the Commission on the animal health and welfare aspects of different housing and husbandry systems for adult breeding boars, pregnant, farrowing sows and unweaned piglets. The EFSA Journal 572, 1-13.

EFSA (European Food Safety Authority), 2007c. Scientific Opinion of the Panel on Animal Health and Welfare on a request from the Commission on the risks associated with tail-biting in pigs and possible means to reduce the need for tail-biting considering the different housing and husbandry systems. The EFSA Journal 611, 1-13.

EFSA (European Food Safety Authority), 2007d. Scientific Opinion of the AHAW Panel on the welfare aspects of the killing and skinning of seals. The EFSA Journal 610, 1-122.

EFSA (European Food Safety Authority), 2008a. Scientific opinion of the Panel on Animal Health and Welfare on animal welfare aspects of husbandry systems for farmed Atlantic salmon, The EFSA Journal 736, 1-31.

EFSA (European Food Safety Authority), 2008b. Scientific opinion of the Panel on Animal Health and Welfare on animal welfare aspects of husbandry systems for farmed trout. The EFSA Journal 796, 1-22.

EFSA (European Food Safety Authority), 2008c. Scientific opinion of the Panel on Animal Health and Welfare on animal welfare aspects of husbandry systems for farmed European eel. The EFSA Journal 809, 1-17.

EFSA (European Food Safety Authority), 2008d. Scientific opinion of the Panel on Animal Health and Welfare on animal welfare aspects of husbandry systems for farmed European sea bass and sea bream. The EFSA Journal 844, 1-21.

EFSA (European Food Safety Authority), 2008e. Scientific opinion of the Panel on Animal Health and Welfare on animal welfare aspects of husbandry systems for farmed common carp. The EFSA Journal 843, 1-28.

EFSA (European Food Safety Authority), 2009a. Scientific Opinion of the Panel on Animal Health and Welfare on species specific welfare aspects of the main systems of stunning and killing of farmed Atlantic salmon. The EFSA Journal 2012, 1-77.

EFSA (European Food Safety Authority), 2009b. Scientific Opinion of the Panel on Animal Health and Welfare on species specific welfare aspects of the main systems of stunning and killing of farmed rainbow trout. The EFSA Journal 1013, 1-55.

EFSA (European Food Safety Authority), 2009c. Scientific Opinion of the Panel on Animal Health and Welfare on species specific welfare aspects of the main systems of stunning and killing of farmed carp. The EFSA Journal 1013, 1-37.

EFSA (European Food Safety Authority), 2009d. Scientific Opinion of the Panel on Animal Health and Welfare on species specific welfare aspects of the main systems of stunning and killing of farmed eels. The EFSA Journal 1014, 1-42.

EFSA (European Food Safety Authority), 2009e. Scientific Opinion of the Panel on Animal Health and Welfare on species specific welfare aspects of the main systems of stunning and killing of farmed seabass and seabream. The EFSA Journal (2009) 1010, 1-52.

EFSA (European Food Safety Authority), 2009f. European Food Safety Authority. Scientific Opinion of the Panel on Animal Health and Welfare on species specific welfare aspects of the main systems of stunning and killing of farmed turbot. The EFSA Journal 1073, 1-34.

EFSA (European Food Safety Authority), 2009g. Scientific Opinion of the Panel on Animal Health and Welfare on species specific welfare aspects of the main systems of stunning and killing of farmed tuna. The EFSA Journal 1072, 1-53.

EFSA (European Food Safety Authority), 2009h. Scientific Opinion of the Panel on Animal Health and Welfare on the impact of housing, nutrition and feeding, management and genetic selection on leg and locomotion problems in dairy cows. The EFSA Journal (2009) 1142, 1-57.

EFSA (European Food Safety Authority), 2009i. Scientific Opinion of the Panel on Animal Health and Welfare on the impact of housing, nutrition and feeding, management and genetic selection on udder problems in dairy cows. The EFSA Journal (2009) 1141, 1-60.

EFSA (European Food Safety Authority), 2009j. Scientific Opinion of the Panel on Animal Health and Welfare on the impact of housing, nutrition and feeding, management and genetic selection on metabolic and reproductive disorders in dairy cows. The EFSA Journal (2009) 1140, 1-75.

EFSA (European Food Safety Authority), 2009k. Scientific Opinion of the Panel on Animal Health and Welfare on the impact of housing, nutrition and feeding, management and genetic selection on behavioural problems in dairy cows. The EFSA Journal (2009) 1139, 1-66.

EFSA (European Food Safety Authority), 2009l. Scientific Opinion of the Panel on Animal Health and Welfare on the overall assessment of dairy cows welfare. The EFSA Journal (2009) 1143, 1-38.

EU (European Commission), 1997. Treaty of Amsterdam – Protocol on protection and welfare of animals. Official Journal C 340, 10/11/1997, p. 110.

Gerritzen, M.A. and Raj, M.A.B., 2009. Animal welfare and killing for disease control. In: Smulders, F.J.M. and Algers, B. (Eds.). Food safety assurance and veterinary public health, Vol.5, Welfare of production animals: assessment and management of risks. Wageningen Academic Publishers, Wageningen, the Netherlands, pp. 191-203.

Anil, H., and Lambooij, B., 2009. Stunning and slaughter methods. In: Smulders, F.J.M. and Algers, B. (Eds.). Food safety assurance and veterinary public health, Vol.5, Welfare of production animals: assessment and management of risks. Wageningen Academic Publishers, Wageningen, the Netherlands, pp. 169-190.

Hartung, J., and Springorum, A.C., 2009. Animal welfare and transport. In: Smulders, F.J.M. and Algers, B. (Eds.). Food safety assurance and veterinary public health, Vol.5, Welfare of production animals: assessment and management of risks. Wageningen Academic Publishers, Wageningen, the Netherlands, pp. 149-168.

Manteca, X., Velarde. A., and Jones. B., 2009. Animal welfare components. In: Smulders, F.J.M. and Algers, B. (Eds.). Food safety assurance and veterinary public health, Vol.5, Welfare of production animals: assessment and management of risks. Wageningen Academic Publishers, Wageningen, the Netherlands, pp. 61-77.

Müller-Graf, C., Candiani, C., Barbieri, S., Ribó, O., Afonso, A., Aiassa, E., Have, P., Correia, S., De Massis, F., Grudnik, T. and Serratosa, J., 2008. Risk assessment in animal welfare – EFSA approach. AATEX 14, Special Issue, March 31, 789-794.

OIE, 2004a. Handbook on Import Risk Analysis for Animals and Animal Products. Volume 1. Introduction and qualitative risk analysis, pp. 57.

OIE, 2004b. Handbook on Import Risk Analysis for Animals and Animal Products. Volume 2. Quantitative risk assessment, pp. 126.

Ribó, O., Candiani, D., Barbieri, S., Grudnik, T., Afonso, A., Berthe, F., De Massis, F., Dhollander, S., Correia, S., Have, P. and Serratosa, J., 2008. EFSA approach on risk assessment in animal welfare for the identification of scientific indicators in different livestock species. OIE, 2nd Global Conference on Animal Welfare, Cairo, Egypt, 20-22 October 2008. Poster.

Ribó, O., Candiani, D. and Serratosa, J., 2009a. Role of the European Food Safety Authority (EFSA) in providing scientific advice on the welfare of food producing animals. Ital. J. Anim. Sci. vol. 8 (Suppl. 1) 9-17, 2009.

Ribó, O., Candiani, D. and Serratosa, J., 2009b. EFSA's scientific assessment on the effects of current farming and husbandry systems on dairy cow welfare. Annual Meeting of the European Association of Animal Production (EAAP), Barcelona, 24-27 August 2009, p.325.

SCAHAW, 2002. Scientific opinion of the Scientific Committee on Animal Health and Animal Welfare (SCAHAW) on the welfare of animals during transport (details for horses pigs sheep and cattle). Adopted on 11 March 2002.

Smulders, F.J.M. 2009. A practicable approach to assessing risks for animal welfare – methodological considerations. In: Smulders, F.J.M. and Algers, B. (Eds.). Food safety assurance and veterinary public health, Vol.5, Welfare of production animals: assessment and management of risks. Wageningen Academic Publishers, Wageningen, the Netherlands, pp. 239-274.

SVC, 1995. E.U. Scientific Veterinary Committee (Animal Welfare Section) Scientific Report 'The welfare of calves' (9 November 1995).

SVC, 1996a. E.U. Scientific Veterinary Committee (Animal Welfare Section) Scientific Report 'The slaughter and killing of animals' (30th October 1996).

SVC, 1996b. E.U. Scientific Veterinary Committee (Animal Welfare Section) Scientific Report 'The welfare of laying hens' (30th October 1996).

SVC, 1997. E.U. Scientific Veterinary Committee (Animal Welfare Section) Scientific Report 'The welfare of intensively kept pigs' (30th September 1997).

Vose, D., Acar, J., Anthony, F., Gupta, R., Nicholls, T., Tamura, Y., Thompson, S., Threlfall, E.J., van Vuuren, M., White, D.G., Wegener H.C., and Costarrica, M.L., 2001. Antimicrobial resistance: risk analysis methodology for the potential impact on public health of antimicrobial resistant bacteria of animal origin. Rev. Sci. Tech. Off. Int. Epiz. 20: pp. 811-827.

WHO, 1999. Codex Alimentarius Commission (CAC). http://www.who.int/foodsafety/publications/micro/cac1999/en/.

# Animal welfare and legislation

**Donald M. Broom**
Centre for Animal Welfare and Anthrozoology, Department of Veterinary Medicine, University of Cambridge, Madingley Road, Cambridge CB3 0ES, United Kingdom; dmb16@cam.ac.uk

## Summary

Animal welfare has been more and more widely accepted as a scientific discipline during the last 25 years and our knowledge of animal functioning, including the extent to which they are sentient, has increased. One result of this has been that the public have demanded more legislation to protect animals and this has been passed in increasing numbers of countries but there are still many countries that do not have adequate laws protecting animals. One of the key ways to improve animal welfare in the long term is for all those who use or have responsibility for animals to receive adequate education and training about the biological functioning of those animals, including ways in which their welfare might be made better or worse. To date, retailers' codes of practice have had the major effect on the welfare of farm animals although both laws and codes are needed. Further scientific studies on animal welfare are also needed and it is important to develop better methodologies for the enforcement of laws and codes and to provide adequate manpower to do this. Where the impact of different factors on animal health, or any other aspect of animal welfare, is being reviewed, careful analysis should involve not only risks but also benefits. Legislators are not just risk managers and a balance has to be struck between risks and benefits in every area of legislation.

**Keywords**: attitudes to animals, risk-benefit analysis, enforcement

## 1. Introduction: do people know what animal welfare is?

The term animal welfare is familiar to many people in Europe and other parts of the world. To most, the concept indicates the individual being happy, not suffering and not having serious problems in life. For some it relates only to their pets. For others it relates to a wider range of animals, including farm and laboratory animals, and to the most knowledgeable it concerns a group of animals with a particular level of biological complexity linked to the idea of sentience. The EU Treaty of Amsterdam is of substantial importance in relation to the number of people who are in this last category because it refers to animals as being sentient.

Poor welfare is one of the factors that can lead to an animal husbandry system or procedure being unacceptable to the public and hence unsustainable. However, which animals should be protected and what degree of protection is required? Animals vary in the extent to which they are aware of themselves (De Grazia, 1996) and of their interactions with their environment, including their ability to experience pleasurable states such as happiness and aversive states such as pain, fear and grief (Panksepp, 1998). This capacity may be referred to as their degree of sentience. A sentient being is one that has some ability to evaluate the actions of others in relation to itself and third parties, to remember some of its own actions and their

consequences, to assess risk, to have some feelings and to have some degree of awareness (Broom, 2006b). Awareness is defined here as a state in which complex brain analysis is used to process sensory stimuli or constructs based on memory (Broom, 1998).

Despite the concerns about which animals are sentient, the term welfare applies to all animals, both sentient and non-sentient so welfare assessment does not necessitate determination of sentience. However, for most people, in a sequence of animals of different complexity the threshold at which animals should be protected by legislation is linked to sentience. Whilst few people would want to protect an amoeba or a nematode worm, increasing proportions of people would be concerned about operations without anaesthetic when earthworms, snails, bees, crabs, octopus, goldfish, frogs, lizards, robins, dogs and chimpanzees are considered. The functioning of some aquatic animals in relation to protection is discussed by De Grazia (1996), Mather (1995, 2004), Chandroo *et al.* (2004) and Broom (2007b). The remarkable abilities of some birds (Clayton and Dickinson, 1999; Pepperberg, 2000) now force many people to put them in the same category as the most complex mammals.

It is generally accepted by animal welfare scientists and those considering human welfare that the concept of welfare refers to the measurable state of the individual on a scale from very good to very poor. It was suggested by Hughes (1982), following the writings of Lorca, that welfare means being in harmony with nature. This has proved to be a valuable idea but being in harmony is a single state and could not vary so it is not a definition of welfare that is usable in welfare assessment. Broom (1986) defined welfare in a way that includes feelings and health: 'The welfare of an individual is its state as regards its attempts to cope with its environment.' The definition refers to the extent to which the individual is failing to cope, or is having difficulties in coping, or is easily coping. The origin of the concept is how well the individual is faring or travelling through life and the definition refers to state at a particular time (Broom and Fraser, 2007; Broom and Johnson, 2000). Since feelings are a part of function, it is not correct to say that this definition of welfare is a functional one, rather than one that refers to suffering and other feelings (Broom, 1991, 1998, 2003; Broom and Fraser, 2007). The definition takes account of biological function but does not imply that a natural environment is necessary for good welfare.

The concept of welfare and the methods for its scientific study developed for non-human animals applies to humans. The discussions by Irwin (2001), Lutgendorf (2001) and Carter (2001) are amongst those that emphasise this. Welfare is a measurable biological variable that is just as important to humans as it is to animals used by, or not used by, humans. The concept of quality of life is essentially the same as the concept of welfare except that welfare can refer to very short-lasting effects while quality of life would not be used unless a period of a few days or more is being considered (Broom, 2007a).

## 2. Do people want animal welfare legislation?

Moral actions are directed more towards those identified as 'us' than towards those considered to be 'them'. Categories included as us may be: (a) individuals readily recognised as close relatives, (b) all of those who know who I am, (c) those who might have access to the same

information that I have, or (d) sentient beings who share characteristics with me (Broom, 2003, 2006b). Increased communication efficiency is revolutionising our degree of concern for other humans and extending our area of moral concern to other species. Companion animals will be in category (a) for some people as Serpell and Paul (1994) found that many pet owners stated that they regarded their pets as part of their family. Most people who have regular contact with individual animals would include them in category (b) and many who know about the functioning, especially the behaviour, of animals would include some or many other species of animals in categories (c) or (d). In many societies now, education levels are high and there is easy access to good quality information about animals whose abilities are complex. Hence people are becoming less likely to cause poor welfare or tolerate it in animals perceived to be aware. Changed attitudes to animals appear to be linked more closely to the educational level of people than to their affluence. In countries where the majority of people are relatively poor but well educated, interest in animal welfare is such that many people are willing to incur some degree of financial loss rather than benefit from poor welfare in animals.

If we use a living animal in a way that gives us some benefit, we have some obligations to that animal. One obligation is to avoid causing poor welfare in the animal except where the action leads to net benefit to that animal, or to other animals including humans, or to the environment. A utilitarian approach is not sufficient to determine all obligations, however, and a deontological approach is also needed because there are some degrees of poor welfare which are never justified by benefit to others.

It is my view (Broom, 2003, 2006a) that all human behaviour and laws should be based on the obligations of each person to act in an acceptable way towards each other person and to each animal that is used. It is better if strategies for living are based on our obligations rather than on perceived rights. This is because many so-called rights can result in harm to others. There are occasions when people state that they have a right to say what they want, or drive as fast as they want, or carry a gun, or select the sex or genetic make-up of their children. In each of these cases the action could cause harm to others which would be accepted by very few people. Whilst arguments based on a concept of rights may sometimes be clearly wrong, arguments based on the obligation of one individual towards others do not suffer from such problems. Hence my conclusion that the concept of rights is not the best to use and that each person should always focus on how they ought to behave. As far as animal rights are concerned, there are no legal rights stated but there are many statements, codes of conduct, or unwritten rules which explain the obligations towards animals of those people who use the animals. This argument is also applicable to references to the freedom which an individual asserts or which it is said should be given. Efforts to list the freedoms which should be allowed to the animals which we keep have been of use as a general guide to management but with the development of information about the needs of animals, it is now possible to be more precise in laws or guidelines for animal care and all of these should refer to needs rather than to freedoms.

It is very old and widespread idea in human society that animals used by people should not be treated like inanimate possessions but should be protected from actions that might cause suffering (Broom, 2006b; Dawkins, 1980; Fraser, 2008). Many people have condemned those

perceived as being cruel to animals, whether or not this is stated in a law. On the other hand, cruelty is part of some forms of human entertainment.

The treatment of animals is an area in which codes of conduct and descriptions of good practice exist. Even amongst groups of people whose objective is to kill animals, there have long existed unwritten codes of conduct concerning what actions are or are not permissible. For example, as discussed by Serpell (1986, 1989), people using guns and dogs to hunt mammals or birds would expend energy and resources trying to ensure that, firstly animals were shot in a way likely to kill quickly and secondly that shot animals were found and killed rather than being left to die slowly. More recently, codes of practice relating to animals kept for food production and other purposes have been produced by various organisations.

Public concern about animal welfare has increased in many countries during the last forty years and especially in the last twenty years. Evidence for this is summarised in Table 1.

Members of the public exert influence by letters to government, other public bodies and commercial organisations and by statements that appear in the media. Members of the European parliament report that they receive more letters about animal welfare than about any other topic. Politicians respond by raising the issues, including them in manifestos, seeking scientific information, encouraging further research and teaching, and passing laws.

The views of the general public are largely made known to farmers and others involved in animal usage via the media. There is frequent coverage of animal welfare issues in newspapers, on radio and on the television and this, by bringing scientific knowledge about animal complexity to the attention of people, affects their attitudes. Farmers and some other animal users may see themselves portrayed as uncaring. Some such portrayals are unfair but others are correct and the farmer cannot hide from them by putting animals in buildings and associating only with other farmers. When public demonstrations about animal welfare issues occur, the people who use the animals need to take note of them. The demonstrations by great numbers of largely orderly and apparently normal people against the shipping of calves to conditions that were illegal within the United Kingdom, had a big influence on farmers and politicians alike. The greatest influence on animal users or politicians is exerted by moderate people who represent a groundswell of public opinion, rather than by the most vociferous people, who may be rather extreme in their views. For some years now, in many surveys in Europe animal welfare has been shown to be an important issue for the general

*Table 1. Evidence for increased concern about animal welfare (Broom, 1999).*

| | |
|---|---|
| 1. | Letters from the public, media coverage |
| 2. | References in parliamentary discussions and government statements |
| 3. | Requests for scientific evidence concerning animal welfare |
| 4. | Activity of scientific and other advisory committees |
| 5. | Funding of scientific research on animal welfare |
| 6. | Increased teaching and conferences |
| 7. | More legislation |

public. For example about three quarters of people questioned in France regarded animal welfare as a problem affecting their purchasing of veal or eggs (Ouedraogo, 1998) and 34% of 420 schoolgirls questioned in Dublin stated that they avoided eating meat, principally (53%) for animal welfare rather than nutrition (29%) reasons (Ryan, 1997).

Laws intended to prevent cruelty to dogs and horses were passed in Europe as long as 200 years ago and were gradually extended to other kinds of animals (Fraser, 2008; Radford, 2001). Most early laws referred to companion animals and working animals but not to farm animals. Some laws protected animals against the forms of animal experimentation which were considered likely to cause substantial pain to the animals. Laws were also passed which proscribed some forms of entertainment involving animals as being cruel but others were still permitted. Laws aimed at preventing poor welfare in animals have become more wide-ranging, both in terms of species and the different animal uses, and have been passed in more and more countries.

As explained by Radford (2001), members of the public have expectations that legislation will result in the worst causes of poor welfare in animals being prevented. Where the law is based upon a necessity for unnecessary suffering to be proved, many practices that result in poor welfare can occur without successful prosecution of the perpetrators. The U.K. Animal Welfare Act, 2006, imposes a duty of care on all who are responsible for animals. This wording can result in a prosecution if the animal has not been cared for in a reasonable way. However, the wording of this Act still refers to unnecessary pain, suffering and distress and this retains the problem that some actions might be deemed necessary, not for the benefit of the animals but for the convenience or profit of the animal user.

The expectations of people about the efficacy of laws vary somewhat from country to country. In the United Kingdom, Germany or Sweden, most people would assume that animal protection laws would protect animals but in Italy or Mexico, many people would be dubious about whether laws would have much impact. The doubts are related to ideas about whether or not the average person would obey the law and the extent to which there would be effective enforcement. However, the differences among countries in these respects appear to be lessening.

## 3. What actually affects the welfare of production animals?

The welfare of animals can be very poor because of the design of animal accommodation or of facilities on farms, for transport or for slaughter. Management systems also have substantial effects on welfare. If the building or management system causes unavoidable problems, as some do, however good the care given to the animals, welfare will not be good. For example, the welfare of social animals deprived of social contact, or of confined animals deprived of exercise or opportunity to groom, cannot be good at all times however well the animals are cared for in those systems.

People who run farms, work on farms, or otherwise use animals are influenced by a variety of factors when they are deciding on their housing and management policies and when

they are executing these policies. Attitudes of animal users depend upon early training, traditional practices, acquisition of knowledge from others subsequent to any training, personal experience and general beliefs and philosophy (Broom, 1999, 2004). Training did not, until recently, include much information about animal welfare except where it impinged on profitability. Even diseases were often mentioned in agriculture training only in relation to effects on growth, offspring production or product quantity and quality. However, education of people who use animals is of key importance in encouraging the improvement of animal welfare in the future and animal welfare courses are becoming more widespread (Broom, 2005; Hewson *et al.*, 2005; Webster, 1994).

Farmers and other animal users have to live with their families, friends and neighbours. If these people are critical of the effects on the welfare of animals of the methods used, the farmer may change these methods. The views of the general public are largely made known to farmers and others involved in animal production via the media. There is frequent coverage of animal welfare issues in newspapers, on radio and on the television and this, by bringing scientific knowledge about animal complexity to the attention of most people, affects the attitudes of people and then comes to represent it. Farmers see themselves portrayed as uncaring in some respects. In many recent surveys in Europe, animal welfare has been shown to be an important issue for the general public.

## 4. The relative importance of legislation and standards in affecting animal welfare

Legislation is generally initiated by pressure from voters on elected politicians. The politicians would usually seek advice from civil servants before formulating new legislation. In recent years in Europe, whenever any legislation on animal welfare is proposed, advice is sought from committees or working groups of scientists. The non-governmental organisations who lobby the politicians, whether they are producers' organisations or animal protection organisations, will have some access to scientific advice so it is important for the legislators to know the latest state of scientific knowledge on the subject. As a consequence, the European Union has set up, *via* the European Commission, scientific committees on a range of subjects. The former committees were the Scientific Veterinary Committee, Animal Welfare Section and the Scientific Committee on Animal Health and Animal Welfare. The present committee is the European Food Safety Authority Scientific Panel on Animal Health and Welfare. Legislation within European countries and E.U. Directives and Regulations have often been preceded by Recommendations from Council of Europe committees such as the Standing Committee of the European Convention on the Protection of Animals Kept for Farming Purposes.

On a world-wide scale the Office International des Épizooties (OIE) or World Organisation for Animal Health is now producing sets of recommendations which are likely to be treated as laws by most nations in the world, just as OIE recommendations on animal disease are respected.

Farmers often sell animals or their production of milk, eggs etc. to single purchasers who represent large retail chains or wholesale distribution companies. The increase in direct

selling to supermarket chains has led to considerable power being placed in the hands of these supermarket companies. It is possible for these purchasers to lay down conditions for animal production and to enforce these by inspection. The standards set by the supermarket chains are determined by what people will buy and by their reputation with the public.

The public image of large companies which retail food, including supermarket chains and fast-food companies, is of great importance to them. Bad publicity because of a risk to public health, a risk to the environment or the occurrence of poor welfare at any stage of the production process can be very damaging. Hence it is in the interest of such food companies to avoid any scandal which might threaten their good image (Broom, 1999). When these companies receive many letters from consumers complaining about a product which they sell, they have to take notice of the points which are being made.

As a consequence of consumer pressure, food retail companies are adopting animal welfare standards which they impose on their suppliers. In some cases, these standards are quite simple, for example Albert Hein in the Netherlands and elsewhere limited their sales of eggs to 'scharreleieren' which meant that the hens were reared in conditions where they could scratch in litter. Marks and Spencer in the U.K. and elsewhere stopped selling eggs from battery cages. In other cases, elaborate standards have been described in detail and sent to suppliers. One of the first systematic attempts to provide comprehensive information about the conditions under which animals were kept in the course of food production was the 'Freedom Foods' scheme run by the R.S.P.C.A. in the U.K. In this scheme, the standards for housing and management have been set by a widely respected animal protection society and farms are inspected by Freedom Foods staff. Retailers who subscribe to the scheme are allowed to use the Freedom Foods logo which is accepted as honest by the purchasing public. Acceptance, by the public, of products which are produced in such a way that the welfare of the animals is good, depends upon trust in the organisation which is carrying out the labelling and inspection. Some large supermarket chains and other food retailers are trusted because it is thought that they could not afford to be found out if they were not labelling and policing adequately.

The enforcement of standards by food retailers has led to substantial changes in the welfare of animals on farms because every producer has to conform to the standards in order to sell their products. The rapid development of such schemes in several countries has, in general, been based on scientific evidence about animal welfare.

As trade in farm products becomes more and more international, and consumers in increasing numbers of countries demand good welfare standards during animal transport, the necessity for international standards for animal welfare during housing, transport and slaughter becomes more important. The OIE standards are likely to be a minimum that most countries will follow. However, the standards enforced by food retail companies in some countries will be higher. Hence many countries will choose to use the higher standards. In general, retailer standards are likely to have a more substantial effect on the welfare of farm animals than legislation.

## 5. The role of risk management in legislation

The risks that a toxic substance will be in a foodstuff, or that a pathogen will enter an animal and proliferate in it, or that a management procedure will result in poor welfare in a farm animal have always been considered in scientific reports like those produced by the European Food Safety Authority (EFSA) or its predecessors in the EU. However, in recent years the methodology for assessing risk has become more sophisticated and systematic so it is now used in many reports on topics like those exemplified above. This has resulted in scientists being more rigorous in their analyses of potential problems. If a quantitative or qualitative risk analysis is carried out, it is less likely that factors that affect the harm under consideration will be missed. Also, the relative importance of the factors involved will often be estimated more accurately. This approach was stimulated by the initial failures in the late 1980s and early 1990s to properly evaluate the risks associated with the outbreak of bovine spongiform encephalopathy (BSE) in cattle and Creutzfeldt-Jakob Disease in humans who had consumed some cattle products. A reluctance to harm the cattle and cattle-feed industries led to inadequate analysis of risk to cattle and humans. This might have resulted in the deaths of millions of people, although fortunately the number will be very much lower than this, and actually caused much more harm to the industries than there would have been if a proper risk analysis had been conducted at an early stage.

The inclusion of risk analysis in scientific reports and opinions produced by EFSA and other organisations has helped decision makers to take appropriate action, for example to minimise animal disease and improve animal welfare. It is desirable that this approach should be continued, with suitable modification according to the limitations associated with the relevant data.

Some factors that affect animals have beneficial effects rather than leading to a greater risk of a harm. This is most obvious when the wide-ranging components of animal welfare are considered. Food, access to other resources, human contact, social interactions and many other factors can result in benefit to the individual. Any one of these factors may also stimulate the immune system of an animal and hence confer benefit by reducing the likelihood of clinical disease. Hence every scientific review of welfare in general or of a component of health, such as the occurrence and effects of a pathological condition, should consider the possible beneficial effects of factors as well as their impact on risk. It is never sufficient in such reports to merely conduct a risk analysis.

Those who formulate laws, such as government ministers or the staff of the European Commission, have to take account of all factors in determining the best course of action so they are never just risk managers. Much of what they are trying to achieve is benefit, not just reduction of risk. This is true in legislation to reduce disease and to promote good health as well as, more obviously, in the animal welfare area. Legislation can promote good welfare and it often does.

## 6. Who has to consider the risk of poor welfare and factors contributing to good welfare?

Where two or more systems for housing and managing animals exist and they differ in their effects on welfare, many people may have an interest in action being taken to provide the best welfare for the animals. As a consequence, a wide range of people may wish to consult accurate scientific reports on the factors that affect welfare. Some members of the general public will inform themselves about the facts so that they can adopt a position on the subject to use in discussion and, on occasions, when voting. They will often be informed by the media, sometimes accurately and sometimes in a sensationalist way. The aims of the media include providing information so they will also wish to consult authoritative scientific reports on animal welfare or those who have thoroughly studied them.

Documents explaining the way in which various factors contribute to good or poor welfare will also be used by stockpersons, individual producers and producer organisations. Many of those directly involved in caring for and managing animals have a personal concern to try to improve the welfare of those animals so they will wish to inform themselves by reading reports or attending courses or lectures by those who have. Those who own or manage businesses centred on animal use and those who work for producer organisations will be aware of the public interest in animal welfare so they will need to obtain good information on factors affecting welfare so that they can act to modify methodologies or building design as necessary. Those who act early are less likely to be criticised. Action to improve animal welfare will, in some cases, allow that animal production area to flourish rather than decline. If a significant number of consumers consider a product to be produced in a morally unacceptable way, they may boycott it. Hence preservation of the good name of an industry into the future may depend on acquiring and acting on information about animal welfare.

There has been a substantial change over the last twenty five years in the way in which animal protection societies act when trying to achieve change. Most societies, like most government agencies and some producer groups, now commence any campaign or the formulation of any document by endeavouring to obtain good quality scientific information about the issue under consideration. They do this because they are more likely to be listened to by government and by the public if what they say is accurate. Hence scientific reports by EFSA are read carefully and where there is no good review of a subject, the animal protection society may commission one from a scientifically-based group or may produce a report from within their own staff.

## 7. The role of animal welfare science in deciding on legislation

Animal welfare scientists provide objective information about the welfare of animals. Most welfare indicators will help to pinpoint the state of the animal wherever it is on the scale from very good to very poor. Some measures are most relevant to short-term problems, such as those associated with human handling or a brief period of adverse physical conditions, whereas others are more appropriate to long-term problems. Tests of avoidance and positive preference help in the design of better conditions and procedures. In all welfare assessment

it is necessary to take account of individual variation in attempts to cope with adversity (Koolhaas *et al.*, 1983; Mendl, *et al.*, 1992) and in the effects which adversity has on the animal. The general methods for assessing welfare are summarised by Broom and Fraser (2007).

The central role of animal welfare scientists is to further develop their discipline and to carry out studies in which they evaluate the welfare of animals housed or treated in different ways. The collection and analysis of data by animal welfare scientists should be carried out in an objective way that is independent of any ethical view about the outcome of the research. After the results have been obtained, scientists, like any member of the public, may make judgements about what should be done.

When scientific reports on animal welfare matters are produced, these are easier for legislators and other informed persons to use and have greater effect if the primary scientific literature is quoted giving full references. However, evaluation of the quality of the scientific information is also important. The conclusions from the data reviewed should be quoted and recommendations made based on the evidence available. Where there is little scientific information, conclusions and recommendations should still be made but should make this clear. In some cases, it is valid to use information from related species but in other cases it is not. For example, every social animal will be adversely affected by being tied up and prevented from showing social as well as normal maintenance behaviour so a report on a social animal that has not been studied directly in this respect could refer to studies of other species. On the other hand, a pathogen that causes infection in one species may not cause infection in another so extrapolation from species to species is not reliable.

Laws can have a significant effect on the ways in which people manage animals. As an example of events leading to an EU Directive, the welfare of calves may be considered. From 1960 onwards there was some public concern that close confinement and inadequate diet lead to poor welfare in calves reared for veal production. This was a focus of the book *Animal Machines* by Harrison (1964) which led to the setting up of a national committee on farm animal welfare in the U.K. In the 1970s and 1980s there were research results giving evidence for serious welfare problems in closely confined calves. In 1988 the recommendation concerning the welfare of cattle from the Council of Europe Standing Committee on the Protection of Animals Kept for Farming Purposes stated that cattle should be able to make all normal movements for grooming, exercise etc. Some European countries passed legislation banning calf-crates. The 1990 report by a group of scientists coordinated by the European Commission was followed in 1991 by Directive 91/629/EEC laying down minimum standards for the protection of calves. This allowed the use of crates of a minimum size but required a report from E.U. Scientific Veterinary Committee by 1/10/97 which Ministers would act on. In the 1990s there was further welfare research on the effects of diet, confinement, space in groups etc. on calf welfare and in 1994-1995 much public pressure for action. The scientific report was requested early so in 1995 the 'Report on the welfare of calves' was produced by the E.U. Scientific Veterinary Committee, Animal Welfare Section. In 1996 there was a proposal for legislation from European Commission staff and the scientific report was considered by Ministry staff from each member state. A revised proposal was discussed by Ministers from each Member state and the 1997 Directive 97/2/EC, phasing out the use of veal crates and inadequate diets, was passed.

Similarly, within the European Union (EU), the Council Regulation (EC) No 1/2005 'On the protection of animals during transport and related operations' takes up some of the recommendations of two separate reports: (1) the E.U. Scientific Committee on Animal Health and Animal Welfare Report 'The welfare of animals during transport [Details for Horses, Pigs, Sheep and Cattle]' (March 2002) and (2) the European Food Safety Authority 'Report on the welfare of animals during transport' (2004) which deals with the other species. There are now many other examples of legislation based on information from scientific reports in the EU and elsewhere.

## 8. Enforcement and animal welfare science

The actual effect of legislation on the welfare of animals depends upon the responses of those owning and managing the animals. This response, in turn, depends upon the nature of any enforcement. Some systems for farm animal production will not continue if they are made illegal because they depend upon large manufacturers who are easily forced to change to a legal system. Other aspects of legislation can be enforced only by checks on farm, transport vehicles, markets, slaughterhouses etc. and the extent of law-breaking will be significantly affected by the frequency and quality of the checks. In most aspects of animal welfare law, little good information on the extent of compliance exists. For many transgressions, unannounced inspections are necessary if transgressors are to be discovered. There are regional and national differences in the extent to which legislation is viewed seriously by those involved in the animal production business. The general direction of movement within the European Union in this respect is towards better enforcement in all member states because it is manifestly unfair for there to be significant differences in the extent of compliance with the law.

Enforcement of 'retailers' codes of practice is likely to be better than enforcement of many national laws because, in order to protect the company's reputation, the retailers will have more inspectors than the government is willing to pay.

Where inspection of animals occurs, as a part of the enforcement process, scientific papers can help to determine which are the best measures of welfare outcomes to monitor. For example, general principles are explained by Botreau *et al.* (2007) and a protocol devised by Whay *et al.* (2003) is appropriate for aspects of dairy cow welfare.

## 9. Conclusions

### 9.1 What has been achieved?

Animal welfare has been more and more widely accepted as a scientific discipline during the last 25 years and our knowledge of animal functioning, including the extent to which they are sentient, has increased. One result of this has been that the public have demanded more legislation to protect animals and this has been passed in increasing numbers of countries. Welfare is affected by the education and attitudes of those responsible for the animals,

as well as by codes of practice and laws. To date, retailers' codes of practice have had the major effect on the welfare of farm animals although both laws and codes are needed. The introduction of risk-assessment methodologies has led to more rigorous scientific analyses of the dangers of toxicity in food or the spread of disease.

### 9.2 What has been neglected?

There are still many countries, including large countries like the United States, that do not have adequate laws protecting animals. The dissemination of information relevant to the improvement of animal welfare has sometimes been blocked, or at least not promoted, by those who view it as having an adverse effect on their business. When laws and codes exist, their efficacy is greatly affected by the quality of the enforcement system. There has been a tendency in many countries to reduce the numbers of staff who check on farm animal welfare, including health. For all analyses of the welfare of humans or other animals, including the impact of disease, the beneficial effects of the various factors should be carefully considered but benefit analysis is seldom carried out.

### 9.3 What needs to be done?

One of the key ways to improve animal welfare in the long term is for all those who use or have responsibility for animals to receive adequate education and training about the biological functioning of those animals, including ways in which their welfare might be made better or worse. Further scientific studies of animal welfare are needed but it is also important to develop better methodologies for the enforcement of laws and codes and to provide adequate manpower to do this. Where the impact of different factors on animal health, or any other aspect of animal welfare, is being reviewed, careful analysis should involve not only risks but also benefits. Legislators are not just risk managers and it is seriously misleading to describe them in that way without reference to the balance that has to be struck between risks and benefits in every area of legislation.

## References

Botreau, R., Veissier, I., Butterworth, A., Bracke, M.B.M. and Keeling, L., 2007. Definition of criteria for overall assessment of animal welfare. Animal Welfare 16, 225-228.

Broom, D.M., 1986. Indicators of poor welfare. British Veterinary Journal 142, 524-526.

Broom, D.M., 1991. Assessing welfare and suffering. Behavioural Processes 25, 117-123.

Broom, D.M., 1998. Welfare, stress and the evolution of feelings. Advances in the Study of Behavior 27, 371-403.

Broom, D.M., 1999. Welfare and how it is affected by regulation. In: Kunisch, M. and Ekkel, H. (Eds.). Regulation of animal production in Europe. Darmstadt: K.T.B.L., pp. 51-57.

Broom, D.M., 2003. The evolution of morality and religion. Cambridge: Cambridge University Press, pp. 259.

Broom, D.M., 2004. Future food animal production efficiency and acceptability. In: Shamsuddin, Z.H. (Ed.). Agriculture congress: innovation towards modernized agriculture. Serdang, Malaysia: Universitiy Putra Malaysia, p. 17-21.

Broom, D.M., 2005. Animal welfare education: development and prospects. Journal of Veterinary Medical Education 32, 438-441.

Broom, D.M., 2006a. Introduction: concepts of animal protection and welfare including obligations and rights. In: animal welfare. Ethical eye series, Strasbourg: Council of Europe Publishing, pp. 13-28.
Broom, D.M., 2006b. The evolution of morality. Applied Animal Behaviour Science 100, 20-28.
Broom, D.M., 2007a. Quality of life means welfare: how is it related to other concepts and assessed? Animal Welfare 16 (suppl.), 45-53.
Broom, D.M., 2007b. Cognitive ability and sentience: which aquatic animals should be protected? Diseases of Aquatic Organisms. 75, 99-108.
Broom, D.M. and Fraser, A.F., 2007. Domestic animal behaviour and welfare, 4th edn. Wallingford: C.A.B.I., pp 437.
Broom, D.M. and Johnson, K.G., 2000. Stress and animal welfare. Dordrecht: Kluwer, pp. 211.
Carter, C.S., 2001. Is there a neurobiology of good welfare? Coping with challenge: welfare in animals including man, D.M. Broom (Ed.). Berlin: Dahlem University Press, pp. 11-30.
Chandroo, K.P., Duncan, I.J.H. and Moccia, R.D., 2004. Can fish suffer? Perspectives on sentience, pain, fear and stress. Applied Animal Behaviour Science 86, 225-250.
Clayton, N.S. and Dickinson, A.D. 1999. Scrub jays (*Aphelocoma coerulescens*) remember when as well as where and what food items they cached. Journal of Comparative Psychology, 113, 403-416.
Dawkins, M.S., 1980. Animal suffering: the science of animal welfare. London: Chapman and Hall.
De Grazia, D., 1996. Taking animals seriously: mental life and moral status. Cambridge University Press: New York, USA.
European Commission, 1991. Council Directive 91/629/EEC of 19 November 1991 laying down minimum standards for the protection of calves. OJ L 268, 24. 9. p. 56.
European Commission, 1996. Council Directive, 97/2/EC of 20 January 1997 amending Directive 91/629/EEC laying down minimum standards for the protection of calves. OJ No C 85, 22. 3. p. 19.
European Commission, 2005. Main points of Regulation (EC) 1/2005 on the protection of animals during transport. Found on http://ec.europa.eu/food/animal/index_en.htm. Last visited in 2009.
Harrison, R., 1964. Animal machines. London: Vincent Stuart.
Hewson, C.J., Baranyiova, E., Broom, D.M., Cockram, M.S., Galindo, F.A., Hanlon, A.J., Hanninen, L., Lexer, D., Mellor, D.J., Molento, C.F.M., Odberg, F.O., Serpell, J.A., Sisto, A.M., Stafford, K.J., Stookey, J.M. and Waldau, P., 2005. Approaches to teaching animal welfare at 13 veterinary schools world wide. Journal of Veterinary Medical Education 32, 422-437.
Hughes, B.O., 1982.The historical and ethical background of animal welfare. In: Uglow, J. (Ed.). How well do our animals fare? Proc. 15th Annual Conference of the Reading University Agricultural Club, 1981, 1-9.
Fraser, D., 2008. Understanding animal welfare: the science in its cultural context, Oxford: Wiley Blackwell, pp. 324.
Irwin, M., 2001. How are stress and depression inter-related? Coping with Challenge: Welfare. In: Broom, D.M. (Ed.). Animals including man. Berlin: Dahlem University Press, pp. 271-288.
Koolhaas, J.M., Schuurmann, T., Fokema, D.S., 1983. Social behaviour of rats as a model for the psychophysiology of hypertension. In: Dembrowski, T.M., Schmidt T.H. and Blumchen, G. (Eds.). Biobehavioural bases of coronary heart disease. Basel (Switzerland), Karger, pp.391-400.
Lutgendorf, S.K., 2001. Life, liberty and the pursuit of happiness: good welfare in humans. Coping with challenge: welfare. In: Broom, D.M. (Ed.). Animals including man. Berlin: Dahlem University Press, pp. 49-62.
Mather, J.A., 1995. Cognition in cephalopods. Advances in the Study of Behavior 24, 316-353.
Mather, J.A., 2004. Cephalopod skin displays: from concealment to communication. In: Oller, K. and Greibel, U. (Eds.). Evolution of communication systems. The MIT Press: Cambridge, USA, pp. 193-213.
Mendl, M., Zanella, A.J. and Broom, D.M., 1992. Physiological and reproductive correlates of behavioural strategies in female domestic pigs. Animal Behaviour 44, 1107-1121.

Ouedraogo, A.P., 1998. Ethical consumers? Social representations of stock farming in France. In: Veissier, I. and Boissy, A. (Eds.). Proceedings of the 32nd Congress of the International Society for Applied Ethology, 204. Clermont Ferrand: I.N.R.A.

Panksepp, J. 1998. Affective neuroscience: the foundations of human and animal emotions. New York: Oxford University Press.

Pepperberg, I., 2000. The Alex Studies: cognitive and communicative abilities of grey parrots. Harvard University Press: Cambridge Massachusetts.

Radford, M., 2001. Animal welfare law in Britain: regulation and responsibility. Oxford: Oxford University Press.

Ryan, Y.M., 1997. Meat avoidance and body weight concerns: nutritional implications for teenage girls. Proceedings of the Nutrition Society 56, 519-524.

Serpell, J.A., 1986. In the company of animals. Cambridge: Cambridge University Press. pp.215.

Serpell, J.A., 1989. Attitudes to animals. In: Paterson, D. and Palmer, M. (Eds.). The status of animals: ethics education and welfare. Wallingford: C.A.B International 162-166.

Webster, J., 1994. Animal welfare: a cool eye towards Eden. Oxford: Blackwell, pp. 273.

Whay, H.R., Main, D.C.J., Green, L.E. and Webster, A.J.F., 2003. Assessment of the welfare of dairy cattle using animal-based measurements: direct observations and investigation of farm records. Veterinary Record 153, 197-202.

# Animal welfare: the risk manager's perspective

**Andrea Gavinelli and Maria Ferrara**
European Commission, Directorate General for Health and Consumers, Unit D5, Animal Welfare, Office: Rue Froissart 101, 1040 Brussels, Belgium; Andrea.Gavinelli@ec.europa.eu, Maria.Ferrara@ec.europa.eu

## Summary

In the European Union animal welfare today is conceived by many citizens as an attribute of an overall 'food quality concept' where animal farming is not seen just a means of food production but it is strictly related to the achievements of other key social goals such as animal health, food safety and quality, safeguarding environmental protection, sustainability and enhancing the quality of life in rural areas. In particular, the impact of food production on the life of rural populations has led to increased awareness of the need to develop animal farming systems able to interface the production process with the social and cultural aspects of the region they are located and with the specific product characteristics (Blokhuis *et al.*, 2008). This quite vast area of interaction where animal welfare is placed is reflected in one of the main principles agreed in the contest of the World Organisation for Animal Health (OIE) defining: 'animal welfare is a complex, multi-faceted public policy issue that includes important scientific, ethical, economic and political dimensions'. Policy making and the risk manager's perspectives in the animal welfare field are not just strictly limited to risk management, understood as reduction and prevention of risks. Instead the European policy making in animal welfare, including its international dimension, integrates the prevention and reduction of risks together with the promotion of benefits in terms of good animal and human life as well as in terms of livestock related livelihood.

Several policy options can be combined to provide parallel and complementary approaches to prevention/reduction of animal welfare risks as well as to the promotion of good quality of life. In this context, the overall aim of the European Commission in its policy making process is to initiate a broad debate to shape a widely accepted policy on animal welfare at European and international level that could continue to lead globally.

**Keywords**: animal welfare policies, ethics, socio-economic evaluations, outcome-based approach

## 1. Introduction

In the European Union (EU), legislation on animal welfare has a longstanding tradition in many Member States, going back to the 19$^{th}$ century. Looking back, the first European Community legislation on animal welfare was adopted in 1974 (European Community, 1974) indicating the importance that was then already attached to this issue and in particular to

the prevention of unnecessary suffering. Since that time until the adoption of the Community Action Plan on the Protection and Welfare of Animals 2006-2010 (European Commission, 2006), the EU Commission policy making has led the way to integrate AW in other areas of public concern, such as animal health, food safety and quality, and sustainability. Through an open and constructive debate with global stakeholders, based on science and ethical considerations, European policies and legislation have been evolving from simply ensuring animal protection, intended as avoiding unnecessary pain and sufferance to animals, to the promotion of 'good life' for animals.

## 2. Key elements for risk managers in animal welfare

Risk Management (RM) may be defined as the process of weighing policy alternatives in the light of the results of risk assessment and, if necessary, selecting and implementing appropriate control options, including regulatory measures. The fact that the use of Risk Assessment (RA) has considerably spread in biosciences in the latest years has affected the RM in the European Union (EU) in particular in relation to Animal Welfare (AW) policies both in the areas of farming animals, in the areas of laboratory animals and regarding certain species such as seals.

Having verified the existence and the nature of the risk at stake, the Competent Authorities (CA) are supposed to decide whether and how to act. Before doing that though, both national and EC CA are called upon to manage the risk threshold acceptable for the whole society (Alemanno, 2008). But the difficulties derived from the absence of international guidelines to define RA methodologies in AW and in particular to weight the social compound are clearly affecting the RM methodology.

First the establishment of a commonly recognized RA and RM methodology and then the identification and analysis of problems and their roots are therefore a crucial step towards addressing issues with proper policy objectives and means. Such evaluation has to be based on science and risk assessment analysis on one hand which needs to be evaluated while considering also socio-economic, environmental, traditional and ethical implications on the other hand. In the European approach to RM, such 'other legitimate factors' need to be integrated with the RA in order to design acceptable and efficient risk managers' decisions.

The need for more formal frameworks for animal welfare RA and RM is also influenced by recent international developments. The World Organization for Animal Health (OIE) has traditionally developed international animal health standards. The OIE has now moved to develop and publish international standards for animal welfare as well (OIE, 2008). Such international standards in animal welfare are based on scientific evidence. Accordingly, improvements in our ability to analyze animal welfare under varying scenarios will assist both in the development and further refinement of standards, as well as in their verification (Paton *et al.*, 2009).

As stated in the general principles guiding the work of the Farm Animal Welfare Council[1], knowledge based on scientific studies of the welfare of animals is increasing rapidly. The term animal welfare is employed frequently in scientific and legal documents and in public statements. In our view, welfare encompasses the animal's health and general physical condition, its mental state, its biological fitness and its ability to cope with any adverse effects of the environment in which it finds itself.

In the areas of policy making related to AW the Community has recently developed a methodology for the identification and analysis of problems which take into account both the new scientific information and the information derived from governmental authorities and stakeholders interacting with the animals. This approach is specific to AW and differs with the one adopted in the area of animal health policies where the ethical compound is not considered at the same level of the scientific assessment.

The reason of this particular approach mainly is provided by the Treaty of the European Union (European Community, 1997). The Treaty states that when formulating policies in several areas related to animals several ethical, cultural and religious elements have to be strictly considered. This is an important definition of the borders within which the managers of AW have to move. The question is mainly if the borders defined by the Treaty correspond to 'enough' AW for the animals or more clearly if these borders are responding to what managers are expected to do after the results given by the RA process.

The process should establish clear objectives that correspond to problems and their causes as well as possible improvements. Objectives may apply at different levels and it is often necessary to define a clear hierarchy and a particular logic to create linkages among them. The objectives established should also be consistent with other policy considerations. When we come to consider AW, one of the major challenges in defining objectives is to make them precise enough to prevent varying interpretation and allow an evaluation of the results achieved (Gavinelli *et al.*, 2008) that could be recognized and evaluated at international level.

In risk management and policy making in the animal welfare field, the risk manager has to give proper attention to those 'other legitimate factors', particularly when it comes to European policies concerning the ethical perspective of animals as 'sentient beings' as identified in the EC Treaty, and to the moral duty society has, to take care of animals' needs respectfully.

But more specifically, what is AW intended to mean to the Europeans and which are the objectives of policy options? Are they merely aimed at ensuring the respect of animal needs while preventing welfare risks or are they aimed at ensuring more broadly a good life to animals? And consequently how AW is assessed to evaluate the results of policy actions?

AW and its management has always been in the last 30 years a cross-disciplinary and multidisciplinary science where both assessment schemes and an epidemiological approach are being developed and where the evolving of the society is playing an important role.

---

[1] http://www.fawc.org.uk/default.htm

When identifying and analysing the problems, risk managers are aware that many factors of the animal itself (e.g. age and breed), its housing (e.g. type of system) and in the management (e.g. feeding routine and handling) can affect the risk of reduced welfare. RA in this way can be considered as measuring the 'output', that is whether or not these risks were actually realised (Blokhuis *et al.*, 2008). Therefore one possible option to risk managers in the policy area is to develop standards to protect the animals and to ensure that their welfare should be based on welfare assessment as well as on the reduction of identified risks and promotion of benefits to welfare, through preventive and corrective actions when there is a problem. However this equation does not have the capacity to evaluate the impact derived by the cultural diversities.

Farm animal welfare measures may have an important impact on animal productivity on one hand and on production costs, on the other hands. Therefore the objectives of such policies need to be based also on socio-economic evaluations, and should be measurable through an overall welfare assessment system integrating both the output of reduced risks and improved benefits. Given the complexity of animal welfare dimension and the possibility to complement welfare assessment and risk assessment policy options in different production systems, animal-based outcomes seem to offer the most appropriate way to monitoring and evaluating a policy option in this field (Figure 1).

Several policy options, such as the regulatory approach or a voluntary standards-market driven approach, can be undertaken or combined to achieve identified animal welfare objectives which are strictly linked and integrated also into animal health and food safety goals.

The EU has traditionally taken a regulatory approach, given the longstanding tradition of animal welfare legislation in many Member States going back to the 19th century. A broader public debate in the earlier 1970s revealed the need to harmonise the production conditions

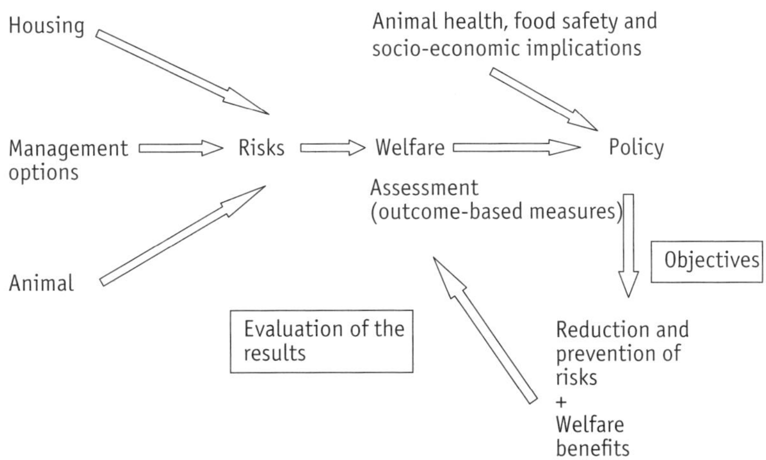


*Figure 1. Key elements for the policy making in animal welfare: relationship between risk assessment, welfare assessment, socio-economic implications and policy options.*

in farms of the Member States and to respond to a growing societal demand for improving the welfare of animals. On this basis, a relevant body of legislation was developed over the last 30 years driven by public concern about specific topics and addressing the protection of animals on the farm, during transport and at the time of killing (Gavinelli *et al.*, 2007a).

However, higher welfare standards are now more related to private certification schemes. In this context, it should therefore be considered whether the legislative approach today represents the best opportunity to achieve higher animal welfare standards or if there are other means such as encouraging stakeholders to put in place voluntary mechanisms that stretch across Europe and abroad and that could extend outside EU borders (Gavinelli *et al.*, 2008).

If it is presumed that the required levels of animal welfare are determined by society as guided by science and ethics, then at what level should the law stop and, for example, be replaced by market driven standards?

The Community Action Plan on the Protection and Welfare of Animals 2006-2010 (European Commission, 2006) indeed emphasised for the first time the need for a more market-oriented approach to animal welfare, encouraging operators to develop their own certification schemes and guidelines always maintaining a close link to science. This would imply a new regulatory approach where legislation provides for a legal framework in support of voluntary approach through a common methodology without providing comprehensive requirements. In particular, this approach could be developed and marketed in a fairly consistent way for systems of production that are well integrated.

The European legislation on the protection of laying hens represents in this sense a case study where all these different approaches have been integrated. Council Directive 1999/74/EC of 19 July 1999 (European Community, 1999) lays down minimum standards for the protection of laying hens and provides that the rearing of laying hens in conventional cages will be prohibited from 1 January 2012. In January 2008 the Commission published an official report (European Commission, 2008) that reflects how the different animal health and welfare, environmental and public health aspects together with socio-economic implications and effects on the Community's economic partners need to be all combined to outweigh the overall benefits of the ban of battery cages by 2012 versus the economic implications of this decision.

Science has shown that conventional cages cause several welfare problems and alternatives such as free range systems, despite the concerns highlighted by the experts for certain aspects in the area of public health, have the capacity to attract citizens' preference and with proper management to advance all the main aspects related to 'food quality'. Today the consumption trends in significant EU 15 Member States reports a regular increase of citizens' interest in non-cage systems and a number of retailers and food service operators are marketing predominantly non-caged eggs.

Therefore, based on scientific information and socio-economic evaluations, the abovementioned analysis led the Commission to not recommend the postponement of the ban of the battery cages but on the other hand to accelerate the introduction of higher animal

welfare standards. Consumer information campaigns, cooperation between stakeholders and the development of a transparent mechanism to make animal welfare labelling trusted by citizens are a powerful mean to characterize and step up EU production in the eyes of sustainable consumers.

## 3. The interaction between ethics, consumers' attitudes and the achievements of benefits versus risk management elements

The question is, which areas of concern should be addressed first by risk managers, when the issue is to give to 'sentient beings', as animals are defined in the Treaty of the European Union (European Community, 1997), their capacity to enjoy a life worth living?

AW is about how animals feel and suffer, therefore while taking risk manager decisions, should priority be given to hazards/factors with low prevalence but which cause very severe adverse effects such as considerable suffering and death of animals than to others that are prevalent but less significant to the animals? Should a hazard/factor leading to death of the animal be considered as more unacceptable than a hazard/factor leading to prolonged stress and suffering of the animal? While addressing the welfare of farmed animals, should animal welfare be assessed and risks evaluated on individual bases or targeted on animal population? In farming of veal calves, is it sufficient to evaluate average blood haemoglobin level or should it be considered as absolute value for each single animal? Is it ethical in a given production system such as broiler farming (or aquaculture) to take care just of population and averages regardless of the state of each single individual? In an integrated approach to animal welfare such as the one undertaken by the EU, up to which level is it justified to compromise animal welfare versus over-production, food safety and animal health aspects?

While accepting that AW is all about the animal, protecting animals involves some form of input from society, or society as represented by governments, to weigh different aspects [e.g. animal welfare, food safety, economy, environment] and interests of different groups [e.g. consumers, farmers, retailers] (Blokhuis *et al.*, 2008). All these aspects have to be considered and pondered by risk managers while replying to the abovementioned questions and designating the best options for addressing identified welfare risks.

However, while replying to those questions, risk managers should also keep in mind that through policy options and legislation they do not just prevent welfare risks and avoid suffering but they do also guarantee welfare benefits and good life for animals and create an opportunity to correspond to social needs.

The adoption of 'prop 2' in California banning several forms of confinement for farm animals is clearly showing that there is a point where animal confinement exceeds what the public will support and *'if industry practice and public tolerance are not synchronized, animal advocates may pursue methods that produce more sudden 'corrections' with the possibility of unintended consequences for both people and animals'* (Emily Patterson-Kane, 2009). These considerations could introduce additional elements to RM going far from the possible recommendations derived from a RA developed into a stricter scientific framework.

Developing the methodology to assess policy making able to guarantee a 'good life' for the animals in a more general term than the pure scientific based assessment, appears today to be a complex but inevitable exercise.

As presented by Peter Sandoe and Stine Christiansen (2008), three different philosophical approaches may well deliver similar conclusions about what it takes to have good life, but for different reasons:
- *Hedonism*: a good life is one in which there is sufficient quantity of positive experiences and sufficiently few negative experiences.
- *Perfectionism*: a precondition of a good life is to be able to realise significant species-specific potentials.
- *Preference theory*: to have a good life it is necessary to achieve what one wants or strives for- that is, to have one's preferences satisfied.

Such approaches, combined in various kinds of hybrid views, could be differently used for defining good life for animals and reflected both in society and in the risk managers' evaluation to be done while designing policy actions. The choice of parameters to measure animal welfare and the interpretation of the results involve ethical reflections both on the requirements of good animal life and on the nature and extent of our duties to animals.

When it comes to defining policy options to be implemented in 27 different countries of the European Union, ethics and inputs from society have to be filtered and interpreted while taking into account and balancing at the same time cultural and traditional differences, based on different political and historical backgrounds and eventually different husbandry systems. Europeans have quite different perceptions of their bounds with animals and their acceptance of intensive farming practices is a lot related to it. Beside what is foreseen in European Directives, some governments in the EU may implement European animal welfare legislation with a different pressure or introduce earlier than in others laws which outlaw, with a transitional period, certain animal production system. The use of conventional cages for laying hens was banned in Sweden already in the late 1980s, Austria has recently banned the use of all types of cage for laying hens, and in the early 1990s the UK banned the keeping of gestating sows (Sandoe and Christiansen, 2008) in stalls more than 20 years earlier compared to the European Directive laying down minimum standards for the protection of pigs.

Although the initial aim of EU decision to create animal welfare legislation has been based on the need to eliminate distortions of competition resulting from varying conditions of animal production, today animal welfare has its place on the EU agenda in relation with the ethical duty as well as the political responsibility of European Institutions towards European citizens to ensure a respectful treatment of farmed animals.

Indeed European citizens in general have a say about animal welfare and their farming, as highlighted in two comprehensive opinion polls on the issue carried out by the Commission, interviewing nearly 25,000 people in the 25 Member States of the EU in 2005 (TNS Opinion & Social, 2005), and more than 29,000 in 2006, already including Bulgaria and Romania and the two candidate countries [Croatia and Turkey] (TNS Opinion & Social 2007). Both surveys demonstrate that there is a considerable interest in animal welfare standards, which is

reflected in the demand for more information and in the awareness that European public commitment is relevant for higher welfare standards. The second survey showed that animal welfare is an issue that Europeans rank highly – giving it 8 out of 10, on average, in terms of importance. The high level of importance given to animal welfare is in line with the results of the first survey, in which 55% of respondents said that they believe animal welfare and protection do not receive enough importance in their country's agricultural policy.

The attitudes towards animals and derived products are probably the most complex parameter to fit in a risk assessment based process and the Commission has made specific efforts in this area developing a long term strategic policy document (the Community Action Plan on the Protection and Welfare of Animals) that combines several of the main components of the risk assessment process.

In the Community Action Plan scientific, ethical and economic elements are considered the tools for policy making. The combination of them in order to assess policy making strategics implies continuous consultation with the stakeholders but at the same time makes information to citizens available in order to adopt a transparent approach.

It is therefore important that everyone understands the ethical value of animal welfare and the importance of respecting the animals needs as the ultimate goal of the European policies. In this context proper communication and in particular risk communication to the sectors concerned demonstrated to play an important role. The farmer's perception of what matters for the welfare of the animals may have a very significant effect to provoke the shift towards an improved level of animal welfare.

The integration of ethics, consumers' concern and risk assessment into risk management in the animal welfare field requires therefore resources combined together in an innovative way. The reasons of this complexity are now clearer but the way to combine them and the efficiency of the proposed solutions are still under study. The EU funded project WelfareQuality® is mainly working to the methodology to solve this complex formula.

## 4. How risk management and the achievement of benefits are integrated into the European policy making

The Protocol on Protection and Welfare of Animals annexed to the EC Treaty in 1999 [the Treaty of Amsterdam amending the Treaty on European Union] (European Community, 1997) represents a milestone for the development of the Community's animal welfare policy, highlighting its ethical dimension. This Protocol spells out the obligation to pay full regard to the welfare of animals as sentient beings when formulating and implementing the Community's policies. The legal recognition of animals as 'sentient beings' has been recently reconfirmed by the Lisbon Treaty.

The link between animal welfare, animal health and food safety was subsequently highlighted in 1999 in the Community White Paper on Food Safety (European Commission, 2000), integrating animal welfare into the food chain policy of the EU as well as in the recent new Animal Health Strategy 2007-2013 (European Commission, 2007). Furthermore, recent reforms of the EU's Common Agricultural Policy strengthened the role of animal welfare considerations by introducing the 'cross-compliance' principle – making compliance with animal welfare requirements a condition for certain farm subsidies.

The way policy making is shaped into the EU, takes into account the growing and global dimension of animal welfare and the related scientific and societal implications.

Nowadays globalisation of trade in animal products in association with increasing demand for animal protein has resulted in a sharp increase in the number of animals slaughtered for meat worldwide. This raises in citizens concerns about the sustainability of the most intensive form of animal production, in particular in relation to their impact on the environment and the countries' social structures (World Society for the Protection of Animals, 2007). As reported by the Food and Agriculture Organisation of the United Nations [FAO] (FAO, 2006), livestock play an important role in the current environmental crisis. They can lead to biodiversity loss and ecosystem changes as a result of habitat change, climate change, overexploitation and pollution, at the local and global level.

It then becomes necessary to consider in the assessment of the welfare aspects of animal farming systems their capacity to interface with different elements mentioned above and if possible to contribute through their sustainability. In this context and on a global scale, farm animal welfare constitutes an important impact on the food chain as it requires visibility of production processes and an understanding of how they affect welfare as well as transparent systems to guarantee welfare quality of animal-based food (Blokhuis *et al.*, 2008).

Moreover, as it has been recently recognised by several international organisations such as OIE and more recently FAO, '*Improvements to animal welfare in food production systems can play a significant role in improving the welfare of people by such means of improving access to food of animal origin, improving economic returns through increased livestock productivity, improving the efficiency of draft animals, and reducing risks to human health through improved food safety and animal health*' (FAO, 2009). Similarly the International Finance Corporation emphasised this potential integration of animal welfare stating that: '*Higher animal welfare standards are also increasingly seen as to be a prerequisite to enhancing business efficiency and profitability, satisfying international markets, and meeting consumer expectations*' (IFC, 2006a,b).

The EU Commission policy making in this field is leading in trying to develop the methodology that could integrate animal welfare in the above mentioned areas of public concern. In order to do so a broad public debate on animal welfare is promoted with major international stakeholders and within multilateral organisations which should allow the shaping in the future of more coherent and widely accepted policies which address animal welfare in all related policy fields such as agriculture, environmental policies and research, while taking into account the socio-economic impact of any new measures.

Risk managers and policy makers in the EU are developing national strategies to approach AW and define their future policy line. It is interesting here to refer to the 'Animal Health & Welfare Strategy's five themes'[2] developed by the UK Department for Environment Food and Rural Affairs (DEFRA).

The five themes of the UK Strategy are defined as important elements to make sure that the Strategy's vision is achieved. These basic management tools include:

1. working in partnership;
2. understanding and accepting roles and responsibilities;
3. 'prevention is better than cure';
   'animals that are cared for appropriately and in accordance with existing welfare standards are more likely to be healthy, and less likely to contract or spread disease';
4. understanding the costs and benefits;
5. delivering and enforcing standards.

DEFRA strategic themes are close to the integrated EU Commission approach, which makes all stakeholders of the food chain (from farmers, to retailers and consumers) actors and part of the process to improve animal welfare, and which has been well received worldwide. Such a coordinated approach through several policy areas needs to be based on sound science, clear identification and analysis of problems and socio-economic aspects as well as shared definition of goals.

Since informed consumers could support, with their purchasing behaviour, the development of sustainable farming practices, the Community Action Plan suggests introducing improved marketing, labelling and communication strategies in order to meet this goal and to enable producers to benefit from market opportunities. Such a system of classification needs to be based on standardised scientific indicators well recognised both in the EU and internationally, and underpinned by risk assessment and research, in order to facilitate the marketing of these products (Gavinelli *et al.*, 2007a).

Such integrated models requires that the specific communication of AW risks, in order to acquire the views of citizens and major stakeholders, has to be performed systematically and profiting of new technologies such as the Eurobarometer polls or web based consultations. However if animal welfare science today is growing to be a recognised and established field of work, the establishment of an harmonised and internationally recognised risk assessment methodology in animal welfare could strengthen the current European policy making in this area and help discussion and negotiation on the issue in multilateral organisation.

Still it is not clear how far the ethical component of the issue could be the limiting factor to make a common risk assessment methodology acceptable internationally. However, if the approach adopted today by many scientists to put the animal at the centre of the assessment will demonstrate its universal validity, there will surely be the opening up of new perspectives in approaching and regulating this issue.

[2] http://www.defra.gov.uk/animalh/ahws/themes.htm.

Fish welfare is an emerging area which has not been studied to the same extent as mammals and birds. Whilst similar measures of welfare developed for other animals are often relevant to fish, clearly defined protocols for fish welfare evaluation are lacking. However, the balance of evidence indicates that some fish species have the capacity to experience pain (EFSA, 2009a) and risk assessment applied to welfare of fish, although not yet harmonized, can certainly help risk managers to identify major welfare risks at farms as well as at killing level to prioritize future policy actions and research fields.

When in such new areas limited amount of quantitative data and published peer-reviewed data are available and risk assessment is mainly based on expert opinion (EFSA, 2009b), this aspect together with the level of uncertainty needs to be reflected in the risk manager evaluation before taking a decision. Although risk assessment and risk management are functionally well separated in the European model, particularly in this case and in the multidisciplinary animal welfare field, communication between risk managers and risk assessors and understanding of needs for and of elements and outcomes of the risk assessment process, are crucial.

Although peer reviewed information on a certain husbandry or slaughter practice is lacking, when practical experience and common sense identify that practice as unacceptable in an animal welfare view point, can the risk manager take action even if the dimension of the risk is not yet totally defined by weighting the implications of non action versus action? In such an integrated European approach to animal welfare and with the different policy tools foreseen in the Community Action Plan, which goes beyond merely a piece of legislation, it is possible to address such concerns, risk assessment, therefore, being an important tool for identification of problems.

## 5. Do we have a golden formula to integrate in a global perspective risk assessment and risk's manager decision making in animal welfare?

*'Animal welfare is a human responsibility that encompasses all aspects of animal well-being, including proper housing, management, nutrition, disease prevention, responsible care, humane handling, and, where necessary, humane euthanasia'* (Canadian Veterinary Medical Association)[3].

*'The health of animals is as important as their well-being. Health and welfare cannot be seen as separate issues. A weakened health status will affect the animals' welfare and when welfare is compromised, the animals will be more susceptible to infections and disease'* (Federation of Veterinarians of Europe)[4].

Are those statements the proper way to explore and develop a global approach to RM in the AW area? Is it relevant and possible to insist on a global approach to RM?

---

[3] http://canadianveterinarians.net/animal.aspx.

[4] http://www.fve.org/news/publications/pdf/policies_actions_08.pdf.

The prototype of approaching animal health issues on a global scale has shown several important and recognized advantages and it is likely that the synergies derived by the development in the future of RM methodologies in AW could facilitate the improvement of the quality of human and animal lives.

RM decision on animal welfare issues in the EU are based on science and on social, economical and ethical evaluations. In particular, based on science and risk assessment, the European Community is building on the vision of integrating the farming of animals in good health and welfare conditions with the respect of other issues such as the safety of food and the respect for the environment in order to bring a real benefit to the animals, to the global society and ensure a sustainable trend of the global trade of animal products.

Legislation, voluntary and international standards, science, communication and information, and capacity building are the main pillar of the European formula of integrating risk assessment into risk management, which has already allowed the EU to take the global initiative in improving farm animal welfare.

However, to make of this approach definitely a golden formula to be tailored and applied globally, some elements still need to be defined such as the development of animal-based indicators and their integration into a harmonised risk assessment methodology. Policy options on animal welfare, including in the international field, will be broadly understood, shared and applicable only if built on an outcome-based approach, allowing a transparent evaluation of animal welfare objectives in terms of 'outcomes' rather than in terms of 'inputs'. In this context, comparison and evaluation of risks in equivalent farming practices will be possible as long as welfare indicators are integrated at the core of the risk assessment process in order to be directly translated into policy options, including the international context.

As Commissioner for Health Androulla Vassiliou stated at the opening of the Conference on Global Trade and Farm Animal Welfare '*The improvement of farm animal welfare is a challenge that the Commission embraced many years ago. However, lasting improvements can only be achieved by common understanding, engagement and international partnerships.*'[5]

When carrying out risk assessment in the animal welfare field, communication and common understanding between risk managers and risk assessors is crucial in order to clearly understand needs and results of risk assessment. In case of uncertainty, transparency in the process needs to be ensured in order to allow risk managers to weigh the risks correctly, even when those are not completely defined. Such transparency in the process is even more relevant when animal welfare risk assessment is to be applied to international policies and to build a constructive international relationship. As said, the global dimension of AW is a key RM dimension and would become even more relevant in the future. This dimension demands even more sophisticated models for RM due to the multidimensional nature of the issue.

The Community Action Plan states that the EU should engage and assist developing countries in implementing animal welfare measures, in order to create a win-win situation for all actors.

[5] http://europa.eu/rapid/middayExpressAction.do?date=19/01/2009.

In particular, the European Commission is committed to work with international organisations such as the OIE for promoting the development and implementation of internationally recognised animal welfare standards as well as with the FAO for supporting capacity building activities on animal welfare in countries with emerging economies.

Such kind of initiatives is based on the vision that, in order to collaborate on the improvements and implementation of animal welfare standards, we need to reflect on equivalent systems rather than identical ones. With this vision, for several years now, the Commission has been actively promoting and discussing animal welfare issues and the prospect of internationally recognized animal welfare standards globally. Several positive achievements have been reached by considering animal welfare into bilateral and multilateral veterinary agreements and by supporting technical initiatives and conferences all around the world, from Latin America, to China and Africa, to reflect on the benefits and opportunities of improving animal welfare practices and standards as well as to consider animal welfare as an added value to trade while delivering food quality and respecting the regional and local dimensions.

As it was highlighted at the 'Conference on Global Trade and Farm Animal Welfare'[6] organised in January 2009 by the European Commission, there are many competing and complimentary issues interacting on farming of animals at global level: starting from remaining competitive in a global marketplace, to the delivering of food for which the demand is projected to double by 2050 and responding to consumers preoccupations on several issues such as animal welfare and last but not least the challenge of climate changes. There are forms of farming in intensive conditions where the welfare for animals is poor to such an extent that are likely to be unacceptable at global level. Moreover, international consensus exists that higher animal welfare standards should be promoted globally, in particular to open new market opportunities for countries with emerging economies, and that '*animal welfare should not be treated as a stand-alone issue but as one among many socially important goals including food safety and security, human and animal health, environmental sustainability, worker safety, rural development, gender equality, and social justice*' (FAO, 2009).

However, international implementation of multilaterally agreed animal welfare standards while respecting the regional and local dimensions as well as international acceptance of animal welfare practices should be based on identification of risks and equivalent systems, good practices and 'outcomes'. As highlighted by a recent FAO initiative, '*As a general approach, improving the welfare of the animals should begin with an assessment of the risks and opportunities in the entire system or production chain, and a search for improvements that will be practical in the given situation. Assessment should include science-based assessment of the needs and welfare of the animals, and risk assessment to identify causes of sub-optimal welfare*' (FAO, 2009).

The development of animal based welfare indicators and of a risk assessment methodology has been indicated as priority to facilitate the assessment of animal welfare in different farming conditions, facilitate their comparison and engage with Third Countries, particularly

[6] http://www.animalwelfareandtrade.com/.

those with emerging economies, to improve animal welfare and related livestock livelihood globally.

As highlighted in the seminar 'Sustainable Agricultural Production and Good Animal Welfare Practices: Trade Opportunities for Developing Countries', organised by the European Commission in the context of the Civil Society Dialogue initiative, extensive and sustainable agricultural systems, with good standards of animal welfare, are still the predominant form of livestock production in many developing countries; products from such systems would readily meet EU animal welfare requirements. The EU should therefore ensure access to its markets for welfare-friendly products, by offering trade-related assistance and capacity building to developing countries together with information, training and mentoring in the development and maintenance of good welfare standards on-farm, during transport, and at slaughter (Gavinelli *et al.*, 2007b).

In shaping international policies and collaborations on animal welfare and while promoting it in the international trade context, risk managers therefore have to consider and evaluate different elements including WTO rules, competitiveness issues for European producers, transparent consumers information on the farming systems applied to produce animal products, food safety, animal health, ethical and non trade concerns, to be integrated with risk assessment and objective comparison of husbandry practices around the world.

## 6. Conclusions

### 6.1 What has been achieved?

Policy formulation in AW is an interactive process where regular revision and analysis is needed as the situation evolves. There has been a general trend in recent years to develop public policies with a perspective that is broader than simply imposing legislation for specific means designed to deliver a specific end.

In the European Union AW today is conceived by many citizens as an attribute of an overall 'food quality concept' where animal farming is not seen just as a means of food production but it is strictly related to the achievements of other key social goals such as animal health, food safety and quality, safeguarding environmental protection, sustainability and enhancing the quality of life in rural areas. In particular, the impact of food production on the life of rural populations has led to increased awareness of the need to develop animal farming systems able to interface the production process with the social and cultural connotations of the regions where they are located and with the specific product characteristics demanded by the market (Blokhuis *et al.*, 2008).

Since the adoption of the first Directive on the protection of animals at the time of slaughter in 1974 until the adoption of the Community Action Plan in 2006, the EU Commission policy making has led the way to integrate AW in the above mentioned areas of public concern and in all related policy fields. Today it is clear that only an open and constructive debate with global stakeholders, based on scientific and ethical considerations could allow this process

to develop. On this basis, European policies and legislation have been evolving from simply ensuring animal protection, intended as avoiding unnecessary pain and sufferance to animals, to the promotion of a specific approach to AW intended to make sure that animals could have a 'good life'.

In this context, animal welfare science, including its societal component, has played a crucial role in supporting such kind of approach and new policies options are studied and developed where human-animal interrelation are meaningful tools to improve animal welfare.

The same multidisciplinary approach and the strong belief in the added value of animal welfare has recently opened new areas of work and collaboration in the international field with main countries trading with the EU as well as with countries with emerging economies where animal welfare can represent an opportunity to improve the production and the related livelihood.

### 6.2 What has been neglected?

Animals have an intrinsic value in our society and in turn, society has a duty to protect the animals and ensure for them a good life. The European approach to animal welfare has already achieved important results and taken the international leadership in this field. However, animal welfare is a policy area which has to evolve according to science and societal concerns. In particular the concerns that AW could be seen as a threat to business as usual indicates that the link between sustainable forms of production and AW and the favourable long term impact on the agricultural activities demands more investments and considerations.

Proper communication and information could contribute to raise awareness in the society of the advantages derived from AW policies encouraging the 'rethinking of our values in relation to the meaning given to quality of life' and accepting the challenge to bring AW into the centre of the sustainable agenda (Rowles, 2006). Nevertheless, the formula to objectively measure the good life for animals regardless of their farming system and its interaction with ethical and social policies is still missing.

### 6.3 What needs to be done?

Legislation, voluntary and international standards, science, communication, information, and capacity building are the main pillars of the European formula of integrating risk assessment into risk management in animal welfare. Driving forces behind AW in different regions of the world are increasing production/efficiency, securing sustainability, and ethical concerns. Probably the best option for assessing AW policies will be to profit and to build around the concept of 'cross-fertilization' intended as the interaction or interchange between more cultures, fields of activity or knowledge in a way to be mutually beneficial and productive.

This approach requires a clear dialogue with all the stakeholders and not only the ones strictly related to the food chain. Such dialogue has to be based on a shared international understanding of AW. Moreover, long terms achievements do also require the development

of long-standing partnerships between international organisations such as the EU, OIE, FAO and the World Bank.

The role of animal welfare in the WTO context also needs to be clarified in order to proper orient AW policies particularly in countries with emerging economies and in order to catalyse resources from international organisations. In particular, international partnership is needed to develop capacity building programmes to implement internationally shared animal welfare standards.

## Note

The views expressed herein are purely those of the authors and may not in any circumstance be regarded as an official position of the European Commission.

## References

Alemanno, A., 2008. The shaping of European risk regulation by community courts, Jean Monnet Working Paper 18/08. Available at: http://www.jeanmonnetprogram.org /papers/08/081801.html. Accessed 2 June 2009.

Blokhuis, H., Keeling, L., Gavinelli, A. and Serratosa J., 2008. Animal welfare's impact on the food chain. Trends in Food Sciences & Technology 19, 79-87.

EFSA, 2009a. Scientific opinion of the panel on animal health and welfare on a request from European Commission on general approach to fish welfare and to the concept of sentience in fish. The EFSA Journal 954, 1-26.

EFSA, 2009b. Scientific opinion of the panel on animal health and welfare on a request from the European Commission on welfare aspect of the main systems of stunning and killing of farmed Atlantic salmon. The EFSA Journal 1012, 1-3.

European Commission, 2000. White Paper on food safety, COM (1999) 719 final, http://europa.eu./comm/dgs/health_consumer/library/pub/pub06_en.pdf. Accessed 2 June 2009.

European Commission, 2006. Communication from the Commission to the European Parliament and the Council on a community action plan on the protection and welfare of animals 2006-2010, COM (2006) 13 final. Available at: http://ec.europa.eu/food/animal/welfare/com_action_plan230106_en.pdf. Accessed 2 June 2009.

European Commission, 2007. Communication from the Commission to the Council, the European Parliament, the European Economic and Social Committee and the Committee of the Regions A new animal health strategy for the European Union (2007-2013) where 'Prevention is better than cure', COM 539 (2007) final. Available at: http://ec.europa.eu/food/animal/diseases/strategy/docs/animal_health_strategy_en.pdf. Accessed 2 June 2009.

European Commission, 2008. Communication from the Commission to the European Parliament and the Council on the various systems of rearing laying hens in particular those covered by Directive 1999/74/EC, COM(2007) 865 final. Available at: http://ec.europa.eu/food/animal/welfare/farm/comm_pdf_com_2007_0865_f_en_acte.pdf. Accessed 2 June 2009.

European Community, 1974. Council Directive 74/577/EEC of 18 November 1974 on stunning animals before slaughter, OJ L 76, 3.7.1974, p. 52.

European Community, 1997. Treaty of Amsterdam amending the Treaty on European Union, the Treaties establishing the European Communities and certain related acts: Protocol annexed to the Treaty on protection and welfare of animals, Official Journal of the European Union, C 340, 10/11/1997, p. 110.

European Community, 1999. Council Directive 1999/74/EC of 19 July 1999 lays down minimum standards for the protection of laying hens, Official Journal of the European Union, OJ L 203, 3.8.1999, p. 53.

European Community, 2007. Treaty of Lisbon amending the Treaty on European Union and the Treaty establishing the European Community, Official Journal of the European Union, C 306, 17/12/2007, p.1.

Food and Agriculture Organization of the United Nations (FAO), 2006. Livestock's long shadow, environmental issues and options. Rome, Italy pp. 181-215. Available at: ftp://ftp.fao.org/docrep/fao/010/a0701e/a0701e00.pdf. Accessed 2 June 2009.

Food and Agriculture Organization of the United Nations (FAO), 2009. Capacity building to implement good animal welfare practices. Executive Summary FAO expert meeting, FAO Headquarters (Rome), 30 September October 2008. Available at: http://www.fao.org/ag/AGAinfo/resources/documents/AW/AW_Exp-meeting_EN.pdf. Accessed on 2 June 2009.

Gavinelli, A., Rhein, C. and Ferrara, M., 2007a. European policies on animal welfare and their effects on global trade. Farm Policy Journal 4, 11-21.

Gavinelli, A. and Park, M., 2007b. Farm animal welfare In: Legislatures, corporate boardrooms, and private kitchens. The state of the animals IV: 2007.

Gavinelli, A., Ferrara, M. and Simonin, D., 2008. Formulating policies for the welfare of animals during long distance transportation. Veterinaria Italiana 44, 71-86.

International Finance Corporation-World Bank Group, 2006a. Animal welfare in livestock operations, good practice note, October 2006, no. 6. Available at: http://www.ifc.org /ifcext/enviro.nsf/AttachmentsByTitle/p_AnimalWelfare_GPN/$FILE/AnimalWelfare_GPN.pdf. Accessed 2 June 2009.

International Finance Corporation-World Bank Group, 2006b, Quick note: creating business opportunity through improved animal welfare. Available at: http://www.ifc.org/ifcext /enviro.nsf/AttachmentsByTitle/p_AnimalWelfare_QuickNote/$FILE/Animal+Welfare+QN.pdf. Accessed 2 June 2009.

Paton, M., Fisher, A., Martin, T. and Giraudo A. Practical uses of risk assessment method in animal welfare. Available at: http://www.daff.gov.au/__data/assets/pdf_file /0004/1046497/37-michael-paton.pdf. Accessed on 2 June 2009.

Rowles, K., 2006. Sustainable development and animal welfare: the neglected dimension, In: Turner, J. and D'Silva, J., (Eds) Animals, ethics and trade – the challenge of animal sentience. Available at: http://www.ciwf.org.uk/includes/documents/cm_docs/2009/r/raw Accessed 16 June 2006.

Sandoe, P. and Christiansen, S.B., 2008. Ethics of animals use. Blackwell Publishing.

TNS Opinion & Social, 2005, Attitudes of consumers towards the welfare of farmed animals, Special Eurobarometer 229/Wave 63.2. Available at: http://ec.europa.eu/food /animal/welfare/euro_barometer25_en.pdf. Accessed 2 June 2009.

TNS Opinion & Social, 2007, Attitudes of EU citizens towards Animal Welfare, Special Eurobarometer 270/Wave 66.1. Available at: http://ec.europa.eu/food /animal/welfare/survey/sp_barometer_aw_en.pdf. Accessed 2 June 2009.

World Organisation for Animal Health (OIE), 2008. Animal welfare, in Terrestrial animal health code, Section 7. Available at: http://www.oie.int/eng/normes/mcode /en_titre_1.7.htm. Accessed 2 June 2009.

World Society for the Protection of Animals, 2007. Industrial animal agriculture-part of the poverty problem. 31 May, London, United Kingdom.

# Animal welfare indicators and their use in society

**Andy Butterworth**
University of Bristol, Clinical Veterinary Science, Langford, N Somerset, BS40 5DU, United Kingdom; andy.butterworth@bris.ac.uk

## Summary

European consumers expect their animal-related food products, to be produced and processed with respect for the welfare of the animals and the focus of EU agricultural policy is increasingly on quality rather than quantity. Traditional price mechanisms do not always allow for considerations like animal welfare to be properly recognised in the prices paid to producers, and, in general, compulsory labelling of food with animal welfare information has not been adopted. Existing assurance schemes tend to assess welfare by examination of the provision of housing or resources (Resource Based Measures RBMs), rather than looking at the animals themselves (Animal Based Measures ABMs). Practically speaking, it is may be possible to combine RBMs and ABMs. For example, if animals are lame (assessed using an ABM) it may be possible to predict lameness if the floor condition is poor (an RBM). Research scientists have for some time suggested that ABMs could provide valid indicators of animal welfare; since welfare is a characteristic of the individual animal, not just of the system in which animals are farmed. The sorts of questions which, in general, are being asked are: (1) Are the animals properly fed and supplied with water?, (2) Are the animals properly housed?, (3) Are the animals healthy?, and (4) Can the animals express a range of behaviours and emotional states? To implement effective use of animal based assessment methods, it seems likely that management support is a critical part of an effective process. Step 1, Measure (ABMs and RBMs) → Step 2, Analyse risk factors → Step 3, Inform (producer, purchaser)→ Step 4, Support management decisions to create improvements in welfare. Once measures have been carried out on a farm, it may be possible to create a range of 'scores'. The individual measures can be combined to give aggregate scores which are considered meaningful, and credible, by the producer and the consumer. The process of combining scores must not devalue the overall meaning of the assessment information, for example, by compensating a very poor score in one area, with a number of 'OK' scores in other areas. Once measures have been made on the farm, the 'natural' next step is to ascribe different weighted values to these measures, to assess the impact of each measure with respect to animal welfare. As well as the potential benefits of the use of animal based measures, there are clearly questions. Who will carry the cost, can the measures be made in a repeatable and reliable way within the timescale of an routine assessment, how would they work in relation to changing seasons, can a single farm-based score provide useful information, and can ABM's fit into existing assessment frameworks?

**Keywords:** farm assessment, animal based measures, outcome based measures, resource based measures, veterinary advice, management support

## 1. Introduction

European consumers expect their animal-related food products, to be produced and processed with respect for the welfare of the animals (Veissier, 2008). In a recent Euro Barometer study (http://ec.europa.eu/food/animal/welfare/euro_barometer25_en.pdf) 54% of respondents said that they had difficulties in finding adequate information on the animal welfare standards applied in producing food. The body of EU legislation on animal welfare has increased steadily in recent years and it would appear that this trend is likely to continue as the Treaty of Amsterdam has raised the ambitions of EU institutions to do more to raise welfare standards. The EU commission community action plan on the Protection and Welfare of Animals 2006-2010 outlines a range of co-ordinated actions in this area (http://ec.europa.eu/food/animal/welfare/actionplan).

There are differences in individual people's emphasis with regard to animal welfare (Fraser, 2004), but in simple terms, good animal welfare is regarded by many as good physical and mental health. Welfare is a multidimensional concept. It comprises both physical and mental aspects and includes physical comfort, hunger, thirst, disease, injuries, and the extent to which an animal experiences positive and negative affective states and events (Broom, 1996; Duncan, 1996). Affective state may also include positive experiences such as those produced by the expression of motivated behaviours (FAWC, 1992). Farming practices have changed rapidly in the last 50 years (Faure *et al.*, 2003), and this, combined with very specific genetic selection for production characteristics may have created animals where welfare-related traits have been, and are, ignored. Indeed, it is increasingly recognised that selective breeding is a powerful tool for alleviating welfare problems (Jones, 1997; Grandin, 1998) and that appropriate environmental enrichment, including positive human contact, can dramatically enhance welfare (Hemsworth and Coleman 1998; Jones, 2001; Mench, 1994).

The focus of EU agricultural policy is increasingly on quality rather than quantity. Traditional price mechanisms do not always allow for considerations like animal welfare to be properly recognised in the prices paid to producers. Labelling regimes, whether voluntary or mandatory, may have an important role to play as consumers seem to be increasingly insistent on higher standards and ways need to be found to provide them with information. Voluntary labelling appears to be on the increase worldwide. Organic farming is probably the most widely adopted voluntary labelling scheme, and is based on Regulated EU standards, but its impact on animal welfare is uncertain, as the preservation of the environment and sustainability of agriculture issues can potentially conflict with animal welfare objectives.

Numerous trade groups including producers, distributors, retailers and chain restaurants (Integrale Keten Beheersing in the Netherlands, Swedish Broiler Control, Filières Qualité Carrefour in France, Mc Donald's Europe, RSPCA Freedom Food in the UK) have created assessment and certification schemes which include elements of animal welfare. These schemes rely heavily on assessment of welfare by examination of the provision of housing or resources (Resource Based Measures RBMs), rather than looking at the animals themselves (Animal Based Measures ABMs) and the emphasis in these schemes is on private, voluntary initiatives. At present, there is no harmonised standard to assess animal welfare on farms within Europe, or globally. Farming bodies and legislators in the EU have recognised that

improvements in farm animal health and welfare can be promoted by standardised assessment techniques and provision for animal health and welfare information to be made available to farmers to support management decisions, and also to consumers so that they can make purchasing choices.

In general, compulsory labelling of food with animal welfare information has not been adopted, to date only one EU instrument makes animal welfare labelling (system description) compulsory (EC, 2001b).

Another example of the way in which the EU describes the minimum standards is Directive 2001/88/CE which describes standards for pig production (EC, 2001a). Some member-states have made additional requirements beyond the directive, for example specific substrate requirements are made in Sweden, and in Norway castration without anaesthesia is banned.

In studies carried out within the WelfareQuality® project (Keeling and Bock, 2007) it was found that participation in animal schemes affects the farmers' definition of animal welfare, the importance they attach to it, and their willingness to further increase animal welfare standards. Three groups of farmers were identified, farmers who participate in specific animal welfare schemes or organic production, farmers who participate in quality assurance schemes, and those who are part of no scheme at all. The first group of farmers was found to be the most concerned about animal welfare and were driven by ethical consideration and their ambition to carry out alternative agricultural production with the potential to receive a premium price as an important potential incentive. On the one hand farmers hope that animal-friendly products will offer opportunities for new markets, but they may also feel threatened by the potential import of cheaper food and the resulting competition on price. For all farmers, a secure market is an important driving factor. Some farmers do not participate in specific animal welfare schemes because they have no faith in the financial benefits promised and see only an increase in bureaucracy and work load and fear a loss of their independence. As part of the WelfareQuality® project, work on societal attitudes to animal welfare reporting, welfare information and welfare assessment has taken place. The spread of opinion amongst farmers can be summarised as being characterized by belief, hope and doubt.

Existing farm assurance standards are beginning to ask 'animal centered' questions, and it is apparent that repetitive resource based assessment can fail to fully answer questions about animal welfare. A number of current agricultural standards in use in Europe use basic 'welfare assessment tools' in that they incorporate the inspectors view and interpretation of the severity of issues observed. Some examples of assessment standards can be found at:

- Globalgap CCCP (Control points and compliance criteria) for Cattle (and sheep), dairy cows, pigs and poultry [http://www.globalgap.org].
- Assured Chicken Production (ACP) [http://www.assuredchicken.org.uk/chicken].

Within these standards, the majority of the standard requirements are written as short descriptions of what is required (clauses), and are almost all 'farm record' or 'resource' based (RBM). However, some of the existing standard 'clauses' are very close to being animal based (ABM) tools, for example:

- Are growing pigs kept in stable social groups? (This question can be answered by the stockman on the basis of establishment of groups of pigs, but also could be answered by observing the social interactions between the animals.)
- Where tail, flank, ear biting or fighting, which goes beyond normal behaviour becomes apparent, is an effective plan agreed...? *(*What is 'normal behaviour', what are the thresholds and what are the welfare implications for the animals of biting and fighting?)
- Is the number of birds found dead on arrival at the slaughterhouse after transportation above 0.25%? (Birds dying between catching at the farm and arrival at the slaughterhouse may reflect poor transport conditions or birds which were in a condition not suitable for transport in the first instance).

Across Europe, the use of assurance schemes in promoting higher levels of farm animal welfare is highly variable. 'Private' standard owners and assurance bodies can be flexible in both driving standards upwards and in responding to local conditions. Private assurance schemes (not government driven), whether linked to NGOs (RSPCA Freedom Food for example) specifically seeking to promote higher standards of farm animal welfare, or participants in retailer driven standards appear to offer most apparent current mechanism for improving farm animal welfare at the present time. In some countries, scheme membership has now shifted from being voluntary, to the current position where many producers now view membership to be an entry requirement (effectively non voluntary) to retailer shelves. In the overall assessment picture, animal welfare is perceived as a component of broader ranging assessments, which also contain environmental, Control of Substances Hazardous to Health (COSHH), animal medicine use, work environment safety, food safety and retailer specific requirements. Moves toward welfare based assessment do have the potential to result in costs to producers. The Commission estimated that the costs arising from the abolition of individual sow stalls was between € 0.006 and € 0.02 per kilogramme of pig meat (European Parliament, 2001) and an increase in egg production costs of a predicted 16 per cent is anticipated by 2012 with the implementation of the ban on battery cages.

Assurance certification can only be applied to farms which operate within the applicable legal framework of the country. As most assurance schemes make demands above (sometimes far above) the legal minimums, the net effect of assurance is to generally promote a voluntary raising of standards. Improvements in animal welfare can positively affect aspects of 'quality' through, for example, a reduction in bruising, bone breaks and blood spots or through increased disease resistance (Jones, 2001; Faure *et al.*, 2003). If there is an increasing emphasis of assessment of ABMs, this may create a shift in the way that farms are inspected – moving the inspection of 'provision' to assessment based on assessing the 'quality of animal lives'. This is arguably a paradigmatic shift that sees the assessment of farm animals from first, protection against cruelty to welfare and, second, from welfare to life quality (Buller and Morris, 2002; Buller and Roe, 2009). One of the possible effects of this shift could be a move towards the notion of farm animal welfare as a 'public good', something enriching the public sphere and so to be supported centrally rather than as a 'private initiative' articulated through privately owned commercial assessment systems.

## 2. What should we 'measure' to assess welfare?

Animals differ in their experience, temperament and the way in which their genetic makeup interacts with their environment. The influence of management and the stockperson can also dramatically influence not only 'productivity' measures, but also the animals experience of a particular situation. Thus resource-based measures (RBMs like type of housing, allocation of resources, stocking density etc.) or management-based measures (like breeding strategies, health plans etc.) may be a poor guarantee of high levels of animal welfare in a particular situation.

Research scientists have for some time suggested that ABMs could provide valid indicators of animal welfare; since welfare is a characteristic of the individual animal, not just of the system in which animals are farmed. A comprehensive assessment of welfare should take as many different potential welfare 'factors' into account and should also be based on scientific knowledge of animal welfare where possible rather than on conjecture based purely on anthropomorphised positions. However, the 'societal position' on animal welfare should not be forgotten in the rush to promote robust scientific methods as, in the end, consumers make the purchasing decisions which influence what is purchased and so, what is finally farmed (Kjaernes and Larvik, 2007).

Some single OBM's have been suggested as being capable of providing an integrated assessment of animal welfare including: corticosteroids, acute phase proteins, and longevity (Barnett and Hemsworth, 1990; Geers *et al.*, 2003; Hurnik, 1990). However, none of these single measures can cover all the dimensions of welfare. It seem probable that several measures are necessary to obtain a comprehensive view of any particular animal's welfare (Dawkins, 1990; Rutter, 1998; Webster, 1997). Recognising the difficulties of 'single measure approaches', grouped measures have been used to advise farmers (Sørensen, 2001), in 'branded' welfare certification schemes [e.g. Freedom Food Scheme] (Main *et al.*, 2001, 2003), and to compare systems to provide information during the creation of new legislation (Bracke *et al.*, 2002) and to check compliance with legislative requirements (Keeling and Svedberg, 1999).

Practically speaking, it is may be possible to combine RBMs and ABMs. For example, if animals are lame (assessed using an ABM) it may be possible to predict lameness if the floor condition is poor (an RBM). However, the philosophy of ABM based assessment is that, if there is a measure from the animal which fulfils the conditions described above (practical, valid, repeatable, robust) then this should be used in preference to the RBM alone – because a given floor condition may be very good for one animal but very poor for another. Additionally, it is clear that some simple questions, for example, 'is the animal thirsty' are not so easy to answer in a farm situation. There is no current feasible animal-based measure for dehydration which can be carried out on a farm (though a blood measure could be carried out at the slaughterhouse, or it could be theoretically 'possible' to offer water to animals in a choice test).

When comparing what scientists may consider to be areas of animal welfare that can be assessed with what farmers think about the needs (and assessment of their welfare) it may be useful to consider some (fabricated) possible viewpoints:

- A practical farmer might say, '*yes... isn't it great, we have made farmed animals lives so much easier, they don't need to worry much about predators, they are protected from disease, food is only a step or two away, and the day is pretty much free from any unexpected challenges. The animals can mostly kick back and PRODUCE!*'
- An ethologist may say, '*here is a list of things I know that these farm animals can do when given a chance – scratching, foraging, dust bathing, social interaction, sexual interaction, taking shade, seeking heat, avoiding wet etc... and I've just been watching the animals on the farm – they don't seem to, or can't do, quite lot of these activities – why? Is this a problem? mostly the animals seem to do little, and they continue to PRODUCE!*'
- A third person may come along and ask, '*it's clear, these animals are well looked after – we can slaughter them earlier because they have PRODUCED so well! – all I want to know is – how many animals survive, and how much do they weigh, i.e. how do they PRODUCE?*'
- And a fourth person may say, '*well this PRODUCTION thing is OK, but what about the LIFE of these animals – if we give them choices, what do the animals choose? We find that some animals choose to spend a lot of their time outside foraging and engaging in social and sexual behaviours and some don't, and that on warm dry days many do, and on cold wet days, many don't. I'd like someone to give me an objective idea of whether the animals have a good life or not so that I can decide what to buy.*'

We may thus have some complex decisions – do we as a society rate growth rate and PRODUCTION as so overwhelmingly important that we forget choice, body conformation, 'naturalness' and the ability to express physiological capacities? Recently there have been moves in Europe, Australasia and North America to create assessment systems for farm animal welfare which use grouped assessment measures to address health and welfare criteria. The sorts of questions which, in general, are being asked are; (1) Are the animals properly fed and supplied with water?, (2) Are the animals properly housed?, (3) Are the animals healthy?, and (4) Can the animals express a range of behaviours and emotional states?

Of course, these questions quite closely resemble the elements of the five freedoms (Brambell, 1965), and WelfareQuality® (Blokhuis *et al.*, 2003), through collaboration between 41 institutes across Europe, has been building on these principles to develop on-farm and slaughter assessment systems to address the key areas of feeding, housing, health and disease and behaviour. WelfareQuality® has created grouped assessment measures for pigs, poultry and cattle which address twelve health and welfare criteria (somewhat like the well known five freedoms), and are testing them on a large number of farms across Europe.

One of the significant aims of the WelfareQuality® project is to develop tools for monitoring animal welfare from farm to slaughter. These tools can be used for a number of purposes:
- to allow inspection and scoring of farms, to inform consumers about the welfare status of the animals from which they buy products;
- to give 'advice' back to the farmer based on the data collected on the farm;
- the assessment protocols could (perhaps) be used by statutory bodies in their assessment of farms.

The tools above address the following key areas:

1 Hunger
This occurs when animals are denied a sufficient and appropriate diet and can lead to poor body condition, reduced ability to fight disease or to cope with thermal and metabolic demands. For most farmed animals, hunger *per se* would not be common, as provision of food is an essential part of farming activity – however, it is possible for individual animals to fail to receive sufficient food through, for example, bullying by other animals, design of feeders which prevent (usually small) animals from readily accessing food – this is particularly a problem in poultry. For ruminants, poor grazing quality – due to drought or cold weather, can result in hunger, and eventually starvation. Additionally, although sufficient volume of food may have been provided, it is possible that the food is inadequate from a compositional viewpoint – and poor balance of protein, energy and trace elements and vitamins can result in functional starvation despite sufficient volume of feed being available.

2 Thirst
Provision of water is a fundamental part of farming – however, animals can still fail to receive sufficient water if drinkers are poorly placed, poorly maintained, if dominant animals control their use, or if the water in drinkers is contaminated with faeces. For animals extensively grazed, drought and reduced water availability from forage can lead to dehydration. For birds, if drinkers are placed at too great a height, small birds can dehydrate – and for all species, if animals become sick and are unable to get to water – this will reduce their ability to cope with disease, and in some animals, for example poultry in intensive conditions, this may contribute to their death.

3 Comfort around resting
Animals can suffer both short term and also chronic discomfort and have problems lying down, getting up and standing. This can occur when they are kept in inappropriately designed housing (e.g. insufficient space, poor ventilation, unsuitable flooring and bedding) or when they are transported in poorly designed or poorly ventilated vehicles.

4 Thermal comfort
Many animals are completely under the control of man in terms of the thermal environment in which they live – they cannot take shade, wallow in mud or create nests or beds to keep warm, because these capacities are reduced or eliminated in many housing systems. For these reasons, the thermal environment, and the quality of its control by the producer, has significant potential effect on animals' quality of life.

5 Ease of movement
Many animals are kept in environments in which their capacity to move (for instance to fly, or to run) is very much restricted. Additionally, some animals are tethered, and other are very closely confined for periods of life – for example sows in farrowing crates. For all animals, even those in extensive systems where the capacity to move about is much greater, there is potential for damage and injury through falling or through poorly designed gates, fences, bedding areas (for example, dairy cows on slippery concrete surfaces).

6 Health: injuries
Animals can suffer physical injuries, such as mutilations, broken bones, bruises or skin lesions, due to factors such as; uneven or slippery flooring, enclosures with poorly

maitained pens or gates and fences with sharp metal or broken wood or concrete edges and environments that promote aggressive behaviours between animals.

7 Health: disease
Animals can suffer a range of diseases- for example, mastitis and metabolic disorders in cattle, lameness in pigs and poultry, respiratory disease in calves. Disease is a 'common' experience for producers – and most producers take practical steps to reduce the incidence of disease through biosecurity measures, vaccination and housing environment control – but when disease does occur it is the step which the producer then makes which can profoundly affect the welfare of animals – is disease rapidly diagnosed and then effectively managed? Poor hygiene, irregular monitoring and insufficient treatment speed can amplify the effects of disease on animals.

8 Pain (not related to injuries or disease)
In addition to suffering pain from injuries and disease, animals can experience intense or prolonged pain due to inappropriate management, handling, slaughter, or invasive management procedures (e.g. castration, dehorning, tail docking) and as a result of aggressive encounters between animals, and also due to the use of goads (electrical or not) used to move them.

9 Normal/natural social behaviours
Animals can be denied the opportunity to express natural, non-harmful, social behaviours, such as grooming, social licking and huddling for warmth. Separating females from their offspring very early in production, and preventing sexual behaviour are examples of more established restrictions on 'normal behaviour' which are common in farming systems.

10 Normal/natural other behaviours
Animals can be denied the possibility of expressing other intuitively desirable natural behaviours, such as exploration and play. The denial of these possibilities may lead to abnormal and/or harmful behaviours such as tongue rolling in cattle, bar biting and tail biting in pigs and feather pecking in chickens.

11 Human-animal relationship
Poor relationships with humans can be reflected in increased avoidance distances and fearful or aggressive animal behaviours. This can occur due to inappropriate handling techniques (e.g. slapping, kicking and the use of electric goads), or when producers, animal transporters or slaughterhouse staff are either insufficiently skilled or possess unfavourable attitudes towards animals.

12 Positive emotions & avoidance of negative emotions
Animals can experience emotions such as fear, distress, frustration or apathy, when they are kept in inappropriate physical or social environments (e.g. where there is a lot of mixing of groups of animals, or not enough space to avoid an aggressive partner). These emotions can be reflected in behaviours such as panic, flight, social withdrawal and aggression and behavioural disorders. Poor management routines and a lack of environmental stimulation may prevent animals from expressing positive emotions. Positive emotions are not easy to assess but may be reflected in certain behaviours, such as play (especially in young animals) and positive social interactions (social licking). Some people consider that promoting 'positive emotional states' in animals is beyond the normal requirements for farmers who are business people trying to make a living from rearing animals – whereas other people consider these exact reasons to be good justification for producers to show 'care' for all aspects of the needs of the animals which

provide their livelihood. Practically speaking, 'care' for the positive needs of animals could be considered an indicator of farming systems ability to consider the whole picture of what farmed animals need to live and to thrive.

Note: the numbers relate directly to the criteria in Table 1.

*Table 1. Twelve health and welfare criteria.*

| Principle | Welfare criteria | | Examples of potential measures |
|---|---|---|---|
| Good feeding | 1 | Absence of prolonged hunger | Body condition score |
| | 2 | Absence of prolonged thirst | Access to water |
| Good housing | 3 | Comfort around resting | Frequencies of different lying positions, standing up and lying down behaviour |
| | 4 | Thermal comfort | Panting, shivering |
| | 5 | Ease of movement | Slipping or falling, possibility of exercise |
| Good health | 6 | Absence of injuries | Clinical scoring of integument, carcass damage, lameness |
| | 7 | Absence of disease | Enteric problems, downgrades at slaughter |
| | 8 | Absence of pain induced by management procedures | Evidence of routine mutilations such as tail docking and dehorning, stunning effectiveness at slaughter |
| Appropriate behaviour | 9 | Expression of social behaviours | Social licking, aggression |
| | 10 | Expression of other behaviours | Play, abnormal behaviour |
| | 11 | Good human-animal relationship | Approach and/or avoidance tests |
| | 12 | Positive emotional state | Novel object test |

## 3. How could an assessment tool which includes ABMs be used to raise the overall level of animal welfare?

There are a number of 'levels' to application of an ABM. First, a standardised description is needed, then a description of the practical application of the method and any linked RBMs is required, then, finally, the ways in which the information that is produced from the use of the ABMs and RBMs can be used to promote and support management decisions and practices to create improvements in welfare.

Step 1: Measure (ABMs and RBMs) →
Step 2: Analyse risk factors →
Step 3: Inform (producer, purchaser)→
Step 4: Support management decisions to create improvements in welfare

### 3.1 Measure

Foot pad dermatitis, description: foot pad dermatitis (or pododermatitis) is a contact dermatitis found on the skin of the foot, most commonly on the central pad, but sometimes also on the toes. The skin is turned dark by contact with litter and deep skin lesions can result. The scoring scale (Figure 1) allows an assessment of the severity of these lesions.

*3.1.1 How many should be scored?*

At least 100 broiler chickens per flock should be assessed: 10 birds taken from 10 areas of the house including 2 areas located near to drinkers, 2 located areas near to feeders, 3 areas located near a wall, 3 areas located away from drinkers and feeders (resting area). At the slaughter plant it is possible to score a large number of feet as the birds pass on the slaughter line. If the line speed is 120 birds per minute, then observing for 10 minutes will enable scoring of 1200 birds using the same scoring scale as used on farm.

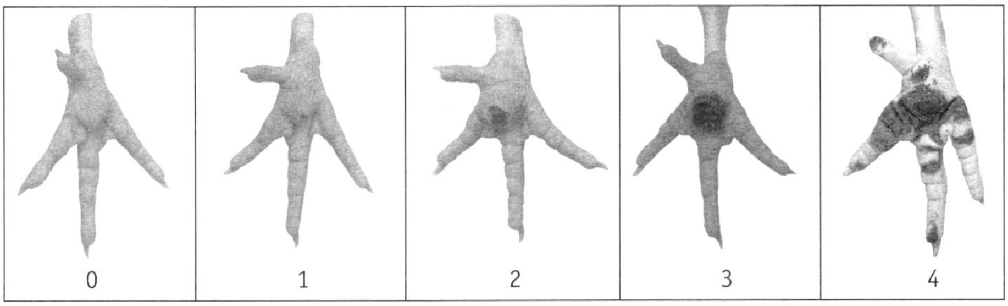


*Figure 1. Scoring scale for pododermatitis (foot pad disease) in broiler chicken (Butterworth, University of Bristol).*

### 3.2 Step 1: measure →

*3.2.1 Training assessors*

- Training in scoring for pododermatitis by means of on-farm visits and examination of photographs.
- The inspectors will require to be assessed during the training course until they develop a uniform scoring.
- Assessors will also be asked to periodically carry out a validation or reference audit to check whether they continue to score in a repeatable way.

*3.2.2 Questionnaire*

As well as recording the ABM (Figures 2 and 3) (Foot pad scores) we will need to complete a farm questionnaire or standard inspection report with the assistance of a representative of the farm, which provides a description of the farm, house and flock (the RBM's) which give

*Figure 2. Handling and penning of birds. Pen approximately 25 birds in an area of the house: birds should not be herded into the pen; the pen should be placed quietly around a group of birds with minimal disturbance; if birds show any signs of distress or overheating, they should be released and a new group penned.*

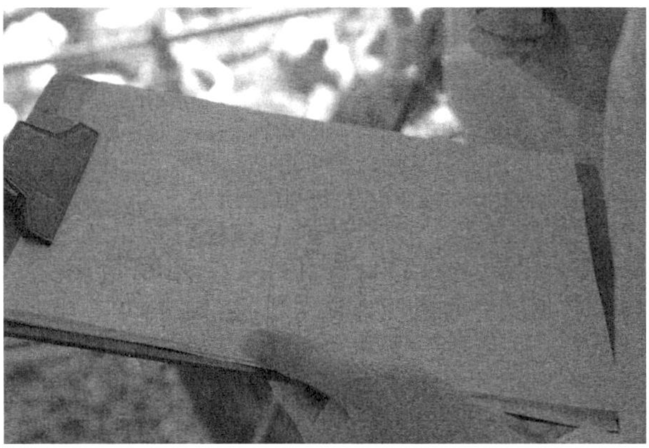

*Figure 3. Paper based recording: pick up individual birds carefully and score them – place them outside of the pen; create a simple tabulation of the numbers of birds scored in each foot pad category.*

the background and risk factors required to provide information to help the farmer resolve his foot pad dermatitis issues.

- Broiler breeder information including genotype/strain, broiler breeder history and age.
- Hatchery information including hatchery, distance/time transported and hatchery vaccination programme for chicks.
- General information including number and weight of chicks placed, sex, time of year, age at assessment and slaughter.
- Specific husbandry practices including stocking density and thinning practice, brooding conditions, nutritional profile (detailed), vitamin/mineral levels, litter substrate, feeder

and drinker design/type, lighting programme, age of house, construction details, target ventilation profile, diseases and medication history, vaccination programme and water source.
- Performance information including growth profile, weekly mortality pattern, weekly leg cull pattern, other culls weekly pattern.
- Background information about the site/company including size of houses, number of birds on site and biosecurity measures.

(Many of these RBMs questions form the basis of existing farm assurance inspection).

### 3.3 Step 2: analyse risk factors →

The company (or the assurance or advisory service) can (in this specific example) analyse the ABM and RBM information and
- Find out the prevalence and severity of foot pad dermatitis in the flocks within the company. Include in this an assessment of the economic impact of small birds, moribund birds and culled birds which result from lameness. In general, significant improvements in profitability can be made if pad problems are tackled within a company, which results in an overall bird welfare: inspection bodies in some countries are now beginning to focus on foot pad health as a marker for company welfare performance.
- Make comparisons between 'good' and 'poor' farms (with respect to foot pads) within the company to help identify management, house environment, feeding, medication, stockmanship and genotype factors which differ between these farms.
- Investigate the use of water in different farms – farms with increased water use per bird (in equal weather conditions) may have systematic problems with leaking drinkers. Small amounts of water leaked chronically into the litter can severely damage the management of litter and foot pad health.
- Carry out an investigation of the bacteriological pathologies linked with foot pad dermatitis and identify whether these bacteria can originate in the hatchery, the transportation or by lapses in farm biosecurity.

### 3.4 Step 3: inform → and step 4: support management decisions to create improvements in welfare

The farmer can be informed about the extend of foot pad dermatitis on his farm, and, with time, and after analysis, a pattern of risk factors may emerge which allow him to make decisions which can reduce this. Factors which, in real farm experience, have been shown to be risk factors for lameness include: growth rate, the age of the birds at slaughter, the use of whole cereals in the diet, the type of feed, the quality of biosecurity measures, litter condition (an important factor), the genotype of the birds. Also, the gender of the birds, levels of feed restriction, the lighting pattern and light intensity, bird activity levels and the stocking density have, in real examples of farms, been altered an manipulated to control levels of foot pad dermatitis in the chickens.

## 4. Turning assessments into 'scores'

Once measures have been carried out on a farm, it may be possible to create a range of 'scores'. The first level of scores would be the results for each measure, provided initially as 'benchmarking' or 'initial position' scores, which tell the producer how he is positioned at the very start of the assessments, and also allows him to compare himself with some known baseline values, and also, with his farming peers. The individual scores for each measure made can be fed back to the farmer for his own information, so that he can see how he did, and also to support any areas that he might wish to develop or focus on (if he has a problem). For example, he might find that the assessment shows that he has an unusually high number of tail bitten pigs in this particular group (when compared to his previous assessments). He might use the periodic information provided by an assessment as a barometer for any changes he chooses to make – if they cause reductions in the 'area of concern', then he can use this information to assess the degree of 'success' that he has achieved – and conversely, if there is no improvement or even an increase, he may be able to use the risk factor assessment to target other areas of change. Von Borell *et al.* (2005) created hazard analysis and critical control points (HACCP) based on recording systems for pig housing, with the potential to feed back results of the effects of altered management at these critical control points. The concept of feeding back individual 'non compliant' assessment points is the basis of all existing farm assurance and certification schemes, and so this approach (detect areas of concern and then request rectification of specific points) is familiar to many farmers- and is seen as the basis of a reasonable approach to ensuring compliance with a recognised 'standard'.

Some schemes, such as the National Dairy Farm Assurance Scheme in the United Kingdom (www.ndfas.org.uk), have two categories of standards, those that 'must' be met (within a given time frame) and those that 'should' be met (i.e. strongly recommended but not imposed) (NDFAS, 2009).

The second level of 'scoring' would be to combine the individual assessment results – perhaps into linked groupings – for example, in the area of housing. A combined score for related areas is possible – but only if:

- The individual measures can be combined to give aggregate scores which are considered meaningful, and credible, by the producer and the consumer.
- The process of combining scores must not devalue the overall meaning of the assessment information, for example, by compensating a very poor score in one area, with a number of 'OK' scores in other areas. If this occurs, then the power of discrimination can be lost, and the credibility of the combined score brought into doubt. Spoolder *et al.* (2003) indicated that when welfare scores are simply summated to give an overall score, a welfare disadvantage can be compensated for by a number of minor advantages, but this effect can be limited (or even eliminated) if minimal requirements are set below which specific score cannot be 'permitted' (Botreau *et al.*, 2007).

Once raw 'values' for measures have been made on the farm, the 'natural' next step is to ascribe different weighted values to these measures, to give impact factors for each measure with respect to animal welfare (Bracke *et al.*, 2002). This type of weighting system is seen in

the Austrian TierGerechtheitsIndex [TGI] (Bartussek, 1999). Weighted sums of scores appear intuitive and the principle is usually readily understood by users.

## 5. Potential veterinary use of animal base assessment

Veterinarians in Europe play some important roles in relation to animal welfare issues. Vets who work in practice interact on a day-to-day basis with their farming clients, diagnosing and treating animal disease and advising farmers on disease prevention and reduction, animal behaviour, welfare issues, breeding and nutrition (Butterworth, 2008). Some vets are responsible for both animal and public health considerations at slaughterhouses, and help to ensure that animals are treated well at the time of slaughter. Others carry out surveillance for both animal and zoonotic disease in farm and companion animals, and during transport and at international borders. As well as assessment of health and welfare, vets can assist farmers to implement of strategies to improve welfare through improvements in the quality of stockmanship, improvements in housing and husbandry, and through genetic selection.

Despite a large amount of co-operation between veterinary organisations and government agencies, there is comparatively little 'benchmarking' information shared between countries when it comes to animal welfare parameters. In all countries of the EU, the state, usually through a state veterinary service carries out some visits to farms to ensure that they can deliver compliance with animal care and welfare legislation. It is possible that demonstration of adequate (or even high) welfare standards would allow targeted farm inspection, and the state inspection load could be optimised using information provided by inspection schemes which included a harmonised welfare assessment. The potential for streamlining of animal health visits could be a genuine benefit to farmers and to the state.

Let us consider an example. A dairy farmer has a problem with lameness in his dairy cows. A structured assessment will tell him how many lame cows he actually has when referenced to some standard definitions. Once he has a tool to gauge whether the level of lameness is increasing (or decreasing) then this can be a barometer against which to judge practical steps he may make to reduce the problem. ABMs may help him to identify practical ways of trying to reduce lameness. For example, information on the type of floor and the farmer's hoof care strategy could be used to help advise on remedial solutions to improve the problem can be both an economic cost (lameness in dairy cows costs the farmer in terms of lost productivity) and also a cost to the animals in terms of disability or discomfort. Targeted improvement may be able to help both the farmer and the animal. To be viable, remedial strategies must satisfy both welfare and economic requirements, and they must be practicable, i.e. affordable and easy to implement by the farmer and/or breeding company.

Alongside this advisory information, the vet is also able to tell the farmer how his animals are doing with regards to early calf mortality, to respiratory disease, to udder lesions and with respect to the feeding of his young cattle. When this information is linked to economic information that the farmer is likely to share with his vet, then the vet may have a powerful tool to help and support his farming clients and to promote best management in animal health and welfare. This productive relationship between the vet and the farmer relies on

several things – trust, information and economic sense. Trust comes from the professional interaction between vet and client, information is part of the armoury of skills that makes the vet a valuable part of the farming system, knowledge of what is common and uncommon, what disease looks like, the specialist techniques of surgery and medicine and knowledge of how his clients farms, and economic, disease and welfare performance relates to other similar farms. Farm veterinary services are not alone in benefiting from benchmarked indices on which to base management decisions, the use of Key Performance Indicators (KPI's) (or indices) is a cornerstone of good business management in many commercial practices. The feedback of information to the farmer, often with veterinary involvement, and his/her uptake of recommendations and remedial measures represent a potential direct advantage of this approach. Veterinarians who work directly with their farm clients, those involved in assurance schemes, those carrying out disease surveillance and import and export work may find that a series of harmonised welfare assessment tools are of value to their business. These tools may be of use for a number of purposes:

- to allow inspection and scoring of farms;
- to inform consumers about the welfare status of the animals of which they buy products;
- to provide advice and support for the farmer based on the data collected on the farm;
- additionally technical specifications may be used by veterinary inspection agencies in their statutory assessment of farms.

## 6. The importance of assurance schemes and their inspectors in the evolving welfare assessment picture

Assurance schemes in a number of countries have the potential to influence and improve farm animal welfare through their programmes of inspection to standards. Higher welfare animal products, sometimes carrying the label of an assurance scheme, are generally sold to consumers as 'quality items' or as 'best' or 'finest', thereby generating a higher price (Buller and Roe, 2009).

Critical to the audit process is the relationship between the farmer and the assessor. Assessors are 'not the police' and, depending on the nature of the assurance scheme, their relationship to farmers usually remains at the level of professional detachment, in which advice may not be given (schemes which operate under a harmonised certification standard EN45011 may not offer advice associated directly with the assessment). However, some schemes, most notably the soil association in the UK have set up a distinct advisory arm (separate from the assessment process) which can support farmers overall to achieve the standards required for certification. Additionally, in a number of countries it appears common for a 'farm advisor' to offer the producer support and advice in preparation for an assessment audit (this person is distinct from the assessor).

However, the practical application of ABMs within existing assurance scheme could be problematic in a number of ways. For example:

- Are existing schemes able to assess ABMs in a repeatable and reliable way within the timescale of a routine assessment?
- Who will carry the cost of assessing ABMs?

- How will assessment of ABMs work in terms of periodicity and seasonality of assessment i.e. will the seasonal changes in production systems make interpretation of the findings difficult?
- Can reduction of ABMs to a single farm-based score actually work?
- ABMs conventionally assess negative, findings (lameness, skin lesions, hunger), can they also be used to convey 'positive' information to consumers?

Alongside these (and other) considerations, the incorporation ABMs into an existing assessment framework which mostly uses RBMs will include practical questions such as:
- To 'integrate ABMs into existing standard farm assessments' or 'to keep them separate'?
- Can use be made of slaughterhouse derived ABMs when creating farm assessment reports?
- Can the vets assess ABMs?

## 7. Conclusions

In conclusion, the tools being developed in animal based measure (ABM) or outcome based measure (OBM) systems now under development in Europe evolve the trend toward inclusion of assessment techniques which reflect what can be measured 'on the animal'. Whilst accepting that many areas of animal assessment are complex, and some are very subtle, it seems likely that some ABMs are likely to find their way into farm assessment schemes, and, to a certain degree, into the farm animal legislature. This raises some fears in the 'community' (producers, scientists, NGO's, legislators), if ABMs are of use, how far will people want to go with them? Should (or even could) they replace some of the prescriptive resource based measures found in existing legislation. Some people appear to be concerned that the good work already carried out in animal protection which is now enshrined in existing legislation could be 'swept away' on the a wave of enthusiasm for ABMs. The inclusion of ABMs is a ripple or evolution (not a wave of revolution) and the stability and conservatism of the legislative and assessment community will ensure that if ABMs do begin to make their way into farm assessment methods and the legislation, then this will take place gradually and without the removal of the existing frameworks which protect farmed animals. In my opinion, animal based measures are 'complementary to existing frameworks until proven otherwise'.

### 7.1 What has been achieved?

Research scientists have for some time suggested that ABMs could provide valid indicators of animal welfare; since welfare is a characteristic of the individual animal, not just of the system in which animals are farmed. The sorts of questions which, in general, are being asked are; (1) Are the animals properly fed and supplied with water?, (2) Are the animals properly housed?, (3) Are the animals healthy?, and (4) Can the animals express a range of behaviours and emotional states?

In current and ongoing work now being carried out around the world, it is clear that a 'bringing together' of expertise and the start of a consensus amongst scientists is taking place, and that it is considered possible to use animal based measures to provide information about the quality of life of farmed animals and about how well animals are being farmed

within the existing 'systems' which are prescribed by legislation and established farming practice. This animal based information may find its ways into retailer and consumer choices, and also may be used to provide advice and support to producers.

### 7.2 What has been neglected?

A number of farmed animal species are clearly not represented fully in this work on ABM's. For example, fish, small ruminants, poultry species other than laying hens and broilers, fur farmed animals. These species have received a large amount of specific focussed scientific study, but currently, there has not been the collaborative work to combine knowledge in these species to produce combined animal welfare assessment systems.

### 7.3 What needs to be done?

At present, the use of ABM's in the commercial setting and on a significant scale outside of controlled studies and for a realistic period has not taken place. Animal based measures are, in my opinion, 'complementary to existing frameworks until proven otherwise' and so, if the use of ABMs in commercial assessment systems can be initiated, most likely in collaboration with existing certification and assurance bodies, then it will be possible to determine whether the claims that ABMs offer real value to the consumer and producer are realistic. The consumer could potentially benefit through informed choice, and the producer may find that he/she can make better management decisions based on knowledge of what happens to lameness, skin disease, animal/human and animal social interaction etc. when things on the farm are changed and 'managed'.

## References

Barnett, J.L. and Hemsworth, P.H., 1990. The validity of physiological and behavioural measures of animal welfare. Applied Animal Behaviour Science, 25.

Bartussek, H, 1999. A review of the animal needs index (ANI) for the assessment of animals' well-being in the housing systems for Austrian proprietary products and legislation. Livestock Production Science 61, 179-192.

Blokhuis, H.J., Jones, R.B., Geers, R., Miele, M. and Veissier, I. 2003. Measuring and monitoring animal welfare: transparency in the food product quality chain. Animal Welfare 12, 445-455.

Borell, E. von and Schaffer, D., 2005. Legal requirements and assessment of stress and welfare during transportation and pre-slaughter handling of pigs. Livestock Production Science 97, 81-87.

Botreau, R., Veissier, I., Butterworth, A., Bracke, M.B.M. and Keeling, L., 2007. Definition of criteria for overall assessment of animal welfare, Animal Welfare 16, 225-228.

Bracke, M.B.M., Spruijt, B.M., Metz, J.H.M. and Schouten, W.G.P., 2002. Decision support system for overall welfare assessment in pregnant sows: a model structure and weighting procedure. Journal of Animal Science 80, 1819-1834.

Brambell, R., 1965. Report of the technical committee to enquire into the welfare of animals kept under intensive livestock husbandry systems. H.M.S.O., London.

Broom, D.M., 1996. Animal welfare defined in terms of attempt to cope with the environment. Acta Agriculturae Scandinavica, Section A, Animal Science 27, 22-29.

Buller, H. and Morris, C., 2002. Farm animal welfare aspects of the CAP, RSPCA, London.

Buller, H. and Roe, E.J., 2009. Constructing quality: negotiating farm animal welfare in food assurance schemes. Proceedings of the WelfareQuality® Advisory Commitee meeting, September 24th – 25th 2008, Copenhagen.

Butterworth, A., Veissier, I., Manteca, J.X. and Blokhuis, H.J., 2008. Welfare trade. European Union 15, Public Service Review 18:54:39; 456-459.

Dawkins, M.S., 1990. From an animal's point of view: motivation, fitness, and animal welfare. Psychological science 13, 1-61.

Duncan, I.J.H., 1996. Animal welfare defined in terms of feelings. Acta Agriculturae Scandinavica. Section A, Animal Science 27, 29-36.

European Parliament, 2001. Communication from the Commission to the Council and the European Parliament on the welfare of intensively kept pigs in particularly taking into account the welfare of sows reared in varying degrees of confinement and in groups, COM(2001) 20 final, 16.01.2001, http://europa.eu.int/eur-lex/en/com/pdf 2001/ com2001_0020en01.pdf.

European Community, 2001a. Council Directive 2001/88/CE of 23 October 2001 amending Directive 91/630/EEC laying down minimum standards for protection of pigs. Official Journal L316, 01/12/2001:0001-0004. Available at http://europa.eu.int/eur-lex/index.html.

European Community, 2001b. Council Regulation 5/1/2001 amending certain marketing standards for eggs. OJ L2 5/1/2001.

Farm Animal Welfare Council, 1992. Report on the welfare of broiler chicken. Farm animal welfare council, Block D, Government buildings, Hook Rise South, Surbiton, Surrey KT6 7NF.

Faure, J.M., Bessei, W. and Jones, R.B., 2003. Direct selection for improvement of animal well-being. In: Muir, W. and Aggrey, S. (Eds.). Poultry breeding and biotechnology. CAB International, pp. 221-245.

Fraser, D. 2004. Applying science to animal welfare standard. In: Proceedings of global conference on animal welfare: an OIE initiative, Paris 23-25 February, 121-135.

Geers, R., Petersen, B., Huysmans, K., Knura-Deszczka, S., De Becker, M., Gymnich, S., Henot, D., Hiss, S. and Sauerwein, H., 2003. On-farm monitoring of pig welfare by assessment of housing, management, health records and plasma haptoglobin. Animal Welfare 12, 643-647.

Grandin, T., 1998. Genetics and the behavior of domestic animals. San Diego, USA, Academic Press.

Hemsworth, P.H. and Coleman, G.J., 1998. Human-livestock interactions. The stockperson and the productivity and welfare of intensively farmed animals. Wallingford, UK, CAB International.

Hurnik, J.F., 1990. World's poultry science association invited lecture: Animal welfare: ethical aspects and practical considerations. Poultry Science 69, 1827-1834.

Jones, R.B., 1997. Fear and Distress. In: Appleby, M.C. and Hughes, B.O. (Eds.). Animal Welfare. Wallingford, U.K, CAB International, pp. 75-87.

Jones, R.B., 2001 Environmental enrichment for poultry welfare. In: Wathes, C.M., (Ed.). Integrated management systems for livestock. British Society for Animal Science, Occasional Publication, 28, pp. 125-131.

Keeling, L. and Bock, B., 2007. Turning welfare principles into practice. Proceeding of the 2nd Welfare Quality Stakeholder Conference. 3-4 May 2007, Berlin. pp 25-28 ISBN 9789078240020.

Keeling, L. and Svedberg, J., 1999. Legislation banning conventional battery cages in Sweden and a subsequent phase-out programme. In: Kunisch, M. and Eckel, H. (Eds.). Proceedings of the Congress Regulation of Animal Production in Europe, pp. 73-78.

Kjaernes, U. and Larvik, R., 2007. Attitudes of consumers, retailers and producers to farm animal welfare. Welfare quality report 2, Cardiff University Press, ISBN 1 902647 734 ISSN 1749.

Main, D.C.J., Webster, J. and Green, L.E., 2001. Animal welfare assessment in farm assurance schemes. Acta Agriculturae Scandinavica, Section A, Animal Science Supplementum 30, 108.

Main, D.C.J., Kent, J.P., Wemelsfelder, F., Ofner, E. and Tuyttens, F.A.M., 2003. Applications for methods of on-farm welfare assessment. Animal Welfare 12, 523-528.

Mench, J.A., 1994 Environmental enrichment and exploration. Laboratory Animal, February, 38-41.
National Dairy Farm Assured Scheme NDFAS Standards & Guidelines. 2009. http://www.ndfas.org.uk/.
Rutter, S.M., 1998. Assessing the welfare of intensive and extensive livestock. In: proceedings of the workshop pasture ecology and animal intake, pp. 1-9.
Sørensen, P., 2001. Breeding strategies in poultry for genetic adaptation to the organic environment. Proceedings, 4th NAHWOA Workshop, Wageningen, the Netherlands, pp. 9.
Spoolder, H., De Rosa, G., Horning, B., Waiblinger, S. and Wemelsfelder, F. 2003. Integrating parameters to assess on-farm welfare. Animal Welfare 12, 529-534.
Veissier, I., Butterworth, A., Bock, B. and Roe, E., 2008. European approaches to ensure good animal welfare. Applied Animal Behaviour Science 113, 279-297.
Webster, J., 1997. Applied ethology: what use is it to animal welfare? Advances in Ethology 32, Supplements to Ethology, 10.

# The scientific basis of animal welfare indicators

**Jeffrey Rushen and Anne Marie de Passillé**
Pacific Agri-Food Research Centre, Agriculture and Agri-Food Canada, P.O. Box 1000, 6946 Highway 7, Agassiz, BC, Canada, V0M 1A0; Jeff.Rushen@agr.gc.ca

## Summary

We describe some of the undeniable advances that science has made in addressing animal welfare issues, and some of the limits on our scientific understanding of, and ability to measure animal welfare. There is always an element of uncertainty when estimating the magnitude of the various hazards to animal welfare. The term 'welfare' refers to the actual state of an animal and involves concerns about the animal's ability to function biologically, its feelings and its ability to live a reasonably natural life. Animal emotions and behavioural deprivation are central issues of animal welfare and are widely accepted to be amenable to scientific investigation. Most often these concerns overlap and only occasionally appear to conflict. Unfortunately, we do not know how to weight the relative importance of the different types of threats to animal welfare, e.g. health versus behavioural expression. Current animal welfare assessment involves a mix of input based criteria (which are easier to audit and can prevent welfare problems from occurring) and outcome-based criteria (which more accurately describe the welfare of the animals). However, there are practical problems with using outcome-based measures on-farm, and all animal welfare indicators show only the presence or absence of specific threats to welfare rather than measuring overall welfare itself. On-farm welfare assessment should involve a combination of health, and behavioural measures of welfare as well as information on the housing, feeding and management of the animals. Measures of production and physiological measures of stress have proven less useful. Assessing the effects of short-term procedures on animal welfare has proven easier than assessing the effects of chronic conditions, such as due to housing or management, which often involve trading off different advantages and disadvantages for animal welfare.

**Keywords**: animal welfare, on-farm assessment, scientific research

## 1. Introduction

According to Webster (2001), a full understanding of animal welfare issues requires that we have (1) a scientific understanding of the factors that affect the welfare of farm animals, (2) an understanding of the ethical or moral reasons why we should respect the welfare of farm animals and (3) an understanding of the economic forces that led us to treat animals the way we do. This chapter is focussed exclusively on the first of these. For governments and inter-governmental organizations, it is important that issues of animal welfare be treated 'scientifically' (OIE, 2008). Research in animal welfare has made tremendous progress in developing measures to assess animal welfare, identifying the main hazards for animal welfare, and developing solutions to remove or reduce the impact of these hazards. Fraser (2008) provides a good recent summary of much of this field. While we do not wish to down

play these achievements, we believe it is essential to avoid hubris and recognize that there are real limits on what science can achieve in dealing with this issue. In this article, we seek to describe some of the successes that science has had in addressing animal welfare issues but also some of the limits on the extent to which we do have a scientific understanding of factors affecting animal welfare and particularly on the extent that we can 'objectively' measure animal welfare.

The European Food Safety Authority (EFSA), which provides independent scientific advice to the European Community governments on matters relating to the health and welfare of animals, has begun to develop a science-based risk analysis approach to animal welfare (EFSA, 2006). This is described as a 'scientifically-based process' which firstly involves identifying and characterizing the major hazards for animal welfare, and estimating the magnitude of the impact upon the animal. A second step is to assess the likelihood of an animal being exposed to these hazards. However, there is always an element of uncertainty both of the estimates of the magnitude of the hazard and of the likelihood of animals being exposed to the hazard. The uncertainty regarding exposure estimates arises primarily from limits to the accuracy of the data collected. However, the uncertainty about the characterization of the hazard and the magnitude of its impact upon the animal involves more profound issues concerning scientific knowledge of animal welfare. It is the latter than we focus upon.

## 2. Definition of animal welfare

It is generally accepted that the term 'welfare' refers to the actual state of an animal (Broom, 2008) rather than to the ethical obligations that people have to care for animals. Thus, an animal in the wild suffering from disease or malnutrition can be said to have poor welfare even if people were not responsible for this and have no ethical obligations to care for the animal. Nevertheless, the concept of animal welfare when applied to farm animals is used most frequently in the context of the ethical obligations people have towards the animals in their care. Thus, to define the concept of farm animal welfare, we do need to understand the concerns that people have about animals. In a series of publications, David Fraser and his colleagues (Fraser, 2008; Fraser *et al.*, 1997) have examined the most common statements made about farm animal welfare in an effort to clarify what are the most important concerns that people have. They have been critical of the notion that animal welfare can be defined in a purely 'objective' fashion without reference to the particular ethical concerns that people have about the way we treat animals. Three broad types of questions are typically raised about the effects of modern farming systems upon animal welfare. Is the animal healthy and producing well (i.e. concerns about the animal's ability to function biologically)? Is the animal happy or is it suffering from pain or other undesirable emotions? (i.e. concerns about the animal's feelings or emotions). Is the animal able to perform its normal behaviour and live a reasonably natural life? (i.e. concerns about the naturalness of how the animal lives). Different stakeholders tend to put the emphasis on different aspects of animal welfare. For example, veterinarians and farmers generally focus on disease, injury, poor growth rates and reproductive problems. In contrast, many members of the public focus upon whether the animals are suffering from unpleasant feelings, such as pain, fear or hunger. For many people interested in animal welfare, especially consumers of organic produce (Lund and Röcklinsberg,

2001), a key concern is whether the animal is able to live a relatively natural life or express its natural behaviour. Fraser *et al.* (1997) argued that none of the three approaches to animal welfare alone adequately dealt with the range of ethical concerns being expressed and that some combination of the three was required. These three aspects of animal welfare are usually included in official definitions. For example the World Animal Health Organization (OIE) defines an animal as being in good animal welfare if it is: 'healthy, comfortable, well nourished, safe, able to express innate behaviour, and ... is not suffering from unpleasant states such as pain, fear, and distress.' (OIE, 2008).

It is beyond the basis of this chapter to argue for or against any particular concept of animal welfare. However, we must briefly discuss the role that science plays in deciding upon the best definition of animal welfare.

In practice, these different concerns about animal welfare usually overlap. For example, a lame dairy cow is suffering pain, her normal behaviour is inhibited and her biological functioning, in terms of milk production and reproduction, is threatened. Indeed, it is more accurate to say that these concerns focus on different aspects of animal welfare rather than providing competing definitions of animal welfare. From an animal welfare perspective, much of the concern about disease arises from the suffering that the animal undergoes as a result of illness, while an inability to perform natural behaviours that the animal is highly motivated to perform can result in frustration and perhaps in some stress-induced immuno-suppression. The fact that animal welfare appears to involve three aspects may simply reflect our lack of knowledge of the relative degree of suffering caused by behavioural deprivation and poor health. Many veterinary text books include psychological well-being as part of overall health (Gunnarsson, 2006), thus emphasizing the link between emotional-based concepts of animal welfare and those based on good functioning. Where the different views of welfare overlap, there is little disagreement that animal welfare is at risk. In some cases, however, these different aspects of animal welfare may appear to be in conflict. For example, housing animals in groups allows them to engage in natural social interactions, but when poorly managed can lead to increased incidence of certain diseases or aggressive interactions. Clearly the best solutions for animal welfare will be those that address all concerns, for example, by creating group-housing systems that avoid competition and allow animals to behave freely and to remain healthy. There are a number of example where new housing systems that have been adopted to provide better behavioural opportunities to animals (such as loose housing for farrowing sows) have been found to reduce health (through increased piglet mortality) but where subsequent research has been able to provide solutions to reduce this problem (Wechsler and Weber, 2007). But sometimes this may not be easy to do.

In such cases, we may need to decide which threat to animal welfare is the most serious. Unfortunately, one of the greatest unresolved issues in research in animal welfare is how to weight the relative importance of the different threats to animal welfare, particularly where these are of a very different type e.g. health versus behavioural expression. Developing widely accepted methods to weight different components of welfare and to combine scores to achieve an overall welfare measure is an area of on-going debate that has yet to be resolved (Botreau *et al.*, 2007). We suggest that it will not be resolved until we have a scientific concept of how the different components of welfare relate to the overall welfare of the

animal. Consequently, for the immediate present, we are limited to being able to develop a qualitative judgment about the impact of different hazards on animal welfare rather than being able to provide a strictly quantitative measure (EFSA, 2008).

Although, such obvious conflicts between the different aspects of welfare are not common, they are responsible for the majority of disagreements between the different stakeholders. What role does science play in deciding which aspects of animal welfare are the most important? Experience has shown that in some cases different scientists can reach opposite conclusions about the relative advantages of different housing systems by favouring different welfare indicators. For example, conflicting conclusions about the relative merits of stalls or group housing for gestating female pigs resulted from different groups of scientists placing the emphasis on different aspects of animal welfare, with some emphasizing the apparent health advantages of gestating stalls, while others placed more emphasis on the greater behavioural opportunities arising from group housing (Fraser, 2003). It seems likely these differing emphases reflect the particular training and discipline of the scientists involved, with veterinarians being more aware of the impact of poor health on animal welfare, while ethologists would be more aware of behavioural problems. Thus, it is essential that any overall scientific review of animal welfare make clear the assumptions as to what are the most important components of animal welfare, and ensure a wide range of scientific disciplines are involved.

Some scientists have reacted to the diversity of views on animal welfare by trying to develop a scientific concept that is most amenable to scientific investigation. However, an important point made by Fraser *et al.* (1997) is that the relative importance given by different stakeholders to different aspects of animal welfare can reflect their personal values, and that science alone cannot determine which one is the 'correct' definition. The OIE (2008) recognized this point when it stated as one of its guiding principles that 'the scientific assessment of animal welfare involves diverse elements which need to be considered together, and that selecting and weighing these elements often involves value-based assumptions which should be made as explicit as possible.' For our purpose, the message (if we accept Fraser's argument) is that science appears to have only a limited role to play in determining the 'correct' definition of animal welfare. However, science does have a very important role in explaining the biology of each species of farm animals so that the public's values and views on animal welfare are well-grounded in accurate biological knowledge.

However, we must not over-estimate the differences between the different stakeholders or scientists in what they consider as good animal welfare. Discussions about the different definitions of animal welfare often give the impression that there is little agreement between scientists on what constitutes good welfare. However, despite occasional disagreements, considerable consensus exists. Some recent examples show how consensus can be achieved on what constitutes good or poor welfare using methods such as the Delphi technique (Anonymous, 2000; Hegelund and Sørensen, 2007; Leach *et al.*, 2008; Whay *et al.*, 2003). Whay *et al.* (2003) found considerable agreement between experts (veterinarians and behavioural scientists) as to which of two dairy farms had the higher level of welfare, when they were presented with a variety of information about the state of the cows on the two farms. The results show that with appropriate techniques, people with experience in animal welfare are

capable of integrating a variety of information about animal health, housing and management to achieve a fair consensus on the level of welfare on the farm. Such techniques have heuristic values in making decisions about animal welfare even in the absence of a clear definition of welfare.

Where genuine disagreement does exist, it often occurs when poorly understood measures of welfare are used. Most often this involves disagreements about the role of animal emotions and of natural behaviour. It is relatively uncontroversial that the presence of a disease indicates poor welfare and that good welfare requires that the animal be in good health. As we discuss later (section 4.1), the main issue here is in measuring the relative impact of different diseases on animal welfare. However, while veterinary medicine is relatively good a diagnosing the presence of individual diseases, the question of deciding the overall health status of animals is a far more difficult procedure, especially since veterinarians appear to lack a clear, consensual definition of what good health is for an animal (Gunnarsson, 2006). Thus, while it may be conceptually easy from a scientific point of view (even if not technically easy) to show that an animal is suffering from a particular disease, it is not scientifically easy to determine if the animal is in good overall health or the extent that its overall welfare is affected by disease.

Suffering by animals is a central aspect of poor welfare (Duncan, 2005). The importance of animals' feelings or emotions to the general public is shown by the acknowledgement of animal sentience in European animal welfare legislation (Camm and Bowles, 2000). A commonly voiced (but somewhat dated) criticism of the emphasis placed on animals' sentience or emotions is that these are beyond the reach of science. However, many recent publications show that animal consciousness is now widely accepted to be amenable to scientific investigation (Brydges and Braithwaite, 2008; Dawkins, 2008; Duncan 2006; Mendl *et al.*, 2009), and developing valid measures of animal emotion remains one of the most interesting and challenging problems in animal welfare science. Considerable progress has been made on understanding and measuring animal pain, and a large and rapidly developing scientific literature on pain assessment and prevention in now available for farm animals (for review see Weary *et al.*, 2006), and more recent research is beginning to examine pain in a much wider range of species, such as invertebrates (Elwood *et al.*, 2009). However, it is a long step from accepting in principle that animals have feelings and emotions, which can be investigated scientifically, to being able to actually measure the mental state of animals in order to determine the effect of farming conditions on their welfare.

The 'natural living' aspect of animal welfare is the issue that causes scientists the most difficulty. Many scientists have difficulty accepting definitions of animal welfare that include the animal being able to perform natural behaviour, even where they have been happy to include animal feelings and emotions (Broom, 2008; Spinka, 2006). However, the 'unnaturalness' of modern housing conditions is one of the greatest sources of concern by the public for the welfare of the animals, and animal welfare science needs to find an effective way of dealing with this issue. These we discuss in more detail in section 4.2.

In summary, the question of what is the best definition of animal welfare is not a scientific question but rather a question about what are the main concerns expressed by all of the

stakeholders. Science has only a limited role to play in deciding what should be encompassed in a definition of animal welfare. However, while it is essential that the concept of animal welfare link up to the full range of concerns people have about welfare, it must also be amenable to scientific study.

## 3. Input versus outcome standards for animal welfare

The criteria for assessing animal welfare are conventionally divided into those that describe the environment of the animal (variously called input-based, engineering or design criteria) and those that describe the actual state of the animal [outcome-based or animal-based criteria] (Mench, 2003; Webster, 2009). Input-based criteria generally describe the aspects of the housing or management system that are thought to have an impact on animal welfare, and include such aspects as the use of cages, space allowances, group size, and tethering etc. Outcome-based criteria include behavioural, physiological and immune measures, incidence of health problems, and production levels (see section 4).

### 3.1 Pros and cons of input-based standards

Input-based criteria are usually favoured in animal welfare standards since they are easier to audit (Mench, 2003). For example, most legislation on animal welfare defines which types of housing or management are considered as unacceptable. A potential advantage of well-chosen input-based criteria is that these should prevent welfare problems from occurring. Identification of hazards to animal welfare that may arise from certain housing systems or management practices is an essential first step in a HACCP approach to preventing animal welfare problems from occurring (Hegelund and Sørensen, 2007). However, to use input-based criteria with any confidence, we need scientific research that shows fairly unequivocally that any given housing or management practice is indeed a hazard leading to reduced welfare.

Usually, research has provided the clearest evidence when assessing acute threats to animal welfare, such as the painfulness of dehorning, tooth clipping, tail docking etc. For example, research has been fairly consistent in showing that dehorning of cattle is painful and that a combination of local anaesthetics and longer lasting analgesics is needed to control both the pain of the operation itself and the post-operative pain (reviewed in Stafford and Mellor, 2005; Rushen *et al.*, 2008). In this case, we can be fairly certain that an input-based animal welfare standard that required the use of anaesthetics and analgesics would improve the welfare of the animals.

However, it has been a far more difficult task for science to show the disadvantages of complex housing systems or management practices that have long lasting, chronic affects on animal welfare. It is beyond the scope of this article to describe all of the research that has attempted to show welfare advantages or disadvantages with battery cages, gestating sow stalls, veal crates etc. But the reviews that have been done show how difficult it is to clearly show that one type of housing system is better, in all respects than another. Usually, the differing housing or management systems have both disadvantages and advantages for animal welfare, and research has been very successful in identifying these. For example, a

current debate is whether furnished cages or non-cage systems provide the best alternative to battery cages for laying hens. One comprehensive study showed that laying hens in non-cage systems make better use of the resources available, appear less fearful and have stronger bones than hens in furnished cages; in contrast, the latter have lower mortality rates and typically are exposed to better air quality (Rodenburg *et al.*, 2008). However, it is difficult to draw clear conclusions about which system is better for the overall welfare of the birds. To do this, we would need to be able to rank the different advantages and disadvantages on a single scale. The inability to unequivocally show that any housing system is a hazard (rather than having both advantages and disadvantages) to animal welfare is an important difference when HACCP or risk analysis approaches, which were developed to deal with unequivocal hazards for food safety, are applied to animal welfare (Hegelund and Sørensen, 2007). It is essential to base our judgements of the relative impact of alternative housing systems on animal welfare on a full analysis of all the advantages and disadvantages within each system.

Recently, some studies of commercial animal welfare standards that rely strongly on input-based criteria have thrown some doubt on their effectiveness at promoting animal welfare: Main *et al.* (2003) showed that many animal-based measures of welfare did not show a clear advantage to the RSPCA Freedom Food standards compared to normal commercial practice. Thus, there is good scientific support for some input based standards (especially those dealing with acutely painful procedures), but less support for others (especially those dealing with housing or management systems).

### 3.2 Pros and cons of outcome-based standards

Because of the problems with using input-based criteria to describe animal welfare, there has been interest in developing outcome-based methods that can be used to assess the actual state of welfare of the animals on a particular farm (Main *et al.*, 2007; Webster, 2009). Certainly there are advantages in increasing the use of outcome-based measures of animal welfare, and science has been successful in developing a wide range of validated measures that can be used in the assessment of animal welfare (Fraser, 2008). But there are difficulties. One problem involves the extent that we can reliably and validly measure the outcomes. Much of section 4 of this article discusses these issues for the various types of measures that have been used.

The second problem involves the real practical difficulties in being able to take the necessary measures (discussed in section 4) from a farm in a relatively short period of time. For example, outcome-based criteria to control pain from dehorning calves would require recording blood concentrations of cortisol, or fine grained behaviours such as ear flicking or head shaking during the hours following dehorning (Rushen *et al.*, 2008).While this can be easily done in well-equipped laboratories, it is unlikely that it could be done during routine inspections of farms. On-farm animal welfare monitoring using outcome-based measures does cost money and so there is always pressure to reduce costs by limiting the number of welfare indicators used, the duration of the visit or the frequency of visits, or the number of animals observed (Sørensen *et al.* 2007). Any audit or visit to a farm involves a measure of time sampling, since it is rare to be able to obtain data continuously. Problems of time sampling are most apparent when using behavioural outcome measures. Many of the behaviours that we would like to

use to assess welfare (discussed in section 4) occur only for short periods of time and do not occur equally throughout the day. Therefore a brief visit to a farm at a time chosen most often for the auditor's convenience is unlikely to obtain a reliable measure of the occurrence of abnormal behaviours throughout the day. Furthermore, the extent that the occurrence of behaviours is distributed over time can be affected by management factors. For example, in broiler flocks, reduced intensity of the lighting tends to reduce the diurnal rhythm in the occurrence of behaviours so that the distribution of different activities occurs more evenly throughout the day compared to when light intensity is high (Alvino *et al.*, 2009). Thus, observations of broiler behaviour at two farms that used different light intensities at fixed times of the day could produce different estimates of the occurrence of the behaviour even if the daily occurrence of the behaviour was the same.

Our increasing ability to automate recording of animal behaviour, either through special technology (Huzzey *et al.*, 2005) or by using technology normally used to manage animal production (Borderas *et al.*, 2008), is likely to aid the use of behavioural measures in on-farm animal welfare monitoring schemes.

Even with measures of illness or injury there are considerable difficulties in obtaining accurate estimates of incidence. Measures of incidence involve the collection of data on the number of animals that become ill over a period of time and hence represent a more complete sample than do measures of prevalence. However, researchers must often rely on data collected by others, such as herdsmen, and any problems in record keeping or diagnosis will reduce the value of the data. Some researchers rely upon records kept of veterinary treatment, and Scandinavian countries typically have a centralized data bank recording all veterinary treatments of farm animals. However, such systems are not in use in other countries. Measures of treatment may give a biased estimate of disease incidence since not all diseases are likely to be treated. For example, because mastitis in dairy cows has immediate effects on milk revenue, cows with mastitis are more likely to be treated than cows suffering from lameness. Thus measures of treatment likely underestimate the incidence of some diseases and likely reflect the perceived economic return on the cost of treatment. Given the relative rarity of veterinary treatments for many diseases, most on-farm assessments of welfare rely upon the farmer's diagnosis. However, the accuracy of these will depend upon the farmer's skill, diligence, and the amount of time available and the effectiveness of the recording method used. Where farms are large, there is also usually only a sample of animals observed. When estimating injury prevalences on large pig farms, it is difficult to observe all pigs and so a sample of pigs or pens is chosen as being representative of the whole farm. Mullan *et al.* (2009) compared estimates of prevalence of injuries using different sampling strategies and found that even samples of a large percentage of the pens could provide inaccurate estimates of the overall prevalence of injury, due to the large variability between pens. Consequently, it is essential to have a reliable and tested sampling strategy in order to obtain reliable estimates. Unfortunately, many studies do not test the adequacy of their sampling strategy.

The third problem is based on the fact that the animal welfare measures do not measure overall welfare but show the presence or absence of specific threats to animal welfare. Measures of illness and injuries in swine and dairy cattle do not measure their welfare; they

measure the incidence of illness and injuries. The absence of illness and injuries does not show that animal welfare is good, merely that those particular threats to animal welfare are absent. However, other threats to animal welfare may still be present. This problem is apparent in some of the attempts that have been done using outcome-based measures of welfare to assess the effectiveness of welfare improvement schemes that use mainly input-based measures. Main *et al.* (2003) showed that many animal-based measures of welfare did not show a clear advantage to the RSPCA Freedom Food standards compared to normal commercial practice. However, the outcome-based measures used focussed very strongly on diseases of dairy cows, and it is questionable whether many of the input-based standards of the RSPCA scheme (such as pain control for acutely painful procedures) would be reflected in altered incidence of these diseases. Similarly, measures of disease incidence give only limited information on the relative welfare status in organic herds (Lund and Algers, 2003) since organic standards place much value on the abilities of animals to express normal behaviour, which may not be apparent in altered health status.

At present, it is unlikely that outcome-based standards can completely replace input-based standards. Given that there are pros and cons for both input- and outcome base criteria, the most sound conclusion is that for the foreseeable future, animal welfare assessments will require both. Indeed, experts tend to choose a mix of such criteria when developing animal welfare assessment schemes (Leach *et al.*, 2008).

## 4. Animal welfare indicators

### 4.1 Health, production and reproduction

#### *4.1.1 Animal disease*

It seems evident that poor health can be a major cause of poor welfare in farm animals and that the occurrence of illness can be used to assess animal welfare. Farm animals of all ages, both in traditional and modern housing systems, and under intensive and extensive management, suffer from a variety of endemic diseases, both infectious and non-infectious, as well as unpredictable outbreaks of epidemic diseases, such as foot and mouth. They can also suffer from various forms of physical injury as a result of poor housing or management. Health measures have long been recognized as potentially useful indicators of animal welfare. Wide scale assessments of the incidence of illness on farms can provide very useful information to judge animal welfare. For example, measures of dairy calf morbidity helped identify large group sizes as being a hazard for the welfare of group-housed calves (Svennson *et al.*, 2003). Measures of leg health of broilers showed that specific housing conditions had a greater impact on welfare than did stocking density, even though proposed legislation focussed on the latter (Dawkins *et al.*, 2004).

However, while it may seem obvious that good welfare is dependent on good health, in order to be able to use measures of the occurrence of different diseases to assess overall welfare, we need to judge the relative impact on animal welfare of different forms of illness or injury. This is not easy. In animal agriculture, the importance of a disease is often judged

by its direct economic impact, but this does not help us decide what impact a disease has upon animal welfare. To understand how different diseases affect animal welfare, we need to know how much suffering they cause the animal (Wells *et al.*, 1998), either in the past, the present or the future. Ideally, we would be able to directly measure the amount of suffering caused to the animal by a disease, but at present this is not feasible. An indirect approach is to compare different diseases by the severity of the symptoms. Severity can be assessed by the duration of the disease and/or the likelihood of the disease causing death. For less serious diseases, some data is available to let us judge the relative impact of different diseases of dairy cattle, by examining their impact on milk production and feed intake (Rushen *et al.*, 2008). This type of data has some potential in evaluating the relative effects of different diseases on cattle welfare, but there are limits to the extent that we can assume that changes in milk production and feed intake really capture the full impact on animal welfare. Another promising approach is to examine any common symptoms that different diseases may have. For example, reductions in feeding behaviour and general activity and increased time spent resting are behavioural changes that accompany a wide number of illnesses in many species (reviewed in Millman, 2007; Weary *et al.*, 2009), such that the relative severity of illnesses might be judged by comparing the relative magnitude of these changes in behaviour. Assessing the degree of change in these common 'sickness' behaviours may help us get a better understanding of the relative impact of different diseases on animal welfare (Reiner *et al.*, 2009). Unfortunately, we still have too few direct analyses of the impact of different diseases on animals in order to judge their affect on animal welfare.

An alternative means of assessing the impact of illness or injury upon animal welfare is to ask the experts, which is a means forward when it is difficult or impossible to obtain more objective information. A number of procedures have been developed to determine what consensus there is among experts. Whay *et al.* (2003) recently conducted a Delphi exercise, which involved repeated consultation with a panel of animal welfare researchers and veterinarians who were asked to rank various animal welfare parameters, including health problems, in terms of their importance for animal welfare. The results of this exercise were then used to assess a number of dairy farms. The results show that experts can come to some consensus about the relative severity of different diseases for animal welfare. However, these results are limited by what the experts know; the importance of some diseases is likely to be underestimated. Until we have better data on the duration and severity of the full range of symptoms of each disease, we will probably need to continue to rely on sophisticated techniques for obtaining 'expert opinion' on the importance of various diseases for animal welfare.

### *4.1.2 Animal productivity and reproduction*

The use of measures of productivity (e.g. growth rates of swine, beef cattle and broiler chickens, or egg or milk production) to assess the welfare of an animal is controversial. Changes in productivity can be powerful indicators of poor welfare, but only if we understand the cause of the change, and can relate this to the fact that the animal is suffering in some way.

There are good reasons why problems causing poor welfare can lead to lower productivity. Activation of the immune system during an illness uses a considerable amount of metabolic energy (Colditz, 2002) and illness often results in reduced feed intake, so that during

illness resources may be limited and diverted into immune function rather than egg or milk production, growth, or reproduction. Drops in production may indicate acute stress or developing illness; for example, reductions in feed intake or milk production of dairy cows can indicate illness. However, this does not mean that all or even most variation between farms in animal productivity is related to differences in animal welfare. Variation in animal productivity can be affected by a host of factors (nutritional, genetic and environmental) that are welfare-neutral. In addition, there is increasing evidence that a high level of production may itself produce a threat to animal welfare. High growth rates among pigs and poultry and high levels of milk production in dairy cattle are associated with an increased occurrence of health problems (Ingvartsen *et al.*, 2003; Rauw *et al.*, 1998). Thus, there is no simple relationship between good productivity and good welfare. Sudden and inexplicable drops in egg production, growth rates, or body condition, may indicate illness, and hence can be used to assess welfare where the causes of these changes are understood. However, the level of production of individual animals and especially of farms cannot be used with any confidence to assess welfare.

Similarly, the fact that poor welfare can sometimes affect reproduction does not mean that reproductive success can be used to assess welfare in practical circumstances. This is especially true when considering differences between farms. When reproductive problems arise because of poor health, this is undoubtedly a sign of poor welfare. However, differences among farms in reproductive failure could be due to many factors that are not related to the welfare of the animals, such as success at oestrus detection, effective artificial insemination strategies, as well as general reproductive management. In short, although poor welfare may reduce reproductive success of animals, measures of reproductive success alone, especially at the herd level, appear to have little potential in welfare assessment.

In summary, changes in productivity or reproduction due to stress or poor health can serve as an indicator of changes in animal welfare but overall levels of productivity of a farm are of little value in welfare assessment. Indeed, few animal welfare experts consider production measures such as milk yield or egg production to be a useful measure in on-farm welfare assessments (Hegelund and Sørensen, 2007; Main *et al.*, 2003; Whay *et al.*, 2003), for example.

### 4.2 Animal behaviour

Some of the most difficult issues in animal welfare deal with the behaviour of the animal, and the study of farm animal behaviour has made major contributions to identifying and helping to solve some of the key problems in the welfare of farm animals. However, there can be difficulties in interpreting animal behaviour.

#### *4.2.1 Animal preferences as input-based criteria for animal welfare*

Designing appropriate environments for animals is much aided by knowing what the animal prefers and these preferences can be assessed directly. Testing animals' preferences is one of the most obvious and direct ways of asking what it is that an animal wants in its environment. However, there can be problems with interpretation. A number of excellent discussions of the pros and cons of preference testing are available (Kirkden and Pajor, 2006). The results

of choice tests can be difficult to interpret when the options presented to an animal differ in a number of properties, e.g. when animals are asked to choose between different housing systems. The general conclusion of recent reviews is that preference tests are most successful when there are only relatively small differences between the options e.g. choosing between flooring with different degrees of softness.

One important limitation in preference tests is that the animals' choices are restricted to the options provided, so a strong preference for one option only means that it is better than the alternative and cannot be taken to mean that it is ideal for the animal. Another important methodological consideration in choice tests is that animals may initially prefer the option that is familiar, yet if encouraged to try the alternative or if given sufficient time, the animals may switch preferences. Choice tests need to ensure that animals have been given sufficient experience with all the options available. In addition, preference tests show how animals rank various options, but provide little information on the strength of their preferences. One way of assessing whether access to a preferred option is really important to the animal is to impose some cost on the animal's choice (Jensen and Pedersen, 2008; Kirkden and Pajor, 2006). There are a number of ways of imposing costs (Jensen and Pedersen, 2008). For example, in a choice situation, animals may have to forgo access to one valued resource, such as food, in order to access another resource. As another option, animals can be trained to perform an 'operant' task such as pressing a bar or pecking a key, in order to access a resource. Once animals have learnt to do this, the 'price' of the item can be increased by, for example, making it press the bar more often in order to gain the reward. The extent that the animal values the item can then be assessed by determining how much work it is willing to expend. An alternative approach is to restrict the amount of time that animals have available to spend in performing different behaviours. As the time available decreases, only the most valued behaviours should continue to be performed. In these different ways, we can estimate the value of different resources to animals. A number of studies have now used various forms of operant testing to determine the extent that animals value different resources (Jensen and Pedersen, 2008; Patterson-Kane *et al.*, 2008).

Most importantly, choice tests reflect the animals' immediate response to the alternatives, but we do not know to what extent animals are capable of judging the long-term effect of a housing treatment on their overall welfare. For example, dairy cows tend to choose wider stalls even though wider stalls can become dirtier potentially increasing the risk of mastitis (Rushen *et al.*, 2008). Animals simply may not have the capacity (or the experience) to judge the long-term consequences of their choices. Some studies show that poultry are able to forgo an immediate small reward in order to obtain a larger reward later (Abeyesinghe *et al.*, 2005), so some farm animals may be able to make some judgements about their longer term welfare (Mendl and Paul, 2008). However, we should not assume that the animals are perfect judges of their own welfare, any more than people are perfect judges of theirs.

Given these caveats, preference tests can provide some evidence of the animal's own perceptions of what are the hazards for its welfare. The underlying assumption behind choice tests is that an animal's welfare will be improved if it is housed under the conditions it prefers, but this assumption has not been adequately explored.

### *4.2.2 Natural behaviour*

The issue of behavioural deprivation is a crucial one for animal welfare and the possibility that farm animals are suffering because they cannot perform behaviour that they normally would perform is one of the enduring concerns that the public has about the welfare of animals in modern husbandry. Housing systems and management practices for farm animals have been criticized on the grounds that they prevent the animals from performing their natural behaviours. The problems with the concept of natural behaviour have been discussed many times (Bracke and Hopster, 2006; Dawkins, 2005; Spinka, 2006). Briefly, there is no reason to think that an animal will inevitably suffer simply because it does not perform all the behaviour patterns shown by its wild ancestors. Indeed, allowing animals to perform some natural behaviour, such as aggressive behaviour or infanticide, may lead to reductions in animal welfare. Furthermore, even within a population of animals in natural environments, there will be marked differences between individual animals in the types of behavioural strategies adopted, raising the question of what should be considered as 'the' natural behaviour (Spinka, 2006). Finally, a multitude of detailed studies on different species has revealed how much artificial selection has altered the behaviour of domestic animals (Jensen, 2006), so we are uncertain as to how much of the behavioural repertoire of their wild ancestors domestic animals have retained. References to natural behaviour in legislation or official definitions of welfare can cause problems for science. For example, the OIE (2008) refers to 'innate behaviour' but there is no clear scientific definition of innateness (Mameli and Bateson, 2006). Despite these criticisms, the longevity and the ubiquity of this concept suggests that it does capture some of the disquiet that modern farming systems provoke in many people. Furthermore, comparisons between farm animals and their wild ancestors or feral counterparts may give some clues as to where to look for welfare problems that can arise from behavioural deprivation (Spinka, 2006) or how to improve housing systems by allowing the animals a greater degree of behavioural expression (Lidfors *et al.*, 2005).

The more recent research examining the genetic basis of behaviour in farm animals holds great promising in being able to pin down the natural behaviour of most farm animal species (Jensen *et al.*, 2008). But in the absence of such information scientists have tended to avoid the question of whether or not a behaviour is 'natural' and instead have translated the notion of natural behaviour into questions about behavioural need and behavioural motivation that are more amenable to scientific investigation.

### *4.2.3 Behavioural needs and motivational analysis*

One alternative approach to asking about natural behaviour is to ask which behaviours are important for the animal to perform. The principle of this approach is that the inability to perform a behaviour will cause greater suffering when the animal is highly motivated to perform the behaviour and when the factors that stimulate the behaviour are internal to the animal (rather than being a feature of the environment) and where the performance of the behaviour itself is necessary to reduce the underlying motivation. There are some particularly good discussions of the concepts and issues underlying this approach (Dawkins, 2005; Jensen and Pedersen, 2008; Spinka, 2006). The importance of external (that is, triggered by stimuli in the environment of the animal) versus internal (that is, triggered by events occurring

within the animal) causes of behaviour can be understood with some simple examples. The natural behaviour of most species includes anti-predator responses that are performed when potential predators are detected. An obvious question is whether farm animals 'need' to express this anti-predator behaviour. Should we expose cattle to wolves or dogs so that they can do this? Most would say no, and the most common reason is that such anti-predator behaviour is stimulated by the sight of the predator itself. In the absence of any predators, which is generally the case in most modern farming systems, the animals are simply not motivated to perform anti-predator behaviour, and so the fact that they do not perform the behaviour in modern husbandry settings is not a problem for their welfare. This situation is often captured by the phrase 'out of sight, out of mind'.

This does not mean that we should ignore the anti-predator behaviour of farmed animals when designing housing for animals. For example, heightened vigilance is the most common method animals use to avoid predation and it may be advantageous to allow animals opportunities to view their surroundings more fully. Foxes appear to favour elevated platforms in their cages because they allow them to monitor their surroundings (Korhonen *et al.*, 2008).

In terms of the factors that are responsible for stopping the behaviour or turning off the motivation, the question is whether these result from the animal achieving the functional goals of the behaviour or whether it is necessary to perform the behaviour itself. As a simple example, the functional goal of eating is the consumption of various nutrients. This is usually achieved by the animal eating, but it need not be: the nutrients could be supplied by a stomach tube, for example. The question in this case is whether feeding motivation is turned off simply by the ingestion of the nutrients, or whether the animal also needs to perform the behaviour of eating. Certainly most farm animals absorb nutrients by eating, but there are many components of feeding behaviour that modern housing systems do not permit animals to perform. For example, many milk fed calves do not suck to ingest their milk, and many dairy cows no longer graze. How much is the performance of these particular behaviour patterns necessary to reduce feeding motivation, even when nutrient intake is adequate? Among calves, the performance of sucking behaviour itself, irrespective of milk intake, does play a role in postprandial satiety (Rushen *et al.*, 2008), and an inability to perform much of their normal feeding behaviours does seem to be a factor underlying the development of stereotypic behaviours in many ungulate farm animals (Bergeron *et al.*, 2006). Thus, we should not underestimate the importance of animals being able to perform some 'natural' behaviours.

A large body of research has thus focussed on understanding the motivation or causes of different behaviours, with a particular focus on whether the main factors are internal or external to the animal and whether or not performance of the behaviour itself is necessary to fulfil the underlying motivation. However, one of the main findings is that the theories of how behaviour is caused are far too simple. Control of behaviour is complex and often involves a mix of internal and external factors. As a result of the complexity of behaviour control, interest in the concept of behavioural needs appears to have waned. However, the underlying issues posed by the concept of behavioural needs have not been resolved. While we have good information on the motivational basis of some key behaviours (e.g. dust bathing and nest building in poultry – Weeks and Nicol, 2006; sucking behaviour of calves – Rushen *et*

*al.*, 2008; nest building by swine – Wischner *et al.*, 2009), research still has not adequately detailed which other key behaviours are the most important for animals to perform.

### *4.2.4 Examining the consequences of behavioural deprivation*

An alternative approach is to assess the effects or consequences of preventing animals from performing certain behaviours. This approach can be used to confirm results from other testing methods. For example, a number of studies have shown that preventing milk-fed calves from sucking to obtain milk can result in cross-sucking, hormonal changes and reduced signs of relaxation after the meal, while reducing the time that dairy cows can rest likely increases the risk of lameness (Rushen *et al.*, 2008). Again, we still have insufficient evidence of how deprivation of many of the other behaviours thought to be important can affect farm animals.

### *4.2.5 Behaviour as an outcome-based criterion for animal welfare*

The performance of certain behaviours by the animal has been used as outcome-based measures to assess the actual welfare status of the animals. Most obvious (and least controversial) are behaviours that directly reduce the welfare of the animals. These include injurious behaviours, such as fighting. Such behaviours are a clear threat to the welfare of the recipient animals, so measures of their occurrence are useful in assessing animal welfare. These we do not discuss further.

The second category is more problematic and includes behaviours that do not themselves reduce the animal's welfare but that may be an indirect sign that the animal's welfare is threatened. These can include behaviours indicating fear or anxiety and so-called abnormal behaviours, such as stereotypic behaviour. These behavioural indicators of poor welfare are indirect in that the behaviour can only be taken as a sign of some problem. For example, animals in pain or frightened express this in some way in their behaviour, and behavioural indicators of pain or fear can indicate the existence of a threat to their welfare. Unfortunately, often purported behavioural signs of pain or fear are not adequately validated and consequently are not easily interpreted (Rushen, 2000).

Some of the issues can be illustrated by attempts to assess animals' fear of people, which has proved particularly problematic. The quality of stockmanship is one of the most important factors affecting the welfare and the productivity of farmed animals (Hemsworth, 2003). One very important aspect of stockmanship that has been extensively researched is the way that animals are handled, especially the use of aversive handling techniques that may result in animals becoming fearful of their caretakers. There is little doubt that rough handling of animals and their resulting fear of people can have markedly deleterious effects on their welfare and their productivity (Hemsworth, 2003; Waiblinger *et al.*, 2006). It would, therefore, seem logical to include measures of the fear that animals show to humans in on-farm welfare audits (Rousing and Waiblinger, 2004). However, de Passillé and Rushen (2005) recently reviewed much of the research in this area and suggested that insufficient attention has been paid to assessing the reliability and the validity of the tests used within on-farm welfare assessments or audits. Reliability usually refers to the degree that a measure can be

taken repeatedly (either at different times or by different people) and give the same answer. Validity refers to the degree that the measure really measures what we want it to i.e. the animals' fear of people and the impact this has on their welfare.

De Passillé and Rushen (2005) concluded that the reliability of many measures used is often low, even if statistically significant, and it is questionable whether it is adequate for on farm assessment that may result in economic losses to farmers. Perhaps more importantly is the lack of adequate validation of the tests as measuring the effects on the animals' welfare. The fact that the measures can be affected by extraneous variables such as the clothing worn, the location of the test etc., reduces confidence in both their validity and reliability. Some more recent research has supported these conclusions. Although some studies do report reasonable degrees of repeatability (Windschnurer *et al.*, 2008), others do not. For example, a recent comparison of tests of pigs' fear of people (Scott *et al.*, 2009) found only moderate reliabilities. The highest test-retest reliabilities showed that only 60% of the variance between pigs on one test was related to the variance when the test was repeated. Different tests of the animals' fear were either not correlated or weakly correlated. Gibbons *et al.* (2009) found that cows were consistent in how they reacted to people, but the reported reliability estimates were only moderate. There was no evidence that the cows' responses to people reflected fearfulness and the cows' responsiveness to people was quite unrelated to other tests of cows' general fearfulness, showing that the tests do not measure animals' general temperament. Some studies have attempted to validate the measures by showing that extra human contact can reduce the distance hens keep from people (Graml *et al.*, 2008). However, while this test may give some information on the 'human-animal relationship' it does not validate these as measures of 'fear', which raises doubt as to their importance for animal welfare. Animals' responses to people are likely to reflect the animals' familiarity with people, as well as their degree of hunger, if the animals have been fed by people. We should not assume that the animals' fear of the people is the most important factor affecting the distance they keep from people. In some cases, studies have found that an increased distance kept between animals and people is associated with physiological signs of stress, which strengthens the argument that the animals' fear of people is an underlying factor. However, this does not mean that all instances where animals avoid people can be taken as a reliable measure of fear. There is little doubt that rough handling by people can increase animals' fear of people and that this can have negative impact on their welfare. Measures of the distance animals keep from people can be used to measure their degree of fear in experimental situations where the fear has been induced by rough handling. However, this does not mean that every time we see animals avoiding people, we can infer that they were handled badly, or that they are afraid of people or that this affects their welfare. Furthermore, good stockmanship comprises more than just appropriate handling of the animals, and includes a much broader range of activities necessary for caring for the animals. Thus, an assessment of the adequacy of stockmanship on a farm should consider more than just the degree of fear that the animals feels towards people (De Passillé and Rushen, 2005).

Communicative signals, such as vocalizations and displays, can also provide information about the animal's state that may be useful in welfare assessment. Animal vocalizations in particular been promoted as a means of assessing animal welfare (Manteuffel *et al.*, 2004). However, one assumption is that communicative behaviour is essentially an honest and

accurate expression of the signaller's state. The extent that animal signals would have evolved to be both is a matter of on-going debate, which is beyond the realms of this article (Bradbury and Vehrencamp, 2000). Furthermore, communication is likely to be performed with the intent of influencing the behaviour of others, rather than just expressing an animal's internal state, and there is evidence that vocal signals are affected by the social context and not only the state of the signaller. For example, calls by poultry thought to indicate frustration are influenced by the presence of an audience (Zimmerman *et al.*, 2003). The important point is that we cannot just assume that animal vocalizations will be honest or accurate reflections of the animal's emotional state. These measures of welfare must be validated like all other measures (Weary and Fraser, 1995). Weary and Fraser (1995) present some good examples of how this can be done.

Farm animals, and animals in captivity, show a variety of behaviours that are often considered 'abnormal' and these have figured prominently in attempts to assess welfare. Sucking on buckets or posts by calves, repetitive pacing by laying hens, tongue playing by cows, and stereotypic rooting at the concrete floor or bars of a pen by pigs are some examples. A considerable body of research has now revealed the causes of these behaviours in a wide number of species (Mason and Rushen, 2007). However, we need to be cautious in interpreting these behaviours in terms of animal welfare. The use of the word 'abnormal' to describe such behaviours implies that they are associated with poor welfare, but the word can have many meanings. First, there is the purely statistical meaning of rare or not usual. This meaning is of little relevance in the present case since such behaviours can often be performed by a large majority of animals in some housing systems (Mason and Latham, 2004). Second is the meaning of unnatural, in the sense that performance of these behaviours is not a part of the natural behaviour of the species. However, this meaning is of little help in discussing animal welfare largely because of the problems with the concept of natural behaviour discussed above. In many cases, such behaviours are labelled abnormal because we find it difficult to imagine why the animal is performing the behaviour. In other words, the function of the behaviour is not clear to the observer. This is more a recognition that we do not understand the behaviour, rather than a property of the behaviour itself. Our own point of view is that the labelling of such behaviours as 'abnormal' is not helpful. Instead, in order to be able to use the occurrence of such behaviours to assess welfare, we need first to understand why animals are performing the behaviours. Research into the causes of such abnormal behaviours has given some surprises. For example, many of the stereotyped behaviours of farmed ungulates appear to be related to feeding problems, such as the use of highly concentrated, low fibre diets for sows (Bergeron *et al.*, 2006) rather than housing problems, as was once thought. More importantly, we must also be able to validate these behaviours as an indicator of welfare by linking performance of the behaviour to reduced welfare. Mason and Latham (2004) and Mason (2007) provide a discussion of how abnormal behaviours are related to animal welfare. They concluded that the environmental conditions that lead to animals performing stereotypic behaviour are associated with poorer welfare. However, within a given environment, it does not seem that individual animals that perform stereotypic behaviour have lower welfare than those that do not.

### 4.3 Stress and physiological indicators of animal welfare

A common approach when assessing animal welfare is to look for physiological signs of stress. The two main physiological systems that respond to stress are the sympathetic nervous system (SNS) and the hypothalamic-pituitary adrenal (HPA) axis. Increased activity of the HPA axis is one of the most commonly reported physiological responses an animal makes in response to stress. Measures of physiological activity, such as changes in HPA axis activity, have been promoted and used to assess animal welfare many times (Mormède *et al.*, 2007). Research using these measures has shown quite clearly that some of the normal management practices do cause pain to the animal, and these measures have proven useful in assessing the most effective pain control procedures (e.g. dehorning and castration of cattle, Stafford and Mellor, 2005). However, others have questioned the extent to which physiological measures of stress really do reflect the level of animal welfare, especially in relation to longer lasting, chronic stressors due to housing (Rushen *et al.*, 2008).

First, there is little evidence of a general stress response that occurs irrespective of the type of stressor. Early in stress research it was assumed that increased HPA activity was a common response to all stressors, which led to the idea that increased HPA activity was the defining indication that an event was a stressor. Consequently, it was assumed that the relative impact of all management and housing conditions on animal welfare could be assessed by measuring the relative change in HPA activity. We now know that this assumption is not correct: not all stressors lead to increased HPA activity. The animal's behavioural, neurophysiological, and peripheral physiological responses to stressors are, in most cases, defence reactions – attempts by the animal either to deal with or avoid the stressor. Since different stressors will differ in many respects, the ways that the animal tries to deal with stressors will also vary from one stressor to another (Rushen, 2000).

One serious obstacle to the use of physiological data to assess animal welfare involves the extent that changes in physiological variables have been validated as indicators of animal welfare. Indicators of animal welfare need to be validated, that is a link between the measures of welfare chosen and an acceptable definition of animal welfare must be shown. Unfortunately, this has rarely been done for many physiological measures of stress. The dangers of not having a clear prediction of how poor welfare will affect HPA activity is illustrated in a recent study of the use of corticosterone concentrations in hen eggs as a means of assessing welfare in cages and free range systems (Bulmer and Gil, 2008). The authors hypothesised that cages would induce stress, which would be apparent in higher corticosterone concentrations. In contrast to their predictions, they found that caged hens had lower corticosterone levels. However, they were able to marshal sufficient evidence that chronic stress could reduce corticosterone output to conclude that the low levels of corticosterone did indeed indicate poor welfare!

We suggest that physiological changes, such as altered activity of the HPA axis, can be taken to indicate changes in animal welfare if these physiological changes indicate that the animal is suffering from some aversive emotional experience e.g. is suffering from pain, fear, anxiety etc. Numerous studies have shown increases in plasma cortisol concentrations following apparently painful procedures such as branding, castration and dehorning in

cattle. Furthermore, these increases are effectively reduced by the use of analgesics or local anaesthetics (reviewed in Rushen *et al.*, 2008). Thus, there seems little dispute that pain can result in marked increases in HPA activity in cattle.

However, a number of apparently painful treatments do not increase HPA activity and plasma concentrations of corticosteroids often appear to reach an upper boundary or 'ceiling' such that they are little value in discriminating between pains of different intensities (Mormède *et al.*, 2007). For sheep, measures of plasma corticosteroids are less effective than behavioural changes at detecting pain responses (Molony *et al.*, 2002). Most importantly, increased plasma concentrations of cortisol also occur following events that do not, on the surface, appear to be painful or cause suffering, for example, after sexual behaviour, feeding and milking (Lindstrom *et al.*, 2001; Norby *et al.*, 2007). In other cases, treatments known to be stressful, such as increased ambient temperatures, can result in reduced cortisol concentrations (Ronchi *et al.*, 2001). In short, increases in plasma cortisol concentrations alone cannot be taken as sufficient evidence that the animal is suffering and their measurements do not provide good information about the degree of suffering the animal is undergoing.

Blood sampling for cortisol assays may be an appropriate means of detecting HPA responses to acute stress that have a well-defined beginning and end. However, this method is far more problematic when dealing with prolonged stress (such as occurs with the effects of housing) or where the end of the stress is not well defined (for example, where there may be prolonged effects, such as chronic pain following de-horning or tail docking) (Mormède *et al.*, 2007). There are important methodological issues that impact on our ability to assess chronic changes in HPA activity from changes in blood cortisol concentrations. Due to the pulsatile secretion of cortisol from the adrenal gland, and the diurnal rhythms, plasma cortisol concentrations can change many-fold in a few minutes or a few hours (Hänninen *et al.*, 2006; Mormède *et al.*, 2007). This makes it difficult to assess effects of housing on HPA activity by taking single blood samples. Automated blood sampling that allows collection of a large number of blood samples over a prolonged period of time (Følnss and Munksgaard, 2008) is essential. However, even then, there is no certainty about whether measures of average concentrations or measures of the nature of the pulsatile release are the most appropriate to determine the degree of stress (Hänninen *et al.*, 2006). Because of the difficulties in assessing prolonged or chronic stress in terms of plasma concentrations of hormones, there has been interest in developing tests in which the HPA axis is challenged in some way (Mormède *et al.*, 2007). However, inconsistent findings on how chronic stress affects adrenal sensitivity make the ACTH test somewhat difficult to interpret at the present time (Rushen *et al.*, 2008). This places a major limitation on our ability to use such data to assess the effects of long-term challenges, such as housing, on animal welfare.

A second way of validating physiological measures as indicators of animal welfare is to show that the physiological changes have biological effects that cause suffering, for example, through stress-induced immuno-suppression leading to disease. Early research emphasized the suppressive effects of stress on the immune system, and the idea that stress can render animals and people more susceptible to disease has entered the public consciousness. We do not have the space to document the relationship between HPA activity and the immune system; fortunately a number of excellent recent reviews are available (Dhabhar, 2008).

Recent research has now shown the complexity of the relation between the HPA axis and the immune system, and there is now ample evidence that activation of the HPA axis can also have either no effect on the immune system, or can have immuno-enhancing effects. It would be foolish to argue that HPA axis activity does not have suppressive effects on the immune system or that changes in immune function and in HPA axis activity are unrelated. However, there is as yet little convincing evidence that assessment of HPA axis activity will reliably predict immune function.

Some evidence that increases in HPA activity may indicate reduced welfare comes from the metabolic consequences of increased HPA axis activity, most notably reduced feed intake, negative energy balance, an increased metabolic rate and subsequently loss of body weight or reduced growth (Elsasser *et al.*, 2000). However, we have little information to predict the effect of these metabolic changes on overall animal welfare just from measures of HPA activity. Thus, the complex relationship between HPA activity, immune system effectiveness and metabolism make it difficult to use changes in HPA activity to predict the nature or magnitude of the metabolic and immune consequences. Direct assessment of metabolism and immune function would seem preferable, although these too need to be validated as predictors of altered susceptibility to disease.

In conclusion, physiological measures of stress have proven useful in examining the relative effects of short-term, acute husbandry procedures on the animals and in testing out methods of reducing pain and distress from these procedures. However, they have proven far less useful in assessing the effects of longer lasting threats to welfare such as that come from the way that animals are housed. Thus, we recommend considerable caution when we attempt to use physiological data to assess the effects of chronic stress on the welfare of animals.

## 5. Conclusions

### 5.1 What has been achieved?

It is undeniable that science has made tremendous advances in dealing with the issue of farm animal welfare. Success has been most evident when examining the effects of short-term painful or frightening management procedures, and in finding ways of improving these to reduce their impact on animal welfare. Research has also developed a range of reliable and validated animal welfare indicators that have been used to document the advantages of alternative housing and management systems for animal welfare. Complex issues of animal emotions and behavioural deprivation have been shown to be amenable to scientific research.

### 5.2 What has been neglected?

However, there are limits to what science has achieved. Alternative housing systems have different advantages and disadvantages for animal welfare housing systems and these must be balanced to obtain an understanding of their overall impact on animal welfare. We still lack a widely-accepted scientific theory of animal welfare that addresses all the concerns of the various stakeholders, and which allows us to balance the various threats to animal

welfare. This complicates our decisions regarding the best housing systems for animal welfare. Many of the behavioural measures of animal welfare that have been used successfully in the laboratory are more problematic to use in farm inspections, although developments in automated monitoring of behaviour hold great promise.

### 5.3 What needs to be done?

We suggest that more attention needs to be paid to validating measures of welfare before using them in on-farm audits. We still lack information on many of the issues concerning which behaviours are the most important for animals to perform and adequate methodology for measuring their occurrence on farms.

## Acknowledgements

We sincerely thank all of you who have discussed these issues with us, with special thanks to our co-authors, Dan Weary, Nina von Keyserlingk and Elsa Vasseur.

## References

Abeyesinghe, S.M., Nicol, C.J., Hartnell, S.J. and Wathes, C.M., 2005. Can domestic fowl, *Gallus gallus domesticus*, show self-control? Animimal Behaviour. 70, 1-11.

Alvino, G.M., Archer, G.S. and Mench, J.A., 2009. Behavioural time budgets of broiler chickens reared in varying light intensities. Applied Animal Behaviour Science. 118, 54-61.

Anonymous, 2001. Scientists' assessment of the impact of housing and management on animal welfare. Journal of Applied Animal Welfare Science. 4, 3-52.

Bergeron, R., Badnell-Waters, A.J., Lambton, S. and Mason, G., 2007. Stereotypic oral behaviour in captive ungulates: foraging, diet and gastrointestinal function. In: Mason, G.J. and Rushen, J., 2006. (Eds.). Stereotypic animal behaviour: fundamentals and implications for animal welfare., second edition. CABI, Wallingford, UK, pp. 19-57.

Borderas, T.F., Fournier, A., Rushen, J. and De Passillé, A.M.B., 2008. Effect of lameness on dairy cows' visits to automatic milking systems. Canadian Journal of Animal Science 88, 1-8.

Botreau, R., Bonde, M., Butterworth, A., Perny, P., Bracke, M.B.M., Capdeville, J. and Veissier, I., 2007. Aggregation of measures to produce an overall assessment of animal welfare. Part 1: a review of existing methods. Animal 1, 1179-1187.

Bracke, M. and Hopster, H., 2006. Assessing the importance of natural behavior for animal welfare. Journal of Agricultural and Environmental Ethics 19, 77-89.

Bradbury, J.W. and Vehrencamp, S.L., 2000. Economic models of animal communication. Animal Behaviour 59, 259-268.

Broom, D.M., 2008. Welfare assessment and relevant ethical decisions: key concepts. Annual Review of Biomedical Sciences 10, 79-90.

Brydges, N.M. and Braithwaite, V.A., 2008. Measuring animal welfare: what can cognition contribute? Annual Review of Biomedical Sciences 10, 91-103.

Bulmer, E. and Gil, D., 2008. Chronic stress in battery hens: measuring corticosterone in laying hen eggs. International Journal of Poultry Sciences 7, 880-883.

Camm, T. and Bowles, D., 2000. Animal welfare and the treaty of Rome – a legal analysis of the protocol on animal welfare and welfare standards in the European Union. Journal of Environmental Law 12, 197-202.

Colditz, I.G., 2002. Effects of the immune system on metabolism: Implications for production and disease resistance in livestock. Livestock Production Science 75, 257-268.

Dawkins, M.S., 2008. The science of animal suffering. Ethology 114, 937-945.

Dawkins, M.S., Donnelly, C.A. and Jones, T.A., 2004. Chicken welfare is influenced more by housing conditions than by stocking density. Nature 427, 342-344.

De Passillé, A.M. and Rushen, J., 2005. Can we measure human-animal interactions in on-farm animal welfare assessment? Some unresolved issues. Applied Animimal Behaviour Science 92, 193-209.

Dhabhar, F.S., 2008. Enhancing versus suppressive effects of stress on immune function: Implications for immunoprotection versus immunopathology. Allergy, Asthma and Clinical Immunology 4, 2-11.

Duncan, I.J.H., 2005. Science-based assessment of animal welfare: farm animals. OIE Revue Scientifique et Technique 24, 483-492.

Duncan, I.J.H., 2006. The changing concept of animal sentience. Applied Animimal Behaviour Science 100, 11-19.

European Food safety Authority, 2006. Opinion on 'The risks of poor welfare in intensive calf farming system. An update of the Scientific Veterinary Committee report on the welfare of calves'. The EFSA Journal 366, 1-36.

Elsasser, T.H., Klasing, K.C., Filipov, N. and Thompson, F., 2000. The metabolic consequences of stress: targets for stress and priorities of nutrient use. In: Moberg, G.P. and Mench, J.A. (Eds.). The biology of animal stress. CABI, Wallingford, UK, pp. 77-110.

Elwood, R.W., Barr, S. and Patterson, L., 2009. Pain and stress in crustaceans? Applied Animimal Behaviour Science 118, 128-136.

Fønss, A. and Munksgaard, L., 2008. Automatic blood sampling in dairy cows. Computers and Electronics in Agriculture 64, 27-33.

Fraser, D., 2003. Assessing animal welfare at the farm and group level: the interplay of science and values. Animal Welfare 12, 433-443.

Fraser, D., 2008. Understanding animal welfare. Wiley-Blackwell, pp. 336.

Fraser, D., Weary, D.M., Pajor, E.A. and Milligan, B.N., 1997. A scientific conception of animal welfare that reflects ethical concerns. Animal Welfare 6, 187-205.

Gibbons, J., Lawrence, A. and Haskell, M., 2009. Responsiveness of dairy cows to human approach and novel stimuli. Applied Animimal Behaviour Science 116, 163-173.

Graml, C., Waiblinger, S. and Niebuhr, K., 2008. Validation of tests for on-farm assessment of the hen-human relationship in non-cage systems. Applied Animimal Behaviour Science 111, 301-310.

Gunnarsson, S., 2006. The conceptualisation of health and disease in veterinary medicine. Acta Veterinaria Scandinavica. 48, 20.

Hänninen, L., Løvendahl, P., De Passillé, A.M. and Rushen, J., 2006. The effect of floor type or relocation on calves' pulsatile growth hormone and cortisol secretion. Acta Veterinaria Scandinavica. A: Animal Sciences 56, 99-108.

Hegelund, L. and Sørensen, J.T., 2007. Developing a HACCP-like system for improving animal health and welfare in organic egg production based on an expert panel analysis. Animal 1, 1018-1025.

Hemsworth, P.H., 2003. Human-animal interactions in livestock production. Applied Animimal Behaviour Science 81, 185-198.

Huzzey, J.M., Von Keyserlingk, M.A.G. and Weary, D.M., 2005. Changes in feeding, drinking, and standing behavior of dairy cows during the transition period. Journal of Dairy Science. 88, 2454-2461.

Ingvartsen, K.L., Dewhurst, R.J. and Friggens, N.C., 2003. On the relationship between lactational performance and health: is it yield or metabolic imbalance that cause production diseases in dairy cattle? A position paper. Livestock Production Science 83, 277-308.

Jensen, M.B. and Pedersen, L.J., 2008. Using motivation tests to assess ethological needs and preferences. Applied Animimal Behaviour Science 113, 340-356.

Jensen, P., 2006. Domestication – from behaviour to genes and back again. Applied Animimal Behaviour Science 97, 3-15.

Jensen, P., Buitenhuis, B., Kjaer, J., Zanella, A., Mormède, P. and Pizzari, T., 2008. Genetics and genomics of animal behaviour and welfare-Challenges and possibilities. Applied Animimal Behaviour Science 113, 383-403.

Kirkden, R.D. and Pajor, E.A., 2006. Using preference, motivation and aversion tests to ask scientific questions about animals' feelings. Applied Animimal Behaviour Science 100, 29-47.

Korhonen, H.T., Jauhiainen, L. and Rekilä, T., 2008. Coping and production performance of blue foxes (*Alopex lagopus*) in a new hall environment. Annals of Animal Science 8, 289-303.

Leach, M.C., Thornton, P.D. and Main, D.C.J., 2008. Identification of appropriate measures for the assessment of laboratory mouse welfare. Animal Welfare 17, 161-170.

Lidfors, L., Berg, C. and Algers, B., 2005. Integration of natural behavior in housing systems. Ambio 34, 325-330.

Lindström, T., Redbo, I. and Uvnäs-Moberg, K., 2001. Plasma oxytocin and cortisol concentrations in dairy cows in relation to feeding duration and rumen fill. Physiology and Behaviour. 72, 73-81.

Lund, V. and Algers, B., 2003. Research on animal health and welfare in organic farming – A literature review. Livestock Production Science. 80, 55-68.

Lund, V. and Röcklinsberg, H., 2001. Outlining a conception of animal welfare for organic farming systems. Journal of Agricultural and Environmental Ethics 14, 391-424.

Main, D.C.J., Whay, H.R., Green, L.E. and Webster, A.J.F., 2003. Effect of the RSPCA Freedom Food scheme on the welfare of dairy cattle. The Veterinary Record 153, 227-231.

Main, D.C.J., Whay, H.R., Leeb, C. and Webster, A.J.F., 2007. Formal animal-based welfare assessment in UK certification schemes. Animimal Welfare 16, 233-236.

Mameli, M. and Bateson, P., 2006. Innateness and the sciences. Biology and Philosophy. 21, 155-188.

Manteuffel, G., Puppe, B. and Schön, P.C., 2004. Vocalization of farm animals as a measure of welfare. Applied Animimal Behaviour Science 88, 163-182.

Mason, G., 2007. Stereotypic behaviour in captive animals: fundamentals and implications for animal welfare and beyond. In: Mason, G.J. and Rushen, J. (Eds.). Stereotypic animal behaviour: fundamentals and implications for animal welfare. Second edition. CABI, Wallingford, Oxford. Pp. 325-356.

Mason, G.J. and Latham, N.R., 2004. Can't stop, won't stop: is stereotypy a reliable animal welfare indicator? Animimal Welfare 13, 57-69.

Mason, G.J. and Rushen, J., 2007. (Eds.). Stereotypic animal behaviour: fundamentals and implications for animal welfare. Second edition. CABI, Wallingford, Oxford.

Mench, J.A., 2003. Assessing animal welfare at the farm and group level: a United States perspective. Animal Welfare 12, 493-503.

Mendl, M. and Paul, E.S., 2008. Do animals live in the present? Current evidence and implications for welfare. Applied Animimal Behaviour Science 113, 357-382.

Mendl, M., Burman, O.H.P., Parker, R.M.A. and Paul, E.S., 2009. Cognitive bias as an indicator of animal emotion and welfare: emerging evidence and underlying mechanisms. Applied Animimal Behaviour Science 118, 161-181.

Millman, S.T., 2007. Sickness behaviour and its relevance to animal welfare assessment at the group level. Animal Welfare 16, 123-125.

Molony, V., Kent, J.E. and McKendrick, I.J., 2002. Validation of a method for assessment of an acute pain in lambs. Applied Animimal Behaviour Science 76, 215-238.

Mormède, P., Andanson, S., Aupérin, B., Beerda, B., Guémené, D., Malmkvist, J., Manteca, X., Manteuffel, G., Prunet, P., van Reenen, C.G., Richard, S. and Veissier, I., 2007. Exploration of the hypothalamic-pituitary-adrenal function as a tool to evaluate animal welfare. Physiology and Behaviour 92, 317-339.

Mullan, S., Browne, W.J., Edwards, S.A., Butterworth, A., Whay, H.R. and Main, D.C.J., 2009. The effect of sampling strategy on the estimated prevalence of welfare outcome measures on finishing pig farms. Applied Animimal Behaviour Science 119, 39-48.

Norrby, M., Madsen, M.T., Alexandersen, C.B., Kindahl, H. and Madej, A., 2007. Plasma concentrations of cortisol and PGF2? metabolite in Danish sows during mating, and intrauterine and conventional insemination. Acta Veterinaria Scandinavica 49, 36.

OIE, 2008. Introduction to the recommendations for animal welfare. Chapter 7.1 Terrestrial Animal Health Code 2008. World Organisation for Animal Health. (OIE), Paris.

Patterson-Kane, E.G., Pittman, M. and Pajor, E.A., 2008. Operant animal welfare: productive approaches and persistent difficulties. Animal Welfare 17, 139-148.

Rauw, W.M., Kanis, E., Noordhuizen-Stassen, E.N. and Grommers, F.J., 1998. Undesirable side effects of selection for high production efficiency in farm animals: A review. Livestock Production Science 56, 15-33.

Reiner, G., Hübner, K. and Hepp, S., 2009. Suffering in diseased pigs as expressed by behavioural, clinical and clinical-chemical traits, in a well defined parasite model. Applied Animimal Behaviour Science 118, 222-231.

Rodenburg, T.B., Tuyttens, F.A.M., de Reu, K., Herman, L., Zoons, J. and Sonck, B., 2008. Welfare assessment of laying hens in furnished cages and non-cage systems: an on-farm comparison. Animal Welfare 17, 363-373.

Ronchi, B., Stradaioli, G., Supplizi, A.V., Bernabucci, U., Lacetera, N., Accorsi, P.A., Nardone, A. and Seren, E., 2001. Influence of heat stress or feed restriction on plasma progesterone, oestradiol-17α, LH, FSH, prolactin and cortisol in Holstein heifers. Livestock Production Science 68, 231-241.

Rousing, T. and Waiblinger, S., 2004. Evaluation of on-farm methods for testing the human-animal relationship in dairy herds with cubicle loose housing systems – Test-retest and inter-observer reliability and consistency to familiarity of test person. Applied Animimal Behaviour Science 85, 215-231.

Rushen, J., 2000. Some issues in the interpretation of behavioural responses to stress. In: Moberg, G.P. and Mench, J.A. (Eds.). The biology of animal stress: basic principles and implications for animal welfare. Wallingford, UK: CABI, Wallingford, Oxford.

Rushen, J., de Passillé, A.M., von Keyserlingk, M. and Weary, D.M., 2008. The welfare of cattle. Springer, Dordrecht, the Netherlands pp. 303.

Scott, K., Laws, D.M., Courboulay, V., Meunier-Salaün, M.-C. and Edwards, S.A., 2009. Comparison of methods to assess fear of humans in sows. Applied Animimal Behaviour Science 118, 36-41.

Sørenson, J.T., Rousing, T., Møller, S.H., Bonde, M. and Hegelund, L., 2007. On-farm welfare assessment systems: what are the recording costs? Animal Welfare 16, 237-239.

Spinka, M., 2006. How important is natural behaviour in animal farming systems? Applied Animimal Behaviour Science 100, 117-128.

Stafford, K.J. and Mellor, D.J., 2005. Dehorning and disbudding distress and its alleviation in calves. Veterinary Journal 169, 337-349.

Svensson, C., Lundborg, K., Emanuelson, U. and Olsson, S.-O., 2003. Morbidity in Swedish dairy calves from birth to 90 days of age and individual calf-level risk factors for infectious diseases. Preventive Veterinary Medicine 58, 179-197.

Waiblinger, S., Boivin, X., Pedersen, V., Tosi, M.-V., Janczak, A.M., Visser, E.K. and Jones, R. B., 2006. Assessing the human-animal relationship in farmed species: a critical review. Applied Animimal Behaviour Science 101, 185-242.

Weary, D.M. and Fraser, D., 1995. Signalling need: costly signals and animal welfare assessment. Applied Animimal Behaviour Science 44, 159-169.

Weary, D.M., Niel, L., Flower, F.C. and Fraser, D., 2006. Identifying and preventing pain in animals. Applied Animimal Behaviour Science 100, 64-76.

Weary, D.M., Huzzey, J.M. and Von Keyserlingk, M.A.G., 2009. Board-invited Review: using behavior to predict and identify ill health in animals. Journal of Animal Science 87, 770-777.

Webster, A.J.F., 2001. Farm animal welfare: the five freedoms and the free market. Veterinary Journal 161, 229-237.

Webster, A.J.F., 2009. The virtuous bicycle: a delivery vehicle for improved farm animal welfare. Animal Welfare 18, 141-147.

Wechsler, B. and Weber, R., 2007. Loose farrowing systems: challenges and solutions. Animal Welfare 16, 295-307.

Weeks, C.A. and Nicol, C.J., 2006. Behavioural needs, priorities and preferences of laying hens. World's Poultry Science Journal 62, 296-307.

Wells, S.J., Ott, S.L. and Hillberg Seitzinger, A., 1998. Key health issues for dairy cattle – new and old. Journal of Dairy Science. 81, 3029-3035.

Whay, H.R., Main, D.C.J., Webster, A.J.F. and Green, L.E., 2003. Assessment of the welfare of dairy cattle using animal-based measurements: direct observations and investigation of farm records. The Veterinary Record 153, 197-202.

Windschnurer, I., Schmied, C., Boivin, X. and Waiblinger, S., 2008. Reliability and inter-test relationship of tests for on-farm assessment of dairy cows' relationship to humans. Applied Animimal Behaviour Science 114, 37-53.

Wischner, D., Kemper, N. and Krieter, J., 2009. Nest-building behaviour in sows and consequences for pig husbandry. Livestock Production Science 124, 1-8.

Zimmerman, P.H., Lundberg, A., Keeling, L.J. and Koene, P., 2003. The effect of an audience on the gakel-call and other frustration behaviours in the laying hen (*Gallus gallus domesticus*). Animimal Welfare 12, 315-326.